T0184141

MATCHED SAMPLING FOR CAUSAL EFFECTS

Matched sampling is often used to help assess the causal effect of some exposure or intervention, typically when randomized experiments are not available or cannot be conducted for ethical or other reasons. This book presents a selection of Donald B. Rubin's research articles on matched sampling, from the early 1970s, when the author was one of the few researchers involved in establishing the statistical foundations of the field, to recent contributions in this now extremely active area. The articles include fundamental theoretical studies that have become classics, important extensions, and real applications that range from breast cancer treatments to tobacco litigation to studies of criminal tendencies. They are organized into seven parts, each with an introduction by the author that provides historical and personal context and discusses the relevance of the work today. A concluding essay offers advice to investigators designing observational studies.

The book provides an accessible introduction to the study of matched sampling and will be an indispensable reference for students and researchers in statistics, epidemiology, medicine, economics, education, sociology, political science, and anyone doing empirical research to evaluate the causal effects of interventions.

Professor Donald B. Rubin is the John L. Loeb Professor of Statistics in the Department of Statistics at Harvard University. Professor Rubin is a Fellow of the American Statistical Association, the Institute for Mathematical Statistics, the International Statistical Institute, the Woodrow Wilson Society, the John Simon Guggenheim Society, the New York Academy of Sciences, the American Association for the Advancement of Sciences, and the American Academy of Arts and Sciences. He is also the recipient of the Samuel S. Wilks Medal of the American Statistical Association, the Parzen Prize for Statistical Innovation, and the Fisher Lectureship. Professor Rubin has lectured extensively throughout the United States, Europe, and Asia. He has more than 300 publications (including several books) on a variety of statistical topics and is one of the top ten highly cited writers in mathematics in the world, according to ISI Science Watch.

MATCHED SAMPLING
FOR
CAUSAL EFFECTS

DONALD B. RUBIN
Harvard University

CAMBRIDGE
UNIVERSITY PRESS

CAMBRIDGE UNIVERSITY PRESS
Cambridge, New York, Melbourne, Madrid, Cape Town, Singapore,
São Paulo, Delhi, Dubai, Tokyo, Mexico City

Cambridge University Press
32 Avenue of the Americas, New York, NY 10013-2473, USA

www.cambridge.org
Information on this title: www.cambridge.org/9780521674362

© Cambridge University Press 2006

First published 2006
Reprinted 2008

A catalog record for this publication is available from the British Library

Library of Congress Cataloging in Publication data

Rubin, Donald B.
Matched sampling for causal effects / Donald B. Rubin
p.   cm.
Includes bibliographical references and index.
ISBN-13: 978-0-521-85762-8 (hardback)
ISBN-10: 0-521-85762-7 (hardback)
ISBN-13: 978-0-521-67436-2 (pbk.)
ISBN-10: 0-521-67436-0 (pbk.)
1. Sampling (Statistics) 2. Statistical matching.   I. Title.
HA31.2.R82   2006
001.4′33 – dc22   2006011564

ISBN 978-0-521-85762-8 Hardback
ISBN 978-0-521-67436-2 Paperback

Contents

Contributor Acknowledgments

The author would like to thank his co-authors for their contributions to the publications and, where appropriate, their permission to reprint the articles in this book. Current affiliations are listed when known.

William G. Cochran

Paul R. Rosenbaum, The Wharton School, The University of Pennsylvania

Neal Thomas, Pfizer, Statistical Research and Consulting Center

Jennifer L. Hill, Columbia University

Ralph B. D'Agostino, Jr., Wake Forest University School of Medicine

June Machover Reinisch, The Kinsey Institute for Research in Sex, Gender, and Reproduction; Institute of Preventive Medicine, Danish Epidemiological Science Center, Copenhagen University; Vice President of Scientific Affairs, The Museum of Sex

Stephanie A. Sanders, The Kinsey Institute for Research in Sex, Gender, and Reproduction, Indiana University

Martin W. McIntosh, Fred Hutchinson Cancer Research Center, University of Washington

Paul W. Holland, Educational Testing Service

Sarnoff A. Mednick, University of Southern California

Herman A. Witkin
Fini Schulsinger
Eskild Bakkestrøm
Karl O. Christiansen
Donald R. Goodenough
Kurt Hirschhorn
Claes Lundsteen
David R. Owen
John Philip
Martha Stocking
Erik Lykke Mortensen

MATCHED SAMPLING FOR CAUSAL EFFECTS

My Introduction to Matched Sampling

This volume reprints my publications on matched sampling, or more succinctly, matching, produced during a period of over three decades. My work on matching began just after I graduated college in 1965 and has continued to the present, and beyond, in the sense that there are publications on matching subsequent to those collected here, and I have continuing work in progress on the topic. For most of the years during this period, I believe I was one of the few statistical researchers publishing in this area, and therefore this collection is, I hope, both interesting and historically relevant. In the introduction to each part, I attempt to set the stage for the particular articles in that part. When read together, the part introductions provide a useful overview of developments in matched sampling. In contrast to the earlier years, in the last few years, there have been many other researchers making important contributions to matching. Among these, ones by technically adroit economists and other social scientists are particularly notable, for example: Hahn (1998); Dehejia and Wahba (1999); Lechner (2002); Hansen (2004); Hill, Reiter, and Zanutto (2004); Hirano, Imbens, and Ridder (2004); Imbens (2004); Zhao (2004); Abadie and Imbens (2005); and Diamond and Sekon (2005). Some of these have had a direct or indirect connection to a course on causal inference I've taught at Harvard for over a decade, sometimes jointly with Guido Imbens.

My interest in matched sampling started at Princeton University, but not until after I graduated. Nevertheless, it was heavily influenced by my time there. I started my college career in 1961 at Princeton University intending to major in physics, as part of a cohort of roughly 20 kids initially mentored by John Wheeler, who (if memory serves correctly) hoped that we would get PhDs in physics in five years from enrollment as freshman – we were all lined up earlier for an AB in physics in three years. In retrospect, this was a wildly overambitious agenda, at least for me. For a combination of reasons, including the Vietnam War and Professor Wheeler's sabbatical at a critical time, I think that no one succeeded in completing the ambitious five-year PhD from entry. We ended up in a variety of departments – I ended up in psychology, but the others in math, chemistry, physics, economics, et cetera. But I still loved the ways of thought in physics and possessed a few of the technical skills, and had some computational skills, which was relatively unusual at the time. Those early computational skills were my entry into the world of matched sampling, and eventually into statistics.

In the fall of 1965 I entered Harvard University in a PhD program in psychology. This lasted about two weeks, until that department's PhD advisor decided that my undergraduate program was deficient, primarily because I lacked enough courses in statistics. I had to take some "baby" statistics courses – me, a previous physics type at Princeton! Hrumpf! Feeling insulted, and unwilling to take such courses, I switched to an applied math program, principally computer science, which I could do fairly simply because I had the aforementioned skills as well as independent support through a National Science Foundation graduate fellowship. In the spring of 1966, I got my Master's in applied mathematics. That summer, I lived near Princeton with old Princeton University roommates, and we supported ourselves doing consulting, which was my introduction to matched sampling.

At that time, Robert Althauser was a junior faculty member in the Department of Sociology at Princeton working on comparisons of white and black students at Temple University. My memory is a bit vague, but I think the objective was to compare white and black students with very similar backgrounds to see if the groups' academic achievements differed. Thus arose the desire to create matched samples of blacks and whites. I was hired primarily as the "computer guy," programming and implementing algorithms that we created. This work resulted in a methodological publication, Althauser and Rubin (1970), but more important to my career, it stimulated an interest in a topic that has had exciting applications to this day, and fostered interesting statistical theory as well.

After returning to Harvard for my second and third years in applied math, I realized that there was a department called "Statistics" that seemed to include the study of things that I was already doing. By my fourth year at Harvard, I was a PhD student in that department, with a ready topic for my thesis, and with a fabulous PhD advisor who was also a wonderful human being, William G. Cochran. Bill, who was then one of three senior faculty in the Department of Statistics at Harvard University, the others being Arthur Dempster and Frederick Mosteller, had a powerful influence on me. He taught me what good statistics meant: doing something to address an important real problem. If a project didn't have some relevance to the real world, Bill's view was that it might be of interest to some, and that was OK, but it wasn't of interest to him. It might be great mathematics, but then I should convince a mathematician of that, not him – he was neither interested nor able to make such an assessment. Over the years, I've tried to instill the same attitude in my own PhD students in statistics.

I submitted my thesis on matched sampling (Rubin, 1970), written under Bill's direction, in the spring of 1970, and stayed on for one year in the Statistics Department at Harvard, post-PhD, co-teaching the statistics course in psychology (with Bob Rosenthal) that had driven me out of psychology five years earlier. Realizing that life as a junior faculty member was, at that stage, not for me, I moved to the Educational Testing Service (ETS) in the fall of 1971, where I also taught as a visiting faculty member in the new Statistics Department at Princeton. During that time, I continued to spend time at Harvard with Bill, Bob Rosenthal, and Art Dempster (another important influence on my view of statistics), and, of particular relevance here, I continued to refine the ideas and the work in my PhD thesis.

But enough of ancient personal history. Matching, or matched sampling, refers to the following situation. We have two groups of units, for example, people. One group is exposed to some "treatment" (e.g., cigarette smoking) and another is not exposed, the "controls" (e.g., never smokers). Our focus is on the causal effect of the exposure (e.g., smoking versus not smoking) on outcomes such as lung cancer or coronary disease. Of course, smokers and nonsmokers may differ on background characteristics such as education, and at a first pass, it would make little sense to compare disease rates in well educated nonsmokers and poorly educated smokers. Whence the idea of matched sampling: create samples of smokers and nonsmokers matched on education and other background variables. Or in the Althauser example, even though it may not make sense to talk about the "causal" effect of a person being a white student versus being a black student, it can be interesting to compare whites and blacks with similar background characteristics to see if there are differences in academic achievement, and creating matched black–white pairs is an intuitive way to implement this comparison.

This technique of comparing "like with like" is intellectually on-target, and it can be formalized in a variety of ways in the context of drawing causal inferences. The topic of matched sampling is not intrinsically about this formalization, but rather concerns the extent to which such matched samples can be constructed and how to construct them, and precisely how well matching works to make the distributions of the background characteristics in the treatment and control groups the same. The formal structure for causal inference that clarifies the role of matching is now commonly called "Rubin's Causal Model" (Holland, 1986b) for a series of articles written in the 1970s (Rubin, 1974, 1975, 1976a, 1977a, 1978a, 1979a, 1980a). This perspective is presented, for example, in Rubin (2005, 2006) and Imbens and Rubin (2006a), and the first text fully expositing it and many further developments, including the topic of matched sampling, is Imbens and Rubin (2006b).

One of the critically important characteristics of matching is that it is done without access to any outcome data – only covariate data and treatment information are used. Therefore, when the matched samples are created this way, the investigator cannot realistically be accused of creating matches to obtain some result with respect to the outcomes, even unconsciously. This is of real benefit for honest study design, a point emphasized in a variety of places in this volume, including the concluding "Advice to the Investigator."

The reprinted articles are presented in seven parts, each part consisting of between two and five chapters, each of which is a reprinted article. The articles have been reset, with figures, tables, and equations renumbered to reflect the appropriate chapters. The articles reprinted in each part are organized to create a coherent topic, and the parts are organized to reflect a systematic historical development, with the last two parts consisting of examples. A final section provides concluding advice to the investigator.

Printers' and related errors in the original articles have been corrected, when recognized, and some other stylistic modifications made for consistency. There was no attempt to make notation entirely consistent across chapters.

PART I: THE EARLY YEARS AND THE INFLUENCE OF WILLIAM G. COCHRAN

Even though this book is limited to reprinting publications of mine on matched sampling, it seems useful to provide some background history to the topic. As described in the initial introduction, I had already started working on matching before I met Bill Cochran. But Bill had been working on the design and analysis of observational studies for many years before my appearance.

I reviewed Cochran's work on observational studies, including his early papers, in Rubin (1984c), the first chapter in this book. This was originally a chapter written for a volume honoring Cochran's impact on statistics, edited by Rao and Sedransk (1984). My review starts with Cochran (1953a), which focused on stratified and pair matching on a single covariate, X, and their effects on efficiency of estimation. That chapter continued with Cochran (1965), which was a compendium of advice on the design and analysis of observational studies. Also reviewed was Cochran (1968a), on subclassification and stratified matching, an article that I regard as extremely important. It was a departure from his earlier work on matching, as well as other early work on the effects of matching, all of which focused on the efficiency of estimation (e.g., Wilks (1932), which assumed exact matching with a normally distributed variable; Greenberg (1953), which compared mean matching and regression adjustment; Billewicz (1965), which I've always found relatively confusing). Cochran (1968a), however, considered the effect of the matching on the *bias* of estimates resulting from the matched, in contrast to random, samples, as well as matching's effects on the *precision* of estimates (i.e., their efficiency or sampling variance).

In some sense it is obvious that the issue of bias reduction dominates that of sampling variance reduction when faced with an observational study rather than a randomized experiment, but this was the first publication I know that actually studied bias reduction mathematically, despite "everyone" talking and writing about it when giving practical advice: A precise estimate of the wrong quantity can easily lead one astray. Cochran (1968a) was a wonderful article that in many ways set the tone for what I tried to do in all of my academic work. Not only was there clever mathematics, but there was also highly useful practical advice, such as the oft-quoted comment that matching using five or six well-chosen subclasses of a continuous covariate typically removes about 90% of the initial biasing effects of that covariate.

Because of the late stage of Cochran's research career when I began working with him, we wrote only one paper together, Cochran and Rubin (1973), included as the second chapter of Part I. He had been invited to write an article for a special issue of *Sankhyā* in honor of the famous Indian statistician, Mahalanobis, and he proposed that we join forces and write up some of his old notes for a book on observational studies that he had shelved in favor of Cochran (1965), plus some of the results in my thesis (subsequently published as Rubin, 1973a,b, both in Part II), along with some of the newer multivariate matching results that I had been developing (subsequently published as Rubin, 1976b,c, both in Part III). This work on discriminant matching, to some extent, anticipated subsequent work on propensity score methods, introduced in Chapter 10. Bill was not only a wonderful writer but a wonderful critic of others' writing, even teaching me about adding "Cochran's commas" to add clarity: "You wouldn't mind if I added a 'Cochran comma' here, would you?" I of course agreed to the commas and the article, and we wrote this paper together in 1970–71 when I was at Harvard for the postgraduate year noted in the initial introduction.

In the article, we addressed the dangers of relying solely on implicit extrapolation through linear regression adjustment on X to control for bias, and documented the improvements that can be achieved when using both matching and regression adjustment on X, noting that the matching is especially important when outcomes are nonlinearly related to X in subtle ways. We also reviewed some of the older Cochran wisdom about the design and analysis of observational studies, the consequences of errors of measurement from Bill's notes, and discussed multivariate matching methods (e.g., discriminant matching, Mahalanobis-metric matching). Even though the topic of observational studies was not one close to Mahalanobis's own work, the fact that his name could be attached to a procedure, "Mahalanobis-metric matching," made the article seem appropriate then for that issue, and with hindsight, even more so because the method has become quite popular (e.g., even currently in economics, Imbens (2004)). The Mahalanobis metric uses the inverse variance–covariance matrix of the matching variables to define the distance between a pair of possible matches, an important inner-product metric.

At the time, there was very little technical work being done on matching. In fact, the only publication I know, not otherwise referenced in this volume, is Carpenter (1977). This *Biometrics* article had technical results on efficiency, where the implied matching method was Mahalanobis-metric matching.

1. William G. Cochran's Contributions to the Design, Analysis, and Evaluation of Observational Studies

Donald B. Rubin

1. INTRODUCTION

William G. Cochran worked on statistically rich and scientifically important problems. Throughout his career he participated in the design, analysis, and evaluation of statistical studies directed at important real world problems. The field of observational studies is a perfect example of a general topic that Cochran helped to define and advance with many contributions. Cochran's work provides an essential foundation for continuing research in this important area of statistics.

An observational study, for purposes here, is a study intended to assess causal effects of treatments where the rule that governs the assignment of treatments to units is at least partially unknown. Thus a randomized experiment on rats for the effect of smoke inhalation on lung cancer is a controlled experiment rather than an observational study, but an analysis of health records for samples of smokers and nonsmokers from the U.S. population is an observational study. The obvious problem created by observational studies is that there may exist systematic differences between the treatment groups besides treatment exposure, and so any observed differences between the groups (e.g., between smokers and nonsmokers) with respect to an outcome variable (e.g., incidence of lung cancer) might be due to confounding variables (e.g., age, genetic susceptibility to cancer) rather than the treatments themselves. Consequently, a primary objective in the design and analysis of observational studies is to control, through sampling and statistical adjustment, the possible biasing effects of those confounding variables that can be measured: a primary objective in the evaluation of observational studies is to speculate about the remaining biasing effects of those confounding variables that cannot be measured.

Although observational studies always suffer from the possibility of unknown sources of bias, they are probably the dominant source of information on causal effects of the treatments and certainly are an important supplement to information arising from experiments. Among the reasons for the importance of observational studies are the following. First, quite often the equivalent randomized experiment cannot be conducted for ethical or political reasons, as when studying the effects of in utero exposure to radiation in humans. Second, if the equivalent randomized experiment is possible to conduct in practice, it usually will be far more expensive than the observational study, and its results may not be available for many years, whereas the results from the observational study may be at hand: for example, consider studying the relationship between cholesterol levels in diet and subsequent heart disease. Third, the units in observational studies are usually more representative of the target populations because randomized experiments generally have to be conducted in restricted environments, such as particular hospitals with consenting patients. Fourth, ascertaining which treatments should be studied in future randomized experiments should be based on the analysis of the best available data sets, which are, in early stages of investigation, nearly always observational. Fifth, many studies designed as experiments become more like observational studies when protocols are broken, patients leave the study, measurements are missing, and so on. Sixth, even within properly conducted randomized experiments, many important questions may be observational; for example, in an experiment randomizing patients to either medical or surgical treatment of coronary artery disease, the study of which version of the surgical treatment (e.g., which hospital or the number of bypasses) is most effective is an observational study.

The ultimate objective of Cochran's statistical research on observational studies was to provide the investigator with reliable statistical tools and sage advice on their use. On more than one occasion Bill told me that it was better to give the applied researcher a reliable tool that is well understood and will be used properly than a more powerful, but complex and potentially misunderstood tool, that can be easily misapplied. As expected given such an orientation, Cochran seems to have had no a priori favorite methods, good practice being more important than philosophical orientation.

In this chapter I will try to convey the broad themes of his advice and trace the historical development of ideas in his articles on observational studies. Although some theory will be presented, it is not necessary to present a comprehensive review of technical results from his articles on observational studies because many of these have been summarized in our joint paper "Controlling Bias in Observational Studies: A Review," published in *Sankhyā-A* in 1973, here reprinted as Chapter 2. Subsequent reviews of statistical research on observational studies include McKinlay (1975a) and Anderson et al. (1980). This review will focus on Cochran's contributions, especially those aspects that I find particularly influential and important. Because he wrote exceptionally well and with deep understanding of the underlying issues, it is appropriate to make liberal use of quotations from his works.

Section 2 summarizes themes of Cochran's advice on observational studies. These themes appear repeatedly in his articles and deserve this emphasis because of their importance. Memories of his written and verbal advice have been very useful to me when thinking about observational studies such as those on the effectiveness of coaching for the SAT (Messick, 1980), the effect of prenatal exposure to hormones (Reinisch and Karow, 1977), the effectiveness of coronary bypass operations (Murphy et al., 1977), and the effectiveness of private versus public schools (Coleman et al., 1981).

Sections 3 through 9 summarize his seven major articles on the design, analysis, and evaluation of observational studies. These were published in 1953, 1957, 1965,

1968, 1970, 1972, and 1973 and are major in the sense of being accessible and influential. Section 10 very briefly comments on other work: four proceedings papers with limited distribution, an unpublished technical report, two papers dealing with historical aspects of experimentation, and his 1968 Harvard seminar on observational studies. Section 11 discusses his monograph on observational studies, published posthumously at the end of 1983, but, I believe, written in large part prior to 1972. Finally, Section 12 adds a few more personal comments in the context of Cochran's attitude towards the interplay of statistical theory and practice.

2. MAJOR THEMES OF ADVICE FOR OBSERVATIONAL STUDIES

Cochran's opening to "Matching in Analytical Studies," published in 1953, defines very clearly the relevant issues in observational studies:

> Most of the following discussion will be confined to studies in which we compare two populations, which will be called the experimental population and the control population. The experimental population possesses some characteristic (called the experimental factor) the effects of which we wish to investigate: It may consist, for example, of premature infants, of physically handicapped men, of families living in public housing, or of inhabitants of an urban area subject to smoke pollution, the experimental factors being, respectively, prematurity, physical handicaps, public housing, and smoke pollution. I shall suppose that we cannot create the experimental population, but must take it as we find it, except that there may be a choice among several populations that are available for study.

> The purpose of the control population is to serve as a standard of comparison by which the effects of the experimental factor are judged. The control population must lack this factor, and ideally it should be similar to the experimental population in anything else that might affect the criterion variables by which the effects of the factor are measured. Occasionally, an ideal control population can be found, but, more usually, even the most suitable control population will still differ from the experimental population in certain properties which are known or suspected to have some correlation with the criterion variables.

> When the control and experimental populations have been determined, the only further resource at our disposal is the selection of the control and experimental *samples* which are to form the basis of the investigation. Sometimes this choice is restricted, because the available experimental population is so small that it is necessary to include all its members, only the control population being sampled.

> The problem is to conduct the sampling and the statistical analysis of the results so that any consistent differences which appear between the experimental and the control samples can be ascribed with reasonable confidence to the effects of the factor under investigation.

Notice that in this introduction examples are used to make sure the reader knows the kind of study to be discussed. Throughout his publications, Cochran continued to use examples to motivate discussion. In particular, he referred often to studies of smoking and health (e.g., U.S. Surgeon General's Committee, 1964), preventing poliomyelitis (e.g., Hill and Knowelden, 1950), and exposure to radiation (e.g., Seltser and Sartwell, 1965).

The major themes of Cochran's advice on observational studies fit well into three broad categories: design, analysis, and drawing conclusions or evaluation. Although such simplifications are never entirely adequate, this classification serves as a useful guide when reading Cochran's work.

In design, Cochran emphasized the need to measure, as well as possible, important variables, both outcome, y, and disturbing, x. Outcome variables are those that are used to assess the effects of the treatments being studied, for example, the presence of lung cancer or polio in medical studies or the score on an achievement test in educational studies. Generally, the purpose of an observational study is to provide insight into how the treatments causally affect the outcome variables. Disturbing (confounding) variables are those that confound the relationship between treatment and y because they may have a different distribution in the experimental group than in the control group. Examples of commonly controlled disturbing variables in studies of human populations include age and gender, as well as measures of pretreatment health in medical studies and pretreatment achievement in educational studies. If the measurements cannot be well taken or the sample sizes would be inadequate, the study may not be worth attempting, and the researchers should consider this possibility.

A second theme in design is the need for a control group, perhaps several control groups (e.g., for a within-hospital treatment group, both a within-hospital control group and a general-population control group). The rationale for having several control groups is straightforward: If similar estimates of effects are found relative to all control groups, then the effect of the treatment may be thought large enough to dominate the various biases probably existing in the control groups, and thus the effect may be reasonably well estimated from the data. My reading of Cochran's work suggests that his insistence on using control groups rather than, say, relying on before–after studies, became stronger as he became older. Another theme in the design of observational studies is the desire to avoid control groups with large initial biases. No statistical procedure can be counted on to adjust for large initial differences between experimental and control populations. A final theme for the design of observational studies is to use matched sampling or blocking to reduce initial bias. Incidentally, Cochran viewed pair matching as an excellent applied tool because the investigator could easily see how similar (or not) the paired units were, and therefore did not have to understand or trust the statistician's fancy adjustment techniques before making judgments about residual bias due to observed disturbing variables.

In analysis, Cochran studied the properties of two major methods of adjustment: subclassification and covariance (regression) adjustment. With subclassification, the treatment and control samples are stratified (i.e., classified) on the primary disturbing variables, and within-stratum (i.e., within-subclass) estimates of treatment effect are made and then combined to form a global estimate of treatment effect. For example, suppose the primary disturbing variable that is measured in a study of the effect of smoking on lung cancer is age: with subclassification, within each age subclass (e.g., < 25, 25–35, > 35–50, and > 50), the difference between the proportions of smokers and nonsmokers with lung cancer would be the estimated effect specific to that subclass; then these subclass-specific estimates would be weighted by the proportion of the target population in each subclass to form a global estimate of the effect of smoking on lung cancer. With covariance adjustment, a model, usually linear, is fit to the regression of y on x, and is used to create an adjusted estimate of treatment effect of the form $\bar{y}_1 - \bar{y}_2 - \hat{\beta}(\bar{x}_1 - \bar{x}_2)$ where $\bar{y}_1, \bar{x}_1$ and $\bar{y}_2, \bar{x}_2$ are the y and x means in the experimental and control groups, and $\hat{\beta}$ is an estimate of the slope, β, of y on x. Cochran generally recommended using these techniques to control disturbing variables and provided the investigator with guidance on how best to use them. Formally, these techniques can be seen as serving two purposes: to increase the precision of comparisons and to remove initial bias due to x. As Cochran's work on observational studies progressed, he moved from focusing on their effectiveness for increasing precision to their effectiveness for removing bias. There are two interrelated reasons for this shift. First, since it is generally not wise to obtain a very precise estimate of a drastically wrong quantity, the investigator should be more concerned about having an estimate with small bias than one with small variance. Second, since in many observational studies the sample sizes are sufficiently large that sampling variances of estimators will be small, the sensitivity of estimators to biases is the dominant source of uncertainty.

When drawing causal conclusions from observational studies, Cochran had two major pieces of advice. First, speculate about sources and directions of residual bias. What unmeasured disturbing variable might be affecting the conclusions of the study, and how might these conclusions change if the variables were controlled? Second, formulate complex, interrelated causal hypotheses. Generally, if there is a causal effect of the treatment, which outcome variables should be most affected and which least, and with respect to which control groups should the effects be largest? Cochran attributed this advice to both Sir Bradford Hill and Sir Ronald Fisher.

A recent National Academy of Sciences meeting called to make suggestions for additional analysis of recent data on the relative effectiveness of private versus public high schools (Coleman et al., 1981) emphasized to me the value of these pieces of advice. First, the advice on speculating about unmeasured confounding variables was highly pertinent. Even though pre-high school achievement tests and parental socioeconomic status variables were measured, there remained the suspicion that students attending private schools differed in unmeasured ways from those attending public schools. For example, it was felt likely that students in private schools were exposed to more parental pressure to achieve and probably were more self-motivated to achieve. Consequently, if such variables had been measured, adjusting for them probably would have led to private schools appearing less effective than before adjustment, although it is not easy to speculate realistically about the numerical results of such an adjustment. One collection of statistical tools for the study of such adjustment are selection models (Heckman, 1976; Greenlees et al., 1982); this type of model must rely very heavily on an untestable assumption like the joint normality of outcome variables and unmeasured disturbing variables, and consequently should be used to explore sensitivity to assumptions, rather than to provide one answer. Presumably, as more statistical tools are developed for formally performing sensitivity analyses to unmeasured disturbing variables, such speculation will become more focused. A recent example of such work is Rosenbaum and Rubin (1983b), which describes a maximum likelihood approach for studying the sensitivity of the estimated effect of a treatment, having adjusted for an observed categorical covariate, to an unobserved binary covariate; the results are illustrated using data on the efficacy of coronary bypass surgery.

The second piece of Cochran's advice on drawing conclusions in observational studies, to make causal hypotheses complex, is also pertinent to the Coleman et al. (1981) data set. For instance, there were many outcome variables measuring different aspects of achievement, and some of these variables could be ordered by anticipated susceptibility to differences between public and private school treatments. In particular, some outcome variables were based on material not taught in high school and other outcome variables were based on material typically taught in high school. If private schools are more effective than public schools, then the treatment effects should be larger for outcome variables with more high school material than outcome variables with less high school material. Also, various control groups could be defined for students in private schools since the pool of nonprivate schools was very large.

Finally, it is important to recall that for obtaining accurate estimates of treatment effects, Cochran never viewed observational studies as a replacement for randomized experiments. He realized that data from observational studies were less expensive to obtain than data from randomized experiments and sometimes were the only data available due to practical constraints. Nevertheless, he recommended that randomized experiments should be considered whenever feasible and, furthermore, kept in mind as an ideal objective when designing observational studies.

3. MATCHING IN ANALYTICAL STUDIES

"Matching in Analytical Studies" (Cochran, 1953a) is Cochran's first article devoted to the topic of observational studies. A previous one, "The Comparison of Percentages in Matched Studies" (Cochran, 1950), alluded to the possibility that the matched samples arose in the context of an observational study, but did not really address any issues peculiar to such studies. Even though the context in Cochran (1953a) is observational studies, variance reduction rather than bias removal is the primary topic. Later articles on observational studies emphasize the importance of controlling bias.

In "Matching in Analytical Studies" Cochran considers the effectiveness of pair matching in simple cases. Pairing or pair-matching here means finding for each experimental subject a perfect control subject, perfect meaning with exactly the same value of x (x generally being a vector of k disturbing variables). For example, suppose $k = 2$ and x represents gender and age in years; then a perfect match for a male 37 years old in the experimental group is a male 37 years old in the control group. For most of the article Cochran assumes an "ideal control population," which is defined by three characteristics: (a) the outcome variable, y, has the same normal linear regression on x in both populations, (b) the distributions of x are the same in both populations, and (c) x represents all relevant confounding factors – that is, "no unsuspected biases are present."

The article begins by comparing two estimates of the treatment effect, the treatment minus control y-means in random samples, $\bar{y}_1 - \bar{y}_2$, and the treatment minus control y-means in perfectly matched samples, $\bar{y}_d$. He shows that the ratio of the variance of $\bar{y}_1 - \bar{y}_2$ to the variance of $\bar{y}_d$ is $(1 - R^2)^{-1}$, where R^2 is the squared multiple correlation between y and k covariates, x, and concludes that "my impression is that the multiple correlation coefficient is often below 0.5, so that the gain in precision from pairing is often modest." Cochran also uses this result to provide advice on the selection of covariables: "inclusion of a specific x-variable in the pairing is worth-while only if it decreases $(1 - R^2)$ by an appreciable amount (say 10 percent), when this x-variable is added to those already selected." This piece of advice was especially appropriate in the days before computers helped eliminate some of the problems of handling data sets with many variables.

Still assuming an ideal control population, Cochran next compares $\bar{y}_d$ with the covariance (regression) adjusted estimator in random samples, $\bar{y}_1 - \bar{y}_2 - b(\bar{x}_1 - \bar{x}_2)$, where $\bar{x}_1 - \bar{x}_2$ is the difference in x-means and b is the pooled, within-group, least squares estimate of the slope of y on x. He shows that the ratio of the variance of the covariance-adjusted estimator to the pairing estimator is $1 + k/(2n - k - 3)$, which indicates that "in these circumstances there is not much to choose between pairing and covariance." The discussion that follows emphasizes practical problems with matching: the need to have x values readily available on a larger control sample, the time involved in constructing the matches, and the fact that loss of one member of a pair will destroy the simplicity of the paired analysis, which finds the standard error of the estimate from the variance of matched pair differences in y.

When the control population is no longer ideal because x does not have the same mean in both populations, pairing may be more effective. The intuition for this result is that the variance of the covariance-adjusted estimate increases as the experimental and control distributions of x become further apart, whereas the variance of the perfectly pair-matched estimator is independent of the difference between the population distributions (of course, in practice, as the experimental and control distributions become further apart, close matches for each experimental unit will become more difficult to obtain). Specifically, the ratio of the variance of the covariance-adjusted estimate to the variance of the pairing estimate is, for large n and one x-variable, approximately $1 + D^2/4\sigma_x^2$ where D is the difference of means of x in the two populations and σ_x^2 is the variance of x in each population. Cochran concludes that "The increase in precision from pairing relative to covariance is probably not great in practice . . . [because $D/\sigma_x = 1$] . . . implies a fairly drastic selection operating on the x variable." Of course the primary benefits of pair matching relative to covariance adjustment arise when the simple linear model underlying covariance adjustment does not hold, but this point is not emphasized here.

Cochran goes on to consider briefly other methods of controlling x: balancing, which matches the x means in the treatment and control groups, and stratified matching, which creates categories from x such that for each category of x (i.e., in each cell defined by x) there are both experimental and control units. Cochran notes that balancing is sensitive to nonlinearities in the regression of y on x and that stratified matching has a potential advantage over pairing:

> The analysis of the results is slightly more complicated, because we can compare the experimental and control results separately for each cell. In return, this analysis focuses attention on any variation that occurs in the effects of the experimental factor as the levels of the covariables change – in other words, on the interactions of the experimental factor with the covariables. Such information may broaden the results of the study.

In conclusion, Cochran reiterates the assumption that "the control and experimental populations differ at most through selection on certain covariables which are included in the matching or covariance" and states:

> The principal conclusion from the preceding discussion is that the selection of the control population is a more crucial step than the selection of a method of matching or of the covariables on which to match. Matching removes the deficiencies of a poor control to only a limited extent.

The final point in this article is one that repeatedly appears in Cochran's work on observational studies – make causal hypotheses complex:

> In conclusion, in observational studies, where it is not feasible to assign a subject at random to the control or experimental sample, we can never be sure that some unsuspected disturbance does not account, in large part, for the observed difference between the two samples. Consequently, the results of tests of significance must be interpreted with more caution in observational studies than in experiments where randomization can be employed. One good practice in observational studies is to check any theory at as many points and in as many ways as ingenuity can devise. . . . If the same kind of result appears repeatedly when the data are analyzed from widely different points of view, it becomes successively more difficult to imagine any 'disturbance' that will explain away *all* the results. Where it can be employed, this technic does much to overcome the handicap under which we all labor in observational studies.

The Hill and Knowelden (1950) study of the relation between inoculation with pertussis vaccine and poliomyelitis is used to illustrate this final point: not only was there a temporal relationship between inoculation and onset of poliomyelitis but also a strong relationship between site of inoculation and site of paralysis.

4. ANALYSIS OF COVARIANCE: ITS NATURE AND USES

Observational studies are only briefly mentioned in Cochran (1957), in Section 2.2, which describes one of the five principal uses of the analysis of covariance: "To remove the effects of disturbing variables in observational studies." The article is included here as a major contribution because Cochran is explicitly and primarily concerned with the use of covariance in observational studies to remove bias, not to increase precision. This position, which Cochran held in the rest of his writing on observational studies, was not explicitly stated in the previous paper.

Cochran warns of two hazards when using covariance analysis to remove bias:

> Although matching and covariance have been skillfully applied, we can never be sure that bias may not be present from some disturbing variable that was overlooked. In randomized experiments, the effects of this variable are distributed among the groups by the randomization in a way that is taken into account in the standard tests of significance. There is no such safeguard in the absence of randomization.

> Secondly, when the *x*-variables show real differences among groups – the case in which adjustment is needed most – covariance adjustments involve a greater or less degree of extrapolation. To illustrate by an extreme case, suppose that we were adjusting for differences in parents' income in a comparison of private and public

school children, and that the private-school incomes ranged from $10,000–$12,000, while the public-school incomes ranged from $4,000–$6,000. The covariance would adjust results so that they allegedly applied to a mean income of $8,000 in each group, although neither group has any observations in which incomes are at or even near this level.

The further discussion of the second hazard is a model of the kind of straightforward advice Cochran could offer, and I repeat it here in its entirety:

Two consequences of this extrapolation should be noted. Unless the regression equation holds in the region in which observations are lacking, covariance will not remove all the bias, and in practice may remove only a small part of it. Secondly, even if the regression is valid in the no man's land, the standard errors of the adjusted means become large, because the standard error formula in a covariance analysis takes account of the fact that extrapolation is being employed (although it does not allow for errors in the form of the regression equation). Consequently the adjusted differences may become insignificant merely because the adjusted comparisons are of low precision. When the groups differ widely in x, these differences imply that the interpretation of an adjusted analysis is speculative rather than soundly based. While there is no sure way out of this difficulty, two precautions are worth observing.

(i) Consider what internal or external evidence exists to indicate whether the regression is valid in the region of extrapolation. Sometimes the fitting of a more complex regression formula serves as a partial check.

(ii) Examine the standard errors of the adjusted group means, particularly when differences become non-significant after adjustment. Confidence limits for the difference in adjusted means will reveal how precise or imprecise the adjusted comparison is.

With the current availability of computers, regression adjustment in observational studies is computationally rather trivial. In my experience, however, users too often do not attend to the fact that in a multivariate sense, the two samples may be far apart, and thus the regression adjustment may implicitly be reliant upon substantial extrapolation. Two recent examples where the initial reports did not show much awareness of this crucial issue are the Federal Trade Commission's analysis for the effects of coaching programs for the Scholastic Aptitude Test (Messick, 1980, provides references), and the new Coleman report on the relative effectiveness of private and public high schools (Coleman et al., 1981).

5. THE PLANNING OF OBSERVATIONAL STUDIES OF HUMAN POPULATIONS

Cochran's Royal Statistical Society discussion paper (Cochran, 1965) is his most comprehensive article on observational studies. He begins it with two basic pieces of advice on their conduct.

1. When selecting samples for study, make sure that they are large enough and have complete enough data to allow effects of practical importance to be estimated, and avoid treatment and control groups with large initial differences on confounding variables.

2. Use both the statistician and the subject-matter expert in the planning stages; in Cochran's words:

> Effective planning of observational studies calls for considerable mastery of the state of research in the subject-matter field, for instance in forming a judgment as to which potential sources of bias are major and must somehow be controlled and which are minor and can be ignored, in knowing whether a proposed method of measurement has been tested for reliability and validity; in appraising which theories of behaviour are consistent with the results of a completed study, or more generally in deciding which type of study it is most fruitful to attempt next. Nevertheless, statisticians have much to contribute – particularly their training and experience in the conditions needed to provide sound inferences and their ability to develop efficient techniques for the analysis of the untidy data that are hard to avoid in observational studies.

Cochran's introduction ends with a quotation from Dorn (1953): "How would the study be conducted if it were possible to do it by controlled experimentation?" I believe that this question should be asked repeatedly, since it defines the objective of an observational study.

Section 2 of this paper, "Major Difficulties," provides a list of issues to be addressed in observational studies: "Setting up the comparisons" – define the precise questions to be asked. "The handling of disturbing variables" – look for a specialized environment in which the disturbing variables have similar distributions in treatment and control groups and use blocking or matching to stratify or select units into comparable groups. "The steps from association to causation" – the objective of an observational study is "to predict the consequences" of a new treatment if implemented, but to do this we usually must rely heavily on information not supplied by the study. "Inferences from sample to population" – these usually are difficult because the samples used are rarely probability samples from a reasonable target population. "Measurement" – problems of measurement are especially vexing in social science applications where, for instance, "measures of socioeconomic status have been developed, but it is hard to be sure that we have adjusted for the really relevant variable; further, errors of measurement decrease the effectiveness of the adjustment." "Multiple variables" – observational studies typically have many outcome and many disturbing variables, and these can create complications in analysis. Cochran's suggestions here remain equally valid today:

> The old maxim that the outlines of the analysis should be carefully sketched as a part of the research plan has lost none of its force. There is also, I believe, useful work to be done by statisticians in learning what some of the newer multivariate techniques really accomplish when applied to data and in explaining this to investigators, many of whom have no clear understanding of the techniques that they are trying to use. We need good expository papers of this type.

The final issue in this section is called "Long term studies" – longitudinal studies of treatments whose effects are slow to appear; these require "much organizational skill which may be only partially successful."   ·

Section 3 focuses on the crucial issue, "The Handling of Disturbing Variables." The first step is to classify disturbing variables into three types:

(1) Major variables for which some kind of matching or adjustment is considered essential. Their number is kept small in view of the complexities involved in matching or adjusting for many variables simultaneously. (2) Variables for which, ideally, we would like to match or adjust, but content ourselves with some verification that their effects produce little or no bias. (3) Variables whose effects, thought to be minor, are disregarded.

The widespread use of computers and the existence of simple multivariate matching techniques (e.g., such as described in Rubin, 1980b) makes it rather straightforward to match using many variables. Also, computers make multivariate covariance adjustment a simple computational task. Nevertheless, this classification is important because existing sample sizes often limit both (1) the number of matching variables that can be used without diluting the quality of matches on the most important disturbing variables and (2) the number of covariates that can be adjusted for by regression techniques.

Cochran's discussion of controlling for disturbing variables, "Matching and Adjustment," outlines the four basic methods that are available: "1. Matching," whereby matched treatment-control pairs are found with similar values of disturbing variables; "2. Equal Sample Sizes Within Subclasses," later called "stratified matching," whereby both populations are stratified into subclasses by the values of the disturbing variables, and within each subclass, samples of the *same size* are drawn from each population; "3. Adjustment for Subclass Differences," later called "subclassification," which is like "equal sample sizes within subclasses" except that the samples within each subclass represent random samples from the treatment and control populations and thus are generally of unequal size; and "4. Adjustment by Regression," which gives the covariance-adjusted estimator. With discrete disturbing variables, methods 1 and 2 are identical. Although Cochran states that "Adjustment by regression is not possible when x is categorical," this remark should not be taken literally since indicator variables for x and at least some of their interactions can be adjusted for by using standard regression techniques. A potentially misleading statement is that pair-matching totally removes bias due to x (continuous or discrete) "provided that the relation between y and x is the same in both populations" and that "the regression approach has the advantage that separate regressions can be computed in the two populations and used in the adjustment, this being a situation in which bias remains even with individual matching." This restriction to parallel regressions is not needed for perfect matching to remove all bias due to x if average treatment effects are being estimated for a target population for which the values of x in the matched sample are representative. This point is clarified in Cochran and Rubin (1973, Section 3.1). Cochran goes on to illustrate the properties of these methods in the special case of normally distributed x. He concludes this section with the following advice:

> To summarize, the similarities among the methods are greater than their differences. When feasible, matching is relatively effective. Overall, covariance seems superior to adjustment by sub-classification, though the superiority will seldom be substantial. If the original x-distributions diverge widely, none of the methods can be trusted to remove all, or nearly all, the bias. This discussion brings out the importance of finding comparison groups in which the initial differences among the distributions of the disturbing variables are small.

In Section 4, "Setting up the Comparisons," Cochran offers advice on the general design of an observational study:

> In making a choice between different studies he might undertake, the investigator should consider the resources and time needed and the status of each study with regard to the handling of disturbing variables and to the quality of the measurements. Other relevant factors are: (1) The quantities that can be estimated from the study. Sometimes one study yields only a correlation coefficient while another gives an estimate of the response curve of y to variations in x. (2) The range of variation of the suspected causal variable. In general, a study that furnishes a wider range of variation may be expected to give more precise estimates of the effect on y. (3) The relation to previous work. One study may be a new approach to the problem, another a repetition of studies done elsewhere. Both have their uses, but preference would normally be given to a new approach, especially if it seems free of some of the biases thought to be present in previous studies.

After presenting this general advice and some concerning the desirability of using multiple control groups with potentially different kinds of biases, Cochran defines four common types of comparisons. The before–after study is the only tool available when the treatment is effectively applied to everyone (e.g., as with national policies). The *ex post facto*, or retrospective, study creates "groups that differ in the y-variable and examines whether they differ in the suspected causal variables;" such studies are relatively efficient when dealing with y-variables that indicate rare events such as unusual diseases. "Multiple Causal Variables" studies are effectively regression studies, "... with no deliberate attempt to borrow the idea of factorial design. The disadvantage in this approach is that if the variables are highly correlated it becomes difficult to disentangle their effects. Further, if one of two correlated variables has a high error of measurement while the other does not, the regression coefficient on the first variable is an overestimate." The last type of study listed is called "Population Laboratory;" such a study is an attempt to accumulate enormous amounts of background and follow-up information on an entire population in some local area and supplies many future research efforts with data. An example of such a study which is now the basis of an observational study of the effect of in utero exposure to hormones and barbiturates is the PLUM sample (Zachau–Christiansen and Ross, 1975).

The final section of this very important paper, "The Step from Association to Causation," begins with the standard advice to "Make your theories elaborate" and goes on to explain that "... when constructing a causal hypothesis one should envisage as many *different* consequences of its truth as possible, and plan observational studies to discover whether each of these consequences is found to hold. If a hypothesis predicts that y will increase steadily as the causal variable z increases, a study with at least three levels of z gives a more comprehensive check than one with two levels." Other concluding advice on making the causal leap includes: replicating in various environments, discussing possible biases and alternative explanations for results, and speculating about causal mechanisms underlying the results.

6. THE EFFECTIVENESS OF ADJUSTMENT
BY SUBCLASSIFICATION IN REMOVING BIAS IN
OBSERVATIONAL STUDIES

Cochran (1968a) is a comprehensive discussion of the method of adjustment by subclassification presented in Cochran (1965). Before introducing any technical results, Cochran presents Canadian, British, and U.S. data on death rates for nonsmokers, cigarette smokers and cigar/pipe smokers. Without any adjustment for disturbing variables, nonsmokers fare no better than cigarette smokers, and cigar and pipe smokers fare worst. Adjusted death rates for the three treatment groups are then formed by: stratifying on age of person, finding death rates within each stratum (i.e., subclass), and then weighting these stratum-specific death rates by the number of nonsmokers in each stratum. After this adjustment for age by subclassification, it becomes obvious that cigarette smokers fare worst and nonsmokers and cigar/pipe smokers fare about equally.

Having exhibited the potential utility of adjustment by subclassification, Cochran begins his technical analysis. Assuming a disturbing variable x that is normally distributed in treatment and control populations with common variance, Cochran produces a table showing the percent reduction in bias of x as a function of (a) the number of standard deviations between the means of x in the two populations and (b) the number of subclasses used, where the subclass boundaries are chosen to yield subclasses of equal size in the control population:

Number of standard deviations between means δ	Number of subclasses				
	2	3	4	5	6
1	61.8	78.2	85.3	89.1	91.5
$\frac{1}{2}$	63.2	79.1	85.9	89.6	91.8
$\frac{1}{4}$	63.6	79.3	86.0	89.7	91.9

Cochran notes that these results appear insensitive to the number of standard deviations between the means, δ, which suggests that the results for very small δ may be representative of those for moderate δ; this leads to a clever analysis based on a Taylor series expansion about δ in order to study more general situations. The more general work reported by Cochran shows first that for normally distributed x, the maximum reduction in bias when using optimally sized subclasses only slightly exceeds the percentage reduction in bias using equally sized subclasses. Second, with nonnormal distributions and equally sized subclasses, the percentage reductions in bias are usually close to, although sometimes larger than, the corresponding percentage reductions in bias assuming normality. Cochran summarizes as follows:

> ... these calculations suggest that the normal values may serve as a working guide
> to the percentages of bias removed in practice with from 2 to 6 subclasses. There is
> a hint that the normal values may underestimate the effectiveness of adjustment for
> some types of distributions of x.

The table given above and the associated discussion can be used to defend the choice of five or six equally sized subclasses as effectively removing 90% or more of the bias due to a continuous underlying variable. Cochran goes on to consider the reduction in *variance* of comparisons due to stratified matching (which yields equal numbers of treatment and control units within each subclass) and adjustment by subclassification (which generally yields different numbers of treatment and control units in each subclass since the units are random samples from the populations). For normal data the percentage reduction in variance due to stratified matching equals the percentage reduction in bias due to stratified matching, but the percentage reduction in variance due to adjustment by subclassification is less than the percentage reduction in bias due to subclassification.

Cochran concludes the technical presentation with a discussion of the effects of errors of measurement in x. Under a simple linear model for the regression of outcome y on x, such errors "reduce the amount of bias removed to a factor $1/(1 + h)$ of its value, where h is the ratio of the variance of the errors of measurement to the variance of the correct measurements ... what seems most needed in order to appraise the seriousness of errors of measurement are data from which the values of h in different measurement problems can be estimated." This comment remains true today.

The final sentences exhibit the typical Cochran directness with respect to the relation between his work and the investigator's problem: "Further, this paper has considered only adjustments on a single variable x. The effectiveness of two-way and three-way classifications used to adjust simultaneously for two and three x-variables also merits investigation." Rosenbaum and Rubin (1983a) presents, I believe, a useful extension allowing adjustment by subclassification for multivariate x. The central idea there is to replace multivariate x with a scalar function of x, $e(x)$, called the *propensity score* because it gives the probability (i.e., the propensity) of being in the treatment versus control group at each value of x. Theory presented in Rosenbaum and Rubin (1983a) shows that proper adjustment for $e(x)$ alone removes all bias due to x. This result suggests the following procedure: estimate $e(x)$ from the data, form subclasses using the estimated scalar $e(x)$, and estimate treatment effects from these subclasses using the proportions of the target population in the subclasses as weights for global estimates. An example presented in Rosenbaum and Rubin (1983a) illustrates the application of the technique to coronary bypass data.

7. THE USE OF COVARIANCE IN OBSERVATIONAL STUDIES

Cochran's summary of this short but important note (Cochran, 1969) is a model of clarity:

> When two groups of subjects are being compared, one group exposed to some presumed causal force and the other not, adjustment of the difference $(\bar{y}_1 - \bar{y}_2)$ in mean responses by means of a regression on one or more x-variables is sometimes employed in order to remove possible biases that arise when subjects are not assigned to groups at random. In such applications, Belson (1956) has suggested that the adjustments be made by means of the regression for the unexposed group only, whereas the routine user of the analysis of covariance employs the pooled regression coefficients from both groups. This note tries to clarify the situations in which Belson's proposal is preferable.

Belson's article concerned a study of the effectiveness of a BBC series designed to teach tourists to France useful words and phrases, as well as other helpful information. The outcome variable y measured knowledge of the French language as well as facts relevant to France. Viewers were compared with nonviewers after controlling for two disturbing variables, $x_1 =$ the score on a French test not affected by the BBC series and $x_2 =$ a measure of educational level. Belson controlled these two covariates by a regression adjustment where the regression coefficients were estimated from the nonviewing group alone; unknown to Cochran or Belson at the time was the fact that, in a sociological context, Peters (1941) had proposed this same method of regression adjustment earlier, as a method for matching without loss of control subjects.

It is easily seen that if (a) the regression of y on (x_1, x_2) is linear in the nonviewing population, (b) the average effect of viewing is desired for the population of viewers (i.e., those that chose to view the program), and (c) x_1 and x_2 are the only relevant disturbing variables, then the Peters–Belson procedure produces an unbiased estimate of the effect of viewing for the population of viewers. The usual covariance adjustment, which estimates the regression coefficients from both samples, is generally only unbiased for this effect when the regressions of y on (x_1, x_2) in the two groups are parallel. When the regressions are parallel, the usual covariance-adjusted estimator is to be preferred to the Peters–Belson estimator because its variance is smaller. Thus, which estimator is preferable depends upon the differences between the regressions of y on (x_1, x_2) in the two groups as well as on the sample sizes. In conclusion, Cochran states that "a more general moral of this note is that before making a routine application of the analysis of covariance in an observational study, the investigator should examine both the adequacy and the implied meaning of his model."

My assessment of the trade-off between the bias and variance of the Peters–Belson estimator and the bias and variance of the standard analysis of covariance estimator is that, very often, the sample sizes are such that the possible bias overwhelms the potential increased precision of the standard covariance estimator. Consequently, when I designed a reanalysis of the Federal Trade Commission's study of the effects of coaching on the Scholastic Aptitude Test, summarized in Messick (1980), I developed a generalization of the Peters–Belson estimator which allowed for different patterns of missing data among the disturbing variables. A recent analysis of an observational study essentially rediscovering the Peters–Belson procedure is the Coleman et al. (1981) study of the effectiveness of private versus public schools.

A related analysis that I designed compared the criminalities of 28 males with an extra chromosome (12 XYY males and 16 XXY males) with 3738 XY males, controlling for height, parents' socioeconomic status, education, and measured intelligence (Witkin et al., 1976). Here the outcome variable was dichotomous (criminal record or not), but still the control group's regression was used to predict outcomes for the XYY's and the XXY's and thereby to obtain estimates of effects for the XYY and XXY populations having controlled for the four disturbing variables. Specifically, the conditional distribution of criminality given the four disturbing variables was estimated from the 3738 XY's by logistic regression, and a predicted probability of being a criminal was obtained for each XYY and each XXY. The differences between the observed criminalities and these predicted probabilities for the XYY group and the XXY group estimate the effects of the extra chromosome for these groups having adjusted for the four disturbing variables.

8. OBSERVATIONAL STUDIES

Cochran (1972) is an easy to read article that begins with a review of examples of observational studies and notes their increased numbers in recent years. He mentions, as examples of observational studies for treatment effects, the Cornell study of seat belts (Kilhberg and Narragon, 1964), studies of smoking and health (U.S. Surgeon General's Committee, 1964), the halothane study (Bunker et al., 1969), and the Coleman report (Coleman et al., 1966). In many ways this article is an updated, conversational summary of Cochran (1965). Here, even more than before, mature advice to the investigator is the focus of the paper.

First, the investigator should clearly state the objective and hypothesis of the study because such "... statements perform the valuable purpose of directing attention to the comparisons and measurements that will be needed." Second, the investigator should carefully consider the type of study. Cochran seems to be more obviously negative about studies without control groups than he was earlier:

> Single-group studies are so weak logically that they should be avoided whenever possible....

> Single-group studies emphasize a characteristic that is prominent in the analysis of nearly all observational studies – the role of judgment. No matter how well-constructed a mathematical model we have, we cannot expect to plan a statistical analysis that will provide an almost automatic verdict. The statistician who intends to operate in this field must cultivate an ability to judge and weigh the relative importance of different factors whose effects cannot be measured at all accurately.

> Comparison groups bring a great increase in analytical insight. The influence of external causes on both groups will be similar in many types of study and will cancel or be minimized when we compare treatment with no treatment. But such studies raise a new problem – How do we ensure that the groups are comparable?

Cochran still emphasizes the importance of effective measurement:

> The question of what is considered relevant is particularly important in program evaluation. A program may succeed in its main objectives but have undesirable side effects. The verdict on the program may differ depending on whether or not these side effects are counted in the evaluations.... Since we may have to manage with very imperfect measurements, statisticians need more technical research on the effects of errors of measurement.

But the primary statistical advice Cochran has to offer is on controlling bias: "The reduction of bias should, I think, be regarded as the primary objective – a highly precise estimate of the wrong quantity is not much help.... In observational studies three methods are in common use in an attempt to remove bias due to extraneous variables.... Blocking, usually known as matching in observational studies.... Standardization (adjustment by subclassification).... Covariance (with x's quantitative), used just as in experiments." The de-emphasis of efficiency relative to bias removal was evident when I began my thesis work under Cochran in 1968. The results of this thesis (Rubin, 1970), in large part published in Rubin (1973a, 1973b) and summarized in Cochran and Rubin (1973), lead to some new advice on the trade-off between matching and covariance: In order to guard against nonlinearities in the regression of y on x, the combination of regression and matching appears superior to either method alone.

Recent work (Rubin, 1979b) extends this conclusion to more than one x. Specifically, based on Monte Carlo results with 24 moderately nonlinear but parallel response surfaces and 12 bivariate normal distributions of x, and using percentage reduction in expected squared bias of treatment effect as the criterion, it appears quite clear that the combination of matched sampling and regression adjustment is superior to either matching or regression adjustment alone. Furthermore, Mahalanobis metric matching, which defines the distance between a treatment and control unit using the inverse of the sample covariance matrix of the matching variables and then sequentially finds the closest unmatched control unit for each experimental unit, was found superior to discriminant matching, which forms the best linear discriminant between the groups and sequentially finds the closest unmatched control unit for each experimental unit with respect to this discriminant. Moreover, regression adjustment that estimates the regression coefficient from the regression of the matched pair y differences on the matched pair x differences is superior to the standard covariance-adjusted estimator, which estimates the coefficients from the pooled within-group covariance matrix.

Cochran goes on to offer advice on sample sizes, handling nonresponse, the use of a pilot study, the desire for a critical colleague in the planning stages, and the relation between sampled and target populations.

It is not surprising that this article concludes with a short section called "Judgment About Causality." Cochran's views are somewhat more bluntly presented here than in previous writing:

> It is well known that evidence of a relationship between x and y is no proof that x causes y. The scientific philosophers to whom we might turn for expert guidance on this tricky issue are a disappointment. Almost unanimously and with evident delight they throw the idea of cause and effect overboard.... A claim of proof of cause and effect must carry with it an explanation of the mechanism by which the effect is produced. Except in cases where the mechanism is obvious and undisputed, this may require a completely different type of research from the observational study that is being summarized. Thus in most cases the study ends with an opinion or judgment about causality, not a claim of proof.

Cochran closes with the standard advice to make causal hypotheses complex:

> Given a specific causal hypothesis that is under investigation, the investigator should think of as many consequences of the hypothesis as he can and in the study try to include response measurements that will verify whether these consequences follow.

9. CONTROLLING BIAS IN OBSERVATIONAL STUDIES: A REVIEW – 1973

Cochran and Rubin (1973) was prepared for the Mahalanobis Memorial Volume of *Sankhyā* as a review of some of the newer developments relevant to the analysis of observational studies, including those from Cochran (1968a, 1969), my thesis (Rubin, 1970), and my current research on multivariate matching methods. The article's summary highlights the important conclusions:

> This paper reviews work on the effectiveness of different methods of matched sampling and statistical adjustment, alone and in combination, in reducing bias due to confounding x-variables when comparing two populations. The adjustment methods were linear regression adjustment for x continuous and direct standardization for x categorical.

> With x continuous, the range of situations examined included linear relations between y and x, parallel and non-parallel, monotonic non-linear parallel relations, equal and unequal variances of x, and the presence of errors of measurement in x.

> The percent of initial bias $E(\bar{y}_1 - \bar{y}_2)$ that was removed was used as the criterion. Overall, linear regression adjustment on random samples appeared superior to the matching methods, with linear regression adjustment on matched samples the most robust method. Several different approaches were suggested for the case of multivariate x, on which little or no work has been done.

Since this article is itself largely a technical summary of work on matching, subclassification, and regression adjustment without much of the general advice common in Cochran's previous papers, I only outline its content here.

Section 2 summarizes work when both y and x are continuous and the regressions are parallel and linear in the two groups. Results are presented for: regression adjustment, caliper matching (which finds for each experimental unit a control unit whose x is within $\pm a$ of the experimental unit's x, the name "caliper matching" coined in Althauser and Rubin, 1970), nearest available matching (which sequentially matches each experimental subject with the closest unmatched control subject, i.e., with the nearest available match), and mean matching (which chooses the sample of matched controls so that its mean x is closest to the mean x in the experimental group; some early work on this method appears in Greenberg, 1953). Most of this material is presented in more detail in Rubin (1973a), which studies mean matching and three nearest available pair-matching methods with respect to their relative ability to remove bias as well as with respect to the maximum possible bias reduction given fixed distributions and fixed sample sizes, and in Rubin (1973b), which presents Monte Carlo results on the ability of nearest available pair matching and various regression adjustments to remove bias. Incidentally, Cochran was a fan of nearest available matching, because by matching each experimental unit to the nearest available control by computer, the problem of unmatched units is avoided. Discarding unmatched experimental units can be a problem for two reasons. First, if the experimental and control response surfaces are parallel, discarding experimental units results in reduced sample size and thus less precise estimates. Second, and more important, if the response surfaces are nonparallel and average effects are desired for a population represented by the experimental group (as implicit with the Peters–Belson estimator discussed in Section 7), then discarding experimental units generally implies that an estimate based on the remaining units will be biased since they will not be representative of the experimental population. An application of nearest available matching to the study of coronary artery disease is given in Cohn et al. (1981).

Section 3 considers complications: regressions linear but not parallel, regressions parallel but quadratic, regressions nonlinear, errors of measurement in x, and an omitted confounding (i.e., disturbing) variable. Matching, covariance adjustment, and their combination are considered. The general conclusion is that the combination of matching and covariance is the preferred procedure; however, matching on x and covariance on x remove only the component of the original bias that is due to x and not that due to uncorrelated errors in x or the orthogonal-to-x component of an omitted confounding variable.

Section 4 summarizes work on matching and adjustment by subclassification when the matching variable is qualitative. The primary work reviewed is Cochran (1968a), reviewed here in Section 6, and as mentioned there, I believe that recent work in Rosenbaum and Rubin (1983a) presents a useful extension of the method of subclassification.

Section 5 presents some of the obvious multivariate generalizations. One of the not-so-obvious generalizations concerns the effects of errors of measurement; even with parallel linear regressions and uncorrelated errors, the percent reduction in bias due to matching and/or covariance adjustment is not necessarily between 0 and 100%.

Section 6 considers multivariate generalizations of matching methods and is a preliminary presentation of work that subsequently was published in *Biometrics* (Rubin, 1976b, 1976c, 1980b). The multivariate matching methods discussed include caliper matching, categorical or stratified matching, matching on the best linear discriminant, and nearest available metric matching, which uses a measure of distance such as the Mahalanobis metric to define how close experimental and control subjects are to each other. Rubin (1976b) extends this discussion by presenting examples of multivariate matching methods that yield the same percentage reduction in bias for each matching variable for various underlying distributions; such matching methods are called equal percent bias reducing (EPBR) and are attractive when outcome variables are thought to have basically linear regressions on the matching variables. Rubin (1976c) presents analytic results giving the maximum percentage reduction in bias possible due to an EPBR matching method as a function of sample sizes and distributional conditions; these results lead to suggestions for the ratio of initial control to experimental sample sizes needed to obtain well-matched samples. Related work on the ability to obtain well-matched samples with multivariate normal x appears in Carpenter (1977). Rubin (1980b) presents a Monte Carlo study of the ability of nearest available Mahalanobis-metric matching to reduce bias.

In the concluding section of Cochran and Rubin (1973), we remind the reader that our concern has been with bias removal, not increased precision as would be appropriate for randomized experiments where no bias is present. We summarize the advice that regression adjustment alone appears to be superior to matching alone, but that the combination appears to be better than either alone. Finally, we indicate how little is really known about the performance of these methods in practice. Although Rubin (1979b) extended the conclusion regarding the superiority of the combination of matching and regression adjustment to cases with bivariate x (briefly summarized here in Section 8), the study of methods for controlling bias in observational studies remains today a fertile field for good work in applied statistics.

10. RELATED WORK

Cochran's ideas about observational studies appear in places other than the seven major articles reviewed here in Sections 3–9. Three reports by Cochran to the Milbank Memorial Fund and presented at their annual meetings were published from 1952 to 1955. The first, "An Appraisal of the Repeated Population Censuses in the Eastern Health District, Baltimore" (Cochran, 1952), is relevant to the problem of observational studies because it emphasizes that large databases can be an extremely valuable source for supplying control groups. The second, "Analysis of Records with a View to Their Evaluation" (Cochran, 1953b), emphasizes the importance of obtaining comparable groups when trying to estimate effects of treatments from observational data. He concludes with some advice for the conduct of observational studies that I have not seen elsewhere:

> Finally, I would like to mention one piece of advice which is usually given to Scottish boys shortly after they are weaned, at the critical time when father first puts a golf club in the boy's hands: 'Keep your eye on the ball and don't press'. In a study of this kind, all sorts of suggestions are made for interesting sidelines that might be explored, but if the team is to get anywhere it must construct a system of priorities against blandishments and outside suggestions, unless and until it has the resources to cope with its main objectives with something to spare for additional studies.

> And by 'don't press,' I mean this: This kind of study is very expensive: it takes a long time and attracts many visitors. The members of the team may come to feel that they are under pressure to produce results and they may begin to worry when some things inevitably go wrong. It is most important to do anything that can be done to lighten this pressure and encourage team members to be more relaxed.

Cochran's third Milbank article, "Research Techniques in the Study of Human Beings" (Cochran, 1955), is delightful and full of good humor. One clear piece of advice on observational studies is offered: "If nature mixes things up thoroughly, as she sometimes seems to do, statistical methods will not sort them out very well.... Experimentation (in the sense in which I have used it) needs to be exploited as much as possible. The question: 'Why can't I do an experiment?' is always worth asking, even if it sounds unrealistic."

In 1967 Cochran presented a paper at the 12th Conference on the Design of Experiments in Army Research and Development (Cochran, 1967a) with material similar to, but less technical than, that presented earlier (Cochran, 1965) and more extensive than that presented later (Cochran, 1972). An unpublished Harvard University Technical Report (Cochran, 1970a) studies a procedure not uncommon in practice: calculate the t test for the difference of x-means; if it is not significant use $\bar{y}_1 - \bar{y}_2$ to estimate the treatment effect, and if it is significant use the covariance-adjusted estimate. Two additional Cochran articles are relevant to observational studies because they describe the development of the role of randomization in experiments (Cochran, 1974; Cochran, 1978). Any study without randomization usually should be considered to be an observational study because we do not know what hidden biases may have caused some units to be exposed to the experimental treatment and other units to be exposed to the control treatment.

Cochran's thoughts on observational studies were systematically expressed in his seminar on observational studies, Statistics 284, at Harvard University, which some lucky students attended over the years. Looking through my class notes from 1968, I see many of the points and themes reviewed above. There was, in the seminar relative to his papers, somewhat more review of basic material on the benefits of randomization and, it seems, more emphasis on trying to speculate about the possible sizes of biases. But basically the material was largely taken from his first six major articles reviewed here, plus some initial results from my thesis. The list of references handed out by Cochran is given below.

Statistics 284

Reference Material – General Books

Glock (ed.). In *Survey Research in the Social Sciences.* Chapter 1, Survey Design and Analysis. Russell Sage Foundation, 1967.

Campbell and Stanley. *Experimental and Quasi-experimental Designs for Research.* Rand McNally, 1966.

Lerner (ed.). *Cause and Effect.* MIT Press, 1965.

Blalock. *Causal Inferences in Nonexperimental Research.* University of North Carolina Press, 1961.

MacMahon, Pugh, and Ipsen. *Epidemiologic Methods.* Little, Brown, 1960.

Suchman. *Evaluation Research.* Russell Sage Foundation, 1967.

Hyman. *Survey Design and Analysis.* Free Press, Glencoe, 1957.

Large Studies

Kinsey, Pomeroy, and Martin. *Sexual Behavior in the Human Male.* W. B. Saunders, 1948.

Coleman et al. *Equality of Educational Opportunity.* U.S. Government Printing Office, 1966.

Sroel et al. *Mental Health in the Metropolis.* McGraw-Hill, 1962.

Surgeon General's Committee. *Smoking and Health.* P.H.S. Publication 1103, 1964.

11. PLANNING AND ANALYSIS OF OBSERVATIONAL STUDIES

This short book was published posthumously late in 1983, the result of editing efforts by Lincoln Moses and Frederick Mosteller. I was somewhat surprised to see a nearly complete monograph by Cochran on the topic of observational studies. It was my impression that Cochran worked on such a monograph in the early sixties and then again in the late sixties, but shelved the idea of publishing it, first in favor of his *JRSS* discussion paper (Cochran, 1965) and then in favor of our joint *Sankhyā* paper (Cochran and Rubin, 1973). In fact, at the time we were writing the *Sankhyā* paper we had a conversation about the monograph during which he indicated that he felt there was not yet enough straightforward, solid advice to the investigator to warrant the publication of an entire book; Cochran felt that a good book in applied statistics should contain easily followed, precise guidance within a well structured outline, presumably like his books on sampling (Cochran, 1977) and experimental design (Cochran and Cox, 1957). Bill was apparently ambivalent about this point – as related by Moses and Mosteller in their preface: "More research is needed in many spots 'before I'll have something worth saying. I think, however, that if I pay too much attention to this point, it will never be written.' "

Moses and Mosteller have made available to us Cochran's nearly completed monograph on observational studies with minimal editing. As a result, the monograph truly reflects Cochran's views, and the major themes of advice expressed in his articles and the basic Cochran directness are clearly visible. Cochran's distinctive style would have been nearly impossible to preserve if extensive editing had been attempted.

Another result of the minimal editing is that it is possible to guess that much of the monograph was written prior to 1972 and thus does not necessarily reflect Cochran's latest thinking on observational studies. There are several reasons for this conclusion, the principal one involving the referenced work. Most of the references are to publications prior to 1971. Cochran's primary reference to my work on matching and covariance adjustment is my thesis (Rubin, 1970) rather than the later and more accessible *Biometrics* articles or our joint 1973 *Sankhyā* paper, which is nowhere even mentioned in the monograph. In addition, the overly restrictive advice regarding matching with nonparallel response surfaces, which appeared in Cochran (1965) but was clarified in the *Sankhyā* paper, appears in the monograph in its pre-*Sankhyā* version.

Although the monograph is nominally written for the investigator with limited statistical background and his articles were primarily written for the statistician, it is instructive to compare the content of this posthumous monograph with the content in the ensemble of Cochran's work on observational studies previously reviewed in this chapter, as if the ensemble were itself a monograph. First, it is not surprising that the ensemble presents more technical work on matching, standardization, and covariance adjustment, and in greater detail, whereas the monograph presents more basic statistical material (e.g., tests of significance, confidence intervals, sample size calculations, regression using dummy variables). It is also not surprising that the monograph, far more than the ensemble, addresses standard issues that arise in sample surveys (e.g., cluster sampling, design effects, nonresponse, questionnaire design, interviewer training). I do find it interesting, however, that the monograph repeatedly emphasizes the distinction between the sampled and target populations and the bias that can arise when ignoring the distinction. Although this topic is mentioned in his articles, as when discussing the effect of using different weights in subclassification, it never receives the repeated attention it does in the monograph. Perhaps Cochran felt that practical sampling issues should be so well-known to statisticians as to need no special attention in articles written for them.

Finally, the ensemble of articles offers more direct advice on *inference for causal effects*. There is no section in this posthumous publication on the steps from association to causation, perhaps because such a section would naturally have appeared at the end as in Cochran (1953a), Cochran (1965), and Cochran (1972), and Bill never completed a draft of the final chapter.

12. STATISTICAL THEORY AND PRACTICE

I conclude this review of Cochran's work on observational studies with some more personal comments on his attitude towards statistical theory and practice. I think his attitude can be summarized as follows: pass along to the practitioner only those pieces of theory that are sensible and useful. Cochran claimed in conversation with me in 1978 after presenting a lecture on drawing causal inferences that he "wasn't very good at the foundations of statistics." This statement puzzled me, and when I pursued it with him, it turned out that what he really meant was more like: in his experience, many supposed insights arising from foundations often seem unhelpful in practice; consequently, he did not understand the criteria used to judge contributions to the foundations of statistics.

Bill was basically atheoretical, with the good sense to let evidence decide which techniques to recommend. He was a frequentist in the obvious sense that he always wanted to know the operating characteristics of the procedures he studied: procedures (e.g., confidence limits) that he recommended should work as advertised in long-run practice. But he was also a Bayesian, at least in the uncontroversial sense that he wanted to use all of his and the investigator's knowledge when setting up the design and analysis of a problem. As an illustration, once, when as a graduate student I was consulting for DATA-TEXT (a Harvard-based statistical computing package), I asked Bill for advice on which of several choices of default options would be best for an analysis of variance routine. He wanted to know what sort of data set was involved, pigs or children or razor blades. I replied that we didn't know that. He then responded that he realized DATA-TEXT needed an answer, but that since a good answer depended on the properties of the actual data set, DATA-TEXT was asking for a choice among bad answers, and suggested that I ask someone else to provide it. Bill was also rather Bayesian in the more controversial sense of being willing to intentionally misinterpret confidence statements as probability statements about the likely values of unknown parameters, knowing it was very likely that this was the way the practical investigator would view them anyway.

I conclude with a Sir Ronald Alymer Fisher story Cochran told me that exhibits Bill's wonderful sense of humor and that, with some imagination, can be used to illustrate Bill's attitude toward the theory and practice of statistics. Cochran (1967b) relates a very similar version of this story.

Fisher and Cochran were waiting to cross Gower Street in London after a seminar at University College. The street was very busy with four lanes of traffic, and Fisher, due to his very poor eyesight, was to follow Cochran's lead on crossing the street.

But Fisher soon became very impatient with Bill's caution and finally grabbed Bill by the arm, dragged him into the cross traffic at a nearly random moment, saying, "Come on Cochran, a spot of natural selection never hurt anyone."

Bill's comment to me when concluding this story was something like: "Never one to argue with Fisher on theoretical issues, off we went."

Now returning to the point on the interplay of theory and practice: Bill's good sense never led him to proselytise at University College or anywhere else, this particular bit of Fisher theory.

Bill was a wonderful man, and his sense of humor, his wisdom, and his love for his work and his colleagues, all combine to keep Bill Cochran very much alive in many of our minds and hearts.

ACKNOWLEDGMENTS

I wish to thank P. R. Rosenbaum, R. J. A. Little, A. P. Dempster, and the editors, P. S. R. S. Rao and J. Sedransk, for helpful comments on an earlier draft of this chapter.

2. Controlling Bias in Observational Studies: A Review

William G. Cochran and Donald B. Rubin

Abstract: This paper reviews work on the effectiveness of different methods of matched sampling and statistical adjustment, alone and in combination, in reducing bias due to confounding x-variables when comparing two populations. The adjustment methods were linear regression adjustment for x continuous and direct standardization for x categorical.

With x continuous, the range of situations examined included linear relations between y and x, parallel and non-parallel, monotonic non-linear parallel relations, equal and unequal variances of x, and the presence of errors of measurement in x.

The percent of initial bias $E(\bar{y}_1 - \bar{y}_2)$ that was removed was used as the criterion. Overall, linear regression adjustment on random samples appeared superior to the matching methods, with linear regression adjustment on matched samples the most robust method. Several different approaches were suggested for the case of multivariate x, on which little or no work has been done.

1. INTRODUCTION

An observational study differs from an experiment in that the random assignment of treatments (i.e., agents, programs, procedures) to units is absent. As has been pointed out by many writers since Fisher (1925), this randomization is a powerful tool in that many systematic sources of bias are made random. If randomization is absent, it is virtually impossible in many practical circumstances to be convinced that the estimates of the effects of treatments are in fact unbiased. This follows because other variables that affect the dependent variable besides the treatment may be differently distributed across treatment groups, and thus any estimate of the treatment is confounded by these extraneous x-variables.

Given the choice between an observational study and an essentially equivalent randomized experiment one would prefer the experiment. Thus in the Report of the President's Commission on Federal Statistics (1971), Light, Mosteller, and Winokur urge greater efforts to use randomized studies in evaluating public programs and in social experimentation, despite the practical difficulties. Often however, random assignment of treatments to units is not feasible, as in the studies of the effects of smoking on health, complications of pregnancy on children, or long-term exposure to doses of radiation on uranium mine workers. Also, as in these examples, one might have to wait many years for the results of an experiment while relevant observational

Reprinted from *Sankhyā: The Indian Journal of Statistics*, 1973, Series A, Vol. 35, Part 4, 417–466.

data might be at hand. Hence, although inferior to an equivalent experiment, an observational study may be superior to or useful in conjunction with a marginally relevant experiment (e.g., one on the long-term effects of radiation on white rats). In addition, the analysis of data from observational studies can be useful in isolating those treatments that appear to be successful and thus worth further investigation by experimentation, as when studying special teaching methods for underprivileged children.

In dealing with the presence of confounding variables, a basic step in planning an observational study is to list the major confounding variables, design the study to record them, and find some method of removing or reducing the biases that they may cause. In addition, it is useful to speculate about the size and direction of any remaining bias when summarizing the evidence on any differential effects of the treatments.

There are two principal strategies for reducing bias in observational studies. In matching or matched sampling, the samples are drawn from the populations in such a way that the distributions of the confounding variables are similar in some respects in the samples. Alternatively, random samples may be drawn, the estimates of the treatment being adjusted by means of a model relating the dependent variable y to the confounding variable x. When y and x are continuous, this model usually involves the regression of y on x. A third strategy is to control bias due to the x-variables by both matched sampling and statistical adjustment. Notice that the statistical adjustment is performed after all the data are collected, while matched sampling can take place before the dependent variable is recorded.

This paper reviews work on the effectiveness of matching and statistical adjustments in reducing bias in a dependent variable y and two populations P_1 and P_2 defined by exposure to two treatments. Here, the objective is to estimate the difference $(\tau_1 - \tau_2)$ between the average effects of the treatments on y.

Section 2 reviews work on the ability of linear regression adjustment and three matching methods to reduce the bias due to x in the simplest case when both y and x are continuous, there are parallel linear regressions in both populations, and x is the only confounding variable. Section 3 considers complications to this simple case: non-parallel regressions, non-linear regressions, errors of measurement in x, and the effect of an omitted confounding variable. Section 4 extends the above cases to include x categorical or made categorical (e.g., low, medium, high). Section 5 presents some multivariate x results which are simple generalizations of the univariate x results. Section 6 considers some multivariate extensions of matching methods. A brief summary of the results and indications for further research are given in Section 7.

2. y, x CONTINUOUS: UNIVARIATE PARALLEL LINEAR REGRESSIONS

2.1. The Model

We begin with the simple case when y and a univariate x are both continuous, and the regressions of y on x are linear and parallel in both populations. For the j-th observation from population i, the model may be written

$$y_{ij} = \mu_i + \beta(x_{ij} - \eta_i) + e_{ij} \qquad (2.1)$$

with

$$E(e_{ij} \mid x_{ij}) = 0, \quad E\left(e_{ij}^2 \mid x_{ij}\right) = \sigma_i^2$$

where μ_i and η_i are the means of y and x respectively in population i, where $\eta_1 > \eta_2$ without loss of generality. Thus the regressions of y on x differ by the constant

$$E(y_{1j} - y_{2j} \mid x_{1j} = x_{2j}) = (\mu_1 - \mu_2) - \beta(\eta_1 - \eta_2). \tag{2.2}$$

If x is the only variable (besides the treatment) that affects y *and* whose distribution differs in the two populations, (2.2) equals the difference in the average effects of the treatments, $\tau_1 - \tau_2$. Thus, in this case, the treatment difference in (2.2) is constant at any level of x.

From (2.1) it follows that conditionally on the values of x_{ij} in samples chosen either randomly or solely on x,

$$E_c(\bar{y}_1 - \bar{y}_2) = (\mu_1 - \mu_2) + \beta(\bar{x}_1 - \eta_1) - \beta(\bar{x}_2 - \eta_2)$$
$$= \tau_1 - \tau_2 + \beta(\bar{x}_1 - \bar{x}_2). \tag{2.3}$$

Letting E_r be the expectation over the distribution of variables in random samples,

$$E_r(\bar{y}_1 - \bar{y}_2) = \mu_1 - \mu_2 = \tau_1 - \tau_2 + \beta(\eta_1 - \eta_2) \tag{2.4}$$

so that the expected bias in $(\bar{y}_1 - \bar{y}_2)$ from random samples is $\beta(\eta_1 - \eta_2)$.

2.2. Linear Regression Adjustment

Since from (2.3) $\bar{y}_1 - \bar{y}_2$ is conditionally biased by an amount $\beta(\bar{x}_1 - \bar{x}_2)$ in random and matched samples, it is reasonable to adjust $\bar{y}_1 - \bar{y}_2$ by subtracting an estimate of the bias. The adjusted estimate would then be

$$\hat{\tau}_1 - \hat{\tau}_2 = (\bar{y}_1 - \bar{y}_2) - \hat{\beta}(\bar{x}_1 - \bar{x}_2).$$

In practice, $\hat{\beta}$ is most commonly estimated from the pooled within-sample regressions. With this model, however, $E_c(\hat{\beta}) = \beta$ either for the pooled $\hat{\beta}$ or for $\hat{\beta}$ estimated from sample 1 or sample 2 alone. From (2.3) for any of these $\hat{\beta}$,

$$E_c(\hat{\tau}_1 - \hat{\tau}_2) = \mu_1 - \mu_2 - \beta(\eta_1 - \eta_2) = \tau_1 - \tau_2.$$

For this model, the regression adjustment removes all the bias either for random samples or for matched samples selected solely using x.

Before using the regression adjusted estimate, the investigator should satisfy himself that the regressions of y on x in the two populations appear linear and parallel. Standard methods of fitting higher order terms in x and separate β's in the two samples are appropriate for helping to answer this question.

2.3. Caliper Matching

In order to construct matched samples of size n, the investigator needs initial reservoirs of data of sizes $r_1 n$, $r_2 n$ from which to seek matches, where $r_i \geq 1$ with at least one $r_i > 1$. The work to be reported here is for the case $r_1 = 1$ in which there is a random sample of size n from population 1 to which the sample from population 2 is to be matched from a reservoir of size rn $(r > 1)$. This case is appropriate in studies in which population 1 is of primary interest, population 2 being a control population (untreated or with a standard treatment) with a larger reservoir from which a sample matched to sample 1 is drawn. The case of only one reservoir is a fairly severe test for matching since it is easier to obtain close matches with reservoirs from both populations.

With a random sample from population 1 and some kind of matched sample from population 2 chosen using x, relation (2.3) gives the expected bias of matched samples as

$$E_m(\bar{y}_1 - \bar{y}_2) - (\tau_1 - \tau_2) = \beta\{\eta_1 - E_m(\bar{x}_2)\} \tag{2.5}$$

where E_m is the expectation over the distribution of variables in samples from population 2 matched on x.

The criterion to be used in judging the effectiveness of matching will be the percentage reduction in bias. From (2.4) and (2.5) this is

$$\theta = (100)\frac{E_m(\bar{x}_2) - \eta_2}{\eta_1 - \eta_2}.$$

We note that with this model the percentage reduction in expected bias of $(\bar{y}_1 - \bar{y}_2)$ equals that in $(\bar{x}_1 - \bar{x}_2)$.

As a measure of the amount of initial bias in x when appraising methods of matching or adjustment, we chose the quantity

$$B = (\eta_1 - \eta_2) \Big/ \left(\frac{\sigma_1^2 + \sigma_2^2}{2}\right)^{1/2}$$

and examined values of B in the range $(0, 1)$. A value of $B = 1$ is considered large. With this bias, the difference $(\bar{x}_1 - \bar{x}_2)$ has about a 90% chance of being detected as significant (5% level) in random samples of 25 when σ_1^2, σ_2^2 are not too unequal. The values of σ_1^2/σ_2^2 studied were $\frac{1}{2}, 1, 2$.

The first method of matching investigated, often used with x continuous, is paired caliper matching. Each x_{1j} has a partner x_{2j} such that

$$|x_{1j} - x_{2j}| \leqslant c.$$

This method is attractive from two points of view. Although we are assuming at present a *linear* regression of y on x, it is clear that a tight caliper matching should remove nearly all the bias in $(\bar{y}_1 - \bar{y}_2)$ under any smooth regression, linear or non-linear, that is the same in both populations. Secondly, at first sight this method provides convenient data for investigating how $E_c(y_{1j} - y_{2j})$ varies with x, since x is close to constant for any single pair.

In presenting results on the percent reductions in bias for x normal (Table 2.1), we have taken

$$c = a\sqrt{(\sigma_1^2 + \sigma_2^2)/2}$$

where $a = 0.2(0.2)1.0$. Strictly, the results hold for $B < 0.5$ but for B between 0.5 and 1, the percent reductions are only about 1 to $1\frac{1}{2}$% lower than the figures shown. A tight matching ($a = 0.2$) removes practically all the bias, while a loose matching ($a = 1.0$) removes around 75%. The ratio σ_1^2/σ_2^2 has a minor effect, although performance is somewhat poorer as σ_1^2/σ_2^2 increases.

Table 2.1. *Percent reduction in bias of x for caliper matching*
to within $\pm a\sqrt{(\sigma_1^2 + \sigma_2^2)/2}$ *with x normal*

a	$\sigma_1^2/\sigma_2^2 = \frac{1}{2}$	$\sigma_1^2/\sigma_2^2 = 1$	$\sigma_1^2/\sigma_2^2 = 2$
0.2	.99	.99	.98
0.4	.96	.95	.93
0.6	.91	.89	.86
0.8	.86	.82	.77
1.0	.79	.74	.69

A disadvantage of caliper matching in practical use is that unless r is quite large there is a non-negligible probability that some of the desired n matches are not found in the reservoir. Nothing seems to be known about the distribution of the number of matches found as a function of r, a, $(\eta_1 - \eta_2)$ and σ_1^2/σ_2^2. We have not investigated the consequences of incomplete matching as often results in practice. Thus we have no help to give the investigator in estimating the reservoir size needed and the probable percent success in finding caliper matches.

2.4. 'Nearest Available' Matching

This disadvantage is avoided by a method (Rubin, 1973a) in which all n pair matches are easily formed by computer. The n values of x from sample 1 and the rn values from reservoir 2 are entered in the computer. In one variant of the method, the sample 1 values of x are first arranged in random order from x_{11} to x_{1n}. Starting with x_{11}, the computer selects the value x_{21} in reservoir 2 nearest to x_{11} and lays this pair aside. The computer next seeks a 'nearest available' partner for x_{12} from the $(rn - 1)$ remaining in reservoir 2, and so on, so that n matches are always found although the value of a is not controlled.

Two other variants of this 'nearest available' method were examined. In these, the members of sample 1 were (i) first ranked from highest to lowest, (ii) first ranked from lowest to highest, before seeking matches from the ranked samples. For $\eta_1 > \eta_2$, Monte Carlo results with x normal showed that for the percent reductions θ in bias of $(\bar{x}_1 - \bar{x}_2)$, $\theta_{LH} > \theta_{ran} > \theta_{HL}$. If, however, the quality of the matches is judged by the average MSE within pairs, $E_m(x_{1j} - x_{2j})^2$, the order of performance was opposite: $MSE_{HL} < MSE_{ran} < MSE_{LH}$. Both sets of results have rational explanations. The differences in performance were usually small. On balance, random ordering is a reasonable compromise as well as quickest for the computer.

For random ordering, Table 2.2 shows the percent reductions in bias of $(\bar{x}_1 - \bar{x}_2)$ and hence of $(\bar{y}_1 - \bar{y}_2)$ for $r = 2, 3, 4, n = 25, 50$ and different combinations of the initial bias B and the σ_1^2/σ_2^2 ratio. Results for $n = 100$ (not shown) differ by at most one or two percentage points from those for $n = 50$, suggesting that the $n = 50$ results hold also for $n > 50$. With this method, the percent reduction in bias decreases steadily as the bias B increases from 1/4 to 1, so that results are given separately for the four values of B.

Table 2.2. *Percent reduction in bias for random order, nearest available matching;*
x normal

	r	$\sigma_1^2/\sigma_2^2 = \frac{1}{2}$				$\sigma_1^2/\sigma_2^2 = 1$				$\sigma_1^2/\sigma_2^2 = 2$			
		$B=\frac{1}{4}$	$\frac{1}{2}$	$\frac{3}{4}$	1	$\frac{1}{4}$	$\frac{1}{2}$	$\frac{3}{4}$	1	$\frac{1}{4}$	$\frac{1}{2}$	$\frac{3}{4}$	1
$n = 25$	2	97	94	89	80	87	82	75	66	63	60	56	48
	3	99	98	97	93	94	91	86	81	77	72	67	61
	4	99	99	99	97	95	95	92	88	81	79	76	68
$n = 50$	2	99	98	93	84	92	87	78	69	66	59	53	51
	3	100	99	99	97	96	95	91	84	79	75	69	63
	4	100	100	100	99	98	97	94	89	86	81	75	71

As regards the effect of σ_1^2/σ_2^2, matching does best when $\sigma_1^2/\sigma_2^2 = \frac{1}{2}$ and worst
when $\sigma_1^2/\sigma_2^2 = 2$. This is not surprising. Since $\eta_1 > \eta_2$ the high values of sample 1
(the ones most likely to cause residual bias) will receive less biased partners when
$\sigma_2^2 > \sigma_1^2$.

The investigator planning to use 'nearest available' matching can estimate B
and σ_1^2/σ_2^2 from the initial data on x. Knowing the value of r, he can estimate the
expected percent reduction in bias under a linear regression from Table 2.2.

A measure has also been constructed (Rubin, 1973a) of the closeness or quality
of the individual pair matches. If pairing were entirely at random, we would have

$$E_m(x_{1j} - x_{2j})^2 = (\sigma_1^2 + \sigma_2^2) + (\eta_1 - \eta_2)^2$$
$$= (\sigma_1^2 + \sigma_2^2)(1 + B^2/2).$$

Consequently the quantity

$$100 E_m(x_{1j} - x_{2j})^2 / (\sigma_1^2 + \sigma_2^2)(1 + B^2/2)$$

was chosen as the measure. Since results vary little with n, only those for $n = 50$
are shown in Table 2.3.

Except for $\sigma_1^2/\sigma_2^2 = 2$ and $B > \frac{1}{2}$, random ordering gives good quality matches.
In fact, since the computer program (Rubin, 1973a) for constructing the matched
pairs is very speedy, the investigator can try random, high-low, and low-high
ordering. By examining $(\bar{x}_1 - \bar{x}_2)$ and $\Sigma(x_{1j} - x_{2j})^2/n$ for each method, he can
select what appears to him the best of the three approaches.

Table 2.3. *Values of* $100 E_m(x_{1j} - x_{2j})^2/(\sigma_1^2 + \sigma_2^2)(1 + B^2/2)$ *for nearest*
available random order matching with x normal

r	$\sigma_1^2/\sigma_2^2 = \frac{1}{2}$				$\sigma_1^2/\sigma_2^2 = 1$				$\sigma_1^2/\sigma_2^2 = 2$			
	$B=\frac{1}{4}$	$\frac{1}{2}$	$\frac{3}{4}$	1	$\frac{1}{4}$	$\frac{1}{2}$	$\frac{3}{4}$	1	$\frac{1}{4}$	$\frac{1}{2}$	$\frac{3}{4}$	1
2	0	1	3	8	1	3	8	15	7	13	20	26
3	0	0	0	1	0	1	3	6	4	8	12	18
4	0	0	0	0	0	1	2	4	3	5	9	13

Table 2.4. *Percent reduction in bias for mean matching: x normal*

r	$\sigma_1^2/\sigma_2^2 = \frac{1}{2}$				$\sigma_1^2/\sigma_2^2 = 1$				$\sigma_1^2/\sigma_2^2 = 2$			
	$B = \frac{1}{4}$	$\frac{1}{2}$	$\frac{3}{4}$	1	$\frac{1}{4}$	$\frac{1}{2}$	$\frac{3}{4}$	1	$\frac{1}{4}$	$\frac{1}{2}$	$\frac{3}{4}$	1
2	100	100	98	87	100	99	91	77	100	95	82	67
3	100	100	100	100	100	100	99	96	100	100	97	84
4	100	100	100	100	100	100	100	100	100	100	100	95

2.5. Mean Matching

For an investigator who is not interested in pair matching and is confident that the regression is linear, a mean-matching method which concentrates on making $|\bar{x}_1 - \bar{x}_2|$ small has been discussed (Greenberg, 1953). The following simple computer method has been investigated (Rubin, 1973a). Calculate $\bar{x}_1$. Select, from reservoir 2, the x_{21} closest to $\bar{x}_1$, then the x_{22} such that $(x_{21} + x_{22})/2$ is closest to $\bar{x}_1$, and so on until n have been selected. For $n = 50$, Table 2.4 shows the percent reductions in bias obtained.

Except in a few difficult cases, particularly $B = 1$, this method of mean matching removes essentially all the bias. So far as we know, mean matching is seldom used, presumably because it relies heavily on the assumption that the regression is linear. With a monotone non-linear regression of y on x, one might speculate that mean matching should perform roughly as well as a linear regression adjustment on random samples. But with the regression adjustment, one can examine the relations between y and x in the two samples before deciding whether a linear or non-linear regression adjustment is appropriate, whereas with mean matching performed before y has been observed, one is committed to the assumption of linearity, at least when matching the samples.

3. COMPLICATIONS

3.1. Regressions Linear But Not Parallel

For $i = 1, 2$, the model becomes

$$y_{ij} = \mu_i + \beta_i(x_{ij} - \eta_i) + e_{ij}. \tag{2.6}$$

It follows that for a given level of x,

$$E\{(y_{1j} - y_{2j}) \mid x_{1j} = x_{2j} = x\} = \mu_1 - \mu_2 - \beta_1\eta_1 + \beta_2\eta_2 + (\beta_1 - \beta_2)x. \tag{2.7}$$

If this quantity is interpreted as measuring the difference in the effects of the two treatments for given x, this difference appears to have a linear regression on x. At this point the question arises whether a differential treatment effect with x is a reasonable interpretation or whether the $(\beta_1 - \beta_2)$ difference is at least partly due to other characteristics (e.g., effect of omitted x-variables) in which the two populations differ. With samples from two populations treated differently, we do not see how this question can be settled on statistical evidence alone. With one study population

P_1 and two control populations P_2, P_2' both subject to τ_2, a finding that $\hat{\beta}_2$ and $\hat{\beta}_2'$ agree closely but differ from $\hat{\beta}_1$ leans in favour of suggesting a differential effect of $(\tau_1 - \tau_2)$.

As it happens, assuming x is the only confounding variable, this issue becomes less crucial if the goal is to estimate the average $(\tau_1 - \tau_2)$ difference over population 1. From (2.7) this quantity is

$$E_1(\tau_1 - \tau_2) = (\mu_1 - \mu_2) - \beta_2(\eta_1 - \eta_2). \tag{2.8}$$

Since from random samples,

$$E_r(\bar{y}_1 - \bar{y}_2) = \mu_1 - \mu_2, \tag{2.9}$$

the initial bias is $\beta_2(\eta_1 - \eta_2)$. With samples matched to a random $\bar{x}_1$,

$$E_m(\bar{y}_1 - \bar{y}_2) = \mu_1 - \mu_2 - \beta_2 E_m(\bar{x}_2) + \beta_2 \eta_2,$$

so that the reduction in bias is

$$E_r(\bar{y}_1 - \bar{y}_2) - E_m(\bar{y}_1 - \bar{y}_2) = \beta_2 [E_m(\bar{x}_2) - \eta_2].$$

Hence the percent reduction in bias due to matching remains, as before,

$$100[E_m(\bar{x}_2) - \eta_2]/(\eta_1 - \eta_2)$$

so that previous results for matching apply to non-parallel lines also with this estimand.

As regards regression adjustment, it follows from (2.8) and (2.9) that

$$E_r[(\bar{y}_1 - \bar{y}_2) - \hat{\beta}_2(\bar{x}_1 - \bar{x}_2)] = (\mu_1 - \mu_2) - \beta_2(\eta_1 - \eta_2) = E_1(\tau_1 - \tau_2).$$

Consequently, in applying the regression adjustment to random samples, use of the regression coefficient calculated from sample 2 provides an unbiased estimate of the desired $E_1(\tau_1 - \tau_2)$. This property was noted by Peters (1941), while Belsen (1956) recommended the use of $\hat{\beta}_2$ in comparing listeners (P_1) with non-listeners (P_2) to a BBC television program designed to teach useful French words and phrases to prospective tourists.

With $E_1(\tau_1 - \tau_2)$ as the objective, the standard use of the pooled $\hat{\beta}_p$ in the regression adjustment gives biased estimates, though Rubin (1970) has shown that 'nearest available' matching followed by regression adjustment greatly reduces this bias. With matched samples, the standard estimate of β, following the analysis of covariance in a two-way table, is $\hat{\beta}_d$, the sample regression of matched pair differences, $(y_{1j} - y_{2j})$ on $(x_{1j} - x_{2j})$. Curiously, the Monte Carlo computations show that use of $\hat{\beta}_p$ on matched samples performs better than use of $\hat{\beta}_d$ in this case.

If non-parallelism is interpreted as due to a $(\tau_1 - \tau_2)$ difference varying linearly with x, the question whether $E_1(\tau_1 - \tau_2)$ is the quantity to estimate deserves serious consideration. To take a practice sometimes followed in vital statistics, we might wish to estimate $(\tau_1 - \tau_2)$ averaged over a standard population that has mean η_s differing from η_1 and η_2. The estimand becomes, from (2.7)

$$E_s(\tau_1 - \tau_2) = \mu_1 - \mu_2 + \beta_1(\eta_s - \eta_1) - \beta_2(\eta_s - \eta_2).$$

From random samples, an unbiased regression estimate is

$$(\bar{y}_1 - \bar{y}_2) + \hat{\beta}_1(\eta_s - \bar{x}_1) - \hat{\beta}_2(\eta_s - \bar{x}_2) \tag{2.10}$$

where $\hat{\beta}_1$ and $\hat{\beta}_2$ are the usual least squares estimates from the separate regressions in the two samples.

Alternatively, particularly if $\hat{\beta}_1$ and $\hat{\beta}_2$ differ substantially, no single average of $(\tau_1 - \tau_2)$ may be of interest, but rather the values of $(\tau_1 - \tau_2)$ at each of a range of values of x. As a guide in forming a judgement whether use of a single average difference is adequate for practical application, Rubin (1970) has suggested the following. Suppose that in the range of interest, x lies between x_L and x_H. From (2.7) the estimated difference in $(\tau_1 - \tau_2)$ at these two extremes is

$$(\hat{\beta}_1 - \hat{\beta}_2)(x_H - x_L). \tag{2.11}$$

From (2.10), the average $(\tau_1 - \tau_2)$ over the range from x_L to x_H is estimated as

$$(\bar{y}_1 - \bar{y}_2) + \hat{\beta}_1(\bar{x} - \bar{x}_1) - \hat{\beta}_2(\bar{x} - \bar{x}_2) \quad \text{where} \quad \bar{x} = (x_L + x_H)/2. \tag{2.12}$$

The ratio of (2.11) to (2.12) provides some guidance on the proportional error in using simply this average difference.

If it is decided not to use the average difference, the differences $(\tau_1 - \tau_2)$ for specified x can be estimated by standard methods from the separate regressions of y on x in the two samples.

To examine the relation between $(\tau_1 - \tau_2)$ and x from pair-matched samples, it is natural to look at the regression of $(y_{1j} - y_{2j})$ on $\bar{x}._j = (x_{1j} + x_{2j})/2$. However, from the models (2.6) it turns out that

$$E\{(y_{1j} - y_{2j}) \mid \bar{x}._j = x\} = (\mu_1 - \mu_2) - \beta_1\eta_1 + \beta_2\eta_2$$
$$+ (\beta_1 - \beta_2)\bar{x}._j + (\beta_1 + \beta_2)E(d_j \mid \bar{x}._j = x)$$

where $d_j = (x_{1j} - x_{2j})/2$. With $\eta_1 \neq \eta_2$ or $\sigma_1^2 \neq \sigma_2^2$, it appears that $E(d_j \mid \bar{x}._j = x) \neq 0$, so that this method does not estimate the relation (2.7) without bias. The bias should be unimportant with tight matching, but would require Monte Carlo investigation.

3.2. Regression Non-linear

Comparison of the performance of pair-matching with linear regression adjustment is of great interest here, since this is the situation in which, intuitively, pair-matching may be expected to be superior. Use of both weapons – linear regression on matched samples – is also relevant.

Monte Carlo comparisons were made, (Rubin, 1973b), for the monotonic non-linear functions $y = e^{\pm\frac{1}{2}x}$ and $e^{\pm x}$ and the random order nearest available matching method described earlier in Section 2.4. In such studies it is hard to convey to the reader an idea of the amount of non-linearity present. One measure will be quoted. For convenience, the Monte Carlo work was done with $\eta_1 + \eta_2 = 0$ and $(\sigma_1^2 + \sigma_2^2)/2 = 1$. Thus in the average population, x is $N(0, 1)$. In this population the percent of the variance of $y = e^{\pm ax}$ that is attributable to its *linear* component of

Table 2.5. *Percent reduction in bias of* $y(\sigma_1^2/\sigma_2^2 = 1)$; *x normal*

Method*	r	$B = \frac{1}{4}$				$B = \frac{1}{2}$			
		$e^{x/2}$	$e^{-x/2}$	e^x	e^{-x}	$e^{x/2}$	$e^{-x/2}$	e^x	e^{-x}
R		100	100	101	101	101	101	102	102
M	2	83	99	70	106	74	94	60	98
	3	90	101	79	104	87	98	75	100
	4	94	101	87	103	92	99	84	100
RM	2	99	103	100	108	102	100	106	101
	3	100	101	100	103	100	100	102	101
	4	100	101	100	102	100	100	101	101

Method	r	$B = \frac{3}{4}$				$B = 1$			
		$e^{x/2}$	$e^{-x/2}$	e^x	e^{-x}	$e^{x/2}$	$e^{-x/2}$	e^x	e^{-x}
R		101	101	104	104	102	102	108	108
M	2	62	87	47	94	53	82	39	91
	3	81	96	68	99	70	92	55	97
	4	87	98	76	100	79	96	65	99
RM	2	103	99	110	100	104	99	113	99
	3	102	99	105	100	103	100	109	100
	4	101	100	103	100	102	100	106	99

*R denotes *linear* regression adjustment on random samples ($\hat{\beta}_p$).
M denotes 'nearest available' matching.
RM denotes *linear* regression adjustment on matched samples ($\hat{\beta}_d$).

regression on x is $100a^2/(e^{a^2} - 1)$. For $a = \pm\frac{1}{2}, \pm1$, respectively, 12% and 41% of the variance of y are *not* attributable to the linear component. From this viewpoint, $y = e^{\pm\frac{1}{2}x}$ might be called moderately and $y = e^{\pm x}$ markedly non-linear.

With regression adjustments on random samples, the regression coefficient used in the results presented here is $\hat{\beta}_p$, the pooled within-samples estimate. With regression adjustments on matched samples, the results are for $\hat{\beta}_d$, as would be customary in practice. Rubin (1973b) has investigated use of $\hat{\beta}_1$, $\hat{\beta}_2$, $\hat{\beta}_p$ and $\hat{\beta}_d$ in both situations. He found $\hat{\beta}_p$ in the unmatched case and $\hat{\beta}_d$ in the matched case to be on the whole the best choices.

The results were found to depend markedly on the ratio σ_1^2/σ_2^2. Table 2.5 presents percent reductions in bias for $\sigma_1^2/\sigma_2^2 = 1$, the simplest and possibly the most common case. Linear regression on random samples performs admirably, with only a trifling over-adjustment for $y = e^{\pm x}$. Matching is inferior, particularly for $B > \frac{1}{2}$, even with a reservoir of size $4n$ from which to seek matches. Linear regression on matched samples does about as well as linear regression on random samples. Results are for $n = 50$.

Turning to the case $\sigma_1^2/\sigma_2^2 = \frac{1}{2}$ in which better matches can be obtained, note first that linear regression on random samples gives wildly erratic results which call for a rational explanation, sometimes markedly overcorrecting or even (with

Table 2.6. *Percent reduction in bias of y ($\sigma_1^2/\sigma_2^2 = \frac{1}{2}$, the easier case for matching); x normal*

Method	r	$B = \frac{1}{4}$				$B = \frac{1}{2}$			
		$e^{x/2}$	$e^{-x/2}$	e^x	e^{-x}	$e^{x/2}$	$e^{-x/2}$	e^x	e^{-x}
R		298	62	−304	48	146	80	292	72
M	2	95	99	106	100	96	99	93	99
	3	99	100	103	100	98	100	94	100
	4	99	100	102	100	99	100	97	100
RM	2	102	100	96	100	101	100	108	100
	3	100	100	100	100	100	100	101	101
	4	100	100	100	100	100	100	100	100

Method	r	$B = \frac{3}{4}$				$B = 1$			
		$e^{x/2}$	$e^{-x/2}$	e^x	e^{-x}	$e^{x/2}$	$e^{-x/2}$	e^x	e^{-x}
R		123	90	170	88	113	96	139	102
M	2	89	96	85	98	76	91	69	96
	3	97	100	94	100	94	98	90	99
	4	99	100	97	100	97	99	94	100
RM	2	103	99	113	100	105	99	118	99
	3	100	100	102	100	99	99	105	100
	4	100	101	101	100	101	100	102	100

$B = \frac{1}{4}$ for e^x) greatly increasing the original bias.[2] Matching alone does well, on the average about as well as with a linear relation (Table 2.6) when $\sigma_1^2/\sigma_2^2 = \frac{1}{2}$. Linear regression on matched samples is highly effective, being slightly better than matching alone.

With $\sigma_1^2/\sigma_2^2 = 2$ (Table 2.7), linear regression alone performs just as erratically as with $\sigma_1^2/\sigma_2^2 = \frac{1}{2}$, the results being in fact the same if e^{ax} is replaced by e^{-ax}. As expected from the results in Section 2.3, matching alone is poor. In most cases, regression on matched samples is satisfactory, except for failures with $e^{-x/2}$ and e^{-x} when $B = \frac{1}{4}$ or $\frac{1}{2}$.

3.3. Regressions Parallel But Quadratic

Some further insight into the performances of these methods is obtained by considering the model

$$y_{ij} = \tau_i + \beta x_{ij} + \delta x_{ij}^2 + e_{ij}. \tag{2.13}$$

---

[2] The most extreme results follow from the nature of the function $e^{\pm ax}$. Consider e^x. Its mean value in population i is $e^{(\sigma_i^2/2 + \eta_i)}$. For $B = \frac{1}{4}$, with $\eta_1 = \frac{1}{8}$, $\eta_2 = -\frac{1}{8}$, $\sigma_1^2 = \frac{2}{3}$, $\sigma_2^2 = \frac{4}{3}$, the initial bias in y is *negative*. Since $\eta_1 > \eta_2$ and $\hat{\beta}_p$ is positive, the regression adjustment greatly increases this negative bias, giving −304% reduction. For $B = \frac{1}{2}$, the initial bias is positive but small, so that regression greatly overcorrects, giving 292% reduction. For $B = \frac{3}{4}$, 1, the initial biases are larger and the over-correction not so extreme (170%, 139%).

Table 2.7. *Percent reduction in bias of y ($\sigma_1^2/\sigma_2^2 = 2$, the harder case for matching); x normal*

Method	r	$B = \frac{1}{4}$				$B = \frac{1}{2}$			
		$e^{x/2}$	$e^{-x/2}$	e^x	e^{-x}	$e^{x/2}$	$e^{-x/2}$	e^x	e^{-x}
R		62	298	48	−304	80	146	72	292
M	2	48	121	35	−50	45	81	30	123
	3	66	139	51	−48	60	89	43	118
	4	70	121	55	1	65	94	48	126
RM	2	90	177	90	−99	100	111	107	171
	3	93	149	92	−29	100	108	105	147
	4	95	140	94	−5	100	107	104	146

Method	r	$B = \frac{3}{4}$				$B = \frac{1}{2}$			
		$e^{x/2}$	$e^{-x/2}$	e^x	e^{-x}	$e^{x/2}$	$e^{-x/2}$	e^x	e^{-x}
R		90	123	88	170	96	113	102	139
M	2	38	72	23	90	31	67	16	83
	3	55	85	39	98	45	79	28	92
	4	60	89	42	100	50	84	29	94
RM	2	106	102	120	115	109	99	127	104
	3	103	102	111	112	106	100	119	104
	4	103	101	111	97	105	99	119	102

It follows that

$$E_c(\bar{y}_1 - \bar{y}_2) = (\tau_1 - \tau_2) + \beta(\bar{x}_1 - \bar{x}_2) + \delta\left(\bar{x}_1^2 - \bar{x}_2^2\right) + \delta\left(s_1^2 - s_2^2\right) \qquad (2.14)$$

where $s_i^2 = \Sigma(x_{ij} - \bar{x}_i)^2/n$. Hence the initial bias in random samples is, unconditionally,

$$(\eta_1 - \eta_2)[\beta + \delta(\eta_1 + \eta_2)] + \delta\left(\sigma_1^2 - \sigma_2^2\right) \qquad (2.15)$$
$$= (\eta_1 - \eta_2)\beta + \delta\left(\sigma_1^2 - \sigma_2^2\right) \qquad (2.16)$$

where without loss of generality we have assumed $\eta_1 + \eta_2 = 0$.

Even though $\eta_1 > \eta_2$, if $\delta > 0$ (as appropriate for the positive exponential function) (2.16) shows that if $\sigma_1^2 < \sigma_2^2$, the initial bias might be small or even negative. This may indicate why some erratic results appear in the percent reduction in bias with non-linear functions.

From (2.14), the remaining bias in matched samples is

$$(\eta_1 - E_m(\bar{x}_2))[\beta + \delta(\eta_1 + E_m(\bar{x}_2))] + \delta\left\{\sigma_1^2 - E_m\left(s_2^2\right)\right\}. \qquad (2.17)$$

The second term should be minor if the samples are relatively well matched. The first term suggests that in this case the percent reduction in bias should approximate that for parallel-linear regressions if $|\delta/\beta|$ is small. For example, let $\sigma_1^2 = \sigma_2^2 = 1$ and θ be the percent reduction in bias for y linear. From (2.16) and (2.17), the percent reduction in bias for y quadratic works out approximately as

$$-(100 - \theta)\frac{\delta}{\beta}[\eta_1 - E_m(\bar{x}_2)] = \theta\left[1 - \frac{\delta}{\beta}\left(1 - \frac{\theta}{100}\right)B\right].$$

For regression adjusted estimates on random samples, $E_c(\hat{\beta}_p)$ may be expressed as

$$E_c(\hat{\beta}_p) = \beta + \delta\left[\frac{2\bar{x}_1 s_1^2 + 2\bar{x}_2 s_2^2}{s_1^2 + s_2^2}\right] + \frac{\delta(k_{31} + k_{32})}{s_1^2 + s_2^2}$$

where $k_{3i} = \Sigma(x_{ij} - \bar{x}_i)^3/n$ is the sample third moment. From (2.14) it follows that the residual bias in the regression adjusted estimate on random samples is conditionally

$$= E_c[(\bar{y}_i - \bar{y}_2) - \hat{\beta}_p(\bar{x}_1 - \bar{x}_2)] - (\tau_1 - \tau_2)$$

$$= \delta\left(s_1^2 - s_2^2\right) + \delta(\bar{x}_1 - \bar{x}_2)\left[(\bar{x}_1 + \bar{x}_2) - \frac{2\left(\bar{x}_1 s_1^2 + \bar{x}_2 s_2^2\right)}{s_1^2 + x_2^2}\right]$$

$$- \delta(\bar{x}_1 - \bar{x}_2)(k_{31} + k_{32})/\left(s_1^2 + s_2^2\right).$$

For a symmetric or near-symmetric distribution of x in both populations the third term becomes unimportant. The first two terms give

$$\delta\left(s_1^2 - s_2^2\right)\left[1 - (\bar{x}_1 - \bar{x}_2)^2/\left(s_1^2 + s_2^2\right)\right].$$

The average residual bias in large random samples after regression adjustment is therefore, for x symmetric and $(\sigma_1^2 + \sigma_2^2)/2 = 1$,

$$\delta\left(\sigma_1^2 - \sigma_2^2\right)\left(1 - \frac{(\eta_1 - \eta_2)^2}{2}\right).$$

This formula suggests, as we found for $e^{\pm ax}$, that with a symmetric x and $\sigma_1^2 = \sigma_2^2$, linear regression adjustment in random samples should remove essentially all the bias when the relation between y and x can be approximated by a quadratic function. The further indication that with $\sigma_1^2 \neq \sigma_2^2$ the residual bias is smaller absolutely as $\eta_1 - \eta_2$ increases towards 1 is at first sight puzzling, but consistent, for example, with the Monte Carlo results for $e^{x/2}$ and e^x when $\sigma_1^2/\sigma_2^2 = 2$ in Table 2.7.

To summarize for the exponential and quadratic relationships: If it appears that $\sigma_1^2 \simeq \sigma_2^2$ and x is symmetric (points that can be checked from initial data on x) linear regression adjustment on random samples removes all or nearly all the bias. Pair matching alone is inferior. Generally, regression adjustment on pair-matched samples is much the best performer, although sometimes failing in extreme cases. An explanation for this result is given in Rubin (1973b) but is not summarized here because it is quite involved. Further work on adjustment by quadratic regression, on other curvilinear relations, and on the cases $\sigma_1^2/\sigma_2^2 = \frac{3}{4}, \frac{4}{3}$ would be informative.

Before leaving the problem of non-linear regressions, we indicate how the above results can be extended to non-linear response surfaces other than quadratic. Let

$$y_{ij} = \tau_i + g(x_{ij}) + e_{ij}$$

where $g(\cdot)$ is the regression surface. Since $\hat{\beta}_p$ may be written as $\sum_i \sum_j (y_{ij} - \bar{y}_i)(x_{ij} - \bar{x}_i)/\sum_i \sum_j (x_{ij} - \bar{x}_i)^2$ the limit of $\hat{\beta}_p$ in large random samples is

$$[\text{cov}_1(x, g(x)) + \text{cov}_2(x, g(x))]/[\text{var}_1(x) + \text{var}_2(x)]$$

where cov_i and var_i are the covariances and variances in population i. Hence the regression adjusted estimate in large random samples has limiting residual bias

$$E_1(g(x)) - E_2(g(x)) - (\eta_1 - \eta_2)[\text{cov}_1(x, g(x)) \\ + \text{cov}_2(x, g(x))]/[\text{var}_1(x) + \text{var}_2(x)].$$

This quantity can be calculated analytically for many distributions and regression surfaces $g(\cdot)$, (e.g., normal distributions and exponential $g(\cdot)$). In addition, if g is expanded in a Taylor series, the residual bias in random or matched samples may be expressed in terms of the moments of x in random and matched samples.

3.4. Errors of Measurement in x

In this section we assume that y has the same linear regression on the correctly measured x (denoted by X) in both populations, but that matching or regression adjustment is made with respect to a fallible $x_{ij} = X_{ij} + u_{ij}$, where u_{ij} is an error of measurement. As in Section 2.1 the model is

$$y_{ij} = \mu_i + \beta(X_{ij} - \eta_i) + e_{ij} \tag{2.18}$$

and the expected bias in $(\bar{y}_1 - \bar{y}_2)$ in random samples is as before $\beta(\eta_1 - \eta_2)$.

To cover situations that arise in practice it is desirable to allow (i) u_{ij} and X_{ij} to be correlated, and (ii) u_{ij} to be a biased measurement, with $E_i(u_{ij}) = v_i \neq 0$. A difficulty arises at this point. Even under more restrictive assumptions (u_{ij}, X_{ij} independent in a given population and $E_i(u_{ij}) = 0$), Lindley (1947) showed that the regression of y_{ij} on x_{ij} is not linear unless the cumulant generating function of the u_{ij} is a multiple of that of the X_{ij}. Lindley's results can be extended to give corresponding conditions when the u_{ij}, X_{ij} are correlated. For simplicity we assume that these extended conditions are satisfied.

The linear regressions of y on the fallible x will be written

$$y_{ij} = \mu_i + \beta^*(x_{ij} - \eta_i - v_i) + e^*_{ij}$$

with $E(e^*_{ij} \mid x_{ij}) = 0$. Hence, from (2.18),

$$\beta^* = \text{cov}(yx)/\sigma_x^2 = \frac{\beta[\sigma_X^2 + \text{cov}(uX)]}{\sigma_X^2 + \sigma_u^2 + 2\text{cov}(uX)}.$$

Unless $\text{cov}(uX) \leqslant -\sigma_u^2$, we have $|\beta^*| < |\beta|$, the slope of the line being damped towards zero. The results in Section 2.2 imply that in random samples or samples matched on x a regression-adjusted estimate $\bar{y}_1 - \bar{y}_2 - \hat{\beta}^*(\bar{x}_1 - \bar{x}_2)$, where $\hat{\beta}^*$ is a least squares estimate of the regression of y on the fallible x, changes the initial bias of $\bar{y}_1 - \bar{y}_2$ by the amount

$$-\beta^*(\eta_1 - \eta_2 - v_1 + v_2).$$

Since the initial bias of $\bar{y}_1 - \bar{y}_2$ in random samples is $\beta(\eta_1 - \eta_2)$, the bias of a regression adjusted estimate is

$$(\beta - \beta^*)(\eta_1 - \eta_2) - \beta^*(v_1 - v_2).$$

The last term on the right shows that biased measurements can make an additional contribution $(+$ or $-)$ to the residual bias. This contribution disappears if the measurement bias is the same in both populations, $v_1 = v_2$. Under this condition the percent reduction in bias due to the regression adjustment is $100\beta^*/\beta$. With the same condition, the percent reduction in bias of $(\bar{y}_1 - \bar{y}_2)$ due to matching on x is easily seen to be

$$\frac{100\beta^*}{\beta} \frac{[E_m(\bar{x}_2) - \eta_2]}{(\eta_1 - \eta_2)}.$$

Thus with this simple model for errors of measurement in X, their effects on matching and adjustment are similar – namely to multiply the expected percent reduction in bias by the ratio β^*/β, usually less than 1. With u, X uncorrelated, this ratio is the quantity σ_X^2/σ_x^2 often called the reliability of the measurement x (Kendall and Buckland, 1971).

If this reliability, say $(1 + a^2)^{-1}$, is known, it can be used to inflate the regression adjustment to have expectation $\beta(\eta_1 - \eta_2)$, (Cochran, 1968b). Thus form the "corrected" regression adjusted estimate

$$\bar{y}_1 - \bar{y}_2 - (1 + a^2)\hat{\beta}^*(\bar{x}_1 - \bar{x}_2),$$

which is unbiased for $\tau_1 - \tau_2$ under this model.

In simple examples in which Lindley's conditions are not satisfied, Cochran (1970b) found the regression of y on the fallible x to be monotone but curved. A thorough investigation of the effects of errors of measurement would have to attack this case also.

3.5. Omitted Confounding Variable

One of the most common criticisms of the conclusions drawn from an observational study is that they are erroneous because the investigator failed to adjust or match for another confounding variable z_{ij} that affects y. He may have been unaware of it, or failed to measure it, or guessed that its effect would be negligible. Even under simple models, however, investigation of the effects of such a variable on the initial bias and on the performance of regression and matching leads to no crisp conclusion that either rebuts or confirms this criticism in any generality.

We assume that y_{ij} has the same linear regression on x_{ij} and z_{ij} in both populations, namely

$$y_{ij} = \mu_i + \beta(x_{ij} - \eta_i) + \gamma(z_{ij} - v_i) + e_{ij}. \qquad (2.19)$$

Hence, assuming x and z are the only confounding variables,

$$\tau_1 - \tau_2 = E(y_{1j} - y_{2j} \mid x_{1j} = x_{2j}, z_{1j} = z_{2j})$$
$$= (\mu_1 - \mu_2) - \beta(\eta_1 - \eta_2) - \gamma(v_1 - v_2)$$

and the initial bias in $(\bar{y}_1 - \bar{y}_2)$ from random samples is now

$$\beta(\eta_1 - \eta_2) + \gamma(\nu_1 - \nu_2). \tag{2.20}$$

Similarly, the bias in $(\bar{y}_1 - \bar{y}_2)$ from samples matched on x is

$$\beta(\eta_1 - E_m(\bar{x}_2)) + \gamma(\nu_1 - E_m(\bar{z}_2)).$$

Thus, depending on the signs of the parameters involved, the presence of z_{ij} in the model may either increase or decrease (perhaps to an unimportant amount) the previous initial bias $\beta(\eta_1 - \eta_2)$. Also, even if $|\eta_1 - \eta_2| > |\eta_1 - E_m(\bar{x}_2)|$ and $|\nu_1 - \nu_2| > |\nu_1 - E_m(\bar{z}_2)|$, the bias of $(\bar{y}_1 - \bar{y}_2)$ may be greater in matched than random samples.

Suppose now that z_{ij} has linear and parallel regressions on x_{ij} in the two populations:

$$z_{ij} = \nu_i + \lambda(x_{ij} - \eta_i) + \varepsilon_{ij}. \tag{2.21}$$

Then (2.19) may be written

$$y_{ij} = \mu_i + (\beta + \gamma\lambda)(x_{ij} - \eta_i) + \varepsilon_{ij} + e_{ij}. \tag{2.22}$$

In (2.22) we have returned to the model in Section 2.2 – same linear regression of y on x in both populations. From Section 2.1, the expected change in bias of $(\bar{y}_1 - \bar{y}_2)$ due to regression adjustment on x in random samples or samples matched on x is therefore

$$-(\beta + \gamma\lambda)(\eta_1 - \eta_2) \tag{2.23}$$

while that due to matching on x is

$$-(\beta + \gamma\lambda)[E_m(\bar{x}_2) - \eta_2]. \tag{2.24}$$

As regards regression, (2.20) and (2.24) lead to the residual bias

$$\gamma[(\nu_1 - \nu_2) - \lambda(\eta_1 - \eta_2)]. \tag{2.25}$$

Thus, adjustment on x alone removes the part of the original bias coming from z that is attributable to the linear regression of z on x. If z has *identical* linear regressions on x in both populations, so that $(\nu_1 - \lambda\eta_1) = (\nu_2 - \lambda\eta_2)$, the residual bias is zero as would be expected. With matching in this situation, the residual bias is

$$(\beta + \gamma\lambda)[\eta_1 - E_m(\bar{x}_2)]$$

matching being less effective than regression.

With regressions of z on x parallel but not identical, the final bias with either regression or matching could be numerically larger than the initial bias, and no simple statement about the relative merits of regression and matching holds under this model.

If the regressions of z_{ij} on x_{ij} are parallel but non-linear, investigation shows that in large samples, regression and matching remove the part of the bias due to z that is attributable to the linear component of the regression of z on x.

4. MATCHING AND ADJUSTMENT BY SUBCLASSIFICATION

4.1. The Two Methods

When the x-variable is qualitative, e.g., sex (M, F), it is natural to regard any male from population 1 as a match for any male from population 2 with respect to x, or more generally, any two members who fall in the same qualitative class as a match. This method is also used frequently when x is continuous, e.g., age. We first divide the range of ages that are of interest into, say, specified 5-year classes 40–44, 45–49, etc. and regard any two persons in the same age class as a match.

In matching to the sample from population 1, let n_{1j} be the number in sample 1 who fall in the j-th subclass. From the reservoir from population 2, we seek the same number $n_{2j} = n_{1j}$ in the j-th class. The average matched-pair difference, $\Sigma n_{1j}(\bar{y}_{1j} - \bar{y}_{2j})/n$ is of course the difference $(\bar{y}_1 - \bar{y}_2)$ between the two matched sample means, this method being self-weighting.

With random samples from the two populations, the alternative method of adjustment by subclassification starts by classifying both samples into the respective classes. The numbers n_{1j}, n_{2j} will now usually differ. However, any weighted mean $\Sigma w_j(\bar{y}_{1j} - \bar{y}_{2j})$, with $\Sigma w_j = 1$, will be subject only to the residual within-class biases insofar as this x is concerned. In practice, different choices of the weights w_j have been used, e.g., sometimes weights directed at minimizing the variance of the weighted difference. For comparison with matching we assume the weights $w_j = n_{1j}/n$.

4.2. Performance of the Two Methods

If sample 1 and reservoir 2 or sample 2 are random samples from their respective populations, as we have been assuming throughout, the n_{1j}, n_{2j} who turn up in the final sample are a random sample from those in their population who fall in class j under either method-matching or adjustment. Consequently, with the same weights n_{1j}/n, the two methods have the same expected residual bias. (An exception is the occasional case of adjustment from initial random samples of equal sizes $n_1 = n_2 = n$, where we find $n_{2j} = 0$ in one or more subclasses, so that subclasses have to be combined to some extent for application of the 'adjustment by subclassification' method.)

With certain genuinely qualitative classifications it may be reasonable to assume that any two members of the same subclass are identical as regards the effect of this x on y. In this event, both matching and adjustment remove all the bias due to x, there being no within-class bias. But many qualitative variables like socio-economic status, degree of aggressiveness (mild, moderate, severe), represent an ordered classification of an underlying continuous variable x which at present we are unable to measure accurately. Two members of the same subclass do not have identical values of x in this event. For such cases, and for a variable like age, we assume the model

$$y_{ij} = \tau_i + u(x_{ij}) + e_{ij}, i = 1, 2, j = 1, 2, \ldots, c, \tag{2.26}$$

the regression of y on x being the same in both populations, with $\tau_1 - \tau_2$ not depending on the value of x.

From (2.26) the percent reduction in the bias of y due to adjustment by subclassification of u equals the percent reduction in the bias of u. If $u(x) = x$, this also equals the percent reduction in the bias of x. If $u(x)$ is a monotone function of x, a division of x into classes at the quantiles of x will also be a division of u into classes at the same quantiles of u. The percent reductions in bias of u and x will not, however, be equal, since these depend both on the division points and on the frequency distributions, which will differ for u and x. The approach adopted by Cochran (1968a) was to start with the case $u(x) = x$, with x normal, and then consider some non-normal distributions of x to throw some light on the situation with $u(x)$ monotone.

In subclassification, the range of x is divided into c classes at division points $x_0, x_1, \ldots, x_c$. Let $f_i(x)$ be the p.d.f.'s of x in the two populations. The overall means of x are

$$\eta_i = \int x f_i(x) dx$$

while in the j-th subclass the means are

$$\eta_{ij} = \int_{x_{j-1}}^{x_j} x f_i(x) dx / P_{ij}, \quad \text{where} \quad P_{ij} = \int_{x_{j-1}}^{x_j} f_i(x) dx.$$

The initial expected bias in x is $(\eta_1 - \eta_2)$. After matching or adjustment, the weighted mean difference in the two samples is

$$\sum_{j=1}^{c} \frac{n_{1j}}{n} (\bar{x}_{1j} - \bar{x}_{2j}). \tag{2.27}$$

Its average value, the expected residual bias, is

$$\sum_{j=1}^{c} P_{1j}(\eta_{1j} - \eta_{2j}). \tag{2.28}$$

This expression may be used in calculating the expected percent reduction in bias.

If $f_1(x)$, $f_2(x)$ differ only with respect to a single parameter, it is convenient to give it the values 0 and Θ in populations 1 and 2, respectively. Expression (2.28) may be rewritten as

$$\sum_{j=1}^{c} P_j(0)\{\eta_j(0) - \eta_j(\Theta)\}. \tag{2.29}$$

A first-term Taylor expansion about 0, assuming Θ small, seems to work well for biases of practical size (Cochran, 1968a) and leads to a useful result obtained in a related problem. From (2.29) the expected *residual* bias is approximately, expanding about $\Theta = 0$,

$$-\Theta \sum_{j=1}^{c} P_j(0) \frac{d\eta_j(\Theta)}{d\Theta} \tag{2.30}$$

Table 2.8. *Percent reductions in bias with equal-sized class in population 1, x normal*

no. of subclasses	2	3	4	5	6	8	10
% reduction	64%	79%	86%	90%	92%	94%	96%

the derivative being measured at $\Theta = 0$. On the other hand, the expected *initial* bias is

$$\sum_{j=1}^{c} [P_j(0)\eta_j(0) - P_j(\Theta)\eta_j(\Theta)] \simeq -\Theta \sum_{j=1}^{c} \left[P_j(0)\frac{d\eta_j(\Theta)}{d\Theta} + \eta_j(0)\frac{dP_j(\Theta)}{d\Theta} \right].$$

(2.31)

On subtracting (2.30) from (2.31), the expected proportional reduction in bias is approximately

$$\sum_{j=1}^{c} \eta_j(0)\frac{dP_j(\Theta)}{d\Theta} \bigg/ \frac{d\eta(\Theta)}{d\Theta}$$

(2.32)

measured at $\Theta = 0$, where $\eta(\Theta) = \eta_2 = \sum_{j=1}^{c} P_j(\Theta)\eta_j(\Theta)$.

In particular, if $f_1(x) = f(x)$, $f_2(x) = f(x - \Theta)$, the two distributions differing only in their means, we have $\frac{d\eta}{d\Theta} = 1$ and

$$P_j(\Theta) = \int_{x_{j-1}}^{x_j} f(x - \Theta)dx = \int_{x_{j-1}-\Theta}^{x_j-\Theta} f(x)dx$$

with

$$\frac{dP_j(\Theta)}{d\Theta} = f(x_{j-1}) - f(x_j)$$

at $\Theta = 0$. From (2.32), the proportional reduction in bias becomes

$$\sum_{j=1}^{c} \eta_j(0)[f(x_{j-1}) - f(x_j)].$$

(2.33)

If $f(x)$ is the unit normal distribution, (2.33) gives

$$\sum_{j=1}^{c} [f(x_{j-1}) - f(x_j)]^2 / P_j(0)$$

(2.34)

for the proportional reduction in bias. Expression (2.34) has been studied in other problems by J. Ogawa (1951) and by D. R. Cox (1957a). Cox showed that it is 1 minus the ratio of the average within-class variance to the original variance of x when x is normal. For our purpose, their calculations provide (i) the optimum choices of the P_{1j}, (ii) the resulting maximum percent reductions in bias, and (iii) the percent reductions in bias with equal-sized classes $P_{1j} = 1/c$. For $c = 2 - 10$, the maximum percent reductions are at most about 2% higher than those for equal P_{1j}, shown in Table 2.8.

Calculations (Cochran, 1968a) of the percent reductions when x follows χ^2 distributions, t distributions and Beta distributions suggest that the above figures can be used as a rough guide to what to expect in practice when the classification represents an underlying continuous x. To remove 80%, 90% and 95% of the initial bias, evidently 3, 5, and 10 classes are required by this method.

5. SIMPLE MULTIVARIATE GENERALIZATIONS

5.1. Parallel Linear Regressions

We now consider the case of many x-variables, say $(x^{(1)}, x^{(2)}, \ldots, x^{(p)})$. Many of the previous results for one x variable have obvious analogues for p x-variables, with the p-vectors η_i, β_i and x_{ij} replacing the scalars η_i, β_i and x_{ij}. However except in the cases where the adjustment removes all the bias, the conclusions are even less sharp than in the univariate case.

The simplest multivariate case occurs when y has parallel linear regressions on x in both populations

$$y_{ij} = \mu_i + \beta(x_{ij} - \eta_i)' + e_{ij}. \tag{2.35}$$

The regressions of y on x in the two populations are parallel "planes" with a constant difference of height

$$E(y_{1j} - y_{2j} \mid x_{1j} = x_{2j}) = (\mu_1 - \mu_2) - \beta(\eta_1 - \eta_2)'. \tag{2.36}$$

If $(x^{(1)}, \ldots, x^{(p)})$ are the only confounding variables, this constant difference is the treatment difference, $\tau_1 - \tau_2$. From (2.35) it follows that conditionally on the values of the x_{ij} in two samples, chosen either randomly or only on the basis of the x-variables,

$$E_c(\bar{y}_1 - \bar{y}_2) = \tau_1 - \tau_2 + \beta(\bar{x}_1 - \bar{x}_2)'.$$

The expected bias of $\bar{y}_1 - \bar{y}_2$ in random samples is

$$E_r(\bar{y}_1 - \bar{y}_2) - (\tau_1 - \tau_2) = \beta(\eta_1 - \eta_2)'. \tag{2.37}$$

Notice that since β and $(\eta_1 - \eta_2)$ are vectors the initial bias in $(\bar{y}_1 - \bar{y}_2)$ may be zero even if $\beta \neq 0$ and $(\eta_1 - \eta_2) \neq 0$.

In random P_1 and matched P_2 samples, the bias is

$$E_m(\bar{y}_1 - \bar{y}_2) - (\tau_1 - \tau_2) = \beta(\eta_1 - E_m(\bar{x}_2))'. \tag{2.38}$$

Formally, the percent reduction in bias is the natural extension of the univariate result,

$$100\beta(E_m(\bar{x}_2) - \eta_2)'/\beta(\eta_1 - \eta_2)'. \tag{2.39}$$

But the bias in matched samples may be greater than in random samples even with $E_m(\bar{x}_2)$ closer to η_1 in all components than η_2 is to η_1(e.g., $(\eta_1 - \eta_2) = (1, -1)$, $(\eta_1 - E_m(\bar{x}_2)) = (\frac{1}{2}, \frac{1}{2})$, $\beta = (1, 1)$), which give initial bias 0 and matched sample bias 1.

The regression adjusted estimate is

$$\hat{\tau}_1 - \hat{\tau}_2 = (\bar{y}_1 - \bar{y}_2) - \hat{\beta}(\bar{x}_1 - \bar{x}_2)' \tag{2.40}$$

where $\hat{\beta}$ is the vector of estimated regression coefficients of y on x. Under this model, $E_c(\hat{\beta}) = \beta$, for $\hat{\beta}_p$, $\hat{\beta}_1$, $\hat{\beta}_2$. Thus, for any of these $\hat{\beta}$ and samples either random or matched on x, (2.35) and (2.40) show that the regression adjusted estimate is unbiased:

$$E_c(\hat{\tau}_1 - \hat{\tau}_2) = \mu_1 - \mu_2 - \beta(\eta_1 - \eta_2)' = \tau_1 - \tau_2.$$

5.2. Non-parallel Linear Regressions

As in the univariate case, the regressions of y on x may not be parallel. Assume the objective is to estimate $(\tau_1 - \tau_2)$ averaged over some standard population with mean x vector η_s (e.g., $\eta_s = \eta_1$ if P_1 is considered the standard). From the multivariate version of (2.7), assuming x are the only confounding variables we have

$$E_s(\tau_1 - \tau_2) = \mu_1 - \mu_2 + \beta_1(\eta_s - \eta_1)' - \beta_2(\eta_s - \eta_2)'. \tag{2.41}$$

In random samples $\bar{y}_1 - \bar{y}_2$ has expectation $\mu_1 - \mu_2$ and thus the initial bias is

$$-\beta_1(\eta_s - \eta_1)' + \beta_2(\eta_s - \eta_2)'.$$

If $\eta_s = \eta_1$, this initial bias becomes $\beta_2(\eta_1 - \eta_2)'$.

For random samples or samples selected solely on x

$$E_c(\bar{y}_1 - \bar{y}_2) = \mu_1 - \mu_2 + \beta_1(\bar{x}_1 - \eta_1)' - \beta_2(\bar{x}_2 - \eta_2)'$$
$$= E_s(\tau_1 - \tau_2) + \beta_1(\bar{x}_1 - \eta_s)' - \beta_2(\bar{x}_2 - \eta_s)'. \tag{2.42}$$

If $\eta_s = \eta_1$, and sample 1 is a random sample, the bias of $\bar{y}_1 - \bar{y}_2$ is $\beta_2(\eta_1 - E_m(\bar{x}_2))'$ while the initial bias is $\beta_2(\eta_1 - \eta_2)'$. By comparison with (2.37) and (2.38) it follows that when population 1 is chosen as the standard the effect of matching on bias reduction is the same whether the regressions are parallel or not.

Now consider the regression estimate. Since (2.42) gives the conditional bias of $\bar{y}_1 - \bar{y}_2$ it would seem reasonable to estimate this bias using the usual within-sample least squares estimates of β_1 and β_2 (and an estimate of η_s if necessary) and forming the regression adjusted estimate

$$\bar{y}_1 - \bar{y}_2 - \hat{\beta}_1(\bar{x}_1 - \hat{\eta}_s)' + \hat{\beta}_2(\bar{x}_2 - \hat{\eta}_s)' \tag{2.43}$$

which is an unbiased estimate of $E_s(\tau_1 - \tau_2)$ under the linear regression model. If $\eta_s = \eta_1$ and the first sample is random, this estimate is the natural extension of the univariate result,

$$\bar{y}_1 - \bar{y}_2 - \hat{\beta}_2(\bar{x}_1 - \bar{x}_2)'.$$

If a single summary of the effect of the treatment is not adequate, one could examine the estimated effect at various values of x using (2.43) where η_s is replaced by the values of x of interest.

5.3. Non-linear Regressions

If y has non-linear parallel regressions on x, expressed by the function $g(x)$, the initial bias, $E_1(g(x)) - E_2(g(x))$, depends on the higher moments of the distributions of x in P_1 and P_2 (e.g., the covariance matrices Σ_1 and Σ_2 if x is normal) as well as the means. The large sample limit of the pooled regression adjusted estimate in random samples is

$$E_1(g(x)) - E_2(g(x)) - (\eta_1 - \eta_2)[\Sigma_1 + \Sigma_2]^{-1}C'$$

where the k-th component of the p-vector C is $\mathrm{cov}_1(x^{(k)}g(x)) + \mathrm{cov}_2(x^{(k)}g(x))$.

This quantity, as well as similar quantities for the case of parallel, non-linear regressions, can be obtained analytically for many distributions and regression functions. As far as we know, no work has been done on this problem or the more difficult one involving matched samples, in which case the distribution of x in matched samples may not be analytically tractable. Expanding $g(x)$ in a Taylor series would enable one to expand the limiting residual bias in terms of the moments of x in random and matched samples (matched moments for the regression adjusted estimate based on matched pairs).

5.4. Errors of Measurement in x

Assume that y has parallel linear regressions on the correctly measured matching variables X, that X are the only confounding variables, but that matching and regression adjustment are done on the fallible $x = X + u$. Hence

$$y_{ij} = \mu_i + \beta(X_{ij} - \eta_i)' + e_{ij}$$

and the initial bias in random samples is $\beta(\eta_1 - \eta_2)'$. If y has a linear regression on the fallible x (e.g., y, X, u are multivariate normal), let

$$y_{ij} = \mu_i + \beta^*(x_{ij} - \eta_i - \nu_i) + e_{ij}^*$$

where $E(e_{ij}^* \mid x_{ij}) = 0$, $E(u_{ij}) = \nu_i$, and $\beta^* = \Sigma_x^{-1}\mathrm{cov}(y, x)$, where Σ_x is the covariance matrix of the x variables.

A regression adjusted estimate based on random samples or samples matched on x changes the initial bias by the amount $-\beta^*(\eta_1 - \eta_2 - \nu_1 + \nu_2)'$, and thus the bias of a regression adjusted estimate is

$$(\beta - \beta^*)(\eta_1 - \eta_2)' - \beta^*(\nu_1 - \nu_2)'. \tag{2.44}$$

Some simple results can be obtained for the special case when $\nu_1 = \nu_2$ (equally biased measurements in both populations), X and u are uncorrelated, and the covariance matrix of u is proportional to the covariance matrix of X, say $a^2\Sigma_X$. With the latter two conditions β^* becomes $(1 + a^2)^{-1}\beta$. This result and $\nu_1 = \nu_2$ imply that (2.44) becomes

$$\frac{a^2}{1 + a^2}\beta(\eta_1 - \eta_2)'$$

and the percent reduction in bias due to regression adjustment is

$$100/(1+a^2),$$

as in the univariate case, since $1/(1+a^2)$ corresponds to the reliability, which we are assuming to be uniform for all variables. Under this same set of special conditions, the percent reduction in bias due to matching on x would be

$$\frac{100}{1+a^2}\beta(E_m(\bar{x}_2)-\eta_2)'/\beta(\eta_1-\eta_2)'.$$

Under models different from the above special case, clearcut results appear more difficult to obtain, and the percent reduction in bias for regression adjustment or matching is not necessarily between 0 and 100 percent even with $\nu_1 = \nu_2$, X and u uncorrelated, and all u_i independent.

If one knew Σ_u, one could form a "corrected" regression-adjusted estimate that is in large samples unbiased for $\tau_1 - \tau_2$. That is, assuming x and u are uncorrelated, form

$$\bar{y}_1 - \bar{y}_2 - \hat{\Sigma}_X^{-1}\hat{\Sigma}_x\hat{\beta}^*(\bar{x}_1 - \bar{x}_2)' \tag{2.45}$$

where $\hat{\beta}^*$ is the usual least squares estimate of the regression of y on x, $\hat{\Sigma}_x$ is the estimated within group covariance matrix of x and $\hat{\Sigma}_X = \hat{\Sigma}_x - \hat{\Sigma}_u$. In the special case when $\Sigma_u = a^2\Sigma_X$, the estimate simplifies to the analogue of the univariate result if a^2 is known

$$\bar{y}_1 - \bar{y}_2 - (1+a^2)\hat{\beta}^*$$

which is unbiased for $\tau_1 - \tau_2$.

5.5. Omitted Confounding Variables

Assume that y has parallel regressions on (x, z) in the populations but that matching and/or adjustment is done on the x variables alone. Also assume that x and z are the only confounding variables. This multivariate case is very similar to the univariate one of Section 3.5 and the multivariate analogs of all the formulas follow in an obvious manner. The basic result is that if z has a linear regression on x, z can be decomposed into z_a along x and z_0 orthogonal to x, and adjustment on x is also adjustment on z_a but does not affect z_0.

6. SOME MULTIVARIATE GENERALIZATIONS OF UNIVARIATE MATCHING METHODS

6.1. Caliper Matching

Thus far we have not discussed any specific multivariate matching methods. The obvious extension of caliper matching is to seek in reservoir 2 a match for each x_{1j} such that $|x_{1j}^{(k)} - x_{2j}^{(k)}| < c_k$ for $k = 1, 2, \ldots, p$. This method is used in practice, the difficulty being the large size of reservoir needed to find matches.

The effect of this method on $E_m(\bar{y}_1 - \bar{y}_2)$ could be calculated from univariate results if all x were independently distributed in P_2 (this restriction will be relaxed shortly). This follows because selection on x_{ij} from P_2 would not affect the other x variables, and so the percent reduction in the bias of the variate $x^{(k)}$ under this method would be the same as that under the univariate caliper matching $|x_{1j}^{(k)} - x_{2j}^{(k)}| < c_k$. From these p percent reductions, the percent reduction in bias could be calculated for any y that is linear in the x. For example, with $p = 2$, let $B = 0.5$ and $\sigma_1^2/\sigma_2^2 = \frac{1}{2}$ for $x^{(1)}$, while $B = 0.25$ and $\sigma_1^2/\sigma_2^2 = 2$ for $x^{(2)}$. Then if $c_1 = 0.4\sqrt{(\sigma_1^2 + \sigma_2^2)/2}$ and $c_2 = 0.8\sqrt{(\sigma_1^2 + \sigma_2^2)/2}$, the reductions for $x^{(1)}$ and $x^{(2)}$ from Table 2.1 are about 96% and 77%. That for $x^{(1)} + x^{(2)}$, for instance, is about $[(.96)(.5) + (.77)(.25)]/(.75) = 90\%$.

With this approach, an attempt to select the c_k from initial estimates of $\eta_1^{(k)} - \eta_2^{(k)}$ and σ_1^2/σ_2^2 for $x^{(k)}$ so that matching gives the same percent reduction in bias for each $x^{(k)}$ has some appeal, particularly when matching for more than one y or when it is uncertain which $x^{(k)}$ are more important. Whenever this property does not hold, matching can *increase* the bias for some y's linear in x. For instance, in the preceding example matching would increase the bias for $x^{(1)} - 2x^{(2)}$, whose bias is initially zero.

In general of course, the matching variables are not independently distributed in P_2, but if they are normally distributed (or more generally spherically distributed, Dempster, 1969) there exists a simple linear transformation

$$z = xH \quad \text{where} \quad H'H = \Sigma_2^{-1} \tag{2.46}$$

such that the z are independently distributed in P_2. Hence, assuming (1) x normal in P_2, (2) a large sample from P_2 so that H is essentially known and all matches can be obtained, and (3) the caliper matching method defined above is used on the $z = xH$ variables, Table 2.1 can be used to calculate the percent reduction in bias for each of the $z^{(k)}$. Also, from these p percent reductions in bias, the percent reduction in bias can be calculated for any linear combination of the $z^{(k)}$, such as any $x^{(k)}$ or any y that is linear in x.

We consider caliper matching on the transformed variables to be a reasonable generalization of univariate caliper matching to use in practice. Caliper matching on the original x variables defines a fixed p-dimensional "rectangular" neighborhood about each x_{1j} in which an acceptable match can be found. If caliper matching is used on the z-variables, a neighborhood is defined about each x_{1j} that in general is no longer a simple rectangle with sides perpendicular to the x-variables but a p-dimensional parallelopiped whose sides are not perpendicular to the x-variables but to the p linear combinations of the x-variables corresponding to the z. Since the original choice of a rectangular neighborhood (e.g., rather than a circular one) was merely for convenience, the neighborhood defined by the z calipers should be just as satisfactory.

6.2. Categorical Matching

As a second example of a commonly used matching method for which we can apply the univariate results, assume the categorical matching method of Section 4 is used with c_k categories for each matching variable, the final match for each member of the first sample being chosen from the members of the second sample lying in the same categories on all variables. If this matching is performed on the transformed variables z given in (2.46), normality is assumed, and the reservoir is large, Table 2.8 can be used to calculate the percent reduction in bias of each $z^{(k)}$ in the final matched sample, and thus of each $x^{(k)}$ or any y linear in x. Actually Table 2.8 requires the ratio of variances to be 1 and B moderate or small but could be extended to include more cases.

By adjusting the number of categories used per matching variable $z^{(k)}$ as a function of $E_1(z^{(k)}) - E_2(z^{(k)})$ and $\mathrm{var}_1(z^{(k)})/\mathrm{var}_2(z^{(k)})$ one can obtain approximately the same percent reduction in bias of any y that is linear in x.

6.3. Discriminant Matching

As a final example of multivariate matching methods for which some of the previous univariate results are applicable, assume the transformation in (2.46) will be used with H defined so that $(\eta_1 - \eta_2)H \propto (1, 0, \ldots, 0)$. Univariate matches are then obtained on $z^{(1)}$, the best linear discriminant with respect to the Σ_2 inner product, as suggested by Rubin (1970). Note with this method there is no (mean) bias orthogonal to the discriminant (i.e., $E_1 z^{(k)} = E_2 z^{(k)}, k = 2, \ldots, p$); hence, if the x are normal in P_2 (so that $z^{(1)}$ and $(z^{(2)}, \ldots, z^{(p)})$ are independent), the percent reduction in bias for any linear function of the x equals the percent reduction in bias of $z^{(1)}$.

Tables 2.1, 2.4, 2.2, or 2.8 can then be used to calculate the percent reduction in bias for each $x^{(k)}$ when univariate caliper, mean, nearest available or categorical matching is used on the discriminant. In using these tables σ_1^2/σ_2^2 is the ratio of the $z^{(1)}$ variances in P_1 and P_2, $(\bar{x}_1 - \bar{x}_2)\Sigma_2^{-1}\Sigma_1\Sigma_2^{-1}(\bar{x}_1 - \bar{x}_2)'/(\bar{x}_1 - \bar{x}_2)\Sigma_2^{-1}(\bar{x}_1 - \bar{x}_2)'$, and B is the number of standard deviations between the means of $z^{(1)}$ in P_1 and P_2, $(\bar{x}_1 - \bar{x}_2)\Sigma_2^{-1}(\bar{x}_1 - \bar{x}_2)'/\sqrt{\frac{1}{2}(\sigma_1^2 + \sigma_2^2)}$. Note that for many matching variables, this B could be quite large even if the means of each matching variable are moderately similar in P_1 and P_2.

Discriminant matching has several appealing properties:

(1) it is easy to control the sizes of the final matched samples to be exactly of size n;
(2) if x is approximately normal in P_2 the method should do a good job of reducing bias of any y linear in x, even for a modest reservoir; this follows from an examination of Tables 2.2 and 2.4;
(3) if x is approximately normal in both P_1 and P_2 with $\Sigma_1 \simeq \Sigma_2$, pair matching should do a good job of reducing the bias of any type of regression when the reservoir is large and/or when combined with regression adjustment.

The third point follows from the fact that if x is normal in P_1 and P_2 with $\Sigma_1 = \Sigma_2$, orthogonal to the discriminant the distributions of the matching variables are identical in P_1 and P_2 and unaffected by the matching. Hence, for any y, all bias is due to the different distributions of the discriminant, and Tables 2.5–2.7 indicate that with moderate r, matching and regression adjustment remove much of this bias;

also when $r \to \infty$ the distributions of all matching variables will be the same in the matched samples if nearest available matching is used and x is normal with $\Sigma_1 = \Sigma_2$.

In addition, if one had to choose one linear combination of the x along which a non-linear y is changing most rapidly, and thus on which to obtain close pair matches, the discriminant seems reasonable since the matching variables were presumably chosen not only because their distributions differ in P_1 and P_2 but also because they are correlated with y.

Of course, the joint distributions of matching variables are not assured to be similar in the matched samples, as they would be with pair matches having tight calipers or with a large number of categories using the methods of Sections 6.1 or 6.2. However, the ability to find tight pair matches on all matching variables in a highly multivariate situation seems dubious even with moderately large r. The implications of these points require study.

In practice the discriminant is never known exactly. However, symmetry arguments (Rubin, 1976b,c) show that under normality in P_2, matching on the sample-based discriminant still yields the same percent reduction in expected bias for each $x^{(k)}$.

6.4. Other Matching Methods

There are two kinds of problems with the preceding matching methods. First, for those utilizing all the z it is difficult to control the size of the final sample of matches. Thus with the caliper or categorical methods little is known about the actual reservoir size needed to be confident of obtaining a match for each member of the first sample, although an argument suggests that the ratio of reservoir to sample size for p variables i.i.d. in P_1 and P_2 is roughly the p-th power of the ratio for one variable. The use of caliper matching to obtain matched samples in a practical problem is described in Althauser and Rubin (1970).

When using mean and nearest available matching on the discriminant it is easy to control the final matched sample to have size n. However using discriminant matching, individual matched pairs are not close on all variables and they rely on specific distributional assumptions to insure that the samples are well-matched, even as $r \to \infty$.

An alternative is to try to define matching methods more analogous to the univariate nearest available matching method using some definition of "distance" between x_{1j} and x_{2j}. We might choose the n matches by ordering the x_{1j} in some way (e.g., randomly) and then assigning as a match the nearest x_{2j} as defined by some multivariate distance measure. Such methods will be called nearest available metric matching methods.

A simple class of metrics is defined by an inner product matrix, D, so that the distance from x_{1j} to x_{2j} is $(x_{1j} - x_{2j}) D(x_{1j} - x_{2j})'$. Rather obvious choices for D are Σ_1^{-1} or Σ_2^{-1} yielding the Mahalanobis (1927) distance between x_{1j} and x_{2j} with respect to either inner product. If $\Sigma_1 \propto \Sigma_2$ and x is spherical, symmetry implies that either Mahalanobis distance yields the same percent reduction in bias for each $x^{(k)}$.

More generally, symmetry arguments (Rubin, 1976b,c) show that for x spherical and an inner product metric, the same percent reduction in bias is obtained for each $x^{(k)}$ if and only if

(1) The P_i covariance matrices of x orthogonal to the discriminants are proportional:

$$\Sigma_+ = \Sigma_1 - \frac{1}{s_1^2}(\eta_1 - \eta_2)'(\eta_1 - \eta_2) = c\left[\Sigma_2 - \frac{1}{s_2^2}(\eta_1 - \eta_2)'(\eta_1 - \eta_2)\right]$$

where $s_i^2 = $ the variance of the discriminant in $P_i = (\eta_1 - \eta_2)\Sigma_i^{-1}(\eta_1 - \eta_2)'$. (Note that this implies the discriminants with respect to the P_1 and P_2 inner products are proportional).

(2) The inner product matrix D used for matching is proportional to $[\Sigma_+ + k(\eta_1 - \eta_2)'(\eta_1 - \eta_2)]^{-1}$ with $k \geqslant 0$ (if $k = 0$ or ∞, the inverse is a generalized inverse, Rao, 1973).

The choice of $k = \infty$ yields matching along the discriminant, $k = 0$ yields matching in the space orthogonal to the discriminant, $k = s_1^{-2}$ yields matching using the P_1 Mahalanobis distance and $k = cs_2^{-2}$ yields matching using the P_2 Mahalanobis distance. Symmetry arguments also show that under normality and condition (1), using the sample estimates of Σ_+ and $(\eta_1 - \eta_2)$ gives the same percent reduction in bias for each $x^{(k)}$.

There are of course other ways to define distance between x_{1j} and x_{2j}, for example by the Minkowski metric

$$\left[\prod_{k=1}^p \left|x_{1j}^{(k)} - x_{2j}^{(k)}\right|^\gamma\right]^{1/\gamma} \quad \text{for some } \gamma > 0.$$

Nothing seems to be known about the performance of such matching methods.

A final class of methods that has not been explored might be described as sample metric matching. The simplest example would be to minimize distance between the means $\bar{x}_1$ and $\bar{x}_2$ with respect to a metric. More interesting and robust against non-linearity would be to minimize a measure of the difference between the empirical distribution functions.

7. SUMMARY COMMENTS

This review of methods of controlling bias in observational studies has concentrated on the performance of linear regression adjustments and various matching methods in reducing the initial bias of y due to differences in the distribution of confounding variables, x, in two populations; this seemed to us the most important aspect in observational studies. We have not considered the effects of these techniques on increasing precision, as becomes the focus of interest in randomized experiments.

If the x variables are the only confounding variables, linear regression adjustment on random samples removes all the initial bias when the (y, x) relations are linear and parallel. With only one x and parallel monotonic curved relations of the types examined, linear adjustment on random samples again removes essentially all the bias if $\sigma_1^2 = \sigma_2^2$ and the distributions of x are symmetric, but may perform very erratically if σ_1^2/σ_2^2 is not near 1, or if the distributions of x are asymmetric.

Except in studies from past records, like the Cornell studies of the effectiveness of seat belts in auto accidents (Kihlberg and Robinson, 1968), matching must usually be performed before y has been measured. A drawback is the time and frustration involved in seeking matches from the available reservoirs, but this will be alleviated if computer methods like the 'nearest available' are extended to more than one x. The appeal of matching lies in the simplicity of the concept and the intuitive idea that a tight matching should work well whether the relation between y and x is linear or curved. In our studies with one x, however, the matching methods alone did not perform as well as linear regression under either a linear (y, x) relation, or a monotonic non-linear relation with $\sigma_1^2 = \sigma_2^2$ and x symmetric. Regression adjustment on matched samples also removes all the bias in the linear case and is about as effective as regression on random samples in the non-linear case. If the (y, x) relation is non-linear and σ_1^2 and σ_2^2 are very different, matching followed by regression adjustment on matched pairs performs best. Monte Carlo results on more moderate σ_1^2/σ_2^2 and asymmetric x would be helpful.

Overall, linear regression adjustment is recommended as superior to matching alone when x is continuous and only a moderate reservoir is available. In a similar comparison with more emphasis on precision, Billewicz (1965) reports that regression was more effective than matching in this respect also. However, it appears that the approach of pair matching *plus* regression adjustment on matched pairs is generally superior to either method alone.

An obvious approach not considered here is to try adjustment by a quadratic regression if this appears to fit well in both samples; there appears to be no work on this problem.

Indeed, this review has indicated numerous topics on which little or no work has been done. Even with univariate x these include research on the sizes of reservoirs needed to obtain caliper or categorical matches, on the effectiveness of the commonly used technique of incomplete matching in which members of sample 1 that lack good matches are discarded, and in methods of relaxing the restrictive assumptions of linearity and normality as suggested in Section 3.3 (and Section 5.3 for the multivariate case). For the case of a dichotomous dependent variable the only work seems to be that of McKinlay (1973).

In Sections 6.1–6.4 we have suggested several multivariate extensions of the matching methods but very little is known about their effectiveness. In this connection a survey of the commonly used methods of control, reservoir sizes and number of variables that occur in applications would be useful in guiding the scope of further research.

PART II: UNIVARIATE MATCHING METHODS AND THE DANGERS OF REGRESSION ADJUSTMENT

The statistical study of the utility of general methods of matched sampling starts with the simplest setting, that with only one matching variable, X, as in Cochran (1968a). When attempting to control for bias in X, a prominent competitor to matching is regression adjustment, also called covariance adjustment (ANCOVA = analysis of covariance). Therefore, it is important when studying matching to compare the relative merits of the methods, including their combination, that is, regression adjustment on the matched samples. In fact, this topic was *the* focus of my thesis.

Chapters 3 and 4, Rubin (1973a,b), were originally published back-to-back in *Biometrics*, and were improved versions of the material in my PhD thesis. The first considered matching on a single normally distributed covariate, X, where the outcome variable, Y, was monotonely but possibly nonlinearly (e.g., exponentially) related to X with possibly nonparallel regressions in the treatment and control groups. The estimand was the average causal effect for the treated group, and the estimator was the simple difference in Y means. The criterion for comparing methods was "percent reduction in the bias" due to selecting samples by pair matching or mean matching on X, rather than random sampling. The conditions considered involved various treatment sample sizes and ratios of control to treatment sample sizes, as well as various mean differences on X between treatment and control groups, and ratios of variance of X in the treatment and control groups. Some analytic results were derived, for example, for the maximum bias reduction, a result that generalized well to the multivariate situation and is still useful today in the context of propensity score matching techniques. Other analytic results concerning trends were summarized in the Appendix, which may be important when studying matching with small sample sizes. Most results, however, were based on extensive simulations, summarized in various tables. Advice to the investigator was included, which warned of the need for overlapping X distributions, and noted the benefit of having more variance of X in the control group to help create overlap with X values in the treated group because the causal effect to be estimated was the average effect in the treated group. Again, this issue is highly relevant today. For example, see the well-known Dehijia and Wahba (1999) reanalysis of the classic Lalonde (1986) data set.

A somewhat humorous aspect of this article was that, in those days, the copy editor for *Biometrics* set the titles for the tables, but the author supplied camera-ready copy for the interior of the tables; apparently, someone there physically dropped the pages containing the interior of the tables and then haphazardly matched titles and interiors without ever checking with anyone! Not the most accurate matching for an article on matching. The resulting multitude of errors were noted in a printer's correction note the next year, and have been corrected here.

Chapter 4, Rubin (1973b), goes on to consider the use of regression adjustment in the same situations as Chapter 3. Three regression estimators of the treatment effect were studied, in addition to the simple matching estimator of Rubin (1973a): the usual ordinary least squares (OLS) one assuming parallel regressions lines in the treated and control groups, the Peters–Belson (defined at the end of this intro-duction) estimator using the control group regression line, and the one based on matched pair differences. The basic conclusion was simple: the combination of regression and matching was usually superior to either alone, with matching alone more robust across different settings involving nonlinear, nonparallel response sur-faces than regression adjustment alone. Although there were some minor analytic results presented, as with the first article, the concluding advice to the investigator was based largely on simulations.

There is a final piece of Cochran wisdom about these two early *Biometrics* articles of mine. These two were my second and third sole-authored journal submissions, and after waiting months for the editorial reports, I received what I would now view as a "tentative reject with encouragement to resubmit" decision for both, together with very detailed referee reports, which often missed the point. I was very unhappy! How could the editor send my wonderful work to such obvious idiots! Bill calmed me down and pointed out that: (1) they may be idiots, but they are, almost certainly, more informed than most of your intended audience (that's why they were asked to review the work, you idiot); (2) if you ever want to communicate with that audience, you'd better learn how to present your work so it is coherent to them; (3) they almost certainly have, at this point, more experience in this area than you; and (4) they didn't get paid to do this – they spent a long time writing out their confusions to try to help you, not hurt you. So be quiet and fix the papers! Very good advice that I've never forgotten, and that I try to pass on to my own students.

Chapter 5, Rubin (1977a), was written after I had been at ETS as chair of the "Statistical Research Group" for a couple of years – ETS was the location of my first real full-time job following my postgraduate year at Harvard, and essentially my only full-time nonacademic job since then. The article grew out of my frustration with educational researchers at the time who were dealing with interventions designed to help underprivileged kids, such as Upward Bound, and who were enamored of "true scores," which were supposed to measure "true abilities." The "true score" is the underlying "true" value of an observed test score, and the mood at ETS was that there was the need to adjust for true scores in some magic way in order to estimate the causal effect of an educational intervention, even if the rule for the assignment of the intervention was based solely on observed scores. At the time, I thought it was obvious that if assignment to treatment versus control is based solely on a stochastic rule involving X, then X is the only variable that needs to be controlled, even when X is a completely fallible measure of ability. Not so, I was told.

Rubin (1977a) proved this simple result for different estimands, including the average treatment effect for all units and for the subset comprising the treated units. Three general methods were considered: matching on X, subclassification (blocking) on X, and regression adjustment (modeling) of X. The need for overlapping distributions of X was again emphasized, without which the investigator had no choice but to rely on assumptions allowing pure extrapolation. In some ways, the messages of this simple article are still often missed today. In many observational studies, assignments into interventions are partly or mostly based on variables people can observe and do measure, and not on some unobservable metaphysical construct.

Before closing this introduction to the second part, it seems appropriate to note some older references on regression adjustment in observational studies. One of the first is Peters (1941), which proposed using the control group regression line to effectively create a matched control value for each treated person by using that line to "impute" (current terminology, not Peters') the value each treated person would have had if that person had not been treated. This method is called the "Peters–Belson" estimator because Belson (1956) independently proposed the same method, as referenced by Cochran (1969), who was not aware of the Peters reference until I pointed it out to him. This Cochran article, and Cochran (1957) on the analysis of covariance, warned of the implicit extrapolation involved when doing regression adjustment, which can occur with either the standard assumption of parallel lines or the Peters–Belson method. This simple advice is very important and merits constant repetition, even today.

3. Matching to Remove Bias in Observational Studies

Donald B. Rubin

Abstract: Several matching methods that match all of one sample from another larger sample on a continuous matching variable are compared with respect to their ability to remove the bias of the matching variable. One method is a simple mean-matching method and three are nearest available pair-matching methods. The methods' abilities to remove bias are also compared with the theoretical maximum given fixed distributions and fixed sample sizes. A summary of advice to an investigator is included.

1. INTRODUCTION

Matched sampling is a method of data collection and organization designed to reduce bias and increase precision in observational studies, i.e., in those studies in which the random assignment of treatments to units (subjects) is absent. Although there are examples of observational studies which could have been conducted as properly randomized experiments, in many other cases the investigator could not randomly assign treatments to subjects. For example, consider the Kihlberg and Robinson [1968] study comparing severity of injury in automobile accidents for motorists using and not using seatbelts. One would not want to randomly assign subjects to "seatbelt" and "no seatbelt" treatments and then have them collide at varying speeds, angles of impact, etc. Neither, however, would one want to simply compare the severity of injury in "random" samples of motorists in accidents using and not using seatbelts; important variables such as "speed of automobile at time of accident" may be differently distributed in the two groups (i.e., seatbelted motorists are generally more cautious and therefore tend to drive more slowly). Hence, in observational studies, methods such as matched sampling or covariance adjustment are often needed to control bias due to specific variables.

We will investigate matched sampling on one continuous matching variable X (e.g., speed of automobile at time of accident) and two treatment populations, P_1 and P_2 (e.g., motorists in accidents using and not using seatbelts). Several articles have previously considered this situation. However, most of these have assumed that the average difference in the dependent variable between the matched samples

Reprinted with permission from *Biometrics*, 1973, **29**, 159–183. © 1973 International Biometric Society.

is an unbiased estimate of the effect of the treatment and thus were interested in the ability of matching to increase the precision of this estimate. See, for example, Wilks [1932], Cochran [1953a], Greenberg [1953], and Billiwicz [1965]. Here, we will investigate the ability of matched sampling to reduce the bias of this estimate due to a matching variable whose distribution differs in P_1 and P_2 (e.g., to reduce the bias due to "speed at time of accident").

We assume that there is a random sample of size N from P_1, say G_1, and a larger random sample of size $rN, r \geq 1$, from P_2, say G_2. All subjects in G_1 and G_2 are assumed to have recorded scores on the matching variable X. Using these scores, a subsample of G_2 of size N will be chosen according to some "matching method"; we call this subsample G_{2*}. The effect of the treatment will then be estimated from the G_1 and G_{2*} samples both of size N. If r is one, G_{2*} would be a random sample from P_2, and matching could not remove any bias due to X; if r is infinite, perfect matches could always be obtained, and all of the bias due to X could be removed. We will study moderate ratios of sample sizes, basically $r = 2, 3, 4$, although some results are given for $r = 6, 8, 10$.

Following Cochran [1968a], we will use "the percent reduction in the bias of X due to matched sampling" as the measure of the ability of a matching method to reduce the bias of the estimated effect of the treatment; justification for this choice is given in Section 2. Then Section 3 states and proves a theorem giving the maximum obtainable percent reduction in bias given fixed distributions of X in P_1 and P_2 and fixed samples sizes N and rN. In Section 4, the ability of a simple mean-matching method to reduce bias will be compared with the theoretical maximum. In Section 5, we compare three "nearest available" pair-matching methods with respect to their ability to reduce bias. Section 6 serves to present practical advice to an investigator.

2. TERMINOLOGY; PERCENT REDUCTION IN BIAS

Suppose that we want to determine the effect of a dichotomous treatment variable on a continuous dependent variable, Y, given that the effect of a continuous matching variable, X, has been removed.[1] The dichotomous treatment variable is used to form two populations P_1 and P_2. In P_1 and P_2 X and Y have joint distributions which in general differ from P_1 to P_2. In P_i the conditional expectation of the dependent variable Y given a particular value of X is called the response surface for Y in P_i, and at $X = x$ is denoted $R_i(x)$.

The difference in response surfaces at $X = x$, $R_1(x) - R_2(x)$, is the effect of the treatment variable at $X = x$. If this difference between response surfaces is constant and so independent of the values of the matching variable, the response surfaces are called parallel, and the objective of the study is the estimation of the constant difference between them. See Figure 3.1. For linear response surfaces, "parallel response surfaces" is equivalent to "having the same slope."

---

[1] As Cochran [1968a] points out, if the matching variable X is causally affected by the treatment variable, some of the real effect of the treatment variable will be removed in the adjustment process.

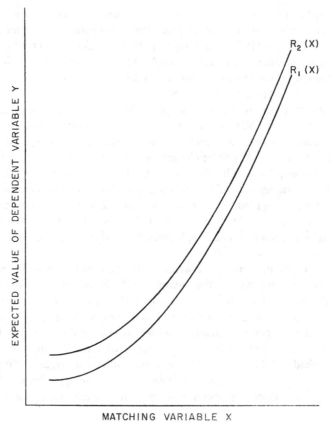

Figure 3.1. Parallel univariate response surfaces.

If $R_1(x) - R_2(x)$ depends on x, the response surfaces are non-parallel and there is no single parameter that completely summarizes the effect of the treatment variable. In this case we will assume that the average effect of the treatment variable (the average difference between the response surfaces) over the P_1 population is desired. Such a summary is often of interest, especially when P_1 consists of subjects exposed to an agent and P_2 consists of controls not exposed to the agent; see for example Belsen's [1956] study of the effect of an educational television program.[2]

---

[2] In other cases, however, this average difference may not be of primary interest. Consider for example the previously mentioned study of the efficacy of seatbelts. Assume that if automobile speed is high seatbelts reduce the severity of injury, while if automobile speed is low seatbelts increase the severity of injury. (See Figure 3.2, where $P_1 =$ motorists using seatbelts, $P_2 =$ motorists not using seatbelts, $X =$ automobile speed, and $Y =$ severity of injury.) A report of this result would be more interesting than a report that there was no effect of seatbelts on severity of injury when averaged over the seatbelt wearer population. Since such a report may be of little interest if the response surfaces are markedly nonparallel, the reader should generally assume "nonparallel" to mean "moderately nonparallel." If the response surfaces are markedly nonparallel and the investigator wants to estimate the effect of the treatment variable averaged over P_2 (the population from which he has the larger sample), the methods and results presented here are not relevant and a more complex method such as covariance analysis would be more appropriate than simple matching. (See Cochran [1969] for a discussion of covariance analysis in observational studies.)

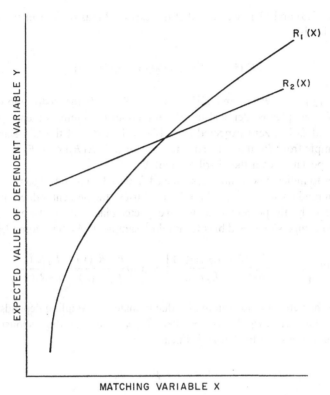

Figure 3.2. Nonparallel univariate response surfaces.

The average difference between non-parallel response surfaces over the P_1 population or the constant difference between parallel response surfaces will be called the (average) effect of the treatment variable or more simply "the treatment effect" and will be designated τ:

$$\tau = E_1\{R_1(x) - R_2(x)\}, \tag{3.1}$$

where E_1 is the expectation over the distribution of X in P_1.

Let y_{1i} and x_{1i} be the values of Y and X for the jth subject in G_1, and similarily let y_{2i} and x_{2i} be the values of Y and X for the jth subject in G_{2^*}, $j = 1, \ldots, N$. Using the response surface notation we can write

$$y_{ij} = R_i(x_{ij}) + e_{ij} \quad i = 1, 2; \quad j = 1, \ldots, N \tag{3.2}$$

where $E_c(e_{ij}) = 0$ and E_c is the conditional expectation given the x_{ij}.

We assume that the difference between dependent variable averages in G_1 and G_{2^*} will be used to estimate τ:

$$\hat{\tau}_0 = \frac{1}{N}\sum y_{1j} - \frac{1}{N}\sum y_{2j} = \bar{y}_{1.} - \bar{y}_{2.} \tag{3.3}$$

Let E represent the expectation over the distributions of X in matched samples and E_{2^*} represent the expectation over the distribution of X in matched G_{2^*} samples.

Then using (3.3) and (3.1) we have that the expected bias of $\hat{\tau}_0$ over the matched sampling plan is

$$EE_c(\hat{\tau}_0 - \tau) = E_1 R_2(x) - E_{2*} R_2(x) \tag{3.4}$$

since $EE_c(\bar{y}_{2.}) = E_{2*} R_2(x)$ and $EE_c(\bar{y}_{1.}) = E_1 R_1(x)$. If the distribution of X in matched G_{2*} samples is identical to that in random G_1 samples then $E_1 R_2(x) = E_{2*} R_2(x)$ and $\hat{\tau}_0$ has zero expected bias. If $r = 1$, that is if the G_{2*} sample is a random sample from P_2, then the expected bias of $\hat{\tau}_0$ is $E_1 R_2(x) - E_2 R_2(x)$ where E_2 is the expectation over the distribution of X in P_2.

In order to indicate how much less biased $\hat{\tau}_0$ based on matched samples is than $\hat{\tau}_0$ based on random samples, Cochran [1968a] uses "the percent reduction in bias" or more precisely "the percent reduction in expected bias": $100 \times (1 - \text{expected bias}$ for matched samples/expected bias for random samples) which is from (3.4)

$$100 \left\{ 1 - \frac{E_1 R_2(x) - E_{2*} R_2(x)}{E_1 R_2(x) - E_2 R_2(x)} \right\} = 100 \frac{E_{2*} R_2(x) - E_2 R_2(x)}{E_1 R_2(x) - E_2 R_2(x)}. \tag{3.5}$$

Notice that the percent reduction in bias due to matched sampling depends only on the distribution of X in P_1, P_2 and matched G_{2*} samples, and the response surface in P_2. If the response surface in P_2 is linear,

$$R_2(x) = \mu_2 + \beta_2(x - \eta_2)$$

where

$$\mu_2 = \text{mean of } Y \text{ in } P_2$$
$$\eta_i = \text{mean of } X \text{ in } P_i$$

and

$$\beta_2 = \text{regression coefficient of } Y \text{ on } X \text{ in } P_2,$$

we have for the denominator of (3.5) $\beta_2(\eta_1 - \eta_2)$ and for the numerator of (3.5) $\beta_2(\eta_{2*} - \eta_2)$, where η_{2*} is the expected value of X in matched G_{2*} samples, $E_{2*}(x)$ (equivalently, η_{2*} is the expected average X in G_{2*} samples, $E(\bar{x}_{2.})$).

Thus, if G_1 is a random sample and the response surface in P_2 is linear, the percent reduction in bias due to matched sampling is

$$\Theta = 100 \frac{\eta_{2*} - \eta_2}{\eta_1 - \eta_2}, \tag{3.6}$$

which is the same as the percent reduction in bias of the matching variable X. Even though only an approximation if the P_2 response surface is not linear, we will use Θ, the percent reduction in the bias of the matching variable, to measure the ability of a matching method to remove bias.

3. THE MAXIMUM PERCENT REDUCTION IN BIAS GIVEN FIXED DISTRIBUTIONS AND FIXED SAMPLE SIZES

Assume that in P_i X has mean η_i (without loss of generality let $\eta_1 > \eta_2$), variance σ_i^2 and $(X - \eta_i)/\sigma_i \sim f_i$, $i = 1, 2$. Define the initial bias in X to be

$$B = \frac{\eta_1 - \eta_2}{\sqrt{\frac{\sigma_1^2 + \sigma_2^2}{2}}} > 0,$$

which if $\sigma_1^2 = \sigma_2^2$ is simply the number of standard deviations between the means of X in P_1 and P_2.

Then if Θ is the percent reduction in bias of X due to some matching method that selects a matched sample, G_{2*}, of N subjects from a random sample, G_2, of rN P_2 subjects, we have

$$\Theta \leq \Theta_{max} = 100 \frac{\Omega_2(r, N)}{B\sqrt{\frac{1 + \sigma_1^2/\sigma_2^2}{2}}}, \tag{3.7}$$

where $\Omega_2(r, N) =$ the expected value of the average of the N largest observations from a sample of size rN from f_2.

Since a reduction in bias greater than 100% is clearly less desirable than 100% reduction in bias, if B, σ_1^2/σ_2^2, and $\Omega_2(r, N)$ are such that $\Theta_{max} \geq 100$ this should be interpreted as implying the existence of a matching method that obtains 100% reduction in expected bias.[3]

This result follows immediately from (3.6): since $\eta_1 > \eta_2$, Θ is the largest when η_{2*} (i.e., $E(\bar{x}_{2*})$) is largest, which is clearly achieved when the N subjects in G_2 with the largest X values are always chosen as matches. The expected value of these N largest values from a sample of rN is $\eta_2 + \sigma_2 \Omega_2(r, N)$. Hence, the maximum value of Θ is

$$\Theta_{max} = 100 \frac{\sigma_2 \Omega_2(r, N)}{\eta_1 - \eta_2} = 100 \frac{\Omega_2(r, N)}{B\sqrt{\frac{1 + \sigma_1^2/\sigma_2^2}{2}}}.$$

The result in (3.7) is of interest here for two reasons. First, for fixed distributions and sample sizes and given a particular matching method, a comparison of Θ and min $\{100, \Theta_{max}\}$ clearly gives an indication of how well that matching method does at obtaining a G_{2*} sample whose expected X mean is close to η_1. In addition, the expression for Θ_{max} will be used to help explain trends in Monte Carlo results. When investigating matching methods that might be used in practice to match finite samples, properties such as percent reduction in bias are generally analytically intractable. Hence, Monte Carlo methods must be used on specific cases. From such Monte Carlo investigations it is often difficult to generalize to other cases or explain trends with much confidence unless there is some analytic or intuitive reason for believing the trends will remain somewhat consistent. It seems clear that

---

[3] A matching method that has as its percent reduction in expected bias min $\{100, \Theta_{max}\}$ may be of little practical interest. For example, consider the following matching method. With probability $P = \min\{1, 1/\Theta_{max}\}$ choose the N G_2 subjects with the largest observations the G_{2*} sample and with probability $1 - P$ choose a random sample of size N as the G_{2*} sample. It is easily checked that the percent reduction in expected bias using this method is min $\{100, \Theta_{max}\}$.

Table 3.1. $\Omega(r, N)$

	$r=$ 2		3		4		6		8		10	
	$N=1$	∞	1	∞	1	∞	1	∞	1	∞	1	∞
$+\chi_2^2$	0.50	0.69	0.83	1.10	1.08	1.38	1.45	1.79	1.72	2.08	1.93	2.30
$+\chi_4^2$	0.53	0.74	0.86	1.13	1.10	1.39	1.43	1.75	1.67	2.00	1.85	2.19
$+\chi_6^2$	0.54	0.76	0.86	1.13	1.09	1.38	1.41	1.72	1.64	1.95	1.81	2.12
$+\chi_8^2$	0.54	0.77	0.86	1.13	1.09	1.37	1.40	1.69	1.61	1.92	1.78	2.08
$+\chi_{10}^2$	0.55	0.78	0.86	1.13	1.08	1.36	1.39	1.68	1.60	1.89	1.76	2.05
Normal	0.56	0.80	0.85	1.09	1.03	1.27	1.27	1.50	1.42	1.65	1.54	1.75
$-\chi_{10}^2$	0.55	0.78	0.78	0.99	0.92	1.11	1.09	1.25	1.19	1.34	1.26	1.39
$-\chi_8^2$	0.54	0.77	0.77	0.98	0.91	1.09	1.06	1.22	1.16	1.30	1.22	1.35
$-\chi_6^2$	0.54	0.76	0.76	0.95	0.88	1.06	1.02	1.17	1.11	1.24	1.17	1.28
$-\chi_4^2$	0.53	0.74	0.73	0.91	0.84	1.00	0.97	1.09	1.04	1.14	1.08	1.17
$-\chi_2^2$	0.50	0.69	0.67	0.81	0.75	0.86	0.83	0.91	0.87	0.93	0.90	0.95

if Θ_{max} is quite small (e.g., 20) no matching method will do very well, while if Θ_{max} is large (e.g., 200) most reasonable matching methods should do moderately well. Hence, we will use trends in Θ_{max} to help explain trends in the Monte Carlo results that follow.

Two trends for Θ_{max} are immediately obvious from (3.7).

(1) Given fixed r, N, f_2 and σ_1^2/σ_2^2, Θ_{max} decreases as B increases.
(2) Given fixed r, N, f_2 and B, Θ_{max} decreases as σ_1^2/σ_2^2 increases.

Given fixed f_2, B, and σ_1^2/σ_2^2 two other trends are derivable from simple properties of the order statistics and the fact that Θ_{max} is directly proportional to $\Omega_2(r, N)$ (see Appendix A for proofs).

(3) Given fixed B, σ_1^2/σ_2^2, f_2 and N, Θ_{max} increases as r increases: $\Omega_2(r, N) \leq \Omega_2(r + a, N)$, $a \geq 0$; N, rN, aN integers.
(4) Given fixed B, σ_1^2/σ_2^2, f_2 and r, Θ_{max} increases as N increases: $\Omega_2(r, N) \leq \Omega_2(r, N + b)$, $b \geq 0$; N, rN, rb integers.

From the fourth trend, we have $\Omega(r, 1) \leq \Omega(r, N) \leq \Omega(r, \infty)$. Values of $\Omega(r, 1)$ have been tabulated in Sarhan and Greenberg [1962] for several distributions as the expected value of the largest of r observations. $\Omega(r, \infty)$ can easily be calculated by using the asymptotic result

$$\Omega(r, \infty) = r \int_w^\infty z f(z), \quad \text{where} \quad \int_w^\infty f(z) = 1/r.$$

Values of $\Omega(r, 1)$ and $\Omega(r, \infty)$ are given in Table 3.1 for $X \sim \pm\chi_\nu^2 (\nu = 2(2)10)$ and $X \sim$ Normal, and for $r = 2, 3, 4, 6, 8, 10$.
Table 3.1 can be summarized as follows.

(a) For fixed r and ν, the results for $+\chi_\nu^2$ are more similar to those for the normal than are those for $-\chi_\nu^2$. This result is expected since the largest N observations

Table 3.2. $\Omega(r, N)$; f normal

	$r = 2$	3	4	6	8	10
$N = 1$	0.56	0.85	1.03	1.27	1.42	1.54
2	0.66	0.96	1.14	1.38	1.53	1.64
5	0.74	1.03	1.22	1.45	1.60	1.70
10	0.77	1.06	1.24	1.47	1.62	1.72
25	0.78	1.08	1.26	1.49	1.64	1.74
50	0.79	1.08	1.27	1.50	1.65	1.75
100	0.80	1.09	1.27	1.50	1.65	1.75
∞	0.80	1.09	1.27	1.50	1.65	1.75

come from the right tail of the distribution and the right tail of $+\chi_\nu^2$ is more normal than the right tail of $-\chi_\nu^2$ which is finite.

(b) Given a fixed distribution, as r gets larger the results differ more from those for the normal especially for $-\chi_\nu^2$. Again this is not surprising because the tails of low degree of freedom χ^2 are not very normal, especially the finite tail.

(c) For $r = 2, 3, 4$, and moderately normal distributions ($\pm\chi_\nu^2$, $\nu \geq 8$) the results for the normal can be considered somewhat representative. This conclusion is used to help justify the Monte Carlo investigations of a normally distributed matching variable in the remainder of this article.

(d) Given a fixed distribution and fixed r, the values for $\Omega(r, 1)$ are generally within 20% of those for $\Omega(r, \infty)$, suggesting that when dealing with moderate sample sizes as might commonly occur in practice, we would expect the fourth trend (Θ_{max} increasing function of N) to be rather weak.

In Table 3.2 values of $\Omega(r, N)$ are given assuming f normal, the same values of r as in Table 3.1, and $N = 1, 2, 5, 10, 100, \infty$. Values were found with the aid of Harter [1960]. For fixed r, the values of $\Omega(r, N)$ for $N \geq 10$ are very close to the asymptotic value $\Omega(r, \infty)$, especially when $r > 2$. Even $\Omega(2, 10)$ is within about 3% of $\Omega(2, \infty)$. These results indicate that the values for $\Omega(r, \infty)$ given in Table 3.1 may be quite appropriate for moderate sample sizes.

4. MEAN-MATCHING

Thus far we have not specified any particular matching method. Under the usual linear model "mean-matching" or "balancing" (Greenberg [1953]) methods are quite reasonable but appear to be discussed rarely in the literature. In this section we will obtain Monte Carlo percent reductions in bias for a simple mean-matching method and compare these with the theoretical maximums given by (3.7).

Assuming linear response surfaces, it is simple to show from (3.3) that the bias of $\hat{\tau}_0$ for estimating τ is $\beta_2(\eta_1 - \bar{x}_{2.}) + \beta_1(\bar{x}_{1.} - \eta_1)$, where β_i is the regression coefficient of Y on X in P_i and $\bar{x}_i$ is the average X in the matched samples. Using $\bar{x}_{1.}$ to estimate η_1 or assuming parallel response surfaces ($\beta_1 = \beta_2$) one would minimize the estimated bias of $\hat{\tau}_0$ by choosing the N G_2 subjects such that $|\bar{x}_{1.} - \bar{x}_{2.}|$ is minimized. A practical argument against using this mean-matching method is that

Table 3.3. Θ_{MN}: *Percent reduction in bias for a simple mean-matching method, X normal*

		$\sigma_1^2/\sigma_2^2 = \frac{1}{2}$				$\sigma_1^2/\sigma_2^2 = 1$				$\sigma_1^2/\sigma_2^2 = 2$			
		$B=\frac{1}{4}$	$\frac{1}{2}$	$\frac{3}{4}$	1	$\frac{1}{4}$	$\frac{1}{2}$	$\frac{3}{4}$	1	$\frac{1}{4}$	$\frac{1}{2}$	$\frac{3}{4}$	1
	$r=$												
$N=25$	2	99	98	95	84	99	97	89	77	97	90	75	63
	3	100	100	100	98	100	100	98	93	100	98	94	81
	4	100	100	100	100	100	100	100	98	100	100	97	91
$N=50$	2	100	100	98	87	100	99	91	77	100	95	82	67
	3	100	100	100	100	100	100	99	96	100	100	97	84
	4	100	100	100	100	100	100	100	100	100	100	100	95
$N=100$	2	100	100	100	88	100	100	96	80	100	98	85	64
	3	100	100	100	100	100	100	100	98	100	100	99	87
	4	100	100	100	100	100	100	100	100	100	100	100	96

finding such a subset requires the use of some time consuming algorithm designed to solve the transportation problem. Many compromise algorithms can of course be defined that approximate this best mean-match.

We will present Monte Carlo percent reductions in bias only for the following very simple mean-matching method. At the kth step, $k = 1, \ldots, N$, choose the G_2 subject such that the mean of the current G_{2*} sample of k subjects is closest to $\bar{x}_{1\cdot}$. Thus, at step 1 choose the G_2 subject closest to $\bar{x}_{1\cdot}$; at step 2 choose the G_2 subject such that the average of the first G_{2*} subject and the additional G_2 is closest to $\bar{x}_{1\cdot}$; continue until N G_2 subjects are chosen.

In Table 3.3 we present Monte Carlo values of Θ_{MN}, the percent reduction in bias for this simple mean-matching method.[4] We assume X normal, $B = \frac{1}{4}, \frac{1}{2}, \frac{3}{4}, 1; \sigma_1^2/\sigma_2^2 = \frac{1}{2}, 1, 2; N = 25, 50, 100$; and $r = 2, 3, 4$. Some limited experience indicates that these values are typical of those that might occur in practice. In addition, values of r and N were chosen with the results of Tables 3.1 and 3.2 in mind – values for percent reduction in bias may be moderately applicable for nonnormal distributions, especially when $r = 2$, and values given when $N = 100$ may be quite representative for $N > 100$. Θ_{MN} exhibits the four trends given in Section 3 for $\Theta_{\max}$.

(1) Given fixed N, r, and σ_1^2/σ_2^2, Θ_{MN} decreases as B increases.
(2) Given fixed N, r, and B, Θ_{MN} decreases as σ_1^2/σ_2^2 increases.
(3) Given fixed $B, \sigma_1^2/\sigma_2^2$ and N, Θ_{MN} increases as r increases.
(4) Given fixed $B, \sigma_1^2/\sigma_2^2$ and r, except for one value (67% for $N = 50, \sigma_1^2/\sigma_2^2 = 2, r = 2, B = 1$) Θ_{MN} increases as N increases.

In Table 3.4 we present values of $\min\{100, \Theta_{\max}\}$ for the same range of N, B and σ_1^2/σ_2^2 as in Table 3.3. Note first that the 67% for $N = 50, \sigma_1^2/\sigma_2^2 = 2$,

---

[4] The standard errors for all Monte Carlo values given in Tables 3.3, 3.5, 3.6, and 3.7 are generally less than 0.5% and rarely greater than 1%.

Table 3.4. *Min* $\{100, \Theta_{\max}\}$; *X normal*

		$\sigma_1^2/\sigma_2^2 = \frac{1}{2}$				$\sigma_1^2/\sigma_2^2 = 1$				$\sigma_1^2/\sigma_2^2 = 2$			
		$B = \frac{1}{4}$	$\frac{1}{2}$	$\frac{3}{4}$	1	$\frac{1}{4}$	$\frac{1}{2}$	$\frac{3}{4}$	1	$\frac{1}{4}$	$\frac{1}{2}$	$\frac{3}{4}$	1
	$r =$												
$N = 25$	2	100	100	100	91	100	100	100	78	100	100	85	64
	3	100	100	100	100	100	100	100	100	100	100	100	88
	4	100	100	100	100	100	100	100	100	100	100	100	100
$N = 50$	2	100	100	100	91	100	100	100	79	100	100	87	65
	3	100	100	100	100	100	100	100	100	100	100	100	89
	4	100	100	100	100	100	100	100	100	100	100	100	100
$N = 100$	2	100	100	100	92	100	100	100	80	100	100	87	65
	3	100	100	100	100	100	100	100	100	100	100	100	89
	4	100	100	100	100	100	100	100	100	100	100	100	100

$r = 2$, $B = 1$ mentioned above is larger than the theoretical maximum and thus suspect. Comparing the corresponding entries in Table 3.3 and Table 3.4 we see that the values for $N = 100$ always attain at least 96% of min$\{100, \Theta_{\max}\}$, {while the values for $N = 50$ always attain at least 91% of min$\{100, \Theta_{\max}\}$, and those for $N = 25$ always attain at least 87% of min$\{100, \Theta_{\max}\}$. Hence this simple method appears to be a very reasonable mean-matching method, especially for large samples.

5. PAIR-MATCHING

Even though a simple mean-matching method can be quite successful at removing the bias of X, matched samples are generally not mean-matched. Usually matched samples are "individually" (Greenwood [1945]), "precision" (Chapin [1947]), or "pair" (Cochran [1953a]) matched, subject by subject. The main reason is probably some intuitive feeling on the part of investigators that pair-matched samples are superior. One theoretical justification is that $\hat{\tau}_0$ based on exactly mean-matched samples has zero expected bias only if the P_2 response surface really is linear, while $\hat{\tau}_0$ based on exactly pair-matched samples has zero expected bias no matter what the form of the response surface. Since an investigator rarely knows for sure that the P_2 response surface is linear, if the choice is between exactly pair-matched samples and exactly mean-matched samples of the same size, obviously he would choose the exactly pair-matched samples.

The ease of constructing confidence limits and tests of significance is a second reason for using a pair-matching method rather than a mean-matching method. Significance tests and confidence limits that take advantage of the increased precision in matched samples are easily constructed with pair-matched data by using matched pair differences, while such tests and limits for mean-matched data must be obtained by an analysis of covariance (Greenberg [1953]).

Another reason for the use of pair-matching methods is that each matched pair could be considered a study in itself. Thus, the investigator might assume

Table 3.5. Θ_{RN}: *Percent reduction in bias for random order, nearest available matching; X normal*

		$\sigma_1^2/\sigma_2^2 = \frac{1}{2}$				$\sigma_1^2/\sigma_2^2 = 1$				$\sigma_1^2/\sigma_2^2 = 2$			
		$B=\frac{1}{4}$	$\frac{1}{2}$	$\frac{3}{4}$	1	$\frac{1}{4}$	$\frac{1}{2}$	$\frac{3}{4}$	1	$\frac{1}{4}$	$\frac{1}{2}$	$\frac{3}{4}$	1
	$r=$												
$N=25$	2	97	94	89	80	87	82	75	66	63	60	56	48
	3	99	98	97	93	94	91	86	81	77	72	67	61
	4	99	99	99	97	95	95	92	88	81	79	76	68
$N=50$	2	99	98	93	84	92	87	78	69	66	59	53	51
	3	100	99	99	97	96	95	91	84	79	75	69	63
	4	100	100	100	99	98	97	94	89	86	81	75	71
$N=100$	2	100	99	96	86	95	90	81	69	67	59	55	49
	3	100	100	99	98	99	96	91	86	81	75	70	64
	4	100	100	100	99	99	98	96	90	85	81	76	71

the response surfaces are nonparallel and use the difference $y_{1j} - y_{2j}$ to estimate the response surface difference at x_{1j}. It follows from (3.2) that the bias of $y_{1j} - y_{2j}$ for estimating $R_1(x_{1j}) - R_2(x_{1j})$ is $R_2(x_{1j}) - R_2(x_{2j})$. Assuming this bias to be some unknown increasing function of $|x_{1j} - x_{2j}|$, one minimizes the bias of each estimate, $y_{1j} - y_{2j}$, by minimizing each $|x_{1j} - x_{2j}|$ rather than $|\bar{x}_{1.} - \bar{x}_{2.}|$.

If each G_1 subject is closest to a different G_2, subject, assigning matches to minimize each $|x_{1j} - x_{2j}|$ is easily done. However, if two or more G_1 subjects are closest to the same G_2 subject, the best way to assign individual matches is not obvious, unless the investigator decides upon some criterion to be minimized, such as one proportional to the average squared bias of the N individual estimates assuming parallel linear response surfaces, $1/N \sum (x_{1j} - x_{2j})^2$. As was already mentioned, in order to find the G_{2*} sample that minimizes any such quantity, some rather time consuming algorithm designed to solve the transportation problem must be used.

Even though more complex pair-matching methods often may be superior, we will investigate three simple "nearest available" pair-matching methods. A nearest available pair-matching method assigns the closest match for $g_1 \varepsilon G_1$ from the yet unmatched G_2 subjects and thus is completely defined if the order for matching the G_1 subjects is specified. The three orderings of the G_1 subjects to be considered here are: (1) the subjects are randomly ordered (random), (2) the subject not yet matched with the lowest score on X is matched next (low-high), and (3) the subject not yet matched with the highest score on X is matched next (high-low). The results will depend on our assumption that $\eta_1 > \eta_2$), for if η_1 were less than η_2, the values for the low-high and high-low pair-matching methods would be interchanged.

In Tables 3.5, 3.6, and 3.7 we present Monte Carlo values for the percent reduction in bias for random ordering (Θ_{RD}), low-high ordering (Θ_{LH}) and high-low ordering (Θ_{HL}). We assume the same range of conditions as given in Table 3.3

Table 3.6. Θ_{HL}: Percent reduction in bias for high-low order, nearest available matching; X normal

		$\sigma_1^2/\sigma_2^2 = \frac{1}{2}$				$\sigma_1^2/\sigma_2^2 = 1$				$\sigma_1^2/\sigma_2^2 = 2$			
		$B=\frac{1}{4}$	$\frac{1}{2}$	$\frac{3}{4}$	1	$\frac{1}{4}$	$\frac{1}{2}$	$\frac{3}{4}$	1	$\frac{1}{4}$	$\frac{1}{2}$	$\frac{3}{4}$	1
	$r=$												
$N=25$	2	78	83	81	75	77	77	72	65	59	58	55	48
	3	92	94	93	90	89	88	84	79	75	71	67	61
	4	96	97	97	95	93	93	90	87	79	78	75	68
$N=50$	2	86	90	86	79	85	84	76	68	63	58	53	51
	3	96	97	96	94	93	93	89	84	77	74	69	63
	4	98	99	98	97	96	96	93	88	84	81	75	71
$N=100$	2	93	94	90	82	90	87	79	69	65	59	55	49
	3	98	98	98	96	96	95	90	85	80	75	70	64
	4	99	99	99	98	98	97	95	90	84	81	76	71

for Θ_{MN}. Θ_{RD} and Θ_{HL} exhibit the four trends given in Section 3 for $\Theta_{\max}$ and exhibited in Table 3.3 for Θ_{MN}.

(1) Given fixed N, r, and σ_1^2/σ_2^2, Θ_{RD} and Θ_{HL} decrease as B increases.
(2) Given fixed N, r, and B, Θ_{RD} and Θ_{HL} decrease as σ_1^2/σ_2^2 increases.
(3) Given fixed B, σ_1^2/σ_2^2, and N, Θ_{RD} and Θ_{HL} increase as r increases.
(4) Given fixed B, σ_1^2/σ_2^2 and r, Θ_{RD} and Θ_{HL} generally increase as N increases.

These same four trends hold for all orderings if "Θ_{RD} and Θ_{HL} increase" is replaced by "Θ_{RD}, Θ_{HL}, and Θ_{LH} get closer to 100%". Values of Θ greater than 100%

Table 3.7. Θ_{LH}: Percent reduction in bias for low-high order, nearest available matching; X normal

		$\sigma_1^2/\sigma_2^2 = \frac{1}{2}$				$\sigma_1^2/\sigma_2^2 = 1$				$\sigma_1^2/\sigma_2^2 = 2$			
		$B=\frac{1}{4}$	$\frac{1}{2}$	$\frac{3}{4}$	1	$\frac{1}{4}$	$\frac{1}{2}$	$\frac{3}{4}$	1	$\frac{1}{4}$	$\frac{1}{2}$	$\frac{3}{4}$	1
	$r=$												
$N=25$	2	118	105	95	82	98	87	76	67	67	62	57	48
	3	106	103	101	97	99	94	88	81	80	73	68	62
	4	103	102	101	99	98	97	93	89	83	80	76	68
$N=50$	2	113	107	99	86	100	91	79	69	69	60	54	51
	3	103	102	102	99	100	97	92	85	81	75	69	63
	4	102	101	101	100	100	98	95	89	87	82	76	71
$N=100$	2	108	106	101	86	100	92	81	69	69	60	55	49
	3	102	101	102	100	101	98	92	86	82	76	70	64
	4	101	101	101	100	99	99	96	90	85	81	76	71

indicate that $\eta_{2*} > \eta_1$ which is of course not as desirable as $\eta_{2*} \simeq \eta_1$ which implies $\Theta \simeq 100$.

Comparing across Tables 3.5, 3.6, and 3.7 we see that given fixed B, σ_1^2/σ_2^2, r and N, $\Theta_{LH} \geq \Theta_{RD} \geq \Theta_{HL}$. This result is not surprising for the following reason. The high-low ordering will have a tendency not to use those G_2 subjects with scores above the highest G_1 score while the low-high ordering will have a tendency not to use those G_2 subjects with scores below the lowest G_1 scores. Since we are assuming $B > 0$ ($\eta_1 > \eta_2$), the low-high ordering should yield the most positive $\bar{x}_{2*}$ followed by the random ordering and then the high-low ordering. When $\sigma_1^2/\sigma_2^2 = \frac{1}{2}$ and $B \leq \frac{1}{2}$, Θ_{LH} can be somewhat greater than 100 (e.g., 113) while $100 \geq \Theta_{RD} \geq 94$. In all other cases ($\sigma_1^2/\sigma_2^2 > \frac{1}{2}$ or $\sigma_1^2/\sigma_2^2 = \frac{1}{2}$ and $B \leq \frac{1}{2}$), Θ_{LH} is closer to 100% than Θ_{RD} or Θ_{HL}. In general the results for Θ_{RD}, Θ_{LH}, and Θ_{HL} are quite similar for the conditions considered.

Comparing the results in this section with those in Section 4, it is easily checked that if $\sigma_1^2/\sigma_2^2 \leq 1$ the three pair-matching methods generally attain more than 85% of $\min\{100, \Theta_{max}\}$ in Table 3.4 indicating that they can be reasonable methods of matching the means of the samples. However, if $\sigma_1^2/\sigma_2^2 = 2$, the pair-matching methods often attain less than 70% of the corresponding Θ_{MN} in Table 3.3 indicating that when $\sigma_1^2/\sigma_2^2 > 1$ these pair-matching methods do not match the means very well compared to a single mean-matching method.

Remembering that pair-matching methods implicitly sacrifice closely matched means for good individual matches, we also calculated a measure of the quality of the individual matches. These results presented in Appendix C indicate that, in general, the high-low ordering yields the closest individual matches followed by the random ordering. This conclusion is consistent with the intuition to match the most difficult subjects first in order to obtain close individual matches.

6. ADVICE TO AN INVESTIGATOR

In review, we assume there are two populations, P_1 and P_2, defined by two levels of a treatment variable. There is a sample, G_1, of size N from P_1 and a larger sample, G_2, of size rN from P_2, both of which have recorded scores on the matching variable X. The objective of the study is to estimate τ, the average effect of the treatment variable on a dependent variable Y over the P_1 population. We assume that $\hat{\tau}_0 = \bar{y}_1. - \bar{y}_2.$ will be used to estimate τ where $\bar{y}_1.$ is the average Y in the G_1 sample and $\bar{y}_2.$ is the average Y in an N-size subsample of G_2 matched to G_1, G_{2*}.

Depending upon the particular study, the investigator may be able, within limits, to control three "parameters".

(a) N, the size of the smaller initial sample (G_1); equivalently, the size of each of the final samples.
(b) r, the ratio of the sizes of the larger initial sample (G_2) and the smaller initial sample (G_1).
(c) The matching rule used to obtain the G_{2*} sample of size N from the G_2 sample of size rN.

Below we present advice for choosing these "parameters" in the order first N, then r, and then the matching method.

(a) Choosing N

We will use a standard method for estimating N (Cochran [1963]) which assumes that the investigator wants $\hat{\tau}_0$ to be within $\pm\Delta$ of τ with probability $1 - \alpha$: Prob $\{|\hat{\tau}_0 - \tau| > \Delta\} = \alpha$. Letting $s/\sqrt{N}$ be the estimated standard error of $\hat{\tau}_0$ we would choose

$$N = z^2 s^2 / \Delta^2, \tag{3.8}$$

where z is the standard normal deviate corresponding to $1 - \alpha$ confidence limits (e.g., if $\alpha = 0.05$, $z \simeq 2$).[5] In order to use (3.8) we must have an estimate of the standard error of $\hat{\tau}_0$, $s/\sqrt{N}$.

Suppose that the response surfaces are linear with slopes β_1 and β_2 and that $\bar{x}_{2.}$ will be exactly matched to $\bar{x}_{1.}$ in the final samples by using one of the matching methods discussed in Sections 4 and 5. Thus, $E E_c(\hat{\tau}_0) = \tau$, and it is easy to show that

$$s^2/N = E E_c(\hat{\tau}_0 - \tau)^2$$
$$= E E_c[\beta_2(\eta_1 - \bar{x}_{2.}) + \beta_1(\bar{x}_{1.} - \eta_1) + \bar{e}_{1.} - \bar{e}_{2.}]^2. \tag{3.9}$$

Setting $\bar{x}_{2.} = \bar{x}_{1.}$ and assuming the usual independent error model where $E_c(e_{ij}^2) = \sigma_{e_j}^2$, $j = 1, 2$, (3.9) becomes

$$s^2/N = \frac{\sigma_{e_1}^2}{N} + \frac{\sigma_{e_2}^2}{N} + \frac{\sigma_1^2}{N}(\beta_1 - \beta_2)^2. \tag{3.10}$$

Rarely in practice can one estimate the quantities in (3.10). Generally, however, the investigator has some rough estimate of an average variance of Y, say $\hat{\sigma}_y^2$, and of an average correlation between Y and X, say $\hat{\rho}$. Using these he can approximate $(1/N)\{\sigma_{e_1}^2 + \sigma_{e_2}^2\}$ by $(2/N)\hat{\sigma}_y^2(1 - \hat{\rho}^2)$.

Approximating $\sigma_1^2/N(\beta_1 - \beta_2)^2$ is quite difficult unless one has estimates of β_1 and β_2. The following rough method may be useful when the response surfaces are at the worst moderately nonparallel. If the response surfaces are parallel $\sigma_1^2/N(\beta_1 - \beta_2)^2$ is zero and thus minimal. If the response surfaces are at most moderately nonparallel, one could assume $(\beta_1 - \beta_2)^2 \leq 2\beta_1^2$ in most uses.[6] Hence, in many practical situations one may find that $0 \leq \sigma_1^2/N(\beta_1 - \beta_2)^2 \leq 2(\sigma_1^2/N)\beta_1^2$, where the upper bound can be approximated by $2(\hat{\rho}^2\hat{\sigma}_y^2/N)$. Hence, a simple estimated range for s^2 is

$$2\hat{\sigma}_y^2(1 - \hat{\rho}^2) < s^2 < 2\hat{\sigma}_y^2. \tag{3.11}$$

If the investigator believes that the response surfaces are parallel and linear, the value of s^2 to be used in (3.8) can be chosen to be near the minimum of this interval. Otherwise, a value of s^2 nearer the maximum would be appropriate.

---

[5] Moderate samples ($N > 20$) are assumed. For small samples $N = t_N^2 - 1 s^2/\Delta^2$ where t_{N-1} is the student-deviate with $N - 1$ degrees of freedom corresponding to $1 - \alpha$ confidence limits.

[6] A less conservative assumption is $(\beta_1 - \beta_2)^2 \leq \beta_1^2$.

(b) Choosing r

First assume that mean-matching is appropriate, i.e., assume an essentially linear response surface in P_2, and that the sole objective is to estimate τ. We will choose r large enough to expect 100% reduction in bias using the simple mean-matching method of Section 4.

(1) Estimate $\gamma = B[(1 + \sigma_1^2/\sigma_2^2)/2]^{1/2}$ and the approximate shape of the distribution of X in P_2. In order to compensate for the decreased ability of the mean-matching method to attain the theoretical maximum reduction in bias in small or moderate samples (see Section 4), if N is small or moderate ($N \leq 100$) increase γ by 5 to 15% (e.g., 10% for $N = 50$, 5% for $N = 100$).

(2) Using Table 3.1 find the row corresponding to the approximate shape of the distribution of X in P_2. Now find approximate values of r_1 and r_∞ such that $\Omega(r_1, 1) \simeq \gamma$ and $\Omega(r_\infty, \infty) \simeq \gamma$. If N is very small ($N < 5$), r should be chosen to be close to r_1; otherwise, results in Table 3.2 suggest that r can be chosen to be much closer to r_∞. r should probably be chosen to be greater than two and in most practical applications will be less than four.

Now assume pair-matches are desired, i.e., the response surfaces may be non-linear, nonparallel and each $y_{1j} - y_{2j}$ may be used to estimate the treatment effect at x_{1i}. We will choose r large enough to expect 95% + reduction in bias using the random order-nearest available pair-matching method of Section 5. Perform steps (1) and (2) as above for mean-matching. However, since in Section 5 we found that if $\sigma_1^2/\sigma_2^2 > 1$ nearest available pair-matching did not match the means of the samples very well compared to the simple mean-matching method, r should be increased. The following is a rough estimate (based on Tables 3.5 and 3.3) of the necessary increase:

if $\sigma_1^2/\sigma_2^2 = \frac{1}{2}$, r remains unchanged
if $\sigma_1^2/\sigma_2^2 = 1$, increase r by about 50%
if $\sigma_1^2/\sigma_2^2 = 2$, at least double r.

(c) Choosing a Matching Method

We assume G_1 and G_2 (i.e., r and N) are fixed and the choice is one of a matching method. If the investigator knows the P_2 response surface is linear and wants only to estimate τ, the results in Section 4 suggest that he can use the simple mean-matching method described in Section 4 and be confident in many practical situations of removing most of the bias whenever $r > 2$.

If confidence in the linearity of the P_2 response surface is lacking and/or the investigator wants to use each matched pair to estimate the effect of the treatment variable at a particular value of X, he would want to obtain close individual matches as well as closely matched means. Results in Section 5 indicate that in many practical situations the random order nearest available pair-matching method can be used to remove a large proportion of the bias in X while assigning close individual matches. The random order nearest available pair-matching is extremely easy to perform since the G_1 subjects do not have to be ordered; yet, it does not appear to be inferior to either high-low or low-high orderings and thus seems to be a reasonable choice in practice.

If a computer is available, a matching often superior to that obtained with the simple mean-matching or one random order nearest available pair-matching may be easily obtained by performing the simple mean-matching and several nearest available pair-matchings (i.e., several random orderings, low-high ordering, high-low ordering) and choosing the "best" matching. There should be no great expense in performing several matchings. Using Fortran IV subroutines given in Appendix B for the simple mean-matching method and nearest available pair-matching methods, a matching of 100 G_1 subjects from 400 G_2 subjects takes about $1\frac{1}{2}$ seconds on an IBM 360/65.

In order to decide which matching is "best," record for all matched samples $\bar{d} = \bar{x}_{1.} - \bar{x}_{2.}$ and $\bar{d}^2 = 1/N \sum(x_{1j} - x_{2j})^2$. Pair-matches (and thus $\bar{d}^2$) for the mean-matched sample can be found by using a nearest available pair-matching method on the final samples. If several matchings give equally small values of $\bar{d}$, choose the matching that gives the smallest value of $\bar{d}^2$. If $\bar{d}$ for one matched sample is substantially smaller than for any of the other matched samples but $\bar{d}^2$ for that sample is quite large, the investigator must either (1) make a practical judgement as to whether closely matched means or close individual matches are more important for his study, or (2) attempt to find matches by a matching method more complex than the ones considered here.

Admittedly, the practical situations and methods of estimating τ covered above are quite limited. The following article extends this work to include regression (covariance) adjusted estimates of τ and nonlinear parallel response surfaces. Rubin [1970] includes extensions to the case of many matching variables. Althauser and Rubin [1970] give a nontechnical discussion of some problems that arise with many matching variables.

ACKNOWLEDGMENTS

I wish to thank Professor William G. Cochran for many helpful suggestions and criticisms on earlier drafts of this article. I would also like to thank the referees for their helpful comments.

APPENDIX A

Proofs of Trends (3) and (4) in Section 3

We prove the intuitively obvious trend (3) by considering a random sample of size $(a + r)N$ from f. Call the order statistics $x_{(1)}, \ldots, x_{(N)}, \ldots, x_{(a+r)N}$ where $x_{(1)}$ is the largest observation. The average of the N largest observations from these $(a + r)N$ is $1/N \sum_1^N x_{(i)}$. By randomly discarding aN of the original observations, we have a random sample of size rN from f. But in any such subset the average of the N largest observations is less than or equal to $1/N \sum_1^N x_{(i)}$. Averaging over repeated random samples we have that $\Omega(r, N) \leq \Omega(r + a, N)$, N, rN, aN positive integers.

We prove trend (4) by a similar but more involved argument. Consider a random sample of size $r(N + b)$ and let $x_{(1)}, \ldots, x_{(N)}, \ldots, x_{(N+b)}, \ldots, x_{(r(N+b))}$ be the order statistics. The average of the $N + b$ largest from these $r(N + b)$ is $1/(N + b) \sum_{i=1}^{N+b} x_{(i)}$. Choosing a random rN-size subset of these observations, we have that the expected value of the average of the N largest from such a subset is

$$\frac{1}{n} \sum_{S \varepsilon S} \frac{1}{N} \text{ (total of largest } N \text{ observations from } S)$$

where

$\mathbf{S}$ = set of all distinct rN size subsets of original $r(N+b)$

$$n = \binom{r(N+b)}{rN} = \text{the number of elements in } \mathbf{S}.$$

This expression can be rewritten as

$$\frac{1}{n}\frac{1}{N}\sum_{i=1}^{r(N+b)}\lambda_i x_{(i)}$$

where λ_i = number of elements of $\mathbf{S}$ in which $x_{(i)}$ is one of the N largest observations, $\sum \lambda_i = Nn$.

For $i = 1, \ldots, N$, λ_i = the number of subsets in which $x_{(i)}$ occurs $= m = \binom{r(N+b)-1}{rN-1}$. For all $i > N$, $\lambda_i \leq m$. Consider the above summation as a weighted sum of the $x_{(i)}$ where the weights are ≥ 0 and add to 1 ($\sum \lambda_i / nN = 1$). Increasing the weights on the largest $x_{(i)}$ while keeping the sum of the weights the same cannot decrease the total value of the sum. Thus,

$$\frac{1}{n}\frac{1}{N}\sum_{1}^{r(N+b)}\lambda_i x_{(i)} \leq \frac{1}{n}\frac{1}{N}\left\{ m\sum_{1}^{N+b-1} x_{(i)} + (nN - m(N+b-1))x_{(N+b)} \right\}$$

$$\leq \frac{m}{nN}\left\{ \sum_{1}^{N+b-1} x_{(i)} + \left(\frac{nN}{m} - (N+b-1)\right)x_{(N+b)} \right\}$$

$$\leq \frac{1}{N+b}\left\{ \sum_{1}^{N+b} x_{(i)} \right\}$$

since $(m/nN) = 1/(N+b)$.

Hence, the expected average of the top N from a random rN-size subset is less than or equal to the average of the top $b+N$ from the original $r(b+N)$; thus averaging over repeated random samples we have

$$\Omega(r, N) \leq \Omega(r, b+N), \quad N, rN, r(b+N) \text{ positive integers.}$$

APPENDIX B

Fortran Subroutines for Nearest Available Pair Matching and Simple Mean Matching

Notation used in the subroutines

$N1 = N$ = size of G_1
$N2 = rN$ = size of initial G_2
$X1$ = vector of length N giving matching variable scores for G_1 sample, i.e., 1st entry is first G_1 subject's score
$X2$ = vector of length rN giving scores for G_2 on matching variable
$AV1 = \bar{x}_{1.}$
D = $\bar{x}_{1.} - \bar{x}_{2.}$; output for matched samples
$D2 = 1/N\sum(x_{1i} - x_{2i})^2$; output for matched samples
$IG1$ = vector giving ordering of G_1 sample for nearest available matching (a permutation of $1 \ldots N1$)
$IG2$ = "current" ordering of G_2 sample. After each call to a matching

```
      SUBROUTINE NAMTCH(D, D2, IG2,IG1, N1,N2,X1,X2)
C     SUBROUTINE TO PERFORM NEAREST AVAILABLE MATCHING
C     NECESSARY INPUTS ARE IG2, IG1, N1, N2, X1, X2
      DIMENSION IG1(1), IG2(1),X1(1),X2(1)
      D=0.
      D2=0.
      DO 200 I=1,N1
      K=IG1(I)
  200 CALL MATCH(D,D2,IG2,X1(K),I,N2,X2)
      D=D/FLOAT(N1)
      D2=D2/FLOAT(N1)
      RETURN
      END

      SUBROUTINE MNMTCH(D, IG2,N1,N2,X2,AV1)
C     SUBROUTINE TO PERFORM SIMPLE MEAN MATCHING TO AV1
C     NECESSARY INPUTS ARE IG2, N1, N2, X2, AV1
      DIMENSION IG2(1), X2(1)
      D=0.
      D2=0.
      DO 200 I=1,N1
      XX=AV1+D
  200 CALL MATCH(D,D2,IG2, XX,I,N2,X2)
      D=D/FLOAT(N1)
      RETURN
      END

      SUBROUTINE MATCH(D,D2,IG2,X1,K1,N2,X2)
C     SUBROUTINE PICKS G2 SUBJECT BETWEEN (INCLUSIVE) K1 AND N2 IN LIST
C     IG2 WHO HAS SCORE (IN X2) CLOSEST TO VALUE X1
C     HIS SUBJECT NUMBER IS PUT IN IG2(K1) AND PREVIOUS ENTRY IN IG2(K1)
C     IS MOVED BEYOND K1 ENTRY
      DIMENSION X2(1),IG2(1)
      LL=IG2(N2)
      IF (K1.EQ. N2) GO TO 410
      DMIN=ABS(X1-X2(LL))
      K2=N2-K1
      DO 400 LK=1,K2
      K=N2-LK
      L=IG2(K)
      IF (ABS(X1-X2(L)).LT. DMIN) GO TO 300
      IG2(K+1)=L
      GO TO 400
  300 IG2(K+1)=LL
      LL=L
      DMIN=ABS(X1-X2(LL))
  400 CONTINUE
  410 CONTINUE
      IG2(K1)=LL
      D=D+X1-X2(LL)
      D2=D2+(X1-X2(LL))**2
      RETURN
      END
```

subroutine, the G_1 subject having subject number $IG1(K)$ is matched to G_2 having subject number $IG2(K)$, $K = 1, \ldots, N1$. Subject number, of course, refers to order in vectors $X1$ and $X2$. Before the first call to a matching subroutine one should set $IG2(K) = K$, $K = 1, \ldots, N2$. After this initialization $IG2$ should be considered output of matching routines.

APPENDIX C

The Quality of Individual Matches:

$$100 \times \frac{E \sum (x_{1j} - x_{2j})^2 / N}{(B^2/2 + 1)(\sigma_1 + \sigma_2)}, \ X \text{ normal}^*$$

			$\sigma_1^2/_2 = \frac{1}{2}$				$\sigma_1^2/_2 = 1$				$\sigma_1^2/_2 = 2$			
			$B=\frac{1}{4}$	$\frac{1}{2}$	$\frac{3}{4}$	1	$\frac{1}{4}$	$\frac{1}{2}$	$\frac{3}{4}$	1	$\frac{1}{4}$	$\frac{1}{2}$	$\frac{3}{4}$	1
Random Order — N = 25		r = 1	47	58	72	81	35	50	63	77	35	48	63	76
		2	01	02	04	10	02	05	10	16	10	13	19	28
		3	00	00	01	02	01	02	05	08	06	09	13	19
		4	00	00	00	01	01	01	03	04	04	06	09	15
N = 50		1	42	55	70	80	30	47	64	76	31	42	57	73
		2	00	01	03	08	01	03	08	15	07	13	20	26
		3	00	00	00	01	00	01	03	06	04	08	12	18
		4	00	00	00	00	00	01	02	04	03	05	09	13
N = 100		1	37	56	69	80	25	46	63	76	25	41	60	73
		2	00	00	02	06	01	02	07	14	06	12	19	27
		3	00	00	00	01	00	01	02	05	03	07	11	17
		4	00	00	00	00	00	00	01	04	03	05	09	13
High-Low Order — N = 25		1	55	42	42	47	30	26	31	40	27	28	34	45
		2	01	01	02	04	02	03	05	09	08	09	13	19
		3	00	00	01	01	01	01	03	05	05	06	09	13
		4	00	00	00	01	01	01	02	03	04	05	07	11
N = 50		1	51	39	38	43	21	19	27	37	20	22	29	40
		2	00	01	01	03	01	02	04	07	05	08	12	16
		3	00	00	00	01	00	01	02	03	03	05	08	11
		4	00	00	00	00	00	00	01	02	02	03	06	09
N = 100		1	49	39	35	41	15	16	24	35	16	19	28	39
		2	00	00	01	02	00	01	03	06	04	07	11	17
		3	00	00	00	00	00	00	01	02	02	03	07	11
		4	00	00	00	00	00	00	01	02	02	03	05	08
Low-High Order — N = 25		1	93	108	122	126	71	92	108	119	55	77	96	110
		2	02	04	10	20	04	09	17	26	13	19	27	39
		3	00	01	02	05	01	03	08	13	07	12	18	26
		4	00	00	01	02	01	02	04	07	05	08	12	20
N = 50		1	95	110	122	127	69	92	110	120	52	71	91	108
		2	00	02	08	18	02	07	16	25	11	19	29	36
		3	00	00	01	03	01	02	05	11	06	11	18	24
		4	00	00	00	01	00	01	03	07	04	07	13	19
N = 100		1	92	111	122	128	62	91	109	120	44	69	92	109
		2	00	01	06	17	01	06	14	25	10	19	28	38
		3	00	00	00	02	00	01	05	10	05	10	17	24
		4	00	00	00	01	00	01	02	07	04	07	13	18

PAIR MATCHING METHOD

*If perfectly matched, equals 00. If randomly matched from random samples, equals 100.

4. The Use of Matched Sampling and Regression Adjustment to Remove Bias in Observational Studies

Donald B. Rubin

Abstract: The ability of matched sampling and linear regression adjustment to reduce the bias of an estimate of the treatment effect in two sample observational studies is investigated for a simple matching method and five simple estimates. Monte Carlo results are given for moderately linear exponential response surfaces and analytic results are presented for quadratic response surfaces. The conclusions are (1) in general both matched sampling and regression adjustment can be expected to reduce bias, (2) in some cases when the variance of the matching variable differs in the two populations both matching and regression adjustment can increase bias, (3) when the variance of the matching variable is the same in the two populations and the distributions of the matching variable are symmetric the usual covariance adjusted estimate based on random samples is almost unbiased, and (4) the combination of regression adjustment in matched samples generally produces the least biased estimate.

1. INTRODUCTION

This paper is an extension of Rubin [1973a] to include regression adjusted estimates and parallel nonlinear response surfaces. The reader is referred to Sections 1 and 2 of that paper for the statement of the general problem and an introduction to the notation.

After presenting the estimates of the treatment effect to be considered in the remainder of Section 1, we go on in Section 2 to present Monte Carlo results for the expected bias of the estimates assuming four exponential response surfaces, normally distributed X, and the random order, nearest available matching method. Section 3 is an attempt to understand the Monte Carlo results in a more general context by examining the bias of the estimates for quadratic response surfaces. Section 4 is a summary of the results.

1.1. The Five Estimates of τ to be Considered Here

We assume that the objective is to estimate the constant difference, τ, between parallel univariate response surfaces in two populations P_1 and P_2:

$$\tau = R_1(x) - R_2(x) \text{ for all } x,$$

Reprinted with permission from *Biometrics*, 1973, **29**, 185–203. © 1973 International Biometric Society.

where $R_i(x)$ is the conditional expectation in P_i of the dependent variable Y given the matching variable $X = x$. Equivalently, we can write

$$R_i(x) = \alpha_i + V(x) \quad i = 1, 2 \tag{4.1}$$

where $V(0) = 0$ and $\alpha_1 - \alpha_2 = \tau$.

We often refer to the function $V(x)$ as "the response surface."

We assume G_1 is a random sample from P_1 of size N and G_2 is a random sample from P_2 of size rN, $r \geq 1$. τ will be estimated from G_1 and G_{2*}, an N size subsample of G_2 "matched" to G_1. For the jth matched pair of subjects in G_1 and G_{2*} with scores y_{ij} and x_{ij} on Y and X we write

$$y_{ij} = \alpha_i + v_{ij} + e_{ij} \tag{4.2}$$

where $v_{ij} = V(x_{ij})$, $E_c(e_{ij}) = 0$, $i = 1, 2$, and E_c is the expectation conditionally given the x_{ij}.

The simplest estimate of τ is average Y difference in G_1 and G_{2*}

$$\hat{\tau}_0 = \bar{y}_1. - \bar{y}_2..$$

The other four estimates of τ we will consider here use an adjustment based on the assumption of a linear model, $V(x) = \beta x$ for a regression coefficient β, which we will temporarily assume to be correct. It is simple to show that the bias of $\hat{\tau}_0$ under this model is $\beta(\bar{x}_1. - \bar{x}_2.)$, and hence if we knew β, the estimate $\bar{y}_1. - \bar{y}_2. - \beta(\bar{x}_1. - \bar{x}_2.)$ would be unbiased. We can obtain an estimate of β, say $\hat{\beta}$, that is conditionally unbiased, $E_c(\hat{\beta}) = \beta$, by fitting a regression model. Thus the estimate $\bar{y}_1. - \bar{y}_2. - \hat{\beta}(\bar{x}_1. - \bar{x}_2.)$ would be an unbiased estimate of τ under the linear model whether we have matched or not.

Probably the most common estimate of β, at least when dealing with random samples, comes from fitting the parallel linear response surface model by least squares. After fitting the means to each group the data are pooled and a pooled estimate of β, $\hat{\beta}_p$, is found. The estimate of τ is then

$$\hat{\tau}_p = (\bar{y}_1. - \bar{y}_2.) - \hat{\beta}_p(\bar{x}_1. - \bar{x}_2.).$$

This method is the standard approach of the analysis of covariance for two groups, and the estimate is of course unbiased under this model of parallel linear response surfaces.

Two more estimates of the regression coefficient are easily found. Assuming that the parallel linear response surface model is correct, the least squares estimate of β found from the G_1 sample, $\hat{\beta}$, is an unbiased estimate of β, as is the estimate found from the G_2 sample, $\hat{\beta}_2$. Hence, we have two more unbiased estimates of the regression coefficient, one estimated from the G_1 data and the other estimated from the G_{2*} data, and so two regression adjusted estimates of τ,

$$\hat{\tau}_1 = (\bar{y}_1. - \bar{y}_2.) - \hat{\beta}_1(\bar{x}_1. - \bar{x}_2.)$$
$$\hat{\tau}_2 = (\bar{y}_1. - \bar{y}_2.) - \hat{\beta}_2(\bar{x}_1. - \bar{x}_2.).$$

These estimates of τ using within group estimates of β are most appropriate when the response surfaces are thought to be non-parallel and an average difference over the G_1 or G_2 sample is desired. See Cochran [1969] and Belsen [1956]. If the

Table 4.1. *Estimates of the response surface difference:*
$$\hat{\tau} = \bar{y}_{1.} - \bar{y}_{2.} - \hat{\beta}(\bar{x}_{1.} - \bar{x}_{2.})$$

Estimate of τ : $\hat{\tau}$	Estimate of β : $\hat{\beta}$
$\hat{\tau}_0$	$\hat{\beta}_0 \equiv 0$
$\hat{\tau}_1$	$\hat{\beta}_1 = \dfrac{S_{xy}}{S_{xx}}; S_{xu} = \displaystyle\sum_j (x_{1j} - \bar{x}_{1.})u_{1j}$
$\hat{\tau}_2$	$\hat{\beta}_2 = \dfrac{S_{xy}}{S_{xx}}; S_{xu} = \displaystyle\sum_j (x_{2j} - \bar{x}_{2.})u_{2j}$
$\hat{\tau}_p$	$\hat{\beta}_p = \dfrac{S_{xy}}{S_{xx}}; S_{xu} = \displaystyle\sum_{1=1}^{2} \sum_{j=1}^{N} (x_{ij} - \bar{x}_{i.})u_{ij}$
$\hat{\tau}_d$	$\hat{\beta}_d = \dfrac{S_{xy}}{S_{xx}}; S_{xu} = \displaystyle\sum_{i=1}^{2} \sum_{j=1}^{N} (x_{ij} - \bar{x}_{i.} - \bar{x}_{.j} + \bar{x}_{..})u_{ij}$

response surfaces are parallel and linear, these estimates will have larger variances than the pooled estimate, $\hat{\tau}_p$, because one is not using half of the data relevant to estimating β.

The last estimate of β to be considered is in some sense the most natural one when dealing with pair-matched data. Forming matched pair differences $y_{dj} = y_{1j} - y_{2j}$ and $x_{dj} = x_{1j} - x_{2j}$, $\hat{\beta}_d$ is the estimate of β found from the regression of y_{dj} on x_{dj}. Equivalently $\hat{\beta}_d$ is the estimate of β found from a two-way analysis of covariance, groups by matched pairs. It is easy to show that $\hat{\beta}_d$ is an unbiased estimate of β under the linear response surface model. The associated estimate of τ,

$$\hat{\tau}_d = (\bar{y}_{1.} - \bar{y}_{2.}) - \hat{\beta}_d(\bar{x}_{1.} - \bar{x}_{2.}),$$

is the constant in the linear regression on matched pair differences.

We have considered five estimates of the difference between parallel response surface, τ, all of the form

$$\hat{\tau} = (\bar{y}_{1.} - \bar{y}_{2.}) - \hat{\beta}(\bar{x}_{1.} - \bar{x}_{2.}). \tag{4.3}$$

The differences between the estimates are thus confined to estimating the regression coefficient and are summarized below in Table 4.1. Note that $\hat{\tau}_d$ is the only estimate that requires $G_1 - G_2$ pairs assigned in the final samples.

1.2. The Percent Reduction in Bias Due to Matched Sampling and Regression Adjustment

We now find the bias of the five estimates presented above. By Table 4.1 and equation (4.3) we have

$$\hat{\beta} = \frac{S_{xy}}{S_{xx}} = \frac{S_{xv}}{S_{xx}} + \frac{S_{xe}}{S_{xx}} \left(\text{for } \hat{\tau}_0, \frac{S_{xv}}{S_{xx}} = \frac{S_{xe}}{S_{xx}} \equiv 0 \right)$$

and from (4.3)

$$\hat{\tau} = \alpha_1 - \alpha_2 + \bar{v}_{1.} - \bar{v}_{2.} + \bar{e}_{1.} - \bar{e}_{2.} - \frac{S_{xv}}{S_{xx}}(\bar{x}_{1.} - \bar{x}_{2.}) - \frac{S_{xe}}{S_{xx}}(\bar{x}_{1.} - \bar{x}_{2.}).$$

Thus the conditional bias of $\hat{\tau}$ given the x_{ij} is

$$E_c(\hat{\tau} - \tau) = \bar{v}_{1.} - \bar{v}_{2.} - \frac{S_{xv}}{S_{xx}}(\bar{x}_{1.} - \bar{x}_{2.}).$$

If the response surface is linear, $(\bar{v}_{1.} - \bar{v}_{2.}) = \beta(\bar{x}_{1.} - \bar{x}_{2.})$ and $S_{xv}/S_{xx} = \beta$ for all estimates except $\hat{\tau}_0$ for which $S_{xv}/S_{xx} = 0$. Hence given parallel linear response surfaces, all estimates except $\hat{\tau}_0$ are unbiased and $\hat{\tau}_0$ will be unbiased if $\bar{x}_{1.} = \bar{x}_{2.}$. However, if the response surface is nonlinear all estimates are in general biased even if $\bar{x}_{1.} = \bar{x}_{2.}$. Thus a mean-matching method or a procedure that concludes unbiased estimates will result if $\bar{x}_{1.} = \bar{x}_{2.}$ is not necessarily appropriate if the response surface is nonlinear. See Cochran [1970a] and Rubin [1973a] for examples of such procedures.

The expected bias of $\hat{\tau}$ over the matched sampling plan is $E\{\bar{v}_{1.} - \bar{v}_{2.} - S_{xv}/S_{xx}(\bar{x}_{1.} - \bar{x}_{2.})\}$ where E is the expectation over the distributions of X in matched samples. Given $r = 1$ (random samples), the expected bias of $\hat{\tau}_0$ is $E_1(\bar{v}_{1.}) - E_2(\bar{v}_{2.})$ where $E_i(\ )$ is the expectation over the distribution of X in P_i. It follows that the percent reduction in expected bias due to matching and/or regression adjustment is

$$100 \left[1 - \frac{E\left[\bar{v}_{1.} - \bar{v}_{2.} - \frac{S_{xv}}{S_{xx}}(\bar{x}_{1.} - \bar{x}_{2.})\right]}{E_1(\bar{v}_{1.}) - E_2(\bar{v}_{2.})} \right]. \tag{4.4}$$

If the matches were exact, $x_{1j} = x_{2j}, j = 1, \ldots, N$ implying that $\bar{x}_{1.} = \bar{x}_{2.}$ and $\bar{v}_{1.} = \bar{v}_{2.}$; hence, the percent reduction in expected bias would be 100% for any response surface. In Section 2 we present Monte Carlo values of the percent reduction in bias due to matching and/or regression adjustment for the estimates $\hat{\tau}_0, \hat{\tau}_1, \hat{\tau}_2, \hat{\tau}_p$, and $\hat{\tau}_d$ for some moderately nonlinear response surfaces and imperfectly matched samples.

2. MONTE CARLO PERCENT REDUCTIONS IN BIAS

When dealing with finite matched samples, the expectations required to calculate the percent reductions in bias are usually analytically intractable. Hence, we will turn to Monte Carlo methods in order to obtain numerical values for the percent reduction in bias of the different estimates in "typical" situations. These numerical values will be used to compare and evaluate the different estimators of τ. After specifying the conditions for the Monte Carlo investigations in Section 2.1, we will present the Monte Carlo results in Sections 2.2, 2.3, and 2.4.

2.1. Conditions of the Monte Carlo Investigation

There are four conditions that must be specified in order to obtain Monte Carlo percent reductions in bias for the five estimates of τ.

1. the distribution of X in P_1 and P_2
2. the sample sizes N and rN
3. the matching method
4. the response surface $V(x)$.

Table 4.2. *Percent reduction in bias of X for random order, nearest available pair-matching: X normal, N = 50*

B	$\sigma_1^2/\sigma_2^2 = 1/2$			$\sigma_1^2/\sigma_2^2 = 1$			$\sigma_1^2/\sigma_2^2 = 2$		
	$r = 2$	3	4	2	3	4	2	3	4
1/4	99	100	100	92	96	98	66	79	86
1/2	98	99	100	87	95	97	59	75	81
3/4	93	99	100	78	91	94	53	69	75
1	84	97	99	69	84	89	51	63	71

We will assume that in P_i, $X \sim \text{Normal}\ (\eta_i, \sigma_i^2)\, i = 1, 2$. Without loss of generality we can assume $\eta_1 = -\eta_2 \geq 0$ and $(\sigma_1^2 + \sigma_2^2)/2 = 1$. Then $B = 2\eta_1$ is the number of standard deviations ($\sqrt{(\sigma_1^2 + \sigma_2^2)/2}$) between the means of X. The choice of X as normal is obvious but restrictive; generalizing the Monte Carlo results to other distributions of X will be considered in Section 3. Some limited experience indicates that the values $B = \frac{1}{4}, \frac{1}{2}, \frac{3}{4}, 1$ and $\sigma_1^2/\sigma_2^2 = \frac{1}{2}, 1, 2$ are representative of the range that might occur in practice and so will be used.

With respect to sample sizes, we will assume $N = 50$ and $r = 2, 3, 4$. Previous work on matching, Rubin [1973a], and preliminary results indicated very consistent trends for moderate and large N so that additional N's were judged unnecessary. Values of r were chosen to represent typical values that might be used in practice.

The matching method must assign to each subject in G_1 a distinct subject in G_2 as a match so that there are N matched pairs. We will assume the random order, nearest available pair-matching method investigated in Rubin [1973a]. First randomly order the G_1 subjects, then for each G_1 subject choose in turn the closest match from those G_2 subjects not yet assigned as matches. This method was chosen for two basic reasons: (1) previous results indicate that it is a relatively intelligent pair-matching method that might be used in practice; and (2) the matching procedure is very fast to perform by computer. Since our study requires extensive Monte Carlo runs, the second point was of practical importance.

Some basic results for this matching method are given in Table 4.2 and Table 4.3, for the values of N and r and distribution of X specified above. Table 4.2 gives Monte Carlo percent reductions in bias for $\hat{\tau}_0$ assuming linear response surfaces, and Table 4.3 gives Monte Carlo values of the ratio of the expected variance of X in

Table 4.3. $(E(s_2^2)/\sigma_2^2) \times 100$ *for random order, nearest available pair-matching: X normal, N = 50*

B	$\sigma_1^2/\sigma_2^2 = 1/2$			$\sigma_1^2/\sigma_2^2 = 1$			$\sigma_1^2/\sigma_2^2 = 2$		
	$r = 2$	3	4	2	3	4	2	3	4
1/4	49	48	48	92	92	96	134	149	157
1/2	49	49	48	82	90	94	116	132	149
3/4	48	48	48	69	83	87	99	114	125
1	45	45	47	59	68	76	76	91	100

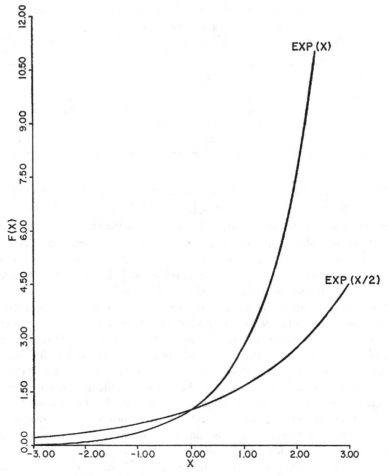

Figure 4.1. Two exponential response surfaces.

matched G_{2*} samples, $E(s_2^2)$, to the expected variance of X in random G_2 samples, σ_2^2. Note that $E(s_2^2)$ is for these conditions always less than σ_1^2, although in "easy conditions" (those in which the percent reduction in the bias of $X > 90\%$) $E(s_2^2)$ is close to σ_1^2.

The last condition to be specified is the range of nonlinear response surfaces, the functions $V(x)$ to be used. Since we are investigating regression adjusted estimates based on the model that the response surface $V(x)$ is linear, we will require the response surface to be moderately linear in the range of interest on the assumption that an alert investigator should be able to detect violent nonlinearity and thus use a more appropriate model for adjusting the estimate (e.g., add a quadratic term).

In Figure 4.1 we have plotted $V(x) = \exp(x/2)$ and $V(x) = \exp(x)$,[1] for $-3 < x < 3$ corresponding roughly to the range of X to be used in the Monte Carlo

---

[1] Strictly by equation (4.1), $V(0) = 0$ so that we really should set $V(x) = \exp(x) - 1$. Since the constant has no effect on result it will be simpler to ignore it in discussion.

samples. Even when disguised by random error, it might reasonably be argued that the nonlinear aspects of exp (x) often would be quite apparent. For this range of X we consider the response surface exp (x) to be an extreme example of what might be termed moderately nonlinear and exp $(x/2)$ a more reasonable example of a moderately nonlinear response surface.

Since $\eta_1 \geq \eta_2$, the response surfaces exp (x) and exp $(x/2)$ are "unfavorable to matching" because they are increasing most rapidly for large X which is where close matches are difficult to find. The response surface exp (x) is, in this sense, more unfavorable to matching than exp $(x/2)$. Also, the response surfaces exp $(-x)$ and exp $(-x/2)$ are "favorable to matching" because they are changing very slowly where the matches are poor. The four exponential response surfaces exp (ax), $a = \pm\frac{1}{2}, \pm1$, were investigated in detail. We consider these to be representative of a range of moderately nonlinear response surfaces that are both favorable and unfavorable to matching.

Since an investigator when deciding whether or not to obtain matched samples or when deciding which estimate of β to use generally has some knowledge of the distribution of the matching variable in P_1 and P_2, it seems logical to present the results of the investigation classified by the distribution of the matching variable. We first present the results when the distribution of X is favorable to matching: $\sigma_1^2/\sigma_2^2 = \frac{1}{2}$, then when only the mean of X differs in P_1 and P_2, $\sigma_1^2 = \sigma_2^2$, and finally when the distribution of X is unfavorable to matching, $\sigma_1^2/\sigma_2^2 = 2$. "Favorable" refers to the previous work on ability to remove the bias of the matching variable X.

2.2. Case I – Distribution of X Favorable to Matching – $\sigma_1^2/\sigma_2^2 = \frac{1}{2}$

The results for case I in which the distribution of X is favorable to matching $(\sigma_1^2/\sigma_2^2 = \frac{1}{2})$ are given in Table 4.4. First consider the results for $\hat{\tau}_0$ (no regression adjustment). $\hat{\tau}_0$ based on samples matched with $r \geq 2$ usually removes most of the bias (90%–110%) of $\hat{\tau}_0$ based on random samples.[2] With the largest initial bias it may remove quite a bit less (e.g., $B = 1, r = 2$, 76% for exp $(x/2)$ and 69% for exp (x)). For practical purposes, the previous results in Table 4.2 for linear response surfaces can be considered slightly optimistic typical values (in general, optimistic by less than 5% for moderately nonlinear response surfaces) when $\sigma_1^2/\sigma_2^2 = \frac{1}{2}$.

Regression adjusted estimates based on random samples seem quite sensitive to even moderate departures from linearity. The extremely wild values for percent reduction in bias (e.g., $-288, 306$) for exp (x) and exp $(x/2)$ when $B = \frac{1}{4}$ and for exp (x) when $B = \frac{1}{2}$ indicate that in these cases the regression adjusted estimates based on random samples are substantially more biased than $\hat{\tau}_0$ based on random samples. Since exp (x) and especially exp $(x/2)$ are not violently nonlinear these results are somewhat surprising.

The explanations seem to be as follows. Since exp (ax) $(a > 0)$ is monotonically increasing (implying that generally $\hat{\beta} > 0$) and $\eta_1 > \eta_2$, the adjustment $-\hat{\beta}(\eta_1 - \eta_2)$ is usually negative in this case. But if $\sigma_1^2 < (\sigma_2^2 - 2B/a)$ we have

---

[2] Values greater than 100% indicate that the expected value of $V(x)$ in matched G_{2*} samples is greater than the expected value of $V(x)$ in P_1, while values less than 0% indicate that the expected value of $V(x)$ in matched G_{2*} samples is less than the expected value of $V(x)$ in P_2.

Table 4.4. Percent reduction in bias, X normal, $N = 50$, case I: $\sigma_1^2/\sigma_2^2 = \frac{1}{2}$

	Estimate	$r = 1$ (random)				$r = 2$				$r = 3$				$r = 4$			
	Response surface	$\exp(x)$	$\exp(x/2)$	$\exp(-x/2)$	$\exp(-x)$	$\exp(x)$	$\exp(x/2)$	$\exp(-x/2)$	$\exp(-x)$	$\exp(x)$	$\exp(x/2)$	$\exp(-x/2)$	$\exp(-x)$	$\exp(x)$	$\exp(x/2)$	$\exp(-x/2)$	$\exp(-x)$
$B = \frac{1}{4}$	$\hat{\tau}_0$	00	00	00	00	106	95	99	100	103	99	100	100	102	99	100	100
	$\hat{\tau}_1$	−288	306	54	32	101	100	100	100	102	99	100	100	101	100	100	100
	$\hat{\tau}_2$	−313	294	65	57	102	100	100	100	102	99	100	100	101	100	100	100
	$\hat{\tau}_p$	−304	298	62	48	101	100	100	100	102	99	100	100	101	100	100	100
	$\hat{\tau}_d$	−228	248	79	77	96	102	100	100	101	100	100	100	100	100	100	100
$B = \frac{1}{2}$	$\hat{\tau}_0$	00	00	00	00	93	96	99	99	94	98	100	100	97	99	100	100
	$\hat{\tau}_1$	326	163	63	38	101	100	100	100	96	99	100	100	98	100	100	100
	$\hat{\tau}_2$	276	138	88	88	101	100	100	100	95	99	100	100	98	100	100	100
	$\hat{\tau}_p$	292	146	80	72	101	100	100	100	96	99	100	100	98	100	100	100
	$\hat{\tau}_d$	226	125	96	104	108	101	100	100	101	100	100	100	100	100	100	100
$B = \frac{3}{4}$	$\hat{\tau}_0$	00	00	00	00	85	89	96	98	94	97	100	100	97	99	100	100
	$\hat{\tau}_1$	220	148	65	38	102	100	101	101	97	99	100	100	98	100	100	100
	$\hat{\tau}_2$	145	111	102	113	103	100	101	101	96	99	100	100	98	100	100	100
	$\hat{\tau}_p$	170	123	90	88	103	100	101	101	97	99	100	100	98	100	100	100
	$\hat{\tau}_d$	137	109	103	114	113	103	99	100	102	100	100	100	101	100	100	100
$B = 1$	$\hat{\tau}_0$	00	00	00	00	69	76	91	96	90	94	98	99	94	97	99	99
	$\hat{\tau}_1$	206	147	63	36	106	101	101	101	98	99	101	101	97	99	100	100
	$\hat{\tau}_2$	106	97	113	136	102	100	101	102	97	99	101	101	97	99	100	100
	$\hat{\tau}_p$	139	113	96	102	105	101	101	101	98	101	101	101	97	99	100	100
	$\hat{\tau}_d$	118	103	108	126	118	105	99	99	105	99	99	100	102	101	100	100

$E_1\{\exp (ax)\} < E_2\{\exp (ax)\}$, so that a *positive* adjustment is needed. Hence for $a = 1, \sigma_1^2 = \frac{2}{3}, \sigma_2^2 = \frac{4}{3}$, and $B = \frac{1}{4}$, the regression adjustment often greatly increases the original bias. Further, for $a = \frac{1}{2}, B = \frac{1}{4}$ as well as for $a = 1, B = \frac{1}{2}$, the regression adjustment is a gross over-adjustment, since the expected bias of $\hat{\tau}_0$ based on random samples is much less than suggested by the difference in means and the "average slope" of $\exp (x/2)$. There is no one regression adjusted estimate based on random samples which is always best, although $\hat{\tau}_d$ appears to be more consistent than any other estimate. Also, $\hat{\tau}_1$ is always very poor, which is not surprising since the range of X in G_1 is limited compared to the range in random G_1 and G_2 samples, implying that $\hat{\tau}_1$ may give a poor linear approximation to the response surface.

Regression adjusted estimates based on samples matched with $r \geq 2$ are far superior to those based on random samples, the difference being more striking when the bias in X is small. In all conditions all of the regression adjusted estimates based on matched samples ($r \geq 2$) can be expected to remove all of the bias (98% – 102%). However, there is some tendency for $\hat{\tau}_d$ to be inferior if the response surface is very unfavorable to matching (i.e., exp (x)). Except for $\hat{\tau}_d$ given the response surface exp (x), there is almost no improvement in using greater than a 2:1 ratio of sample sizes.

2.3. Case II – $\sigma_1^2 = \sigma_2^2$

The percent reductions in bias for case II ($\sigma_1^2 = \sigma_2^2$) are given in Table 4.5. First consider $\hat{\tau}_0$. Given the response surface, B, and $r > 1$, $\hat{\tau}_0$ is more biased in case II ($\sigma_1^2 = \sigma_2^2$) than in case I ($\sigma_1^2 = \sigma_2^2/2$). This result is expected since in case II the distribution of X is less favorable to matching than in case I. The values given in Table 4.2 for linear response surfaces can be considered mildly optimistic typical values (optimistic by 5%–10%) for these nonlinear response surfaces when $\sigma_1^2 = \sigma_2^2$.

The results for the regression adjusted estimates in case II are surprising when compared with the results in case I. First, the regression adjusted estimates based on random samples are much better in all conditions in case II than in case I. In fact, $\hat{\tau}_p$ based on random samples generally removes almost all of the bias (98%–102%). A possible explanation for this result is that since the variances and higher moments of X about the mean are equal in random samples from the two populations, all of the bias of $\hat{\tau}_0$ is due to the difference in the means of X which linear regression should be good at removing. This comment implies that regression adjusted estimates should be approximately unbiased for nonlinear response surfaces (that can be approximated by a polynomial) whenever the distributions of X in G_1 and G_2 are the same except for a difference in means. In Section 3 we will see that this comment is not completely accurate and that the results we are discussing are somewhat dependent upon the symmetry of the normal distribution.

In samples matched with $r \geq 2$, $\hat{\tau}_d$ is the best estimate, in general removing most of the bias (99%–105%). The other regression adjusted estimates based on samples matched with $r \geq 2$ often have values outside the range 90%–110%. $\hat{\tau}_2$ is especially poor for matched samples possibly because of the small range of X in G_{2*} on which to base the regression (see Table 4.2). $\hat{\tau}_1$ and $\hat{\tau}_2$ are somewhat better when based on matched samples than when based on random samples. Surprisingly, $\hat{\tau}_p$ is often worse when $r \geq 2$ than when $r = 1$.

Table 4.5. *Percent reduction in bias, X normal, $N = 50$, case II: $\sigma_1^2/\sigma_2^2 = 1$*

	Estimate	r = 1 (random)				r = 2				r = 3				r = 4			
Response surface		$\exp(x)$	$\exp(x/2)$	$\exp(-x/2)$	$\exp(-x)$	$\exp(x)$	$\exp(x/2)$	$\exp(-x/2)$	$\exp(-x)$	$\exp(x)$	$\exp(x/2)$	$\exp(-x/2)$	$\exp(-x)$	$\exp(x)$	$\exp(x/2)$	$\exp(-x/2)$	$\exp(-x)$
$B = \frac{1}{4}$	$\hat{\tau}_0$	00	00	00	00	70	83	99	106	79	90	101	104	87	94	101	103
	$\hat{\tau}_1$	113	106	94	88	80	91	107	112	84	94	103	106	90	97	102	104
	$\hat{\tau}_2$	88	94	106	113	77	91	107	113	82	94	104	107	90	96	102	105
	$\hat{\tau}_p$	101	100	100	101	79	91	107	113	83	94	104	107	89	96	102	105
	$\hat{\tau}_d$	109	101	107	125	100	99	103	108	100	100	101	103	100	100	101	102
$B = \frac{1}{2}$	$\hat{\tau}_0$	00	00	00	00	60	74	94	98	75	87	98	100	84	92	99	100
	$\hat{\tau}_1$	127	113	88	77	80	91	105	108	83	94	103	104	88	96	102	102
	$\hat{\tau}_2$	77	88	113	127	74	89	107	110	81	93	103	104	87	96	102	103
	$\hat{\tau}_p$	102	101	101	102	77	90	106	109	82	94	103	104	88	96	102	103
	$\hat{\tau}_d$	109	102	105	116	106	102	100	101	102	100	100	101	101	100	100	101
$B = \frac{3}{4}$	$\hat{\tau}_0$	00	00	00	00	47	62	87	94	68	81	96	99	76	87	98	100
	$\hat{\tau}_1$	142	120	82	67	83	92	106	109	88	93	104	105	86	95	102	103
	$\hat{\tau}_2$	67	82	120	142	71	87	109	112	78	91	105	106	83	94	103	104
	$\hat{\tau}_p$	104	101	101	104	79	90	107	110	81	92	104	105	84	94	103	103
	$\hat{\tau}_d$	111	102	104	114	110	103	99	100	105	102	99	100	103	101	100	100
$B = 1$	$\hat{\tau}_0$	00	00	00	00	39	53	82	91	55	70	92	97	65	79	96	99
	$\hat{\tau}_1$	158	127	77	58	88	93	106	109	82	92	105	106	82	93	103	104
	$\hat{\tau}_2$	58	77	127	158	68	85	110	114	72	88	107	109	75	90	105	105
	$\hat{\tau}_p$	108	102	102	108	81	90	107	111	79	91	105	107	79	92	104	105
	$\hat{\tau}_d$	112	102	106	120	113	104	99	99	109	103	100	100	106	102	100	99

As would be expected, the estimates generally become slightly less biased as r increases from two to four.

2.4. Case III – Distribution of Matching Variable Unfavorable to Matching – $\sigma_1^2/\sigma_2^2 = 2$

The results for case III in which the distribution of X is unfavorable to matching ($\sigma_1^2/\sigma_2^2 = 2$) are given in Table 4.6. First consider the results for $\hat{\tau}_0$. As expected from the results for linear response surfaces, $\hat{\tau}_0$ based on matched samples is more biased when $\sigma_1^2 > \sigma_2^2$ than when $\sigma_1^2 \leq \sigma_2^2$. In fact, for exp $(-x)$ and $B = \frac{1}{4}$, $\hat{\tau}_0$ based on samples matched with $4 > r \geq 2$ can be more biased than $\hat{\tau}_0$ based on random samples. This strange result is due to the same circumstances as mentioned previously in case I when discussing regression adjusted estimates based on random samples. Even though exp $(-x)$ is monotonically decreasing, in case III when $B = \frac{1}{4}$, $E_1 \{\exp(-x)\} > E_2 \{\exp(-x)\}$. Matching with $r < 4$ decreases the difference in means in G_1 and G_{2*} more than the difference in variances which tends to increase the bias of $\hat{\tau}_0$. In general, however, the values for percent reduction in bias of $\hat{\tau}_0$ for linear response surfaces given in Table 4.2 can be considered representative optimistic values (optimistic by about 10%) for moderately nonlinear response surfaces when $\sigma_1^2/\sigma_2^2 = 2$.

As already observed in case I, regression adjusted estimates based on random samples are quite biased when $\sigma_1^2 \neq \sigma_2^2$ even for moderately linear response surfaces. Also, as in case I, given the response surface and B, regression adjusted estimates based on samples matched with $r \geq 2$ are less biased than when based on random samples. However, the estimates are more biased in case III than in case I, presumably because the matching is poorer than in case I. As might be expected $\hat{\tau}_2$ is generally the worst regression adjusted estimate based on matched samples since the range of X in the matched G_{2*} samples is small compared to the range of X over both samples. $\hat{\tau}_d$ is the best estimate in case III, with $r \geq 2$ generally removing most (90%–110%) of the bias. In those conditions in which $\hat{\tau}_d$ does poorly the only better estimate is $\hat{\tau}_0$ which in general is not very satisfactory. The advantages of matching with $r = 4$ rather than with $r = 2$ are greater in case III than in case I or II but still are not substantial.

3. LINEAR REGRESSION ADJUSTMENT AND QUADRATIC RESPONSE SURFACES

Somewhat surprisingly there appears to be little literature on the use of linear regression to remove bias when the response surfaces are not exactly linear. A theoretical study of this situation would be especially valuable for interpreting results such as those presented in Section 2. The following discussion is an attempt to understand the preceding Monte Carlo results within a more general framework than provided by normal distributions and exponential response surfaces. It is not intended to be a complete study of linear regression and nonlinear response surfaces. Assuming that the response surface is actually quadratic, we derive expressions for the bias of $\hat{\tau}_1$, $\hat{\tau}_2$ and $\hat{\tau}_p$ in Section 3.1 and for the bias of $\hat{\tau}_d$ in Section 3.2. Since exp (ax) may be approximated by such a response surface in the range of conditions

Table 4.6. Percent reduction in bias, X normal, N = 50, case III: $\sigma_1^2/\sigma_2^2 = 2$

Response surface Estimate	r = 1 (random)				r = 2				r = 3				r = 4			
	exp(x)	exp(x/2)	exp(−x/2)	exp(−x)	exp(x)	exp(x/2)	exp(−x/2)	exp(−x)	exp(x)	exp(x/2)	exp(−x/2)	exp(−x)	exp(x)	exp(x/2)	exp(−x/2)	exp(−x)
B = 1/4																
$\hat{t}_0$	00	00	00	00	35	48	121	−50	51	66	139	−48	55	70	120	01
$\hat{t}_1$	57	65	294	−313	57	73	216	−141	64	79	184	−85	67	82	164	−40
$\hat{t}_2$	32	54	306	−288	47	68	227	−149	57	76	192	−96	61	80	170	−45
$\hat{t}_p$	48	62	298	−304	53	71	220	−144	61	78	187	−89	65	81	167	−42
$\hat{t}_d$	84	83	240	−218	90	90	177	−99	92	93	149	−29	94	95	140	−05
B = 1/2																
$\hat{t}_0$	00	00	00	00	30	45	81	123	43	60	89	118	48	65	94	126
$\hat{t}_1$	88	88	138	276	66	80	132	220	68	83	123	181	68	84	121	177
$\hat{t}_2$	38	63	163	326	47	71	140	237	55	77	128	191	58	79	124	184
$\hat{t}_p$	72	80	146	292	60	77	135	226	63	81	125	185	64	82	122	180
$\hat{t}_d$	108	98	126	240	107	100	111	171	105	100	108	147	104	100	107	146
B = 3/4																
$\hat{t}_0$	00	00	00	00	23	38	72	90	39	55	85	98	42	60	89	100
$\hat{t}_1$	113	102	111	145	76	85	121	152	72	85	116	138	71	85	114	131
$\hat{t}_2$	38	65	148	220	44	70	133	173	54	76	123	150	54	78	119	140
$\hat{t}_p$	88	90	123	170	66	80	125	159	66	82	119	142	65	83	116	109
$\hat{t}_d$	123	106	106	132	120	106	102	115	111	103	102	112	111	103	101	97
B = 1																
$\hat{t}_0$	00	00	00	00	16	31	67	83	28	45	79	92	29	50	84	94
$\hat{t}_1$	136	113	97	106	88	90	114	132	78	87	114	129	74	86	113	124
$\hat{t}_2$	36	63	147	206	40	69	130	158	46	73	125	146	46	74	121	138
$\hat{t}_p$	102	96	113	139	76	85	118	139	69	83	117	134	66	82	115	128
$\hat{t}_d$	133	111	101	114	127	109	99	104	119	106	100	104	119	105	99	102

considered, these expressions will be used to help interpret the Monte Carlo results of Section 2.

3.1. Bias of $\hat{\tau}_0$, $\hat{\tau}_1$, $\hat{\tau}_2$ and $\hat{\tau}_p$

Consider two samples of size N with means $\bar{x}_{i\cdot}$, $i = 1, 2$, sample variances $s_i^2 = \Sigma(x_{ij} - \bar{x}_{i\cdot})^2/N$, $i = 1, 2$ and sample skewness $k_i = \Sigma(x_{ij} - \bar{x}_{i\cdot})^3/N$, $i = 1, 2$. We will assume that the true response surface can be accurately approximated by a quadratic response surface: $V(x) = \beta x + \delta x^2$. Hence, for the samples

$$y_{ij} = \alpha_i + \beta x_{ij} + \delta x_{ij}^2 + e_{ij} \qquad (4.5)$$

where $E_c(e_{ij}) = 0$, and $E_c(\cdot)$ is the expectation conditionally given the x_{ij}. Using the simple results

$$\frac{1}{N} \sum x_{ij}^2 = \bar{x}_{i\cdot}^2 + s_i^2$$
$$\frac{1}{N} E_c \sum [(x_{ij} - \bar{x}_{i\cdot})y_{ij})] = \beta s_i^2 + \delta(2\bar{x}_{i\cdot}s_i^2 + k_i) \qquad (4.6)$$

we can calculate the bias of $\hat{\tau}_0$, $\hat{\tau}_1$, $\hat{\tau}_2$ and $\hat{\tau}_p$ as follows.

$$E_c(\hat{\tau}_0 - \tau) = \beta(\bar{x}_{1\cdot} - \bar{x}_{2\cdot}) + \delta(\bar{x}_{1\cdot}^2 - \bar{x}_{2\cdot}^2) + \delta(s_1^2 - s_2^2), \qquad (4.7)$$

$$E_c(\hat{\tau}_i - \tau) = \delta \left\{ (s_1^2 - s_2^2) \mp (\bar{x}_{1\cdot} - \bar{x}_{2\cdot})^2 - (\bar{x}_{1\cdot} - \bar{x}_{2\cdot})\frac{k_i}{s_i^2} \right\} \qquad (4.8)$$

where the $-$ holds for $\hat{\tau}_1$ and the $+$ for $\hat{\tau}_2$, and

$$E_c(\hat{\tau}_p - \tau) = \delta \left[(s_1^2 - s_2^2) + (\bar{x}_{1\cdot} - \bar{x}_{2\cdot}) \left\{ (\bar{x}_{1\cdot} + \bar{x}_{2\cdot}) - \frac{2(\bar{x}_{1\cdot}s_1^2 + \bar{x}_{2\cdot}s_2^2)}{s_1^2 + s_2^2} \right\} \right.$$
$$\left. - \frac{\bar{x}_{1\cdot} - \bar{x}_{2\cdot}}{s_1^2 + s_2^2}(k_1 + k_2) \right]. \qquad (4.9)$$

We now use expressions (4.8) and (4.9) to interpret the Monte Carlo results of Section 2. First consider random samples. For the results presented $E(\bar{x}_{1\cdot}) = B/2$, $E(\bar{x}_{2\cdot}) = -B/2$, $E(s_i^2) \doteq \sigma_i^2$, $i = 1, 2$, $E(s_1^2 + S_2^2)/2 \doteq 1$, $E(k_1) = 0$ and $E(k_2) = 0$ since the normal is symmetric. Thus for the random samples considered in Section 3 the expected bias of $\hat{\tau}_i$, $i = 1, 2$ is approximately

$$\delta\{(\sigma_1^2 - \sigma_2^2) \mp B^2\}, \qquad (4.10)$$

and for $\hat{\tau}_p$ the expected bias is approximately

$$\delta(\sigma_1^2 - \sigma_2^2)(1 + B^2/2). \qquad (4.11)$$

Hence in case II when $\sigma_1^2 = \sigma_2^2$, $\hat{\tau}_p$ should be approximately unbiased, while both $\hat{\tau}_1$ and $\hat{\tau}_2$ should be biased by an amount $\mp\delta B^2$. This claim is substantiated by the Monte Carlo results presented in Section 3. In case I, $\sigma_1^2 = \frac{2}{3}, \sigma_2^2 = \frac{4}{3}$. Since $1 \geq B > 0$, from (4.10) and (4.11) the least biased estimate should be $\hat{\tau}_2$ followed by $\hat{\tau}_p$ and then $\hat{\tau}_1$. The Monte Carlo results again substantiate this claim. In case III $\sigma_1^2 = \frac{4}{3}, \sigma_2^2 = \frac{2}{3}$ and the ordering implied by (4.10) and (4.11) is $\hat{\tau}_2, \hat{\tau}_p, \hat{\tau}_1$, which is again in agreement with previous results.

Now consider case I and samples matched with $r \geq 2$: $E(s_1^2) \doteq E(s_2^2) \doteq \sigma_1^2$ (see Table 4.3). For $\hat{\tau}_p$, the expected bias is approximately

$$\delta \left\{ -\frac{E(\bar{x}_{1.} - \bar{x}_{2.})^2}{2\sigma_1^2} E(k_1 + k_2) \right\}, \tag{4.12}$$

and for $\hat{\tau}_i, i = 1, 2$ the expected bias is approximately

$$\delta \left\{ \mp E(\bar{x}_{1.} - x_{2.})^2 - \frac{E(\bar{x}_{1.} - \bar{x}_{2.})}{\sigma_1^2} E(k_i) \right\}. \tag{4.13}$$

In case I for matched samples with $(r \geq 2)$, $E(\bar{x}_{1.} - \bar{x}_{2.}) \doteq 0$ (see Table 4.2); hence $\hat{\tau}_p, \hat{\tau}_1$ and $\hat{\tau}_2$ all have approximately zero expected bias. Notice however, that if $E(s_1^2) = E(s_2^2)$ but $E(\bar{x}_{1.} - \bar{x}_{2.})$ is large, $\hat{\tau}_p$, would have approximately zero expected bias only if $E(k_1 + k_2) \doteq 0$ (case II, random samples). Thus a situation in which there is a large bias in the mean of X but very similar higher moments $(s_1^2 \doteq s_2^2, k_1 \doteq k_2)$ would not necessarily be favorable to using the estimate $\hat{\tau}_p$ unless the distribution of X is symmetric $(k_1 \doteq k_2 \doteq 0)$.

In cases II and III for samples matched with $r \geq 2$, $\sigma_1^2 > E(s_2^2)$ (see Table 4.3) and $E(\bar{x}_{1.} - \bar{x}_{2.})$ is not trivial (see Table 4.2). Expressions (4.8) and (4.9) suggest that $\hat{\tau}_2$ should be the worst estimate since $E(s_1^2 - s_2^2) > 0$, $E(\bar{x}_{1.} - \bar{x}_{2.})^2 > 0$, and $E(k_2/s_2^2) < 0$ (results not presented indicate that in the conditions considered with $r > 1$, $E(k_2) < 0$). Also, $\hat{\tau}_1$ should be better than $\hat{\tau}_2$ or $\hat{\tau}_p$ in these cases because $E(k_1) \doteq 0$ and we are subtracting $E(\bar{x}_{1.} - \bar{x}_{2.})^2$ from $E(s_1^2 - s_2^2)$. The Monte Carlo results confirm these trends.

The above discussion has at least to some extent explained the Monte Carlo results in Section 2 for $\hat{\tau}_1, \hat{\tau}_2$ and $\hat{\tau}_p$ and thus generated an understanding of the effect of these estimates which is not tied to normal distributions and exponential response surfaces.

3.2. Bias of $\hat{\tau}_d$

Thus far no explanation has been offered for the often superior performance of the regression adjusted estimate based on matched pair differences, $\hat{\tau}_d$. This omission was intentional because even the intuitive explanation given below is somewhat involved.

We begin by showing that for any set of N matched pairs there is some mth degree $(1 \leq m \leq N + 1)$ non-trivial polynomial response surface for which the pairs are exactly matched. Let $P(x) = \sum_{k=1}^{m} a_k x^k$ be an mth degree polynomial in X; then if all matched pairs are exactly matched with respect to P we have

$$\sum_{k=1}^{m} a_k \left(x_{1j}^k - x_{2j}^k \right) = 0, \quad j = 1, \ldots, N.$$

Such a polynomial always exists if $m = N + 1$ because we can always find a non-trivial solution for $N + 1$ unknowns given N homogeneous linear equations. If the pairs are exactly matched, the minimum m is 1; the samples are exactly matched for linear response surfaces as well as for all higher degree response surfaces.

Intuitively, one might feel that as the matches become better the minimum degree of the response surface for which the pairs are exactly matched should decrease, or at least the degree of the response surface for which the pairs are almost exactly matched should decrease. In this discussion we will assume that the matched pairs are close enough so that they are almost exactly matched for some quadratic response surface:

$$a_1(x_{1j} - x_{2j}) + a_2 \left(x_{1j}^2 - x_{2j}^2\right) = d_j \tag{4.14}$$

where d_j is small for all matched pairs and its average value over matched pairs is zero. Since we require the average value of d_j to be zero we have

$$0 = a_1(\bar{x}_{1.} - \bar{x}_{2.}) + a_2 \left(\bar{x}_{1.}^2 + s_1^2 - \bar{x}_{2.}^2 - s_2^2\right).$$

If $\bar{x}_{1.} \equiv \bar{x}_{2.}$, all regression adjusted estimates considered are identical to $\hat{\tau}_0$. If $\bar{x}_{1.} \neq \bar{x}_{2.}$ without loss of generality, let $a_2 = 1$, so that

$$a_1 = -\frac{\left(\bar{x}_{1.}^2 - \bar{x}_{2.}^2\right) + \left(s_1^2 - s_2^2\right)}{(\bar{x}_{1.} - \bar{x}_{2.})}. \tag{4.15}$$

The bias of $\hat{\tau}_d$ is

$$E_c(\hat{\tau}_d - \tau)$$

$$= E_e \left[(\bar{y}_{1.} - \bar{y}_{2.}) - \frac{\sum \{x_{1j} - x_{2j} - (\bar{x}_{1.} - \bar{x}_{2.})\}(y_{1j} - y_{2j})}{\sum \{x_{1j} - x_{2j} - (\bar{x}_{1.} - \bar{x}_{2.})\}^2}(\bar{x}_{1.} - \bar{x}_{2.}) \right]$$

$$= \delta \left\{ \left(\bar{x}_{1.}^2 - \bar{x}_{2.}^2\right) + \left(s_1^2 - s_2^2\right) \right\} \tag{4.16}$$

$$- (\bar{x}_{1.} - \bar{x}_{2.}) \frac{\delta \sum \{x_{ij} - x_{2j} - (\bar{x}_{1.} - \bar{x}_{2.})\}\left(x_{1j}^2 - x_{2j}^2\right)}{\sum \{x_{1j} - x_{2j} - (\bar{x}_{1.} - \bar{x}_{2.})\}^2}.$$

But from (4.14) and (4.15) assuming $\bar{x}_{1.} \neq \bar{x}_{2.}$, we have

$$\left(x_{1j}^2 - x_{2j}^2\right) = d_j + (x_{1j} - x_{2j}) \frac{\left(\bar{x}_{1.}^2 - \bar{x}_{2.}^2\right) + \left(s_1^2 - s_2^2\right)}{(\bar{x}_{1.} - \bar{x}_{2.})}. \tag{4.17}$$

Inserting (4.17) into (4.16) and taking expectations over matched samples we have

$$E(\hat{\tau}_d - \tau) \doteq \delta A, \quad \text{where} \quad A = -\frac{E(\bar{x}_{1.} - \bar{x}_{2.})}{\text{Var}(x_{1j} - x_{2j})} \, \text{Cov}\,(x_{1j} - x_{2j}, d_j),$$

and Var $(\cdot)$ and Cov $(\cdot)$ are the variance and covariance over the distribution of x_{ij} in matched samples.

If the samples are well matched with respect to some quadratic response surface the d_j should be relatively small and relatively uncorrelated with the $x_{1j} - x_{2j}$, implying a small A. Hence, $\hat{\tau}_d$ should be approximately unbiased if the response surface is basically quadratic and the samples are well matched with respect to some other quadratic response surface. If the true response surface is exactly quadratic and the matches are exact with respect to any quadratic response ($d_i \equiv 0$), $\hat{\tau}_d$ will be conditionally unbiased. A might be large if the samples are very poorly matched as with random samples; thus, it is not surprising that for random samples in case II $(\sigma_1^2 = \sigma_2^2)$ $\hat{\tau}_p$ is superior to $\hat{\tau}_d$. However, even in those situations in which $\hat{\tau}_d$ is not the best estimate, the expected bias of $\hat{\tau}_d$ is rarely substantially larger than that of the least biased estimate. In general it appears that for moderately linear response surfaces and moderately well matched samples, $\hat{\tau}_d$ will be the least biased estimate that we have considered.

4. SUMMARY

We now summarize the Monte Carlo results of Section 2 and the analytic results of Section 3 in the form of advice to an investigator who wants to estimate the constant difference between parallel univariate response surfaces in two populations P_1 and P_2. This constant difference is called the treatment effect and designated τ.

In review, we assume G_1 is a random sample from P_1 of size N and G_2 is a random sample from P_2 of size $rN, r \geq 1$. τ will be estimated using G_1 and a N-size subsample of G_2 matched to G_1 on the matching variable X, G_{2*}. If $r = 1 G_{2*}$ is a random sample from P_2.

Five estimates of τ are considered (see Table 4.1 for explicit definitions).

(1) $\hat{\tau}_0$ – the average difference across matched pairs.
(2) $\hat{\tau}_p$ – the covariance adjusted estimate using the pooled estimate of the regression coefficient from a one-way analysis of variance. This estimate is the natural regression adjusted estimate when dealing with random samples.
(3) $\hat{\tau}_d$ – the regression adjusted estimate using matched pair differences, or equivalently a two-way analysis of variance. This regression adjusted estimate is the natural one when dealing with matched samples.
(4) $\hat{\tau}_1$ – the regression adjusted estimate using only the G_1 sample to estimate the regression coefficient.
(5) $\hat{\tau}_2$ – the regression adjusted estimate using only the matched G_{2*} sample to estimate the regression coefficient.

$\hat{\tau}_1$ and $\hat{\tau}_2$ are most natural when average differences between non-parallel response surfaces are desired.

4.1. No Regression Adjustments – $\hat{\tau}_0$

If no regression adjustments are to be performed, random order, nearest available pair-matching with a ratio of sample sizes $r \geq 2$ generally reduces the bias of the estimate, $\hat{\tau}_0 = \bar{y}_{1.} - \bar{y}_{2.}$, of the response surface difference, especially if the variance of the matching variable is greater in P_2 than in P_1. The values of percent reduction in bias for linear response surfaces given in Table 4.2 are optimistic by less than 10% in most nonlinear cases considered here. However, in some "unfavorable" cases, the values of percent reduction in bias for moderately nonlinear response surfaces are much less than suggested by the values for linear response surfaces. Detailed advice on $\hat{\tau}_0$ for linear response surfaces is given in Rubin [1973a].

4.2. Regression Adjusted Estimates – $\hat{\tau}_1, \hat{\tau}_2, \hat{\tau}_p, \hat{\tau}_d$

A. $\hat{\tau}_d$ vs $\hat{\tau}_p$ – Variances of X Approximately Equal and Distribution of X Symmetric

When the variances of X are approximately equal in P_1 and P_2 and the distribution of X is symmetric in both P_1 and P_2 the Monte Carlo results of Section 2 and the analytic results of Section 3 suggest the following conclusions:

(1) The estimate $\hat{\tau}_p$ based on random samples is approximately unbiased when the response surfaces are approximately linear or quadratic; hence, for these simple and often assumed distributions of X there may be little gain in obtaining matched samples.
(2) If matched samples have been obtained with $4 \geq r \geq 2$, $\hat{\tau}_d$ will be the least biased estimate but generally not less biased than $\hat{\tau}_p$ based on random samples, and hence probably less preferred because of the fewer degrees of freedom used to estimate the regression coefficient.

B. $\hat{\tau}_d$ vs $\hat{\tau}_p$ – Variances of X Unequal and/or Distributions of X Non-Symmetric

When the variances of X are different in P_1 and P_2 and/or the distributions of X are not symmetric in P_1 and P_2, results in Sections 2 and 3 suggest the following conclusions:

(1) Matching with $r \geq 2$ and using the estimate $\hat{\tau}_d$ based on matched pair differences should in most cases be the least biased procedure we have considered, removing most (90–110%) of the original bias of $\hat{\tau}_0$ based on random samples.
(2) Even with random samples ($r = 1$), assigning matches and using $\hat{\tau}_d$ in these cases may often be superior to the usual pooled estimate $\hat{\tau}_p$.
(3) If in the final matched samples the variances of X are approximately equal and the distributions of X appear symmetric, $\hat{\tau}_p$ may be slightly less biased than $\hat{\tau}_d$.
(4) In general, the decrease in bias of a regression adjusted estimate from matching with $r = 4$ rather than $r = 2$ is minor.
(5) If the response surface is linear, all regression adjusted estimates are unbiased, but $\hat{\tau}_p$ will be superior to $\hat{\tau}_d$ because it uses all of the data to estimate the regression coefficient and thus has smaller variance.

C. $\hat{\tau}_1$ and $\hat{\tau}_2$

In general, for the cases considered in Sections 2 and 3, $\hat{\tau}_1$ and $\hat{\tau}_2$ are inferior to either $\hat{\tau}_d$ or $\hat{\tau}_p$ and their use should be avoided when the response surfaces are parallel.

4.3. Other Estimates of τ

It could be argued in those cases in which $\hat{\tau}_d$ was the least biased estimate that the extra $N - 1$ degrees of freedom (D.F.) used to estimate parameters for the N matched pairs when forming matched pair differences could be better used on the pooled data to estimate the response surface more accurately. Thus a reasonable suggestion would be to obtain matched or random samples and whenever the response surfaces are thought to be even slightly nonlinear use the pooled data to estimate a quadratic (or higher order) term in X assuming parallel response surfaces. A possible criticism of this method is that in a multivariate case one may not have a large enough sample to estimate all quadratic terms. The generalization of this work to the multivariate case is currently being studied.

ACKNOWLEDGMENTS

I wish to thank Professor William G. Cochran for many helpful suggestions and criticisms on earlier drafts of this article. I would also like to thank the two referees for their helpful comments.

5. Assignment to Treatment Group on the Basis of a Covariate

Donald B. Rubin

Abstract: When assignment to treatment group is made solely on the basis of the value of a covariate, X, effort should be concentrated on estimating the conditional expectations of the dependent variable Y given X in the treatment and control groups. One then averages the difference between these conditional expectations over the distribution of X in the relevant population. There is no need for concern about "other" sources of bias, e.g., unreliability of X, unmeasured background variables. If the conditional expectations are parallel and linear, the proper regression adjustment is the simple covariance adjustment. However, since the quality of the resulting estimates may be sensitive to the adequacy of the underlying model, it is wise to search for nonparallelism and nonlinearity in these conditional expectations. Blocking on the values of X is also appropriate, although the quality of the resulting estimates may be sensitive to the coarseness of the blocking employed. In order for these techniques to be useful in practice, there must be either substantial overlap in the distribution of X in the treatment groups or strong prior information.

1. INTRODUCTION

In some studies, the experimental units are divided into two treatment groups solely on the basis of a covariate, X. By this we mean that if two units have the same value of X either they both must receive the same treatment or they must be randomly assigned (not necessarily with probability 0.5) to treatments. For example, those units (children) with high scores on X (a reading test) receive the experimental treatment (a compensatory reading program); those with low scores on X receive the standard control treatment; perhaps those with intermediate scores on X are randomly assigned with equal probability to the treatments. The critical point is that the probability that an experimental unit is exposed to Treatment 1 rather than Treatment 2 is a function only of the values of X in the sample. After exposure to treatments, a dependent variable Y (a second reading test) is recorded in both treatment groups.

The central question is: what is the average effect on Y of Treatment 1 vs. Treatment 2 for the relevant population? For simplicity of discussion, we will usually assume the relevant population is the one from which all the units being studied are considered a random sample, say P. The associated effect is called τ. Some researchers might wonder whether to use gain scores, simple posttest scores,

Reprinted with permission from the publisher. *Journal of Educational Statistics*, 1977, **2**, 1–26. © 1977 by the American Educational Research Association and the American Statistical Association.

covariance adjusted scores (possibly adjusted for reliability), or some other device to estimate τ.

We will show that the appropriate estimate of τ is the average value of the difference between the estimated conditional expectations of Y on X in the two treatment groups, the average being taken over all units in the study if the relevant population is P. The conditional expectations (regressions) can be estimated using least squares, robust techniques, blocking, or matching methods. Neither gain scores nor scores adjusted for the reliability of X are generally appropriate (no matter how unreliable X may be).

In the special case of parallel linear regressions of Y on X in the two groups and least squares estimation, the average difference between the estimated regressions in the two treatment groups corresponds to the simple covariance adjusted estimator. There are previous references to the appropriateness of the covariance adjusted estimator in versions of this special case; see, for example, Cox (1951, 1957b), Finney (1957), Goldberger (1972a, 1972b), Greenberg (1953), Kenney (1975), Snedecor and Cochran (1967, pp. 438–439).

However, the results presented here are general and emphasize (a) recording the variable X used to make assignment decisions, (b) estimating the conditional expectations of Y given X in each treatment group, and (c) averaging the difference between the estimated conditional expectations over the estimated distribution of X in the relevant population. These three steps are essential in order to estimate causal effects of treatments in those studies which are not classical randomized designs. Bayesian analogues for these results are presented in Rubin (1978a).

In the following development we use unbiasedness as the criterion indicating the appropriateness of estimators. We do so only to show that the estimator tends to estimate the correct quantity without further adjustment. We do not mean to suggest that all biased estimators are unacceptable (a biased estimator with small mean squared error may of course be preferable to an unbiased estimator whose variance is large).

2. DEFINING THE EFFECTS OF TREATMENTS: ASSUMPTIONS AND NOTATION

The definition of the effect of Treatment 1 vs. Treatment 2 that we will use is standard in the sense that if the population P is essentially infinite, then the average treatment difference in very large randomized experiments on random samples from P will estimate the effect with negligible variance. However, we will explicitly present the assumptions in order to avoid ambiguity. The definition follows that given in Rubin (1974), the basic idea being that for each unit in P there is a value of Y that we could observe if the unit had been exposed to Treatment 1 and another value that we could observe if the unit had been exposed to Treatment 2; an important assumption is that these values of Y do not change as the other units in P receive different treatments. It is also assumed that the values of X are the same no matter which treatments the units received (i.e., X is a proper covariate).

More precisely, first suppose that all units in P were exposed to Treatment 1; let μ_1 be the resulting average value of Y for all units in P, and let $\mu_1(x)$ be the resulting average value of Y for all those units in P with score x on variable X. Second, suppose that all units in P were exposed to Treatment 2; let μ_2 be the resulting average value of Y for all units in P, and let $\mu_2(x)$ be the resulting average value of Y for all those units in P with score x on variable X. Letting $\overset{\text{ave}}{\underset{x \in P}{}}[\cdot]$ denote the average value of the quantity in brackets over the distribution of X in P, we have that

$$\mu_i = \overset{\text{ave}}{\underset{x \in P}{}} [\mu_i(x)].$$

Assume that X is unaffected by the treatments so that a unit's score on X will be the same no matter how treatments are assigned; this will be the case when X is recorded before treatments are assigned. Also assume "no 'interference' between different units" (Cox, 1958, p. 19) so that a unit's Y value given Treatment i is unaffected by which treatments the other units in P received. Without this assumption, even if P were infinite, different infinitely large randomized experiments would estimate different effects, in the sense that the variance of the average treatment difference over all such randomized experiments generally would not be negligible. There are weaker assumptions under which one can estimate causal effects, but we do not consider them here. Note that the usual null hypothesis of no treatment effect assumes that Y given Treatment 1 equals Y given Treatment 2 for all units and assignments of treatments, a very special form of no interference.

Good experimental design often reflects the assumption of no interference between different units. For example, consider a time-consuming compensatory reading treatment. First, suppose each student is a unit with the compensatory and regular reading treatments assigned to different students in the same classroom. In this case the no interference assumption may be suspect because of social interactions among the students and competition for the teacher's time (the effect of the compensatory reading treatment on a student being different when only a few students receive the compensatory reading treatment than when all the students in the class receive the compensatory reading treatment). Now suppose instead classrooms in different schools were the units, and each classroom was assigned either to the regular or compensatory treatment condition (perhaps all students in a classroom receiving the compensatory reading instruction, or a randomly chosen group of ten, or only those in need – these reflect three different compensatory reading treatments being applied to the classroom). With the choice of classrooms as units, the no interference assumption seems quite plausible.

We are now ready to define the average causal effect of Treatment 1 vs. Treatment 2. Consider a unit randomly drawn from P and then exposed to Treatment i (i.e., each unit in P was equally likely to be chosen). Because of the assumption of no interference between units, the average value of Y for such a unit (i.e., averaging over all random draws of one unit from P) is μ_i, no matter what treatments the other units in P received. Hence $\tau = \mu_1 - \mu_2$ is called the average or expected effect of Treatment 1 vs. Treatment 2 on Y in the population P.

Now consider a unit randomly chosen from those units in P with X = x and then exposed to Treatment i (i.e., each unit in P with X = x was equally likely to be chosen). Because of the assumption of no interference between units, the average value of Y for such a unit exposed to Treatment i (i.e., averaging over all random draws of one unit from P with X = x) is $\mu_i(x)$ no matter what treatments the other units in P received. Hence $\mu_1(x) - \mu_2(x)$ is called the effect of Treatment 1 vs. Treatment 2 on Y at X = x in P. (See Figure 5.1.) If $\mu_1(x) - \mu_2(x)$ is constant for all

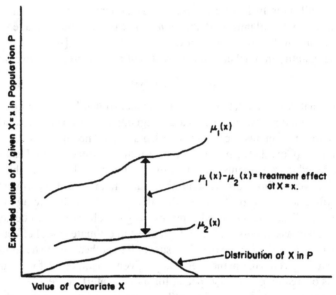

Figure 5.1. The treatment effect in population $P : \tau = \underset{x \in P}{\text{Ave}} \left[\mu_1(x) - \mu_2(x) \right].$

x, $\mu_1(x)$ and $\mu_2(x)$ are called parallel, and the effect of Treatment 1 vs. Treatment 2 is the same for each value of X. Generally, however, the relative effect of the treatments will depend on the value of X.

The $\mu_i(x)$ are called the conditional expectations of Y given X and treatment condition, or the "response functions of Y given X " or the "regressions of Y on X." Often the $\mu_i(x)$ are assumed to be linear in x, but this restriction is not needed for the general results presented here.

It follows from the above definitions that the average effect of Treatment 1 vs. Treatment 2 on Y in P, $\tau = \mu_1 - \mu_2$, is $\mu_1(x) - \mu_2(x)$ averaged over the distribution of X in P:

$$\tau = \underset{x \in P}{\text{ave}} \left[\mu_1(x) - \mu_2(x) \right]. \tag{5.1}$$

This simple relationship is exploited to estimate τ in non-randomized studies. In Figure 5.1, τ is calculated by taking the vertical difference between the $\mu_1(x)$ and $\mu_2(x)$ curves at each x, and finding the average value of this difference weighted by the distribution of X in P.

3. PRELIMINARY RESULTS

Throughout the rest of the paper we will assume the following sampling situation. A random sample of size $n_1 + n_2$ from P is divided into two groups of sizes n_1 and n_2 *solely on the basis of the values of X and perhaps some randomization.* That is, the assignment decisions are such that either all sampled units with the same value of X are in the same treatment group or are randomly divided (not necessarily with equal probability) into the two treatment groups. The first group is exposed to Treatment 1 and the second group is exposed to Treatment 2. Let x_{ij}, y_{ij}, $i = 1, 2; j = 1, \ldots, n_i$ be the values of X and Y in the two samples.

Since the x_{ij} are a random sample from P, Result 1 is immediate from equation (5.1).

Result 1: The quantity

$$\frac{1}{n_1 + n_2} \sum_{i=1}^{2} \sum_{j=1}^{n_i} \left[\mu_1(x_{ij}) - \mu_2(x_{ij}) \right] \tag{5.2}$$

is unbiased for τ.

Notice that the notation $\mu_1(x_{ij})$ and $\mu_2(x_{ij})$ in expression (5.2) means that the functions $\mu_1(x)$ and $\mu_2(x)$ are to be evaluated at the observed values x_{ij}, and that by the phrase "unbiased for τ" we mean that the average value of expression (5.2) over all random samples of size $n_1 + n_2$ from P is τ.

If we had conditionally unbiased estimates of the values $\mu_1(x_{ij})$ and $\mu_2(x_{ij})$, $i = 1, 2; j = 1, \ldots, n_i$, we could substitute them into expression (5.2) to obtain an unbiased estimate of τ. By conditionally unbiased we mean unbiased given the values x_{ij} that occur in the sample (i.e., averaging over all random draws from P that yield the same values for the x_{ij} as observed in our sample). Only the values of x_{ij} in the sample and the conditional expectations of Y given X under the two treatments are needed in order to obtain an unbiased estimate of τ. No matter how "unreliable" X is, no reliability correction is relevant; nor does it matter what functional forms $\mu_1(x)$ and $\mu_2(x)$ take.

Result 2 is the key to obtaining unbiased estimators of τ since it gives us conditionally unbiased estimates of some $\mu_1(x)$ and $\mu_2(x)$ values.

Result 2: The value y_{1j} is a conditionally unbiased estimate of $\mu_1(x_{1j})$, $j = 1, \ldots, n_1$, and the value y_{2j} is a conditionally unbiased estimate of $\mu_2(x_{2j})$, $j = 1, \ldots, n_2$.

In order to prove Result 2, first note that sampled units with X = x were randomly sampled from those units in P with X = x. Next note that since assignment was on the basis of X, sampled units with X = x are either (a) always assigned the same treatment or (b) randomly assigned treatments (not necessarily with probability 0.5 of receiving each treatment). In either case, sampled units with X = x who were assigned Treatment i were randomly chosen from those units in P with X = x. Now by the definition of $\mu_i(x)$ and the assumption of no interference between units, the average value of Y for a unit randomly drawn from those units in P with X = x and then assigned Treatment i is $\mu_i(x)$ no matter what the other sampled values of X or the other treatment assignments. Therefore, y_{ij} (the observed values of Y for a sampled unit with X = x_{ij} given exposure to Treatment i) is a conditionally unbiased estimate of $\mu_i(x_{ij})$.

Note the crucial role in this proof of assignment on the basis of X. If assignment depended on some variable Z other than X, sampled units with X = x who were assigned to Treatment i were not randomly sampled from those units in P with X = x, but rather from those units in P with (a) X = x and (b) Z satisfying the conditions that determined the assignment to Treatment i.

By Result 2, we have conditionally unbiased estimates of points on the Treatment 1-part of $\mu_1(x)$ (i.e., $\mu_1(x_{1j})$, $j = 1, \ldots, n_1$) and points on the Treatment 2-part of $\mu_2(x)$ (i.e., $\mu_2(x_{2j})$, $j = 1, \ldots, n_2$). However, we still lack conditionally unbiased estimates of points on the Treatment 2-part of $\mu_1(x)$ (i.e., $\mu_1(x_{2j})$, $j = 1, \ldots, n_2$) and points on the Treatment 1-part of $\mu_2(x)$ (i.e., $\mu_2(x_{1j})$, $j = 1, \ldots, n_1$). And we need these estimates in order to use Result 1 to obtain an unbiased estimate of τ.

We will discuss two general methods for obtaining conditionally unbiased estimates of these quantities: (a) fitting a model to the data to obtain estimates of the functions $\mu_1(x)$ and $\mu_2(x)$, and (b) grouping Treatment 1 and Treatment 2 units with similar values of X to obtain estimates of the difference $\mu_1(x) - \mu_2(x)$ at particular X values that are representative of the distribution of X in P.

4. ESTIMATING $\mu_1(x_{ij})$ AND $\mu_2(x_{ij})$ BY MODEL FITTING

One method for estimating the values of $\mu_1(x_{2j})$ and $\mu_2(x_{1j})$ is via a model for the functions $\mu_1(x)$ and $\mu_2(x)$. This is most appropriate when X takes on many values (e.g., age, height). Obviously the accuracy of the resulting estimates will be somewhat dependent on the accuracy of the model chosen.

The modelling of $\mu_1(x)$ and $\mu_2(x)$ will be illustrated in the simple case when we assume both are linear in x. The usual least squares estimates are:

$$\hat{\mu}_1(x) = \bar{y}_1 + \hat{\beta}_1(x - \bar{x}_1), \quad \text{and} \tag{5.3}$$

$$\hat{\mu}_2(x) = \bar{y}_2 + \hat{\beta}_2(x - \bar{x}_2), \tag{5.4}$$

where

$$\hat{\beta}_i = \frac{\sum_{j=1}^{n_i}(y_{ij} - \bar{y}_i)(x_{ij} - \bar{x}_i)}{\sum_{j=1}^{n_i}(x_{ij} - \bar{x}_i)^2}, \quad i = 1, 2. \tag{5.5}$$

Result 3: If both $\mu_1(x)$ and $\mu_2(x)$ are of the form $\mu_i(x) = \alpha_i + \beta_i x$, the estimator

$$\bar{y}_1 - \bar{y}_2 - (\bar{x}_1 - \bar{x}_2)\frac{n_1\hat{\beta}_2 + n_2\hat{\beta}_1}{n_1 + n_2} \tag{5.6}$$

is unbiased for τ.

In order to prove Result 3, first note that expression (5.6) equals

$$\frac{1}{n_1 + n_2} \sum_{i=1}^{2} \sum_{j=1}^{n_i} \left[\hat{\mu}_1(x_{ij}) - \hat{\mu}_2(x_{ij})\right], \tag{5.7}$$

where $\hat{\mu}_1(x_{ij})$, $\hat{\mu}_2(x_{ij})$, and $\hat{\beta}_i$ are given by equations (5.3), (5.4), and (5.5). By Result 2, the conditional expectation of $\hat{\mu}_i(x)$ is $\mu_i(x)$: (a) the conditional expectation of $\bar{y}_i$ is $\alpha_i + \beta_i \bar{x}_i$, and (b) the conditional expectation of $\hat{\beta}_i$ is $\sum_{j=1}^{n_i}(\mu_i(x_{ij}) - \mu_i(x_{ij}))(x_{ij} - \bar{x}_i)/\sum_{j=1}^{n_i}(x_{ij} - \bar{x}_i)^2 = \beta_i$. Consequently,

$\hat{\mu}_1(x_{ij}) - \hat{\mu}_2(x_{ij})$ is a conditionally unbiased estimate of $\mu_1(x_{ij}) - \mu_2(x_{ij})$ (for $j = 1, \ldots, n_i; i = 1, 2$). Thus by Result 1, expression (5.7) (and thus (5.6)) is unbiased for τ.

Result 4: If $\mu_1(x)$ and $\mu_2(x)$ are both linear in x and parallel, then the simple analysis of covariance estimator

$$\bar{y}_1 - \bar{y}_2 - (\bar{x}_1 - \bar{x}_2)\hat{\beta} \tag{5.8}$$

where

$$\hat{\beta} = \frac{\sum\limits_{i=1}^{2}\sum\limits_{i=1}^{n_i}(y_{ij} - \bar{y}_i)(x_{ij} - \bar{x}_i)}{\sum\limits_{i=1}^{2}\sum\limits_{j=1}^{n_i}(x_{ij} - \bar{x}_i)^2}$$

is unbiased for τ.

The proof of Result 4 is essentially the same as the proof of Result 3 with the change that now $\beta_1 = \beta_2 = \beta$ and both $\hat{\beta}_1$ and $\hat{\beta}_2$ are replaced by $\hat{\beta}$ which is a conditionally unbiased estimate of β.

Results analogous to Results 3 and 4 follow when $\mu_1(x)$ and $\mu_2(x)$ are polynomial in x or any linear combination of specified functions of x (e.g., e^x). The only change is in the method of estimating the parameters, i.e., by a multiple least squares regression. Methods more robust than least squares (c.f., Beaton & Tukey, 1974) might be appropriate for estimating the conditional expectations $\mu_1(x)$ and $\mu_2(x)$, especially when there is the possibility of outliers or long-tailed distributions.

Of course we never know whether the functions $\mu_1(x)$ and $\mu_2(x)$ are linear in x (or linear in some specified functions of x). But the unbiasedness of the estimators given by (5.6) and (5.8) is dependent upon the accuracy of the linear model. There is evidence (Rubin, 1973b) that in some cases the linear approximation is adequate to remove most of the bias present in the simple estimator $\bar{y}_1 - \bar{y}_2$ but that in other cases it is inadequate even when $\mu_1(x)$ and $\mu_2(x)$ are smooth monotone functions. The troublesome cases are basically those having quite different variances of X in the treatment groups.

If the observed values of X in the treatment groups are similar, it may be possible to check that both $\hat{\mu}_1(x)$ and $\hat{\mu}_2(x)$ are reasonable approximations to $\mu_1(x)$ and $\mu_2(x)$ for the full range of observed values of X. This checking is important because we must average $\hat{\mu}_1(x) - \hat{\mu}_2(x)$ over the full range of observed X values, and therefore must have confidence in both models for most of the values of X that occur in the sample.

If the X values in the two samples do not overlap (e.g., as in the regression discontinuity design, Campbell and Stanley, 1963, pp. 61–64) it is impossible to check the accuracy of either $\hat{\mu}_1(x)$ or $\hat{\mu}_2(x)$ for the full range of observed X values, and we must rely on our *a priori* assumptions. Consequently, in order for the model-fitting efforts described above to be useful in practice, we must either have samples that overlap or strong *a priori* information about the functional forms of the $\mu_i(x)$.

5. ESTIMATING $\mu_1(x_{ij})$ AND $\mu_2(x_{ij})$ BY BLOCKING ON X

When the assignment to treatment group allows the distribution of X in the two treatment groups to overlap substantially, it may be possible to obtain conditionally unbiased estimates of $\mu_1(x_{2j})$ and $\mu_2(x_{1j})$ without fitting a model. The obvious but crucial point is that if $x_{1j} = x_{2j}$, then y_{1j} is conditionally unbiased for $\mu_1(x_{2j}) = \mu_1(x_{1j})$ and y_{2j} is conditionally unbiased for $\mu_2(x_{1j}) = \mu_2(x_{2j})$.

Suppose that in the samples there are only K distinct values of X, say $x_1, \ldots, x_K$, where n_{1k} Treatment 1 units and n_{2k} Treatment 2 units have X values equal to x_k, $k = 1, \ldots, K$. Let $\bar{y}_{1k}$ be the average Y value for the n_{1k} Treatment 1 units whose X value equals x_k; similarly let $\bar{y}_{2k}$ be the average Y value for the n_{2k} Treatment 2 units whose X value equals x_k. If $n_{ik} = 0$ for some i and and k, then the corresponding $\bar{y}_{ik}$ is not defined.

Result 5: If $n_{1k} > 0$ and $n_{2k} > 0$ for all $k = 1, \ldots, K$, then the estimator

$$\frac{1}{n_1+n_2}\left[\sum_{k=1}^{K}(n_{1k} + n_{2k})(\bar{y}_{1k} - \bar{y}_{2k})\right]$$

$$= \frac{1}{n_1+n_2}\left[n_1\bar{y}_1 - n_2\bar{y}_2 + \sum_{k=1}^{K} n_{2k}\bar{y}_{1k} - \sum_{k=1}^{K} n_{1k}\bar{y}_{2k}\right]$$

$$(5.9)$$

is unbiased for τ.

Result 5 follows because by Result 2 $\bar{y}_{1k}$ is an unbiased estimate of $\mu_1(x_k)$, $\bar{y}_{2k}$ is an unbiased estimate of $\mu_2(x_k)$, and so $\bar{y}_{1k} - \bar{y}_{2k}$ is an unbiased estimate of $\mu_1(x_k) - \mu_2(x_k)$, $k = 1, \ldots, K$. That is, the difference between the Y mean for those Treatment 1 units whose X value is x_k and the Y mean for those Treatment 2 units whose X value is x_k is an unbiased estimate of the Treatment 1 vs. Treatment 2 effect at x_k. Hence, from Result 1 we have Result 5.

The advantage of the estimator given by (5.9) is that it does not depend on the accuracy of some underlying model for its unbiasedness. The disadvantage of the estimator is that if X takes on many values, some n_{ik} may be zero and then the estimator is not defined; this occurrence is not unusual in practice.

A common practical method used when some $n_{ik} = 0$ is to aggregate values of the original X variable to define a new variable X^* for which all $n_{ik} > 0$. However, since the assignment process was on the basis of X (not X^*), the estimator given by (5.9) based on X^* is no longer necessarily unbiased for τ. If X^* takes on many values, the bias might be small. For a similar situation, Cochran (1968a) concluded that in many cases blocking on an aggregated version of X with as few as 5 or 6 values was adequate to remove over 90% of the bias present in the simple estimator $\bar{y}_1 - \bar{y}_2$.

Of particular interest is the case in which X^* is chosen with minimum aggregation (i.e., K is maximized subject to the constraint that each $n_{ik} > 0$). It would be of practical importance to investigate the bias of the estimate (5.9) based on this X^* under (a) various underlying distributions of X in P, (b) different assignment processes based on X, and (c) several response functions $\mu_1(x)$ and $\mu_2(x)$.

Another method for handling cases in which some $n_{ik} = 0$ is to discard units. Result 6 is immediate from Result 1.

Result 6: If $\mu_1(x)$ and $\mu_2(x)$ are parallel, then

$$\sum_{k=1}^{K} \delta_k(\bar{y}_{1k} - \bar{y}_{2k}) / \sum_{k=1}^{K} \delta_k \tag{5.10}$$

where

$$\delta_k = \begin{cases} 0 & \text{if } n_{1k} \times n_{2k} = 0 \\ (n_{1k}^{-1} + n_{2k}^{-1})^{-1} & \text{otherwise} \end{cases}$$

is unbiased for τ.

Since $\mu_1(x)$ and $\mu_2(x)$ are parallel, there are many other choices of δ_k in (5.10) that will yield unbiased estimates of τ but they generally yield different precisions. The choice of optimal δ_k depends on conditions we have not discussed, those in (5.10) being optimal when the conditional variance of Y given X is constant. For further discussion, see for example, Kalton (1968) or Maxwell and Jones (1976).

Notice that the estimator given by (5.10) essentially discards those units whose X values are not the same as the X value of some unit who was exposed to the other treatment. This procedure, known as matching on the values of X, makes a lot of sense in some cases. Suppose X has been recorded and there is an additional cost in recording Y even though the treatments have already been given to all of the units. For example, the regular and compensatory reading programs have been given, background variables have been recorded, but there is an additional expense in giving and recording a battery of detailed posttests to each student. In these situations it is appropriate to ask how to choose the units on which to record Y. However, it may not be appropriate to assume the regressions are parallel, so that the estimator given by (5.10) may not be useful for estimating τ. Matching is more applicable when a subpopulation of P is of primary interest, that is, when the parameter of primary interest is the average treatment effect in a subpopulation.

6. GENERALIZING TO A SUBPOPULATION OF P DEFINED BY X

At times, the relevant population will not be P, but rather a subpopulation of P, say P_x defined by values of the covariate X (perhaps supplemented by some randomization). For example, the units exposed to Treatment 1 may be considered to be a random sample from the relevant population, perhaps those in need of extra treatment because of low values of X.

In such cases, all the results presented here generalize to estimating $\tau_x = \underset{x \in p_x}{\text{ave}} [\mu_1(x) - \mu_2(x)]$. The quantity τ_x is the treatment effect in the population P_x because $\mu_1(x) - \mu_2(x)$ is the treatment effect in P_x at $X = x$ as well as in P at $X = x$. That is, the conditional expectation of Y given (a) Treatment i, (b) $X = x$, and (c) X satisfies some criterion that defines membership in P_x is simply the conditional expectation of Y given (a) Treatment i and (b) $X = x$, which is defined to be $\mu_i(x)$.

Hence, Result 1 generalizes to estimating τ_x if the average over all X values in the sample is replaced by the average over X values that are representative of P_x. Result 2 is true as stated for P_x. In Result 3, the corresponding estimator of τ_x is now given by expression (5.7) with the averaging over all units replaced by averaging

Table 5.1. *Raw data for 25 Program 1 children and 47 Program 2 children*

Pretest scores	Posttest scores	
	Program 1	Program 2
10	15	6,7
9	16	7,11,12
8	12	5,6,9,12
7	8,11,12	6,6,6,6,7,8
6	9,10,11,13,20	5,5,6,6,6,6,6,6,6,8,8,8,9,10
5	5,6,7,16	3,5,5,6,6,7,8
4	5,6,6,12	4,4,4,5,7,11
3	4,7,8,9,12	0,5,7
2	4	4
1	–	–
0	–	7

over units representative of P_x. For example, if the units exposed to Treatment 1 are considered a random sample from P_x, this averaging of expression (5.7) leads to

$$(\bar{y}_1 - \bar{y}_2) - (\bar{x}_1 - \bar{x}_2)\hat{\beta}_2 \tag{5.11}$$

as the unbiased estimator of τ_x. This estimator given by expression (5.11) is discussed in some detail by Belsen (1956) and Cochran (1970).

If $\mu_1(x)$ and $\mu_2(x)$ are parallel, $\tau = \tau_x$ so that Result 4 as well as Result 6 apply for obtaining unbiased estimates of the treatment effect for any subpopulation P_x.

The extension of Result 5 to the subpopulation P_x is somewhat more interesting, although equally straightforward. For example, again suppose the units exposed to Treatment 1 are a random sample from P_x; then if $n_{2k} > 0$ whenever $n_{1k} > 0$, the estimator

$$\frac{1}{n_1} \sum_{k=1}^{K} n_{1k}(\bar{y}_{1k} - \bar{y}_{2k}) \tag{5.12}$$

is unbiased for τ_x. This estimator discards those units exposed to Treatment 2 whose X values are not found among the units exposed to Treatment 1. Finding for each unit exposed to Treatment 1 a unit exposed to Treatment 2 with the same X value and forming the estimate (5.12) has been called matched sampling (Rubin, 1973a). As discussed at the end of Section 5, estimators that discard data are most appropriate when one must decide for which units the value of Y should be recorded.

7. A SIMPLE EXAMPLE

Table 5.1 presents the raw data from an evaluation of a computer-aided program designed to teach mathematics to children in fourth grade. There were 25 children in Program 1 (the computer-aided program) and 47 children in Program 2 (the regular program). All children took a Pretest and Posttest, each test consisting of

20 problems, a child's score being the number of problems correctly solved. These data will be used to illustrate the estimation methods discussed in Sections 4, 5, and 6. We do not attempt a complete statistical analysis nor do we question the assumption of no interference between units.

7.1. Assignment on the Basis of Pretest

Suppose first that assignment to Program 1 or Program 2 was on the basis of Pretest, so that children with the same Pretest score were randomly assigned to programs with the probability of assignment to Program 1 increasing with lower Pretest scores. Hence, Pretest is the covariate X in the discussion of the previous sections. Posttest is the dependent variable Y. Figure 5.2 plots Posttest on Pretest for the treatment groups. Notice that although the Program 1 children scored somewhat lower on the Pretest than the Program 2 children (the Pretest means being 5.24 and 5.85 respectively) the Program 1 children scored higher on the Posttest than the Program 2 children (the Posttest means being 9.76 and 6.53 respectively). Consequently, we expect estimates of τ to be positive. Furthermore, notice that the distributions of X in the groups overlap substantially; hence, the methods described in the previous sections are appropriate for estimating treatment effects.

First consider estimating τ without assuming $\mu_1(x)$ and $\mu_2(x)$ are parallel. Using the least squares model-fitting methods described in Section 4, we fit separate linear conditional expectations in the two groups and obtain $\hat{\beta}_1 = 1.22$ and $\hat{\beta}_2 = 0.46$; from equation (5.6), the estimate of τ is 3.81. Fitting a quadratic conditional expectation in each group by least squares

$$\hat{\mu}_i(x) = \bar{y}_i + (x - \bar{x}_i)\bar{\gamma}_{1i} + \left(x^2 - \overline{x_i^2}\right)\hat{\gamma}_{2i}, \, i = 1, 2, \tag{5.13}$$

we have that $\overline{x_1^2} = 31.40$, $\overline{x_2^2} = 38.28$, $\hat{\gamma}_{11} = 1.40$, $\hat{\gamma}_{12} = 0.06$, $\hat{\gamma}_{21} = -0.02$, and $\hat{\gamma}_{22} = 0.04$; since the average difference between $\hat{\mu}_1(x)$ and $\hat{\mu}_2(x)$ over the values of X that occur in the sample is

$$(\bar{y}_1 - \bar{y}_2) - (\bar{x}_1 - \bar{x}_2)\frac{n_1\hat{\gamma}_{12} + n_2\hat{\gamma}_{11}}{n_1 + n_2} - (\overline{x_1^2} - \overline{x_2^2})\frac{n_1\hat{\gamma}_{22} + n_2\hat{\gamma}_{21}}{n_1 + n_2}, \tag{5.14}$$

the resulting estimate of τ under the quadratic model is 3.81. The blocking methods of Section 5 may also be used to estimate τ. After pooling the one child who scored "0" on the Pretest with the two children that scored "2" on the Pretest, we use equation (5.9) and obtain 3.98 for the estimate of τ.

Now, as discussed in Section 6, let us estimate the treatment effect for the population P_x, where the Program 1 children are considered a random sample from P_x. Assuming $\mu_2(x)$ is linear in x and using least squares estimation, equation (5.11) is appropriate and yields 3.51 for the estimate of τ_x. Using a quadratic model for $\mu_2(x)$ and least squares estimation, the appropriate estimate of τ_x is given by

$$(\bar{y}_1 - \bar{y}_2) - (\bar{x}_1 - \bar{x}_2)\hat{\gamma}_{12} - \left(\overline{x_1^2} - \overline{x_2^2}\right)\hat{\gamma}_{22} \tag{5.15}$$

which equals 3.54 for our data. And using the matching estimator given by (5.12) (discarding the data from the Program 2 child who scored "0" on the Pretest), we obtain 3.84 for the estimate of τ_x.

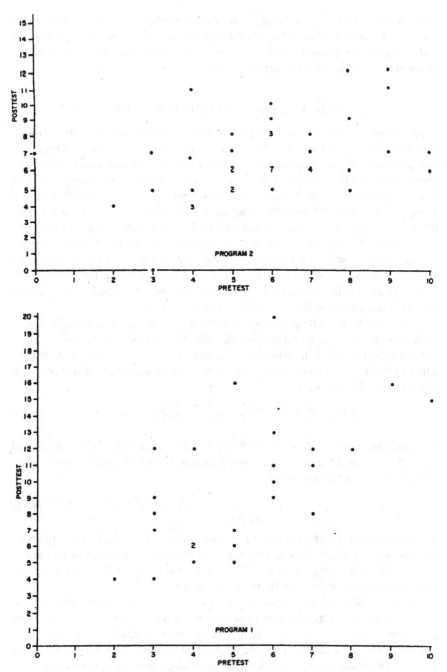

Figure 5.2.   Posttest vs. Pretest.

Table 5.2. *Estimates of treatment effect for data in Table 5.1 (relevant expression numbers in parentheses)*

| | Method of estimation | | |
| | Model-fitting | | |
Parameter being estimated	Linear	Quadratic	Blocking
τ	3.81 (6)	3.81 (14)	3.98[a] (9)
τ_x with Program 1 units a random sample from P_x	3.51 (11)	3.54 (15)	3.84 (12)
$\tau = \tau_x$ assuming $\mu_1(x)$ and $\mu_2(x)$ parallel	3.67 (8)	3.67 (17)	3.97 (10)

[a] The unit with Pretest score of "0" is blocked with those units with Pretest score "2."

Finally, suppose that we assume $\mu_1(x)$ and $\mu_2(x)$ are parallel. Least squares estimation of the linear model gives $\hat{\beta} = 0.7180$, and thus from equation (5.8), 3.67 for the estimate of $\tau = \tau_x$. Using least squares to estimate a parallel quadratic fit,

$$\hat{\mu}_i(x) = \bar{y}_i + (x - \bar{x}_i)\,\hat{\gamma}_1 + \left(x^2 - \overline{x_i^2}\right)\hat{\gamma}_2, i = 1, 2, \tag{5.16}$$

we find $\hat{\gamma}_1 = 0.3855$ and $\hat{\gamma}_2 = 0.0293$; since τ is estimated under this quadratic model by

$$(\bar{y}_1 - \bar{y}_2) - (\bar{x}_1 - \bar{x}_2)\hat{\gamma}_1 - \left(\overline{x_1^2} - \overline{x_2^2}\right)\hat{\gamma}_2, \tag{5.17}$$

we obtain 3.66 for the estimate of $\tau = \tau_x$. The blocking estimate of $\tau = \tau_x$ found by substituting into equation (5.10) is 3.97.

The nine estimates presented for this example are summarized in Table 5.2. The pattern of values for these estimates suggests that $\mu_1(x)$ and $\mu_2(x)$ may not be parallel, since the effect of Program 1 vs. Program 2 appears smaller for the lower values of X that occur more frequently in the Program 1 group. The implication is that the children who scored higher on the Pretest tended to profit more from Program 1. However, the estimates displayed in Table 5.2 exhibit little variability, ranging between 3.5 and 4.0.

Of course in practice, one should be concerned not only with the variability of the estimated treatment effects across different models, but also with the variability of the estimated treatment effect given a particular model (i.e., the standard error of the estimate under the model). For details on calculating standard errors of estimators like these under the normal error model see, for example, Snedecor and Cochran (1967, pp. 268–270, 423).

7.2. Assignment Not on the Basis of Pretest

Suppose now that instead of assignment to treatment group on the basis of Pretest, all Program 1 children came from one school and all Program 2 children came from a different school. That is, suppose that School 1 children had been assigned to Program 1 and School 2 children had been assigned to Program 2. The same plot of Posttest on Pretest as given in Figure 5.2 would be observed, but the estimates given above would not be unbiased for the effect of Program 1 vs. Program 2 because the

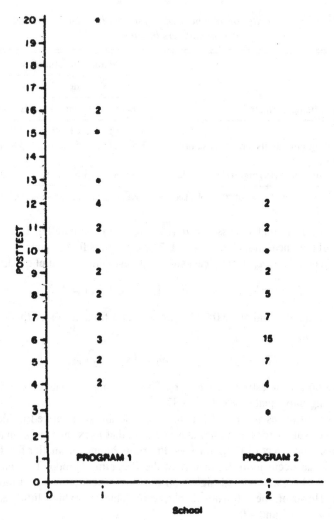

Figure 5.3. Posttest vs. School.

covariate that was used to assign treatments was not Pretest but School. Now School must be included as a covariate in order to apply the results of this paper. However, the plot of Y vs. School looks like Figure 5.3: each treatment group has only one value of the covariate X. We now cannot estimate the response functions $\mu_1(x)$ or $\mu_2(x)$ using the methods discussed in Sections 4, 5, 6 because there is no range of X values in each group. Nor can we block Program 1 and Program 2 children with similar values of X, because there are no such children.

Thus, if assignment to treatment group is on the basis of School, the methods we have presented cannot be directly applied because there is no overlap in the distribution of the covariate School in the two treatment groups. Using the estimates presented in Section 7.1 with Pretest as the covariate makes the implicit assumption that in each treatment group, the expected value of Y given Pretest and School is the same as the expected value of Y given just Pretest. Whether this assumption is reasonable depends, for instance, on how children were assigned to schools.

This simple example brings out two critical points relevant to all nonrandomized studies. First, knowledge of the assignment process is critical to drawing inferences about the effect of the treatments; one cannot simply look at the plot of Posttest on Pretest and properly estimate treatment effects. Second, even when the assignment mechanism is fully understood, the most defensible analysis of the data requires the distribution of the covariate to overlap in the two groups; without overlap, the analysis relies on assumptions that cannot be checked using the data at hand.

An example similar to this one is discussed by Lord (1967), but is used to emphasize the benefits of randomization.

8. DISCUSSION OF NEEDED INVESTIGATIONS

In this paper we have stated the fact that if assignment to treatment group is on the basis of the value of a covariate, X, one must concentrate effort on the essential problem of estimating the conditional expectation of Y given X in each treatment group. One then averages the difference between these conditional expectations over the values of X that are representative of the population of interest.

Two general methods for estimating these expectations were discussed: model fitting and blocking on the values of X. Little relevant work has been done on how well these techniques are likely to do in practice, either alone or in combination. A relevant simulation would include several careful choices of:

(a) the sample size, $n_1 + n_2$
(b) the distribution of X in P
(c) the assignment mechanism
(d) the functional forms for the conditional expectations, $\mu_1(x)$ and $\mu_2(x)$.

One would then find the distribution of estimates resulting from using the model fitting and blocking methods discussed here.

The case of multivariate X is of real interest because in natural settings we may not know the assignment mechanism but may feel that it can be described reasonably well by a particular collection of variables that are recorded. For example, teachers deciding which students should receive compensatory reading treatments presumably use personal assessments of students in addition to background characteristics of the children and test scores (*not* "true scores"), but the assignment mechanism might be adequately approximated by some function of the recorded background variables and tests, personal assessments hopefully being largely determined by the recorded variables.

All the results presented here for univariate X generalize immediately (conceptually at least) to multivariate X (e.g., $\hat{\beta}$ is now a vector of regression coefficients). Some work on multivariate matching methods is given in Althauser and Rubin (1970), Cochran and Rubin (1973) and Rubin (1976b, 1976c), but has received little attention otherwise.

Certainly a serious effort on both the univariate case and the multivariate case is worthwhile, not only in order to improve the analysis of existing nonrandomized studies but also in order to study the possibility of finding designs that are tolerable given social constraints, not randomized in the usual sense, but still allow useful inferences for the effects of treatments.

ACKNOWLEDGMENTS

I wish to thank P. W. Holland, M. R. Novick and a referee for suggestions that substantially improved the presentation, and D. T. Thayer for performing the computations needed in Section 7.

PART III: BASIC THEORY OF MULTIVARIATE MATCHING

Part III begins with another pair of back-to-back *Biometrics* articles, Rubin (1976b,c), which were written shortly after completing my PhD, and were focused entirely on analytic results concerning multivariate matching. Obviously, for practice the multivariate case is the norm; in fact, often there are very many matching variables.

Chapter 6, Rubin (1976b), defines a class of matching methods called "Equal Percent Bias Reducing" (EPBR). EPBR methods have the property that the percent reduction in bias due to the matching is the same for each of the matching variables. There are always linear combinations of the covariates that have the same means in the treatment and control groups before matching, and if a method is not EPBR, some of these will have different means after matching, implying that the matching infinitely increases the bias for them! This is not an attractive feature of a matching method – to increase bias in some directions, especially in the context of outcomes, Y, that are commonly assumed to be approximately linearly related to the X variables.

Chapter 6 goes on to describe classes of matching methods (e.g., caliper methods, inner-product methods) and corresponding distributional conditions on X (e.g., exchangeable, ellipsoidally symmetric) that lead to the methods being EPBR for those distributions. The most generally useful of these methods have turned out to be inner-product methods, including Mahalanobis-metric matching and discriminant matching, which can be a special case of the now widely used propensity score matching. Various technical results are stated and proved, most of them very simply, but there are also a couple of more subtle theorems concerning inner-product matching methods, which are proved in appendices. Sadly, once again, the *Biometrics* printers "dropped the ball" and never inserted the corrections noted on the page proofs, and so, once again, there were numerous printer's corrections made here in the reprinted article.

Chapter 7, Rubin (1976c), presented technical results on the maximum percent bias reduction that can be achieved with fixed sample sizes and fixed distributions. These results and closed form analytic approximations, also given in this chapter, have proven to be very useful in subsequent matching work. Also, because the method of proof considers moving the two distributions maximally toward each other in the direction of the best linear discriminant, the article anticipated the key role that the propensity score plays in matching methods. The more tedious technical results are proven in the appendices. As appeared to be their habit at *Biometrics* at the time, once again, the printer ignored the corrections on the page proofs, and so a printer's correction note was added, but the corrections have been made to the reprinted version here.

Neither of these articles had any simulation results and so were a departure from my earlier *Biometrics* articles. The next one reprinted here, however, Chapter 8, Rubin (1979b), summarized the results of a relatively massive simulation (at least relatively massive for the 1970s) stimulated by the analytic work in the two previous articles. Here X was bivariate normal in both the treatment and control groups, and the conditions considered in my PhD thesis (and thus in Chapters 3 and 4) were generalized to the bivariate setting, including the various linear regression estimators. Obviously, the setting had become more complicated, with now nearest available discriminant matching and Mahalanobis metric both available, and with percent reduction in integrated squared bias replacing percent reduction in bias as criteria of merit for estimators. Also, monotone nonlinear (exponential) response surfaces for Y as a function of bivariate X replaced the previous univariate response surfaces. Because of the wealth of conditions needing summarization, simple tables reporting basic results would have been hopelessly extensive, so the simulation results themselves were analyzed, just like the results of any experiment, using: ANOVA tables, marginal tables to display important one-way and two-way effects, outlier identification, et cetera. Actually, I am still pleased with this article, not only for its results concerning matching methods, but also because in it I analyzed simulation results the way a statistician should, using statistical tools, rather than by reproducing dozens of tables mindlessly listing all numbers produced by the computer to four decimal places.

There were three broad practical conclusions in Rubin (1979b): first, the combination of regression adjustment and matching is usually superior to either alone; second, typically matching alone is more robust than regression adjustment alone; and third, in this two-dimensional setting, Mahalanobis-metric matching works well relative to discriminant matching.

Chapter 9, Rubin (1980b), presents a collection of results from the same large simulation as in Rubin (1979b), but which did not fit logically into that earlier article. The article focused on the comparison of the percent reduction in bias when using Mahalanobis-metric matching with the theoretical maximum percent bias reduction, and it showed that when going from one to two dimensions in the covariates, there was the expected increase in remaining bias. The practical conclusions of both of these simulation-based articles with bivariate X now read to me as somewhat naïve, in that they do not appreciate the beneficial role for pair matching on the discriminant (e.g., linear propensity score) with larger dimensional X. More work with applied problems with many X's brought this message home very clearly, and was one of the forces that led to the development of propensity score methods, addressed in Part IV.

6. Multivariate Matching Methods That Are Equal Percent Bias Reducing, I: Some Examples

Donald B. Rubin

Abstract: Multivariate matching methods are commonly used in the behavioral and medical sciences in an attempt to control bias when randomization is not feasible. Some examples of multivariate matching methods are discussed in Althauser and Rubin [1970] and Cochran and Rubin [1973] but otherwise have received little attention in the literature. Here we present examples of multivariate matching methods that will yield the same percent reduction in bias for each matching variable for a variety of underlying distributions. Eleven distributional cases are considered and for each one, matching methods are described which are equal percent bias reducing. The methods discussed in Section 8 will probably be the most generally applicable in practice. These matching methods are based on the values of the sample best linear discriminant or define distance by the inverse of the sample covariance matrix.

1. INTRODUCTION

In an observational (non-randomized) study the objective is often to determine the effect of a dichotomous treatment variable (e.g., exposure to a specific drug) on several dependent variables (e.g., blood pressure, cholesterol levels). The treatment variable defines two populations of units, P_1 (e.g., subjects given the drug) and P_2 (subjects not given the drug). Since the treatments were not randomly assigned to the units, estimating the effect of the treatment variable on the dependent variables using random samples from P_1 and P_2 may be quite biased. For example, suppose subjects given the drug are typically older than subjects not given the drug; then the simple comparison of post-treatment blood pressures in random samples of subjects given and not given the drug would tend to underestimate the effect of the drug for lowering blood pressures.

Matched sampling attempts to reduce biases like these by finding samples from P_1 and P_2 in which the distributions of the matching variables (e.g., age, initial blood pressure) are nearly identical. More precisely, consider the following matched sampling plan:

1. Obtain random samples from P_1 and P_2, G_1 and G_2, of sizes N_1 and N_2. Without loss of generality let $N_1 \leq N_2$.
2. Record the p matching variables, $\mathbf{X}$, for all units in G_1 and G_2.

Reprinted with permission from *Biometrics*, 1976, **32**, 109–120. © 1976 International Biometric Society.

3. Using some matching method that examines the values of **X**, find "matched" sub-samples of G_1 and G_2, say G_{1*} and G_{2*}, of sizes N_{1*} and N_{2*}. (Often $N_{1*} = N_{2*}$ and the matched samples are organized in matched $G_{1*} - G_{2*}$ pairs of units, each pair having similar values of **X**).
4. Record the dependent variables for all units in G_{1*} and G_{2*} (e.g., via intensive medical examinations).
5. Estimate the effect of the treatment variable on the dependent variables using the matched samples G_{1*} and G_{2*}.

In this paper, we are only concerned with methods of performing the third step: obtaining matched samples from given random samples. Discussion of the more general problem is found in Cochran [1972], Cochran and Rubin [1973] and Rubin [1974, 1973a, b].

We will restrict our discussion of multivariate matching methods in two ways. First, we assume that none of the **X** is strictly categorical (nominal, e.g., religion). If there are some strictly categorical matching variables, we assume that either (i) the matching on **X** takes place for groups of G_1 and G_2 units with the same values on these categorical matching variables, or (ii) each categorical matching variable has been recoded into several non-categorical variables (e.g., polynomial contrasts) and these new variables are included in **X**. The second restriction is that we consider only matching methods that can be "equal percent bias reducing" (EPBR) for the matching variables **X**. By this we mean that the percent reduction in (mean) bias is the same for all p matching variables.

2. EQUAL PERCENT BIAS REDUCING MATCHING METHODS

Let **X** have (finite) mean vector μ_i and (finite) covariance matrix Σ_i in P_i, $i = 1, 2$. Also let μ_{i*} be the expected mean vector of **X** in matched G_{i*} samples obtained by a particular matching method. More precisely, given (a) fixed sizes of the random samples N_1, N_2, (b) fixed distributions of **X** in P_1 and P_2 and (c) a fixed method of obtaining matched samples of sizes N_{1*} and N_{2*}, in repeated drawings of random samples, the average of the mean vectors in the resultant matched samples will converge to μ_{1*} and μ_{2*}.

If $(\mu_{1*} - \mu_{2*}) = \gamma(\mu_1 - \mu_2)$, where γ is a scalar, we will say that the matching method is equal percent bias reducing (EPBR) for **X** because the percent reduction in the (mean) biases of each of the matching variables is the same; i.e., using the superscript j to represent this jth matching variable, the fractional reduction in bias of the jth matching variable is: $1 - (\mu_{1*}{}^j - \mu_{2*}{}^j)/(\mu_1{}^j - \mu_2{}^j) = 1 - \gamma$ for all j if and only if $(\mu_{1*} - \mu_{2*}) = \gamma(\mu_1 - \mu_2)$. Choosing random samples from G_1 and G_2 to be G_{1*} and G_{2*} is EPBR but particularly uninteresting.

A matching method that is EPBR for **X** is appealing basically because one obtains the same percent reduction in bias for any linear function of the **X** if and only if the matching method is EPBR for **X**. That is, given a linear function $\mathbf{X}\beta'$ (all vectors are row vectors), the expected $P_1 - P_2$ bias is $(\mu_1 - \mu_2)\beta'$ while the expected $G_{1*} - G_{2*}$ bias is $(\mu_{1*} - \mu_{2*})\beta'$ and the ratio of these biases is a constant for all β if and only if $(\mu_{1*} - \mu_{2*}) = \gamma(\mu_1 - \mu_2)$. Also notice that if the matching method is not EPBR for $\mathbf{X}$, then the matching substantially increases the bias for some linear functions of $\mathbf{X}$, even if all univariate means are closer in G_{1*} and G_{2*} than in P_1 and P_2; e.g., consider $\beta = (1, 1)$, $\mu_1 = (1, 0)$, $\mu_2 = (0, 1)$, $\mu_{1*} = (.1, .1)$ and $\mu_{2*} = (-.1, -.1)$; the initial bias is zero but the final bias is .4. Hence, if the dependent variables (or reasonable transformations of the dependent variables) are thought to be (approximately) linearly related to $\mathbf{X}$, a matching method that is EPBR for $\mathbf{X}$ is quite appealing.

One might not care about using a matching method that is EPBR if there is one dependent variable and there exists some specific knowledge about its dependence on $\mathbf{X}$ (e.g., $Y \doteq 2X^1 + \ln(X^2)$) or if there is some specific knowledge that for all dependent variables some matching variables are more important than the others (e.g., $|\mu_1{}^1 - \mu_2{}^1| \gg |\mu_1{}^2 - \mu_2{}^2|$ and for all dependent variables $|\beta^1| \gg |\beta^2|$).

The objective here is to introduce terminology useful for describing matching methods and present examples of matching methods that can be EPBR. Whether or not a specific matching method is EPBR generally depends upon the distributions of $\mathbf{X}$ in P_1 and P_2.

It turns out that the distributional conditions required for matching methods to be EPBR are less restrictive when $G_1 = G_{1*}$ than when both G_1 and G_2 are to be subsampled; thus, we consider matching methods in which G_1 is and is not subsampled.

Section 3 describes special distributional cases in which univariate matching is EPBR for $\mathbf{X}$ and Section 4 describes special cases in which "equibias intersection" matching is EPBR for $\mathbf{X}$. Matching variables satisfying one of these special distributional cases even approximately are unlikely to be found in practice. However, as Section 5 demonstrates, they are of interest because if $\mathbf{X}$ is normally distributed and the samples are large, there is a known linear transformation to at least two of these special cases.

Section 6 describes special cases, again unlikely to occur in practice, in which "exchangeable" matching methods are EPBR for $\mathbf{X}$. Section 7 shows that when X is "ellipsoidal" (e.g., normal, multivariate-t) and the samples are large, there exists a known linear transformation to at least one of the special cases of Section 6.

Section 8 shows that if $\mathbf{X}$ is ellipsoidal, univariate matching on a sample estimate of the best linear discriminant, or metric matching using a sample-based inner-product, is EPBR for $\mathbf{X}$, even in small samples. The methods described in Section 8 will probably be the most useful in practice.

3. SPECIAL CASES IN WHICH UNIVARIATE MATCHING IS EPBR

In some distributional cases, matching on one particular variable is EPBR for $\mathbf{X}$ even though matched units will be "close" only with respect to the one variable used for matching.

Some methods of univariate matching with $N_1 = N_{1*} = N_{2*}$ are summarized in Cochran and Rubin [1973] and further discussed in Billewicz [1965], Cochran [1968a] and Rubin [1973 a, b]. "Nearest available matching" orders the G_1 sample in some way and then in turn assigns as a match for each G_1 unit that G_2 unit not yet assigned as a match that is closest on the matching variable; the $N_1 G_2$ units so chosen comprise G_{2*}. As another example, "mean matching" chooses for G_{2*} the N_{2*} sized subsample of G_2 whose mean on the matching variable is closest to the G_1 mean on the matching variable. Methods of univariate matching with $N_{1*} < N_1$ (i.e., that subsample both G_1 and G_2) include (a) nearest available matching followed by retaining only the closest $N_{1*} = N_{2*}$ matched pairs and (b) finding the N_{1*} sized subsample of G_1 and the N_{2*} sized subsample of G_2 whose means are closest on the matching variable.

Case 1:

(a) $(\mu_1 - \mu_2) \propto (1, 0, \ldots, 0)$ (i.e., all mean bias is in the first variable) and
(b) X^1 and $(X^2, \ldots, X^v)$ are independent in P_2.

In Case 1, any univariate matching on X^1 that does not subsample G_1 (i.e., $G_{1*} = G_1$, $N_{1*} = N_1$) is EPBR for **X**.

This result holds because by condition (b), matching on X^1 implies that the distribution of $(X^2, \ldots, X^v)$ in P_2 is the same as it is in matched G_{2*} samples so that by condition (a) $(\mu_1 - \mu_{2*}) \propto (1, 0, \ldots 0)$; since $G_1 = G_{1*}$ implies $\mu_1 = \mu_{1*}$, we thus have $(\mu_{1*} - \mu_{2*}) = (\mu_1 - \mu_{2*}) \propto (1, 0, \ldots, 0) \propto (\mu_1 - \mu_2)$.

Case 2:

(a) $(\mu_1 - \mu_2) \propto (1, 0, \ldots, 0)$,
(b) X^1 independent of $(X^2, \ldots, X^v)$ in P_2 and
(c) X^1 independent of $(X^2, \ldots, X^v)$ in P_1.

In Case 2, any univariate matching method on X^1 is EPBR for X.

This follows because by (b) and (c), any matching on X^1 results in the same distribution of $(X^2, \ldots, X^v)$ in (i) P_1 and matched G_{1*} samples, and (ii) P_2 and matched G_{2*} samples. Hence $(\mu_{1*} - \mu_{2*}) \propto (1, 0, \ldots, 0) \propto (\mu_1 - \mu_2)$.

Section 4 considers transforming normally distributed matching variables to Case 1 or Case 2.

4. SPECIAL CASES IN WHICH "EQUIBIAS INTERSECTION" MATCHING ON X IS EPBR

A class of multivariate matching methods often tried in practice may be called "intersection" matching methods. These methods first define a set of acceptable matches independently on each variable by a univariate rule and then define the set of acceptable matches with respect to all p variables as the intersection of the p sets. Some specific methods are discussed in Althauser and Rubin [1970] and Cochran and Rubin [1973]. A simple example is multivariate "caliper" matching: for each G_1 unit assign as a match a G_2 unit whose value on X^k is within $\pm c_k$ of the G_1 unit's value on X^k for all $k = 1, \ldots, p$. As another example of an intersection matching method consider "categorical" ("quantile") matching: categorize each X^k into g_k categories (e.g., high, medium, low), and for each G_1 unit choose as a match a G_2 unit that falls in the same category on all categorized variables. "Variable caliper" matching defines the set of acceptable matches on X^k to be the m_k closest members of G_2. Intersection methods can also use different rules for different matching variables.

G_1 units without matches can be discarded or retained depending on whether G_1 is to be subsampled. Notice that N_{2*} (and N_{1*} if G_1 is subsampled) is not known exactly until after the matching is completed.

Intersection matching methods that define the set of univariate matches with respect to X^k such that the percent reduction in expected bias is the same for all $k = 1, \ldots, p$ will be called equibias intersection matching methods. For example, when caliper matching, the c_k can be adjusted as a function of the differences in the distributions of X^k in P_1 and P_2 (assuming these distributions are known) so that each of the p univariate caliper matchings yields the same percent reduction in expected bias. Or when categorical matching, the number of categories, g_k, and their boundaries can be adjusted so that each of the p univariate matching rules yields the same percent reduction in expected bias.

Case 3: The X^k are independently distributed in P_2.

In Case 3, any equibias intersection matching that does not subsample G_1 is EPBR for **X**.

This follows because (i) equibias intersection matching with $G_1 = G_{2*}$ and $100(1 - \gamma)$ percent bias reduction in each univariate rule implies that the expected mean for the G_2 units matched by the kth univariate rule is $\mu_1{}^k - \gamma(\mu_1{}^k - \mu_2{}^k)$ and (ii) the independence of all X^k in P_2 implies that the expected mean of X^k in G_{2*} is the same as the expected mean before the intersection of sets of matches; hence $\mu_{2*}{}^k = \mu_1{}^k - \gamma(\mu_1{}^k - \mu_2{}^k)$ or $\mu_{1*}{}^k - \mu_{2*}{}^k = \gamma(\mu_1{}^k - \mu_2{}^k)$ for all k since $G_1 = G_{2*}$ implies $\mu_1{}^k = \mu_{1*}{}^k$.

Case 4:

(a) the X^k are independently distributed in P_2 and
(b) the X^k are independently distributed in P_1.

In Case 4, any equibias intersection matching is EPBR for **X**.

This follows because the independence of **X** in both P_1 and P_2 implies that the percent reduction in bias in each X^k is unaffected by the intersection of the sets of matches.

5. LARGE SAMPLE EPBR MATCHING METHODS FOR X NORMALLY DISTRIBUTED

The range of cases for which it is easy to define EPBR matching methods can be increased by considering normally distributed **X**. For **X** normal and Σ_2 full rank, consider the transformed variables

$$Z = XH, \quad \text{where} \quad HH' = \Sigma_2^{-1} \quad \text{and} \quad (\mu_1 - \mu_2)H \propto (1, 0, \ldots, 0). \quad (6.1)$$

H can be constructed via any triangular decomposition of Σ_2^{-1} followed by an orthogonal rotation. Because the Z^k are uncorrelated and thus independent by the normality, **Z** satisfies at least two of the four special cases presented previously. Since we assume **H** is known, the methods are restricted to large samples. Section 8 presents methods which are EPBR for normal **X** even in small samples.

Case 5: **X** is normal in P_2.

In Case 5,

(a) any univariate matching on Z^1 that does not subsample G_1 or
(b) any equibias intersection matching on **Z** that does not subsample G_1 is EPBR for **X**.

These results hold because **Z** in (6.1) satisfies the conditions of both Case 1 and Case 3 since in P_2 all Z^k are independent and the difference of means in P_1 and P_2 is proportional to $(1, 0, \ldots, 0)$; thus univariate matching on Z^1 with $G_1 = G_{1*}$ and intersection matching with G_1 and G_{1*} are EPBR for **Z** and thus for **X** because **X** is a linear combination of **Z**.

The variable Z^1 is the best linear discriminant with respect to the P_2 inner product, i.e., $\mathbf{X}\Sigma_2^{-1}(\mu_1 - \mu_2)'$. Matching on this discriminant was suggested by Rubin [1970] and further discussed by Cochran and Rubin [1973], Section 6.3. Intersection matching methods using transformed variables are discussed in Cochran and Rubin [1973], Sections 6.1 and 6.2.

Case 6:

(a) **X** is normal in P_2,
(b) **X** is normal in P_1 and
(c) the best linear discriminant with respect to Σ_1 is proportional to the best linear discriminant with respect to Σ_2.

In Case 6,

(a) any univariate matching on Z^1 (the common best linear discriminant) or
(b) any equibias intersection matching on **Z** that does not subsample G_1 is EPBR for **X**.

These results follow because Z satisfies the conditions of Cases 1, 2 and 3, i.e., the first row of the covariance matrix of **Z** in P_1 is $(1, 0, \ldots, 0)\mathbf{H}'\Sigma_1\mathbf{H}$ which by (6.1) is proportional to $(\mu_1 - \mu_2)\Sigma_2^{-1}\Sigma_1\mathbf{H}$ which by condition (c) is proportional to $(\mu_1 - \mu_2)\mathbf{H}$, which by (6.1) is proportional to $(1, 0, \ldots, 0)$.

Case 7:

(a) **X** is normal in P_2,
(b) **X** is normal in P_1 and
(c) the covariance matrices of **X** in P_1 and P_2 are proportional.

In Case 7,

(a) any univariate matching on Z^1 or
(b) any equibias intersection matching on **Z** is EPBR for **X**.

These follow because **Z** satisfies the conditions of Cases 1–4.

6. SPECIAL CASES IN WHICH EXCHANGEABLE MATCHING ON X IS EPBR

Two more special cases in which there exist EPBR matching methods specify that the distribution of **X** is exchangeable (cf. Feller [1966], p. 225). Exchangeable means that the functional form of the distribution remains unchanged under permutation of indices of the variables (e.g., the mean vector is proportional to $\mathbf{1} = (1, \ldots, 1)$ and the covariance matrix is proportional to $(\mathbf{I} + d\mathbf{1}'\mathbf{1})$, $d \geq -1/p$ for the p-variate normal).

Similarly, exchangeable matching methods are ones whose rules remain unchanged under permutation of the indices of the matching variables. For example, a univariate method is exchangeable if the scalar matching variable is the sum of the matching variables, the product of the matching variables, etc.

For a second example, intersection matching methods are exchangeable if each univariate matching uses the same rule (e.g., c_k = one standard deviation of X^k in G_2). In fact, intersection methods whose rules are modified and remodified until a minimum number of G_1 units have matches can be exchangeable, e.g., "guaranteed variable caliper matching" discussed in Althauser and Rubin [1970].

For a third example, "metric" matching methods are exchangeable if distance is defined by an exchangeable function. For instance, if $\mathbf{x}_d$ is the difference between a G_1 unit's and a G_2 unit's values on $\mathbf{X}$, one can use the inner product matrix $\mathbf{Q} \propto (\mathbf{I} + d\mathbf{1}'\mathbf{1})$, $d \geq -1/p$, to define the distance between the units to be $\mathbf{x}_d \mathbf{Q} \mathbf{x}_d'$. Then by ordering the G_1 units, one can nearest available match and retain the closest N_{1*} matched pairs. Or letting $\mathbf{x}_d$ be the difference of means between an N_{1*} sized subset of G_1, and an N_{2*} sized subset of G_2, one can choose for G_{1*} and G_{2*} those subsets that minimize $\mathbf{x}_d \mathbf{Q} \mathbf{x}_d'$.

For a fourth example, methods which choose G_{1*} and G_{2*} to minimize an exchangeable measure of the difference between their empirical cumulative distribution functions (e.g., the Kolmogorof-Smirnoff two-sample statistic) are exchangeable.

Case 8:

(a) $(\mu_1 - \mu_2) \propto \mathbf{1} = (1, 1, \ldots, 1)$ (i.e., the mean bias is the same for each variable) and

(b) the distribution of $\mathbf{X}$ is exchangeable in P_2.

In Case 8, any univariate exchangeable matching on $\mathbf{X}$ that does not subsample G_1 is EPBR for $\mathbf{X}$.

This follows because univariate exchangeable matching with $G_{1*} = G_*$ implies that (i) $\mu_{1*} = \mu_1$ and (ii) the expected G_{2*} mean on the exchangeable scalar matching variable is an exchangeable function of the p components of μ_{2*} (as well as the components of the other moments of $\mathbf{X}$ in G_{2*}) since the distribution of $\mathbf{X}$ in P_2 is exchangeable; hence μ_{2*} (as well as $\mu_{1*} = \mu_1$) must be proportional to $\mathbf{1}$.

Case 9:

(a) the distribution of $\mathbf{X}$ is exchangeable in P_2 and

(b) the distribution of $\mathbf{X}$ is exchangeable in P_1.

In Case 9 any exchangeable matching on $\mathbf{X}$ is EPBR for $\mathbf{X}$.

This result follows because the total symmetry of distribution and matching rule implies that μ_1, μ_2, μ_{1*}, and μ_{2*} are all proportional to $\mathbf{1}$.

7. LARGE SAMPLE EPBR MATCHING METHODS
FOR X ELLIPSOIDAL

The range of cases for which it is straightforward to define EPBR matching methods can be increased by considering **X** to be ellipsoidally distributed. The variable **X** is ellipsoidal if there is constant probability on the surface of ellipsoids, or equivalently, if there is a linear transformation to a spherical distribution (Dempster [1969], Chapter 12). The multivariate normal and t distributions are commonly referenced ellipsoidal distributions.

Consider the transformation (for Σ_2 full rank)

$$\mathbf{W} = (\mathbf{X} - \boldsymbol{\mu}_i)\mathbf{G} \tag{6.2}$$

where $\mathbf{GG'} = \Sigma_2^{-1}$, $(\boldsymbol{\mu}_1 - \boldsymbol{\mu}_2)\mathbf{G} = b\mathbf{1}$, for some scalar b.

Case 10: **X** is ellipsoidal in P_2.

In Case 10, any univariate exchangeable matching on **W** that does not subsample G_1 is EPBR for **X**.

This follows from Case 8 because **W** is exchangeably distributed in P_2 with mean $0 = (0, \ldots, 0)$ in P_1; specifically, in P_2 the mean of **W** is $b\mathbf{1}$ and the covariance of **W** is **I**.

Case 11:

(a) **X** is ellipsoidal in P_2,
(b) **X** is ellipsoidal in P_1 and
(c) $\Sigma_1 = a\Sigma_2 + ac(\boldsymbol{\mu}_1 - \boldsymbol{\mu}_2)'(\boldsymbol{\mu}_1 - \boldsymbol{\mu}_2)$, for some scalars $a > 0, c \geq -[b^2 p]^{-1}$.

In Case 11, any exchangeable matching on **W** is EPBR for **X**.

This follows from Case 9 because **W** is exchangeably distributed in both P_1 and P_2; i.e., **W** is exchangeable in P_2 by the argument in Case 10 and is exchangeable in P_1 because the covariance of **W** in P_1 is

$$\mathbf{G'}\Sigma_1\mathbf{G} = \mathbf{G'}[a\Sigma_2 + ac(\boldsymbol{\mu}_1 - \boldsymbol{\mu}_2)'(\boldsymbol{\mu}_1 - \boldsymbol{\mu}_2)]\mathbf{G} = a[\mathbf{I} + b^2 c\mathbf{1}'\mathbf{1}]. \tag{6.3}$$

Condition (c) in Case 11 implies that there is a common best linear discriminant and that the covariances of **X** orthogonal to the discriminant are proportional (see Appendix A).

8. DISCRIMINANT AND INNER PRODUCT MATCHING

Thus far we have not presented any matching methods that are EPBR in small samples except under very restrictive conditions. That is, the special cases presented in Sections 3, 4, and 6 are unlikely to occur even approximately in practice, and the cases presented in Sections 5 and 7 for normal and ellipsoidal matching variables require knowledge of population parameters in order to perform the suggested transformations. When the sample sizes are very large relative to p, these transformations present no practical problem, but in smaller samples it is not clear what will be the effect of using estimated rather than population moments.

Within the context of ellipsoidal **X**, it is natural to think of linear operators. In fact, we will show that for ellipsoidal **X**, linear operators can generate matching methods based solely on the sample values of $\mathbf{X}$ that are EPBR for $\mathbf{X}$ even in small samples.

In Case 10, the natural, exchangeable, scalar matching variable is $\mathbf{W1}' = (\mathbf{X} - \mathbf{\mu}_1)\mathbf{G1}' \propto (\mathbf{X} - \mathbf{\mu}_1)\mathbf{\Sigma}_2^{-1}(\mathbf{\mu}_1 - \mathbf{\mu}_2)'$, the best linear discriminant between P_1 and P_2 with respect to the P_2 inner product translated by a constant. This suggests using the sample estimate of this discriminant for matching, i.e.,

$$\mathbf{X}\mathbf{S}_2^{-1}(\bar{\mathbf{x}}_1 - \bar{\mathbf{x}}_2)', \tag{6.4}$$

where $\bar{\mathbf{x}}_i =$ sample mean of $\mathbf{X}$ in G_i,

$\quad\quad \mathbf{S}_i =$ sample covariance of $\mathbf{X}$ in G_i (if $\mathbf{S}_2$ is singular, let $\mathbf{S}_2^{-1}$ be the Moore–Penrose pseudo-inverse of $\mathbf{S}_2$).

Theorem 1. *If* $\mathbf{X}$ *is ellipsoidal in* P_2*, then any translation invariant univariate matching on the estimated best linear discriminant, (6.4), that does not subsample* G_1 *is EPBR for* $\mathbf{X}$*.*

By translation invariant univariate matching we mean that the matches are unaffected by the addition of any constant to the matching variable. The proof of Theorem 1 is given in Appendix B.

In Case 11, the natural, exchangeable, multivariate matching method is metric matching using as inner product for $\mathbf{W}$

$$(\mathbf{I} + d\mathbf{1}'\mathbf{1}), \quad d \geq -1/p.$$

In terms of $\mathbf{X}$, this matching inner product is

$$\mathbf{G}(\mathbf{I} + d\mathbf{1}'\mathbf{1})\mathbf{G}' = \mathbf{G}\left[\mathbf{I} - \frac{1}{p}\mathbf{1}'\mathbf{1}\right]\mathbf{G}' + \mathbf{G}\left[\left(d + \frac{1}{p}\right)\mathbf{1}'\mathbf{1}\right]\mathbf{G}'$$

$$= \left[\mathbf{\Sigma}_2^{-1} - \frac{1}{\sigma_2^2}\mathbf{\Sigma}_2^{-1}(\mathbf{\mu}_1 - \mathbf{\mu}_2)'(\mathbf{\mu}_1 - \mathbf{\mu}_2)\mathbf{\Sigma}_2^{-1}\right] \tag{6.5}$$

$$+ \left(d + \frac{1}{p}\right)\frac{p}{\sigma_2^2}\left[\mathbf{\Sigma}_2^{-1}(\mathbf{\mu}_1 - \mathbf{\mu}_2)'(\mathbf{\mu}_1 - \mathbf{\mu}_2)\mathbf{\Sigma}_2^{-1}\right],$$

where $\sigma_2^2 =$ the variance of the P_2 best linear discriminant in P_2. Recall that in Case 11 there is a common best linear discriminant and the covariances of $\mathbf{X}$ orthogonal to the discriminant are proportional in P_1 and P_2. This observation plus the form of expression (6.5) makes it clear that the EPBR matching inner-products for $\mathbf{X}$ calculate squared distance along the discriminant (i.e., the second term in (6.5)) and in the space orthogonal to the discriminant (i.e., the first term in (6.5)) and let the total squared distance be proportional to any convex combination of these two orthogonal components. The choice of the minimum value of d (i.e., $d = -1/p$) corresponds to defining distance by the Mahalanobis metric in the space orthogonal to the discriminant, the intermediate value $d = 0$ corresponds to defining distance by the P_2 Mahalanobis metric, and the maximum value $d = \infty$ corresponds to defining distance along the best linear discriminant. Also an intermediate value for d corresponds to defining distance by the P_1 Mahalanobis metric (see equation (A.5) in Appendix A).

The above discussion suggests using a sample estimate of an inner-product of this form for matching.

Theorem 2. *If*

(a) **X** *is ellipsoidal in* P_2,
(b) **X** *is ellipsoidal in* P_1,
(c) $\Sigma_1 \propto \Sigma_2 + c(\mu_1 - \mu_2)'(\mu_1 - \mu_2)$, *c some scalar,*

then any metric matching using as inner product

$$\left[\mathbf{S}^{-1} - \frac{1}{s^2}\mathbf{S}^{-1}(\bar{\mathbf{x}}_1 - \bar{\mathbf{x}}_2)'(\bar{\mathbf{x}}_1 - \bar{\mathbf{x}}_2)\mathbf{S}^{-1}\right] + k[\mathbf{S}^{-1}(\bar{\mathbf{x}}_1 - \bar{\mathbf{x}}_2)'(\bar{\mathbf{x}}_1 - \bar{\mathbf{x}}_2)\mathbf{S}^{-1}], \qquad (6.6)$$

where $k \geq 0$,
$\mathbf{S} = \lambda_1\mathbf{S}_1 + \lambda_2\mathbf{S}_2$, *for any* $\lambda_1 \geq 0, \lambda_2 \geq 0$ *but* $\lambda_1\lambda_2 > 0$ (*if* **S** *is singular,* $\mathbf{S}^{-1}$ *is the Moore– Penrose pseudo-inverse*)

$$s^2 = (\bar{\mathbf{x}}_1 - \bar{\mathbf{x}}_2)\mathbf{S}^{-1}(\bar{\mathbf{x}}_1 - \bar{\mathbf{x}}_2)'$$

is EPBR for **X**. *The proof of Theorem 2 is given in Appendix C.*

If $\lambda_i = N_i - 1$, **S** is the pooled estimate of the within group covariance matrix and for **X** ellipsoidal is positive definite if $N_1 + N_2 - 2 \geq p$. Then $k = 0$ corresponds to metric matching in the space orthogonal to this estimate of the discriminant, $k = \infty$ corresponds to metric matching along this estimated discriminant, $k = 1/s^2$ corresponds to metric matching with respect to the **S** Mahalanobis metric and intermediate values of k correspond to metric matching with respect to the G_1 and G_2 sample covariance Mahalanobis metrics.

Nearest available metric matching using an inner product of the form (6.6) to define distance between G_1 and G_2 units seems rather natural within the often assumed context of normally distributed **X** and is easy to implement in practice because of the general availability of the required matrix operators. Also, it is easy to control the number of final matches when using such a matching method.

Some general analytic results on the maximum percent reduction in bias for any EPBR matching method given fixed sample sizes and fixed distributions are given in Rubin [1976c]. These results can be used to suggest minimum ratios of sample sizes needed to obtain well-matched samples. However, the evaluation of the utility of specific matching methods, alone and in combination with regression adjustment, probably requires extensive Monte–Carlo work studying dependent variables both linear and nonlinear in **X**.

ACKNOWLEDGMENTS

I would like to thank W. G. Cochran, A. P. Dempster and T. W. F. Stroud for helpful comments on this paper.

APPENDIX A

In the following lemmas, **X** has mean μ_i and full rank (p) covariance Σ_i in $P_i, i = 1, 2$. Let $\Delta_i = (\mu_1 - \mu_2)\Sigma_i^{-1}$ be the coefficient of the P_i best linear discriminant, $i = 1, 2$. Let $\Gamma_i = \Sigma_i - 1/v_i^2(\mu_1 - \mu_2)'(\mu_1 - \mu_2)$ be the covariance of **X** in P_i orthogonal to the P_i discriminant, where

$$v_i^2 = \Delta_i\Sigma_i\Delta_i' = \Delta_i(\mu_1 - \mu_2)' = (\mu_1 - \mu_2)\Sigma_i^{-1}(\mu_1 - \mu_2)' \qquad (A.1)$$

is the variance of the P_i discriminant in $P_i, i = 1, 2$.

Lemma 1: *If the covariance of* $\mathbf{X}$ *in* P_1 *orthogonal to the* P_1 *discriminant is proportional to the covariance of* $\mathbf{X}$ *in* P_2 *orthogonal to the* P_2 *discriminant,*

$$\Gamma_1 = a\Gamma_2 \tag{A.2}$$

then

$$\Sigma_1 = a[\Sigma_2 + c(\mu_i - \mu_2)'(\mu_1 - \mu_2)] \tag{A.3}$$

for some scalar c.

Proof.

$$\Sigma_1 = \Gamma_1 + \frac{1}{v_1^2}(\mu_1 - \mu_2)'(\mu_1 - \mu_2) = a\Gamma_2 + \frac{1}{v_1^2}(\mu_1 - \mu_2)'(\mu_1 - \mu_2)$$

$$= a\left[\Sigma_2 + \left(\frac{1}{av_1^2} - \frac{1}{v_2^2}\right)(\mu_1 - \mu_2)(\mu_1 - \mu_2)\right]$$

Lemma 2: *Equation (A.3) implies that there is a common best linear discriminant,*

$$\Delta_1 \propto \Delta_2. \tag{A.4}$$

Proof. First note that (A.3) implies

$$\Sigma_1^{-1} = \frac{1}{a}\left[\Sigma_2^{-1} - \frac{c}{1 + cv_2^2}\Sigma_2^{-1}(\mu_1 - \mu_2)'(\mu_1 - \mu_2)\Sigma_2^{-1}\right]. \tag{A.5}$$

Now calculate Δ_1 using (A.5):

$$\Delta_1 = (\mu_1 - \mu_2)\Sigma_1^{-1} = \frac{1}{a}\left[1 - \frac{c}{1 + cv_2^2}v_2^2\right](\mu_1 - \mu_2)\Sigma_2^{-1}$$

$$= \frac{1}{a(1 + cv_2^2)}\Delta_2 \tag{A.6}$$

Lemma 3: *Equation (A.3) implies equation (A.2).*

Proof.

$$\Gamma_1 = \Sigma_1 - \frac{1}{v_1^2}(\mu_1 - \mu_2)'(\mu_1 - \mu_2)$$

$$= a\left[\Sigma_2 + c(\mu_1 - \mu_2)'(\mu_1 - \mu_2) - \frac{1}{av_1^2}(\mu_1 - \mu_2)'(\mu_1 - \mu_2)\right]$$

$$= a\left[\Gamma_2 + \left(\frac{1}{v_2^2} - \frac{1}{av_1^2} + c\right)(\mu_1 - \mu_2)'(\mu_1 - \mu_2)\right].$$

But from (A.6) we know that (A.3) implies

$$v_1^2 = \Delta_1(\mu_1 - \mu_2)' = \frac{v_2^2}{a(1 + cv_2^2)},$$

so that (A.3) implies $\Gamma_1 = a\Gamma_2$.

Note that Lemmas 1, 2, and 3 have proved that the conditions given by equations (A.2) and (A.3) are equivalent and either implies a common best linear discriminant, equation (A.4).

APPENDIX B

The proof of Theorem 1 follows by first conditioning on the G_1 sample and showing that the matching is EPBR for $\mathbf{X}$ regardless of the values of $\mathbf{X}$ in G_1; thus the matching is EPBR for $\mathbf{X}$ for any distribution of $\mathbf{X}$ in P_1.

Consider the transformation

$$\mathbf{V} = (\mathbf{X} - \bar{\mathbf{x}}_1)\mathbf{K} \quad \text{where} \quad \mathbf{K}\mathbf{K}' = \Sigma_2^{-1} \quad \text{and} \quad (\bar{\mathbf{x}}_1 - \mu_2)\mathbf{K} \propto \mathbf{1}.$$

The distribution of $\mathbf{V}$ is exchangeable in P_2 and its mean difference in P_1 and P_2 is proportional to $\mathbf{1}$ (given $\mathbf{x}_1$). The sample values of $\mathbf{V}$ are $\mathbf{v}_i = (\mathbf{x}_i - \mathbf{1}'\bar{\mathbf{x}}_1)K$; hence, $\mathbf{x}_i = \mathbf{v}_i\mathbf{K}^{-1} + \mathbf{1}'\bar{\mathbf{x}}_1, (\bar{\mathbf{x}}_1 - \bar{\mathbf{x}}_2) = (\bar{\mathbf{v}}_1 - \bar{\mathbf{v}}_2)\mathbf{K}^{-1}, \mathbf{S}_2 = \mathbf{K}^{-1'}T_2\mathbf{K}^{-1}$, where T_2 is the sample covariance of $\mathbf{V}$ in P_2. Thus

$$\mathbf{S}_2^{-1}(\bar{\mathbf{x}}_1 - \bar{\mathbf{x}}_2)' = \mathbf{K}T_2^{-1}\mathbf{K}'\mathbf{K}^{-1'}(\bar{\mathbf{v}}_1 - \bar{\mathbf{v}}_2)' = \mathbf{K}T_2^{-1}(\bar{\mathbf{v}}_1 - \bar{\mathbf{v}}_2)'. \tag{B.1}$$

If the univariate matching on $\mathbf{X}\mathbf{S}_2^{-1}(\bar{\mathbf{x}}_1 - \bar{\mathbf{x}}_2)'$ is translation invariant, it is thus identical to univariate matching on

$$(\mathbf{X} - \mathbf{1}'\bar{\mathbf{x}}_1)\mathbf{S}_2^{-1}(\bar{\mathbf{x}}_1 - \bar{\mathbf{x}}_2)' = \mathbf{V}T_2^{-1}(\bar{\mathbf{v}}_1 - \bar{\mathbf{v}}_2)'. \tag{B.2}$$

Now the inverse (or if $\mathbf{T}_2$ has rank $< p$, the Moore–Penrose inverse) is an exchangeable function of the variables. Thus given $\mathbf{x}_1$, using (B.2) for matching is a univariate exchangeable matching method based on the variables $\mathbf{V}$. Also given $\mathbf{x}_1$, the distribution of $\mathbf{V}$ is exchangeable in P_2. Hence any matching using (B.2) that does not subsample G_1 is EPBR for $\mathbf{V}$ (by Case 10), and thus EPBR for $\mathbf{X}$, since given $\mathbf{x}_1$, $\mathbf{X}$ is a fixed linear combination of $\mathbf{V}$. Since given $\mathbf{x}_1$ matching on (B.2) is EPBR for $\mathbf{X}$, it must be unconditionally EPBR for $\mathbf{X}$ as well.

APPENDIX C

The proof of Theorem 2 follows the lines of Theorem 1 but is simpler. Consider $\mathbf{Z} = (\mathbf{X} - \mu_1)\mathbf{G}, \mathbf{G}\mathbf{G}' = \Sigma_2^{-1}, (\mu_1 - \mu_2)\mathbf{G} \propto \mathbf{1}$. Then matching using (6.6) to define distance for $\mathbf{X}$ is identical to using (C.1) to define distance for the $\mathbf{Z}$:

$$\left[\mathbf{T}^{-1} - \frac{1}{t^2}\mathbf{T}^{-1}(\bar{\mathbf{z}}_1 - \bar{\mathbf{z}}_2)'(\bar{\mathbf{z}}_1 - \bar{\mathbf{z}}_2)\mathbf{T}^{-1} \right] + k[\mathbf{T}^{-1}(\bar{\mathbf{z}}_1 - \bar{\mathbf{z}}_2)'(\bar{\mathbf{z}}_1 - \bar{\mathbf{z}}_2)\mathbf{T}^{-1}], \tag{C.1}$$

where $\mathbf{T} = \lambda_1\mathbf{T}_1 + \lambda_2\mathbf{T}_2, \mathbf{T}_i$ the sample covariance of $\mathbf{Z}$ in G_i, and $t_2 = (\bar{\mathbf{z}}_1 - \bar{\mathbf{z}}_2)\mathbf{T}^{-1}(\bar{\mathbf{z}}_1 - \bar{\mathbf{z}}_2)' = s^2$. Since metric matching using (C.1) as inner product for $\mathbf{Z}$ is an exchangeable matching method (by symmetry), Case 11 implies that the matching is EPBR for $\mathbf{Z}$ and $\mathbf{X}$ since $\mathbf{X}$ is a fixed linear combination of the $\mathbf{Z}$. Hence, matching using (6.6) to define distance for $\mathbf{X}$ is EPBR for $\mathbf{X}$.

7. Multivariate Matching Methods That Are Equal Percent Bias Reducing, II: Maximums on Bias Reduction for Fixed Sample Sizes

Donald B. Rubin

Abstract: Matched sampling is a method of data collection designed to reduce bias and variability due to specific matching variables. Although often used to control for bias in studies in which randomization is practically impossible, there is virtually no statistical literature devoted to investigating the ability of matched sampling to control bias in the common case of many matching variables. An obvious problem in studying the multivariate matching situation is the variety of sampling plans, underlying distributions, and intuitively reasonable matching methods. This article considers one class of multivariate matching methods which yield the same percent reduction in expected bias for each of the matching variables. The primary result is the derivation of the expression for the maximum attainable percent reduction in bias given fixed distributions and fixed sample sizes. An examination of trends in this maximum leads to a procedure for estimating minimum ratios of sample sizes needed to obtain well-matched samples.

1. INTRODUCTION

This introduction is brief; the reader is referred to Rubin [1976b] for a more detailed explanation of the notation and terminology.

Let G_1 and G_2 be two random samples of sizes N_1 and N_2 from two populations, P_1 and P_2. Matched sampling is an attempt to find subsamples of G_1 and G_2, G_{1*} and G_{2*} of sizes N_{1*} and N_{2*}, such that the distributions of the p matching variables $\mathbf{X}$ are more similar in G_{1*} and G_{2*} than in G_1 and G_2. Let $\mathbf{X}$ have mean μ_i and covariance Σ_i in P_i, $i = 1, 2$. We also use the notation μ_{i*} to represent the expected mean vector of $\mathbf{X}$ in matched G_{i*} samples, $i = 1, 2$. All vectors are row vectors.

We consider a particular class of matching methods that are called "equal percent bias reducing" (EPBR) because they yield the same percent reduction in bias for each matching variable, i.e., $(\mu_1 - \mu_2)$ is proportional to $(\mu_{1*} - \mu_{2*})$. Some EPBR matching methods have been previously presented in Cochran and Rubin [1973] and a more detailed list is given in Rubin [1976b].

The basic reason for studying EPBR matching rules is that they are the easiest multivariate matching rules to evaluate when the dependent variables can be any linear functions of $\mathbf{X}$ since there is only one particular percent reduction in bias of interest. Even though nonlinear functions of $\mathbf{X}$ deserve study (e.g., in the univariate case see Rubin [1973b]), it seems reasonable to begin study of multivariate

Reprinted with permission from *Biometrics*, 1976, **32**, 121–132. © 1976 International Biometric Society.

matching methods in the simpler linear case and then extend that work to the more complex nonlinear case. In this sense then, EPBR matching methods are the simplest multivariate starting point.

Section 2 derives the expression for θ_{max}, the maximum percent reduction in bias due to any EPBR matching method given fixed distributions and fixed sample sizes. Section 3 shows that the "optimal" percent reduction in bias, min $\{100, \theta_{max}\}$, can be attained by an EPBR matching method in some cases. Section 4 points out five trends in this maximum and derives a closed form approximation for it that appears to hold quite well for a variety of distributions. Section 5 indicates how the formula for the maximum can be used to suggest minimum sample sizes for non-randomized studies.

2. THE MAXIMUM PERCENT REDUCTION IN BIAS GIVEN FIXED DISTRIBUTIONS AND FIXED SAMPLE SIZES

Given a matching method that is EPBR for $\mathbf{X}$, an obvious measure of how much more similar the distributions of $\mathbf{X}$ are in matched G_{1*} and G_{2*} samples than in random G_1 and G_2 samples is the percent reduction in bias, which is

$$100[1 - (\mu_{1*} - \mu_{2*})\beta'/(\mu_1 - \mu_2)\beta'] \tag{7.1}$$

for any vector β. This ratio will obviously depend on

(a) the distributions of $\mathbf{X}$ in P_1 and P_2,
(b) the particular matching method,
(c) the sizes of the random samples G_1 and G_2, $N_1 = r_1 N_{1*}$, $N_2 = r_2 N_{2*}$, ratios $r_1, r_2 \geq 1$ and
(d) the sizes of the matched samples G_{1*} and G_{2*}, N_{1*}, N_{2*}.

The following theorem gives the maximum percent reduction in bias for fixed distributions and fixed sample sizes and any EPBR matching method. The "optimal" percent reduction in bias, min $\{100, \theta_{max}\}$, can be attained in some cases. The result is important because it suggests trends in the ability to attain well-matched samples and also suggests minimum ratios of sample sizes needed in order to attain any given percent reduction in bias.

Theorem. *Given*

(a) *fixed distributions of* $\mathbf{X}$ *in* P_1 *and* P_2 *with mean vectors* μ_1 *and* μ_2 *and covariance matrices* Σ_1 *and* Σ_2,
(b) *fixed sample sizes of* G_1 *and* G_2, $N_1 = r_1 N_{1*}$ *and* $N_2 = r_2 N_{2*}$, $r_1 \geq 1, r_2 > 1$, *and*
(c) *fixed sizes of* G_{1*} *and* G_{2*}, N_{1*} *and* N_{2*},

then the maximum (most positive) percent reduction in bias for any matching method that is EPBR for $\mathbf{X}$ *is*

$$\theta_{max} = \frac{100}{B\sqrt{(1 + \sigma_1^2/\sigma_2^2)/2}} \left[\Omega_2{}^+(r_2, N_{2*}) - \frac{\sigma_1}{\sigma_2}\Omega_1{}^-(r_1, N_{1*}) \right], \tag{7.2}$$

$$\sigma_i^2 = \Delta\Sigma_i\Delta', \text{ the variance of the best linear discriminant with respect}$$
to the P_2 inner product in P_i, $\Delta = (\mu_1 - \mu_2)\Sigma_2^{-1}$,

$$B = (\eta_1 - \eta_2)/\sqrt{(\sigma_1^2 + \sigma_2^2)/2}, \text{ the number of "standard deviations"}$$
between the means of $\mathbf{X}\Delta'$ in P_1 and P_2, $\eta_i = \mu_i\Delta'$,

$$\Omega_2^+(r_2, N_{2*}) = \text{ the expectation of the sample average of the } N_{2*} \text{ largest of}$$
$r_2 N_{2*}$ randomly chosen observations from F_2, where F_2 is the distribution of $\mathbf{X}\Delta'$ in P_2 normed to have zero mean and unit variance, i.e., the distribution of $(\mathbf{X} - \mu_2)\Delta'/\sigma_2$ in P_2, and

$$\Omega_1^-(r_1, N_{1*}) = \text{ the expectation of the sample average of the } N_{1*} \text{ smallest of } r_1 N_{1*}$$
randomly chosen observations from F_1, F_1 the distribution of $(\mathbf{X} - \mu)\Delta'/\sigma_1$ in P_1.

Although the notation appears complex, the proof is very straightforward and helps to explain the notation.

For any EPBR matching method, the percent reduction in bias equals the percent reduction in bias of the P_2 best linear discriminant $\mathbf{X}\Delta'$ since it is a fixed linear combination of $\mathbf{X}$; thus (7.1) for an EPBR matching method may be written as

$$100[1 - (\eta_{1*} - \eta_{2*})/(\eta_1 - \eta_2)] \tag{7.3}$$

where $\eta_{i*} = \mu_{i*}\Delta'$, the mean of $\mathbf{X}\Delta'$ in matched G_{i*} samples, $i = 1, 2$. The choice of $\mathbf{X}\Delta'$ rather than some other linear combination of $\mathbf{X}$ is related to the existence of cases discussed in Section 3 in which matching on this discriminant is EPBR for $\mathbf{X}$ and attains min $\{100, \theta_{max}\}$ percent reduction in bias in large samples.

The mean of $\mathbf{X}\Delta'$ is greater in P_1 than in P_2 because $\eta_1 - \eta_2 = (\mu_1 - \mu_2)\Sigma_2^{-1}(\mu_1 - \mu_2) > 0$. For fixed samples one then maximizes the percent reduction in bias of $\mathbf{X}\Delta'$ by choosing the N_{1*} units from G_1 with the most negative $\mathbf{X}\Delta'$ values to comprise G_{1*} and the N_{2*} units from G_2 with the most positive $\mathbf{X}\Delta'$ values to comprise G_{2*}. Thus, the maximum percent reduction in bias (most positive, not necessarily optimal in the sense of closest to 100) is attained with

$$\eta_{1*} = \eta_1 + \sigma_1\Omega_1^-(r_1, N_{1*})$$
$$\eta_{2*} = \eta_2 + \sigma_2\Omega_2^+(r_2, N_{2*})$$

using the notation given in the statement of the theorem. Hence the maximum possible percent reduction in bias is from (7.3)

$$\frac{\sigma_2\Omega_2^+(r_2, N_{2*}) - \sigma_1\Omega_1^-(r_1, N_{1*})}{\eta_1 - \eta_2},$$

which equals (7.2) upon substitution of B.

3. ATTAINING THE "OPTIMAL" PERCENT REDUCTION IN BIAS, MIN $\{100, \theta \text{ MAX}\}$

The bound given by $\theta_{\max}$ is not very interesting unless it can be attained under some conditions by some matching method that is EPBR for **X** (other than the trivial case when **X** is univariate). Actually an "optimal" EPBR matching method would attain min $\{100, \theta_{\max}\}$ percent reduction in bias rather than $\theta_{\max}$ because a percent reduction in bias greater than 100 indicates that $(\mu_{1*} - \mu_{2*})$ has a different sign than $(\mu_1 - \mu_2)$ which is not as desirable as $\mu_{1*} \cong \mu_{2*}$. Hence, we now show that min $\{100, \theta_{\max}\}$ percent reduction in bias can be attained by an EPBR matching method in some cases.

First let $r_1 = 1$ $(G_1 = G_{1*})$. Consider the case in which all mean bias is in X^1 and $(X^2, \ldots, X^p)$ are independent of X^1 in P_2. X^1 is thus the best linear discriminant with respect to the P_2 inner product. The following matching method is EPBR for **X**: $G_1 = G_{1*}$ $(r_1 = 1)$ and the N_{2*} sized subsample of G_2 whose mean on X^1 is closest to the mean of X^1 in G_1, $\bar{x}_1$, is chosen to be G_{2*}. (See Case 1, Rubin [1976b]). In large samples this method attains min $\{100, \theta_{\max}\}$ percent reduction in bias. This follows when $\theta_{\max} \leq 100$ because G_{2*} will consist of the $1/r_2$ portion of G_2 with X^1 values closest to $\bar{x}_1{}^1$ (large samples imply negligible variance in sample means). If $\theta_{\max} > 100$ and the samples are large we now show that there is a subsample of G_2 whose mean is $\bar{x}_1{}^1$ and thus the percent reduction in bias is 100 percent for the mean matching method. Order the G_2 sample on X^1 and then divide G_2 into the top $1/r_2$ fraction and the bottom $(1 - 1/r_2)$ fraction. Consider taking a random sample from each part of G_2 with sizes $(N_2/r_2)w$ for the top and $(N_2/r_2)(1 - w)$ for the bottom, $1 \geq w \geq 0$. $w = 1/r_2$ gives 0 percent reduction in bias while $w = 1$ gives $\theta_{\max} > 100$ percent reduction in bias. Clearly, by adjusting the sampling fraction w we can obtain any percent reduction in bias between 0 and $\theta_{\max}$, e.g., 100. Hence in large samples with $\theta_{\max} > 100$ choosing the N_2/r_2 sized subsample of G_2 whose mean is closest to $\bar{x}_1{}^1$ gives 100 percent reduction in bias.

Also consider the case in which **X** is ellipsoidally distributed in P_2 (i.e., there exists a linear transformation to a spherical distribution, see Dempster [1969]). Letting $r_1 = 1$ and matching the means on the estimated best linear discriminant with respect to the G_2 inner product is EPBR (see Theorem 1, Rubin [1976b]) and in large samples it yields the optimal min $\{100, \theta_{\max}\}$ percent reduction in bias. This follows because the matching method is equivalent to mean matching on the population best linear discriminant, and by the same argument as above, this matching will yield min $\{100, \theta_{\max}\}$ percent reduction in bias.

Now let both r_1 and r_2 be greater than 1 so that both G_1 and G_2 are to be subsampled. Consider the case in which all mean bias is in X^1 and $(X^2, \ldots, X^v)$ are independent of X^1 in both P_1 and P_2. The following matching method is EPBR for X: Choose for G_{1*} and G_{2*} the N_{1*} and N_{2*} sized subsets of G_1 and G_2 whose means are closest to each other on X^1 (see Case 2, Rubin [1976b]). In large samples this matching method attains min $\{100, \theta_{\max}\}$ percent reduction in bias as can be proven by an argument analogous to that given above when $r_1 = 1$ and all mean bias is in X^1.

Also, consider the case in which **X** is normal in both P_1 and P_2 and the best linear discriminants with respect to the P_1 and P_2 inner products are proportional.

In large samples there is a known linear transformation to the previous case in which the first variable is the common best linear discriminant. (See Case 6, Rubin [1976b]). Hence in large samples matching the means on the estimated discriminant is EPBR and attains min $\{100, \theta_{max}\}$ percent reduction in bias. Also, mean matching on the estimated best linear discriminant is EPBR for ellipsoidal **X** even when the samples are small if the covariance matrices of **X** orthogonal to the discriminant are proportional, implying a common best linear discriminant. (See Theorem 2 and Appendix A, Rubin [1976b]). Hence, under these conditions, means matching on the estimated discriminant is EPBR and attains min $\{100, \theta_{max}\}$ percent reduction in bias in large samples.

4. TRENDS IN THE MAXIMUM PERCENT REDUCTION IN BIAS

The above theorem is important not only because it allows one to calculate a theoretical maximum which we have shown can be attained but also because it demonstrates trends that shed light on how difficult it is to obtain well matched samples.

One trend is obvious from (7.2) and is rather intuitive:

(1) For fixed sample sizes and fixed σ_1^2/σ_2^2 and F_1 and F_2 (i.e., fixed distributional form), θ_{max} is linearly increasing in B^{-1}, i.e., everything else being equal, the farther apart the means of **X** in the populations, the more difficult it is to obtain well-matched samples.

Another trend follows from the intuitive fact that $\Omega_2{}^+(r, N)$ is increasing ($\Omega_1{}^-(r, N)$ is decreasing) in r for fixed N. (For proof see Appendix A, Rubin [1973a]).

(2) For fixed distributions and fixed N_{1*}, N_{2*} and r_i ($i = 1$ or 2), θ_{max} is increasing in r_j ($j = 2$ or 1), i.e., the larger the relative sizes of the pools from which to obtain matches, the better the matches will be.

A third trend specifies further how θ_{max} depends on r_1 and r_2. It follows from the result proved in Appendix A that the rate of increase of $\Omega_2{}^+(r, N)$ with respect to r is decreasing in r; that is, for $\delta = 1/N$, N and rN integers:

$$\Omega_2{}^+(r + 2\delta, N) - \Omega_2{}^+(r + \delta, N) \le \Omega_2{}^+(r + \delta, N) - \Omega_2{}^+(r, N).$$

Hence, the third trend in θ_{max}.

(3) For fixed distributions and fixed N_{1*}, N_{2*} and r_i ($i = 1$ or 2), the rate of increase of θ_{max} with respect to r_j ($j = 2$ or 1) is decreasing in r_i, i.e., one gains less and less from successive fixed increases in the ratios r_1 and r_2.

A fourth trend follows from the fact that $\Omega_2{}^+(r, N)$ is increasing in N for fixed r. (For proof see Appendix A, Rubin [1973a]).

(4) For fixed distributions and fixed r_1, r_2, and N_{i*} ($i = 1$ or 2), θ_{max} is increasing in N_{j*} ($j = 2$ or 1), i.e., the larger the samples the better the matches.

This fourth trend is important because it implies that θ_{max} is bounded by

$$\frac{100}{B\sqrt{(1 + \sigma_1^2/\sigma_2^2)/2}}\left[\Omega_2{}^+(r_2, \infty) - \frac{\sigma_1}{\sigma_2}\Omega_1{}^-(r_1, \infty)\right]. \tag{7.4}$$

Previous work (Rubin [1973a]) has suggested that even for small N, $\Omega_2{}^+(r, N)$ is very nearly $\Omega_2{}^+(r, \infty)$ for normal and non-normal distributions where $\Omega_2{}^+(r, \infty)$ is calculated as the expectation in the right $1/r$ tail of the standardized (zero mean, unit variance) distribution.

In order to elucidate further the second and third trends in θ_{max} (i.e., how rapidly θ_{max} increases as a function of r_1 and r_2), we will calculate some values of $\Omega_2{}^+(r, \infty)$ for a simple family of distributions. If $\mathbf{X}$ is ellipsoidal in P_i, $\mathbf{X}\Delta'$ is symmetrically distributed in P_i and thus $\Omega_2{}^+(r, \infty) = -\Omega_1{}^-(r, \infty)$. The family of "standardized" (zero means, unit variance) t-distributions with f degrees of freedom, $\sqrt{f - 2/f} t_f$, $f > 3$ ($t_\infty = $ normal), is a simple family of such univariate distributions.

In Fig. 7.1, graphs of $\Omega_2{}^+(r, \infty) = -\Omega_1{}^-(r, \infty)$ are presented for the normal and four other standardized t-distributions, $f = 3, 4, 5, 10$. The range of r from 1 to 100 is thought to cover all values likely to occur commonly in practice. Also plotted is an approximation for the normal given by

$$2^{\pi/2}(2\pi)^{-1/2} r^{(1-\pi/4)} \left(1 - \frac{1}{r}\right)^{\pi/4}. \tag{7.5}$$

See Appendix B for the derivation of this approximation and the numerical methods used to calculate the values for the other curves.

Notice first that Fig. 7.1 agrees with the implications of trend 2 (i.e., $d/dr\, \Omega_2{}^+(r, \infty) \geq 0$) and trend 3 (i.e., $d^2/dr^2 \Omega_2{}^+(r, \infty) \leq 0$). Also note that the function given by (7.5) is an excellent approximation to $\Omega_2{}^+(r, \infty)$ and that the simple function $(\pi/4) \ln (r)$ is a reasonable approximation for these distributions and this range of r.

If the matching methods are restricted to those with $G_{1*} = G_1$, θ_{max} simplifies to

$$\frac{100}{B} \frac{\Omega_2{}^+(r_2, N_{2*})}{\sqrt{\frac{1}{2}(1 + \sigma_1{}^2/\sigma_2{}^2)}}, \tag{7.6}$$

since $\Omega_1{}^-(r_{1*}, N_{1*}) = \Omega_1{}^-(1, N_1) = 0$. Hence if $r_1 = 1$, in addition to the trends in θ_{max} discussed above there is the fifth trend:

(5) For fixed sample sizes with $r_1 = 1$ and fixed B and distributional forms, θ_{max} is decreasing in $\sigma_1{}^2/\sigma_2{}^2$.

The advantages and disadvantages of subsampling G_1 in addition to G_2 are discussed in Section 5.

5. SUGGESTIONS FOR MINIMAL SAMPLE SIZES

The formula for θ_{max} and the associated trends discussed in Section 4 can be used to suggest minimum sample sizes for non-randomized studies.

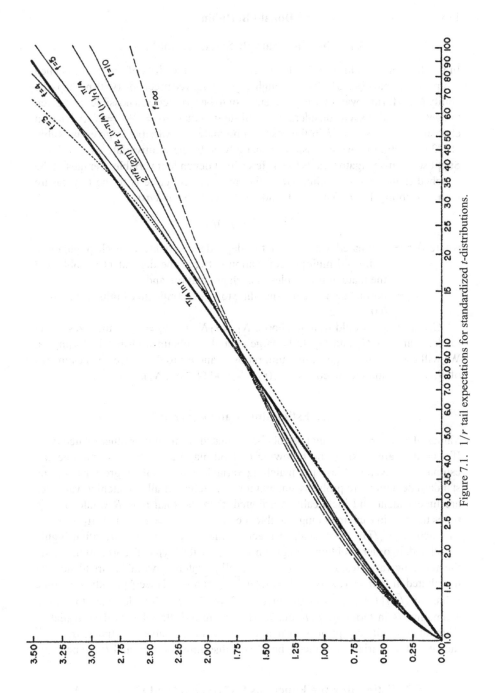

Figure 7.1. $1/r$ tail expectations for standardized t-distributions.

5.1. The Final Sample Sizes, N_{1*} and N_{2*}

In practice, one would want to choose N_{1*} and N_{2*}, the sizes of the final matched samples, G_{1*} and G_{2*}, to be large enough to give a desired precision (with perhaps the ability to perform covariance adjustments), or to be as precise as possible for a given budget. Generally, such considerations will suggest that $N_{1*} = N_{2*} = N$. Advice on choosing $N_{1*} = N_{2*} = N$ in the case of one matching variable is given in Rubin [1973a], Section 6. An extension of that advice to the multivariate case follows. Suppose the investigator wants the difference of dependent variable averages in the matched samples to be within $\pm d$ of the treatment effect he is trying to measure with probability $1 - \alpha$. Then N should be chosen to be *at least*

$$2\hat{\sigma}_y^2 z^2 (1 - \hat{\rho}^2)/d^2$$

where $\hat{\sigma}_y^2 =$ estimated variance of the dependent variable in each population,
$\hat{\rho} =$ estimated multiple correlation between the dependent variable and the matching variables in each population, and
$z =$ normal deviate corresponding to $1 - \alpha$ confidence limits (e.g., if $\alpha = 0.05$, $z \doteq 2$).

Given N, one would want to choose $N_1 = r_1 N$ and $N_2 = r_2 N$, the sizes of the random samples G_1 and G_2, to be large enough to obtain well-matched samples. We will use θ_{max} to suggest minimum ratios r_1 and r_2; to do so, we need estimates of B, σ_1^2/σ_2^2 and the functions $\Omega_2^+(r_2, N_2)$ and $\Omega_1^-(r_1, N_1)$.

5.2. Estimating B and σ_1^2/σ_2^2

Values of B and σ_1^2/σ_2^2 can possibly be estimated from pilot or other related data. If such data are lacking, the following method may be helpful. Assume one can estimate an "average bias" per matching variable, say b; values greater than one standard deviation are probably uncommon in practice. If all p matching variables are independent and identically distributed, then the total bias B would be $\sqrt{p}$ times the bias in each matching variable, i.e., if $\Sigma_1 = \Sigma_2 = \mathbf{I}_{p \times p}$ and $(\mu_1 - \mu_2) = b\mathbf{1}$, then $B = \sqrt{p}b$. Of course, in practice, matching variables are often highly correlated. In order to obtain a rough guide, let ρ be the "typical" correlation among the matching variables, $\rho > -1/(p - 1)$. If all p matching variables are identically distributed in P_i, $i = 1, 2$, then $B = b\sqrt{p/[1 + \rho(p - 1)]}$ (see Appendix C). Some values of $\sqrt{p/[1 + \rho(p - 1)]}$ are given in Table 7.1 which is designed to help in estimating B in terms of p, b and ρ when there is little relevant data available. Without contradictory prior information, it seems reasonable to estimate σ_1^2/σ_2^2, the ratio of the variances of the P_2 best linear discriminant in P_1 and P_2, to be unity.

5.3. Estimating the Functions $\Omega_2^+(r_2, N_{2*})$ and $\Omega_1^-(r_1, N_{1*})$

Section 4 suggested that the functions plotted in Fig. 7.1 may be used to approximate $\Omega_2^+(r_2, N_{2*})$ and $-\Omega_1^-(r_1, N_{1*})$. Also, the simple function $(\pi/4)\ln(r)$ seemed to give quite good approximations. Hence Fig. 7.1, or the function $(\pi/4)\ln(r)$, plus estimates of B and σ_1^2/σ_2^2 can be used to estimate the minimum r_1 and r_2 needed to obtain a given percent reduction in bias. Note that if we combine the approximation for B in terms of the number of matching variables p, the typical bias per matching

Table 7.1. *A rough guide for the ratio of the total bias, B, to the "average bias" per matching variable, b, for p positively correlated matching variables:* $B/b = \sqrt{p/[1 + \rho(p-1)]}$

Number of matching variables, p	Average correlation, ρ				
	0.0	0.2	0.4	0.6	0.8
2	1.41	1.29	1.20	1.12	1.05
3	1.73	1.46	1.29	1.17	1.07
4	2.00	1.58	1.35	1.20	1.08
6	2.45	1.73	1.41	1.22	1.10
8	2.83	1.83	1.45	1.24	1.10
10	3.16	1.89	1.47	1.25	1.10
20	4.47	2.04	1.53	1.27	1.11
50	7.07	2.15	1.56	1.28	1.11
100	10.00	2.19	1.57	1.29	1.11

variable b, and the typical correlation ρ, with the approximation $(\pi/4)\ln(r)$, and let $\tau^2 = \sigma_1^2/\sigma_2^2$ be the typical ratio of P_1 to P_2 variance for each matching variable, we have from (7.2) that $\theta_{\max}$ is approximately

$$\frac{25\pi}{b} \sqrt{\frac{1 + \rho(p-1)}{p(1+r^2)/2}} [\ln(r_2) + r \ln(r_1)].$$

This expression may be useful for obtaining preliminary estimates of sample size ratios. Sections 5.4 and 5.5 use estimates of B and σ_1^2/σ_2^2 plus Fig. 7.1 to obtain estimates of sample size ratios.

5.4. Estimating the Minimum r_2 Needed to Obtain a Given Percent Reduction in Bias When $r_1 = 1$

If $G_{1*} = G_1$ (i.e., $r_1 = 1$) only r_2 need be chosen. Actually this sampling situation is rather common in practice where P_1 is a population exposed to a treatment (cigarette smokers) and P_2 is an unexposed reservoir (nonsmokers). Maintaining G_1 as a random sample then has two simple advantages. First, as Rubin [1976b] demonstrates, EPBR matching methods with $G_1 = G_{1*}$ can be defined under less restrictive distributional conditions than EPBR matching methods with $r_1 > 1$. Second, even when there is an interaction between the treatment and matching variables (i.e., the effect of the treatment on the dependent variable varies with the values of $\mathbf{X}$, equivalently nonparallel regressions or response surfaces), effects estimated from matched samples with $r_1 = 1$ can be thought of as estimating the average effect for the P_1 population without further adjustment (see Rubin [1973a] or Cochran and Rubin [1973] for details).

As an example of determining a minimum ratio r_2 when $r_1 = 1$, assume we estimate σ_1^2/σ_2^2 as 1 and B as 1; then $B\sqrt{(1 + \sigma_1^2/\sigma_2^2)/2}$ is estimated as 1. From expression (7.6) and Fig. 7.1, $r_2 = 3$ should yield a maximum percent reduction in bias close to 100 percent. As a second example, estimate σ_1^2/σ_2^2 as 2 and B as 1.5; then $B\sqrt{(1 + \sigma_1^2/\sigma_2^2)/2}$ is estimated as 1.83 and from Fig. 7.1, $r_2 = 12$

Table 7.2. *Approximate ratio of sample sizes, r_2, needed to obtain a maximum percent reduction in bias close to 100 percent*[*]

r_1	σ_1^2/σ_2^2	Total bias, B					
		1/4	1/2	3/4	1	$1^1/_2$	2
	1/2	1.1	1.4	1.8	2.5	5.0	9.1
	2/3	1.1	1.4	2.0	2.7	5.6	12.0
$r_1 = 1$	1	1.2	1.5	2.2	3.0	6.8	15.0
	$1^1/_2$	1.2	1.6	2.4	3.5	8.5	23.0
	2	1.2	1.7	2.7	4.5	12.0	35.0
$r_1 = r_2$	Any value between 1/2 and 2	1.1	1.2	1.3	1.5	2.2	3.0

[*] As suggested by the curves in Figure 7.1.

should yield a maximum percent reduction in bias close to 100 percent. Table 7.2 summarizes some values of r_2 needed to obtain a maximum percent reduction in bias close to 100 percent when $r_1 = 1$.

5.5. Estimating the Minimum r_1 and r_2 Needed to Obtain a Given Percent Reduction in Bias

If both G_1 and G_2 can be large it seems reasonable to try to subsample both G_1 and G_2 because the third trend as well as intuition suggest that it may be easier to obtain well-matched samples if neither G_{1*} nor G_{2*} is a random sample. If the regressions of the dependent variable on **X** are not parallel in P_1 and P_2, and neither G_{1*} nor G_{2*} is a random sample, results from the matched samples cannot be immediately generalized to P_1 (or P_2). However, the average effect for P_1 can be estimated from (a) an estimate of the distribution of **X** in P_1 obtained from G_1, (b) an estimate of the regression of Y on **X** in P_1 obtained from G_{1*} and (c) an estimate of the regression of Y on **X** in P_2 obtained from G_{2*}. This procedure, of course, relies on the linearity of the regressions of the dependent variable on **X**, which may not be desirable.

Suppose that both matched samples will be of the same size $N = N_{1*} = N_{2*}$ and that the total number of units sampled, $N_1 + N_2$, is fixed, i.e., $r_1 + r_2 = r_0$ a constant ≥ 2. This case is appropriate if units in P_1 and P_2 are equally expensive to sample, the budget allows $(N_1 + N_2) = r_0 N$ units to be sampled and $2N$ units are to be subsampled and studied. The problem then becomes to find r_1, r_2 such that θ_{max} is maximized. Obviously the exact choices will depend not only on $N, r_0, B, \sigma_1^2/\sigma_2^2$ but also on the functional forms of $\Omega_2^+(r_2, N)$ and $\Omega_1^-(r_1, N)$ (i.e., the distributional forms of the best linear discriminant in P_1 and P_2). If $\sigma_1^2 = \sigma_2^2$ and $\Omega_2^+(r, N) = -\Omega_1^-(r, N)$, the second and third trends imply that letting $r_1 = r_2$ (i.e., $N_1 = N_2$) maximizes θ_{max} for fixed $r_1 + r_2$. In order to maximize θ_{max} for fixed $r_1 + r_2$ when $\Omega_2^+(r, N) = -\Omega_1^-(r, N)$ and $\sigma_1^2 > \sigma_2^2$ we should have $r_1 \geq r_2$ and vice versa when $\sigma_1^2 < \sigma_2^2$. Using the approximation $\Omega_2^+(r, N) = -\Omega_1^-(r, N) = \pi/4 \ln(r)$, the ratio of r_1/r_2 that maximizes θ_{max} for fixed $r_1 + r_2 = r_0$ is found to be between σ_1/σ_2 and 1.

Hence, without knowledge that $\sigma_1{}^2$ and $\sigma_2{}^2$ are quite different, $r_1 = r_2 = r$ is a reasonable choice. In this case with $\Omega_2{}^+(r, N) = -\Omega_1{}^-(r, N)$, θ_{max} in equation (7.2) becomes

$$\frac{100\,(1 + \sigma_1/\sigma_2)}{B\sqrt{(1 + \sigma_1{}^2/\sigma_2{}^2)/2}}\Omega_2{}^+(r, N).$$

Therefore, determining minimum ratios $r_1 = r_2 = r$ is no more difficult than when $r_1 = 1$. For example, as in Section 5.3, estimating B as 1 and $\sigma_1{}^2/\sigma_2{}^2$ as 1 yields 0.5 as the estimate of $B\sqrt{(1 + \sigma_1^2/\sigma_2^2)/2}/(1 + \sigma_1/\sigma_2)$. Then from Fig. 7.1, $r_1 = r_2 = 1.5$ should give a maximum percent reduction in bias close to 100 percent. Thus in this case subsampling both G_1 and G_2 has led to a saving of approximately 25 percent in $r_1 + r_2$ (i.e., $1+3$ vs. $1.5 + 1.5$). For a second example, as in Section 5.3, estimate σ_1^2/σ_2^2 as 2 and B as 1.5; and so estimate $B\sqrt{(1 + \sigma_1{}^2/\sigma_2{}^2)/2}/(1 + \sigma_1/\sigma_2)$ as 0.76. From Fig. 7.1, $r_1 = r_2 = 2.2$ should yield a percent reduction in bias close to 100 percent. Hence, in this case subsampling both G_1 and G_2 has resulted in a savings of about 63 percent in sample size. Table 7.2 summarizes some values of $r_1 = r_2$ needed to obtain a maximum percent reduction in bias close to 100 percent.

This advice on choosing sample sizes is only a beginning. Nothing has been said about which matching methods should be used or whether regression adjustments should be made on the matched samples. The issues of robustness to nonlinearity and nonparallelism of regression and nonnormality of distributions are central to such advice. These problems are important, complex and apparently require a great deal of Monte-Carlo study.

ACKNOWLEDGMENTS

I would like to thank W. G. Cochran and T. W. F. Stroud for helpful comments on this paper, and D. H. Saxe and R. Parris for insightful discussions relevant to some of the results proved in the appendix.

APPENDIX A

Claim: $\Omega^+(r + 2\delta, N) - \Omega^+(r + \delta, N) \le \Omega^+(r + \delta, N) - \Omega^+(r, N)$
    where $\delta = 1/N$
    N, rN positive integers
    $\Omega^+(a, b) = $ the expectation of the sample average of the largest b of a random sample of size ab from a specified distribution.
The proof follows from first taking any sample of size $rN + 2$ and showing that $E_2 - E_1 \le E_1 - E_0$, where E_2 is the average of the N largest, E_1 is the expectation of the average of the N largest from all $rN + 1$ sized subsets of the $rN + 2$, and E_0 is the expectation of the average of the N largest from all rN sized subsets. Having proved this conditionally for any sample of size $rN + 2$, the result follows for the unconditional expectation.

Let $X(i)$, $i = 1, \ldots, rN + 2$ be the ordered values, with $X(1) = $ the largest.

$$E_2 = \frac{1}{N} \sum_{i=1}^{N} X(i).$$

Now calculate E_1 by finding the number of $rN + 1$ sized subsets in which $X(i)$ is one of the N largest.

There are $\binom{rN+2}{rN+1}$ subsamples of size $rN + 1$.

In $\binom{rN+1}{rN}$ of these, $X(i)$, $i = 1, \ldots, N$ is one of the N largest since if it's in the sample it's one of the N largest. In N of these subsamples $X(N + 1)$ is one of the N largest since if any one of the top N is excluded $X(N + 1)$ is one of the N largest. If none of these sub-samples will $X(i) i > N + 1$ be one of the N largest. Hence,

$$E_1 = \frac{1}{N}\left[\sum_{i=1}^{N}\left(\frac{rN+1}{rN+2}\right)X(i) + \left(\frac{N}{rN+2}\right)X(N+1)\right].$$

Now calculate E_0. There are $\binom{rN+2}{rN}$ subsamples of size rN. In $\binom{rN+1}{rN-1}$ subsamples $X(i) i = 1, \ldots, N$ is one of the N largest since if it's in the sample, it's one of the N largest. In $N(rN + 2 - 1 - N) + \binom{N}{2}$ subsamples $X(N + 1)$ is one of the N largest since this happens if exactly one of the top N and one of the bottom $rN + 2 - 1 - N$ are not chosen or if exactly two of the top N are not chosen. In $\binom{N+1}{2}$ subsamples $X(N + 2)$ is one of the N largest since if two of the top $N + 1$ are not chosen $X(N + 2)$ is one of the N largest. In no subsamples is $X(i) i > N + 2$ one of the N largest.

Hence,

$$E_0 = \frac{1}{N}\left[\sum_{i=1}^{N}\left(\frac{rN}{rN+2}\right)X(i) + N\frac{2rN - N + 1}{(rN+2)(rN+1)}X(N+1)\right.$$
$$\left. + N\frac{(N+1)}{(rN+1)(rN+2)}X(N+2)\right].$$

Now if $E_2 - 2E_1 + E_0 \le 0$, then $E_2 - E_1 \le E_1 - E_0 \le 0$ and we are done.

$$E_2 - 2E_1 + E_0 = \frac{1}{N}\sum_{i=1}^{N}\left[1 - \frac{2(rN+1)}{rN+2} + \frac{rN}{rN+2}\right]X(i)$$
$$+ \left[-\frac{2}{rN+2} + \frac{2rN - N + 1}{(rN+2)(rN+1)}\right]X(N+1)$$
$$+ \frac{(N+1)}{(rN+1)(rN+2)}X(N+2)$$
$$= \frac{-(N+1)}{(rN+2)(rN+1)}[X(N+1) - X(N+2)] \le 0.$$

APPENDIX B

Letting w be the $(1 - 1/r)$ quantile, simple calculus shows that the expectation in the right $1/r$ tail of the standardized t-distribution with f degrees of freedom is

$$r\frac{\Gamma(f/2 + 1/2)}{\Gamma(f/2)\sqrt{\pi}}\frac{\sqrt{f-2}}{f-1}\left(1 + \frac{w^2}{f}\right)^{-1/2(f-1)} \qquad f \ge 3.$$

The corresponding function for the normal density is

$$r(2\pi)^{-1/2} \exp\left(-\frac{1}{2}w^2\right). \tag{B.1}$$

Of course, in both these expressions, w itself is a function of r. Given a value of r, the value of w can be approximated for the normal from tables of normal integrals or functions given by Johnson and Kotz ([1971a], pp. 53–57) while values for the standardized t-distributions can be found from the normal values and functional approximations such as those given in Johnson and Kotz ([1971b], p. 102). A simple approximation to the tail expectation of the normal can be generated by approximating the integral in

$$\frac{1}{r} = \int_w^\infty (2\pi)^{-1/2} \exp\left(-\frac{1}{2}x^2\right) dx \text{ as}$$

$$\frac{1}{2}\{1 \pm [1 - \exp(-2w^2/\pi)]^{1/2}\}, \ + \text{ if } r \leq 2, \ - \text{ if } r > 2.$$

(See Johnson and Kotz [1971b], p. 57). Solving for $\exp\left(-\frac{1}{2}w^2\right)$ and substituting in (B.1) yields the approximation in (7.5).

APPENDIX C

The Ratio to Total Bias to Average Bias

If all p matching variables are identically distributed

$$(\mu_1 - \mu_2) = a\mathbf{1}$$

$$k_1^{-1}\Sigma_1 = k_2^{-1}\Sigma_2 = [\mathbf{I} + \rho(\mathbf{1'1} - \mathbf{I})].$$

Then $\Delta = q\mathbf{1}$ for $q = ak_2/[1 + \rho(p - 1)]$

$$\sigma_i^2 = q^2 k_i p[1 + \rho(p - 1)]$$

$$B = qpa/\sqrt{(\sigma_1^2 + \sigma_2^2)/2}$$

$$b = \text{bias per matching variable} = a/\sqrt{(k_1 + k_2)/2}$$

Hence $B/b = \sqrt{p/[1 + \rho(p - 1)]}$.

8. Using Multivariate Matched Sampling and Regression Adjustment to Control Bias in Observational Studies

Donald B. Rubin

Abstract: Monte Carlo methods are used to study the efficacy of multivariate matched sampling and regression adjustment for controlling bias due to specific matching variables **X** when dependent variables are moderately nonlinear in **X**. The general conclusion is that nearest available Mahalanobis metric matching in combination with regression adjustment on matched pair differences is a highly effective plan for controlling bias due to **X**.

1. INTRODUCTION

Our objective is to study the utility of matched sampling and regression adjustment (covariance adjustment) for controlling specific matching variables in observational studies. This introduction is brief; we assume that the reader is familiar with the literature on matching and covariance adjustment in observational studies (e.g., Althauser and Rubin 1970; Billewicz 1964, 1965; Campbell and Erlebacher 1970; Cochran 1953a, 1968; Cochran and Rubin 1973; Gilbert, Light, and Mosteller 1975; Greenberg 1953; Lord 1960; McKinlay 1974, 1975a,b; and Rubin 1974, 1977a, 1978a). In particular, this work is a natural extension of earlier Monte Carlo work on one matching variable (Rubin 1973a,b) and theoretical work on multivariate matching methods (Rubin 1976b,c.)

Matched sampling refers to the selection of treatment units (e.g., smokers) and control units (e.g., nonsmokers) that have similar values of matching variables, **X** (e.g., age, weight), whereas regression adjustment refers to a statistical procedure that adjusts estimates of the treatment effects by estimating the relationship between the dependent variable Y (e.g., blood pressure) and **X** in each treatment group. Hence, matched sampling and regression adjustment may be used alone or in combination, that is, samples may be random or matched, and regression adjustment may or may not be performed. Our use of the term *matching* excludes methods that discard units with Y recorded; thus, our matching methods should be thought of as choosing units on which to record Y when Y can be recorded only on a limited number of units (e.g., Y is obtained by an expensive medical examination in the matched samples).

A major problem with matching methods is that in practice it is rare that enough matched pairs of treatment and control units with identical values of $\mathbf{X}$ can be found, and then the matching does not perfectly control for $\mathbf{X}$. A major problem with regression adjustment is that the linear model relating Y to $\mathbf{X}$ may be wrong, and then the adjustment being applied may not be entirely appropriate. We study cases with imperfect matches and Y moderately nonlinear in $\mathbf{X}$.

Cochran and Rubin (1973) summarize work on the efficacy of univariate matching and regression adjustment with quantitative Y and X. The general conclusions of these univariate investigations are that (a) a very simple and easy-to-use pair-matching method known as nearest available pair matching (order the treatment units and sequentially choose as a match for each treatment unit the nearest unmatched control unit) seems to be an excellent matching method; and (b) the combination of regression adjustment on matched samples is usually superior to either method alone.

We extend this work with quantitative Y and X to the case of bivariate $\mathbf{X}$. The two main questions to be addressed are (a) Which of two multivariate nearest available pair-matching methods (discriminant, Mahalanobis metric) is preferable? and (b) Which of three regression adjustments (no adjustment, pooled estimate, estimate based on matched pair differences) is preferable? Section 2 introduces terminology and notation, and Section 3 defines the conditions of our Monte Carlo study. Section 4 presents results on the ability of regression adjustment to control bias in random samples. Section 5 presents results for matched samples, with and without regression adjustment. The broad conclusion is that nearest available Mahalanobis metric pair-matching coupled with regression adjustment on the matched pairs is a quite effective general plan for controlling the bias due to matching variables, and this combination is clearly superior to regression adjustment on random samples.

2. TERMINOLOGY AND NOTATION

Let P_1 be a population of treatment units and let P_2 be a population of control units. Random samples are obtained from P_1 and P_2; these samples, G_1 and G_2, consist of N and rN units ($r \geq 1$) with recorded values of the matching variables, $\mathbf{X}$. Matched samples are created by assigning to each G_1 unit a G_2 unit having similar values of $\mathbf{X}$; the algorithm used to make the assignments is the matching method. The dependent variable Y is then recorded on all $2N$ units in the matched samples, and the effect of the treatment is estimated. Regression adjustments may be performed by fitting a linear model to the conditional expectation of Y given $\mathbf{X}$. These regressions are estimated from the matched samples and not the random samples because Y is only recorded in the matched samples. Of course, if $r = 1$, the matched samples are simply random samples of size N with pairing of G_1 and G_2 units.

2.1. Matching Methods to be Studied

We will study two matching methods: nearest available pair matching on the estimated best linear discriminant and nearest available pair matching using the Mahalanobis metric to define distance. Nearest available pair-matching methods first order the G_1 units and then have each G_1 unit choose in turn the closest match from the yet unmatched G_2 units; that is, the first G_1 unit selects the closest G_2 unit, the second G_1 unit selects the closest G_2 unit from the $rN - 1$ not yet matched, and so on, until all G_1 units are matched. These matching methods are fully defined once we specify the order for matching the G_1 units and the precise meaning of closest. Since previous univariate work (Rubin 1973a) indicated that random ordering is usually satisfactory, we will study random order, nearest available matching methods. Closest is clearly defined for one matching variable but not for more than one.

Let x_i be the $N_i \times p$ data matrix of X in G_i (where $N_1 = N$, $N_2 = rN$), let $\bar{x}_i$ be the $1 \times p$ sample mean vector in G_i, and let

$$S = \left[\left(x_1{}^T x_1 - N_1 \bar{x}_1{}^T \bar{x}_1 \right) + \left(x_2{}^T x_2 - N_2 \bar{x}_2{}^T \bar{x}_2 \right) \right] / (N_1 + N_2 - 2)$$

be the pooled within-sample covariance matrix of X based on the random samples G_1 and G_2. Mahalanobis metric matching calculates the distance between a G_1 unit with score X_1 and a G_2 unit with score X_2 as

$$(X_1 - X_2)S^{-1}(X_1 - X_2)^T. \tag{8.1}$$

Discriminant matching calculates each unit's score on the estimated discriminant as XD^T where $D = (\bar{x}_1 - \bar{x}_2)S^{-1}$, and then matches on this variable; equivalently, it defines the distance between a G_1 unit with score X_1 and G_2 unit with score X_2 as

$$(X_1 - X_2)D^T D(X_1 - X_2)^T. \tag{8.2}$$

We study these two particular matching methods because both matching methods are easy to implement using commonly available computer programs for sorting and calculating a pooled covariance matrix and because both matching methods have an appealing statistical property discussed in Rubin (1976b,c; 1978b).

2.2. The Treatment Effect

For the expected value of Y given X in P_i we write $\alpha_i + W_i(X)$. This expectation is often called the response surface for Y in P_i. The difference in expected values of Y for P_1 and P_2 units with the same value of X is thus $\alpha_1 - \alpha_2 + W_1(X) - W_2(X)$; when P_1 and P_2 represent two treatment populations such that the variables in X are the only ones that affect Y and have different distributions in P_1 and P_2, then this difference is the effect of the treatment at X. If $W_1(X) = W_2(X) = W(X)$ for all X, the response surfaces are called parallel, and $\alpha_1 - \alpha_2$ is the effect of the treatment for all values of the matching variables X.

Nonparallel response surfaces are not studied here in order to limit the number of conditions in the Monte Carlo experiment and because a straightforward argument suggests that matching must have beneficial effects when the response surfaces are nonparallel and the average treatment effect in P_1 is desired. The expected treatment effect over population P_1 is

$$E_1[\alpha_1 - \alpha_2 + W_1(X) - W_2(X)] = E_1[\alpha_1 + W_1(X)] - E_1[\alpha_2 + W_2(X)] \tag{8.3}$$

Table 8.1. *Estimators of the response surface difference:*
$$\hat{\tau} = \bar{y}_1. - \bar{y}_2. - (\bar{x}_1. - \bar{x}_2.)\hat{\beta}$$

Estimators of τ : $\hat{\tau}$	Estimators of β: $\hat{\beta}$	
$\hat{\tau}_o$	$\hat{\beta}_o = 0$	
$\hat{\tau}_p$	$\hat{\beta}_p = \mathbf{S}_{xx}^{-1}\mathbf{S}_{xy};$	$\mathbf{S}_{xu} = \sum_{i=1}^{2}\sum_{j=1}^{N}(\mathbf{x}_{ij} - \bar{\mathbf{x}}_{i.})^{\mathsf{T}}\mathbf{u}_{ij}$
$\hat{\tau}_d$	$\hat{\beta}_d = \mathbf{S}_{xx}^{-1}\mathbf{S}_{xy};$	$\mathbf{S}_{xu} = \sum_{i=1}^{2}\sum_{j=1}^{N}(\mathbf{x}_{ij} - \bar{\mathbf{x}}_{i.} - \bar{\mathbf{x}}_{.j} + \bar{\mathbf{x}}_{..})^{\mathsf{T}}\mathbf{u}_{ij}$

where E_1 is the expectation over the distribution of $\mathbf{X}$ in P_1. An unbiased estimate of the first expectation in (8.3) is simply $\bar{y}_1.$, the average observed Y in G_1. If we knew the P_2 response surface, an unbiased estimate of the second term in (8.3) would be

$$\sum_{j=1}^{N} [\alpha_2 + W_2(\mathbf{x}_{1j})]/N, \tag{8.4}$$

that is, the average value of the P_2 response surface across the values of $\mathbf{X}$ in the G_1 sample. Expression (8.4) implies that in order to estimate the expected treatment effect in P_1, we must extrapolate the P_2 response surface, $W_2(\cdot)$, into the region of G_1 data. When the P_2 response surface is estimated from G_2 data, this extrapolation can be subject to great error unless the sample from P_2 used to estimate $W_2(\cdot)$ has values of $\mathbf{X}$ similar to values in G_1, that is, unless the sample from P_2 is matched to G_1. See Billewicz (1965) and Rubin (1973a, 1977a) for further discussion of nonparallel response surfaces.

Henceforth, we will assume $W_1(\mathbf{X}) = W_2(\mathbf{X}) = W(\mathbf{X})$ so that the treatment effect is $\tau = \alpha_1 - \alpha_2$.

2.3. Estimators of τ

We will consider three estimators of τ, all of the form

$$\hat{\tau} = (\bar{y}_1. - \bar{y}_2.) - (\bar{\mathbf{x}}_1. - \bar{\mathbf{x}}_2.)\hat{\beta}$$

where $\bar{y}_{i.}$ and $\bar{\mathbf{x}}_{i.}$ are the means of Y and $\mathbf{X}$ in the matched samples, and $\hat{\beta}$ is an estimated regression coefficient of Y on $\mathbf{X}$. The differences between the estimators are thus confined to estimating the regression coefficient and are summarized in Table 8.1: $\hat{\tau}_o$ is simply the difference of Y means in the matched samples, $\hat{\tau}_p$ is the analysis of co-variance estimator of τ from the two-group design ignoring the paired structure of the matched samples, and $\hat{\tau}_d$ is the analysis of covariance estimator of τ using the two group by N matched-pair structure of the matched samples (equivalently forming matched-pair differences, $y_{dj} = y_{1j} - y_{2j}$ and $\mathbf{x}_{dj} = \mathbf{x}_{1j} - \mathbf{x}_{2j}$, $\hat{\beta}_d$ is the estimate of β found from regressing y_{dj} on $\mathbf{x}_{dj}$). Note that $\hat{\tau}_d$ is the only estimator that requires matched pairs to be assigned in the matched samples.

Simple algebra shows that the conditional bias of τ given the $\mathbf{x}_{ij}$ is

$$\bar{w}_1. - \bar{w}_2. - (\bar{\mathbf{x}}_1. - \bar{\mathbf{x}}_2.)\left(\mathbf{S}_{xx}^{-1}\mathbf{S}_{xw}\right) \tag{8.5}$$

where $w_{ij} = W(\mathbf{x}_{ij})$, $i = 1, 2$, $j = 1, \ldots, N$, and for $\hat{\tau}_o$, $\mathbf{S}_{xw} \equiv \mathbf{0}$. With multivariate $\mathbf{X}$ and moderate N, the variance of this conditional bias can be substantial, and then the expected value of the conditional bias may not be a good indicator of the utility of a procedure. Hence, we will use the expected value of the squared conditional bias to measure the utility of a procedure:

$$E_* \left[\bar{w}_{1.} - \bar{w}_{2.} - (\bar{\mathbf{x}}_{1.} - \bar{\mathbf{x}}_{2.}) \left(\mathbf{S}_{xx}^{-1} \mathbf{S}_{xw} \right) \right]^2 \tag{8.6}$$

where E_* is the expectation over the distribution of $\mathbf{X}$ in matched samples. When $r = 1$, the expected squared bias of $\hat{\tau}_o$ is

$$[E_1(\bar{w}_{1.}) - E_2(\bar{w}_{1.})]^2 + [V_1(\bar{w}_{1.}) + V_2(\bar{w}_{1.})] = [E_1(W(\mathbf{X})) - E_2(W(\mathbf{X}))]^2$$
$$+ [V_1 W(\mathbf{X}) + V_2 W(\mathbf{X})]/N$$

where E_i is the expectation and V_i the variance over the distribution of X in P_i. It follows that the percentage reduction in expected squared bias resulting from matching and/or regression adjustment is

$$100 \left\{ 1 - \frac{E_* \left[\bar{w}_{1.} - \bar{w}_{2.} - (\bar{\mathbf{x}}_{1.} - \bar{\mathbf{x}}_{2.}) \left(\mathbf{S}_{xx}^{-1} \mathbf{S}_{xw} \right) \right]^2}{[E_1(W(\mathbf{x})) - E_2(W(\mathbf{x}))]^2 + [V_1 W(\mathbf{x}) + V_2 W(\mathbf{x})]/N} \right\}.$$

If the matches were always perfect ($\mathbf{x}_{1j} = \mathbf{x}_{2j}$, $j = 1, \ldots, N$), then $\bar{\mathbf{x}}_{1.} = \bar{\mathbf{x}}_{2.}$ and $\bar{w}_{1.} = \bar{w}_{2.}$; hence, the percentage reduction in expected squared bias would be 100 for any response surface and all of our estimators. If the response surfaces were parallel and linear, then the percentage reduction in expected squared bias would be 100 for the regression adjusted estimates ($\hat{\tau}_p$ and $\hat{\tau}_d$) whether random or matched samples were used. But in general with imperfect matches and nonlinear response surfaces, the percentage reduction in expected squared bias will be less than 100.

3. MONTE CARLO COMPARISONS OF PROCEDURES – CONDITIONS

Except for the cases noted at the end of Section 2.2, the computations of percentage reductions in expected squared bias in matched samples appear to be analytically intractable. Hence we turn to Monte Carlo techniques. Our study can be compactly described as a $2 \times 3 \times 6 \times 4 \times 3 \times 3 \times 8$ factorial study with one summary value (percentage reduction in expected squared bias) per cell. This summary value was in fact obtained by a covariance adjustment on 100 replications using the first, second, and third moments of $\mathbf{X}$ in each of the random samples G_1, G_2 as nine covariates. The resultant precision of the value is roughly equivalent to that obtained with 300 replications and yields standard errors usually less than 1 percent, although larger in cases with smaller percentage reductions in expected squared bias. The Appendix provides details of the design.

    The factors in this study are

Factor 1: matching method: metric matching, discriminant matching.
Factor 2: regression adjustment: $\hat{\tau}_o$, $\hat{\tau}_p$, $\hat{\tau}_d$.
Factor 3: ratio of sample sizes, r: 1, 2, 3, 4, 6, 9 ($N = 50$ for all conditions).
Factor 4: bias along discriminant, B: $\frac{1}{4}, \frac{1}{2}, \frac{3}{4}, 1$.
Factor 5: ratio of variances along discriminant, σ^2: $\frac{1}{2}, 1, 2$.

Factor 6: ratio of variances orthogonal to discriminant, $\xi^2 : \frac{1}{2}, 1, 2$.

Factor 7: response surfaces $W(\mathbf{X})$; curvature along and orthogonal to discriminant:
$++, +0, +-, 0+, 0-, -+, -0, --$; see Equation (8.8).

The first three factors define the procedures that we study. The next three factors specify the distributions of the matching variables in P_1 and P_2; we assume that $\mathbf{X}$ has the following normal distributions:

$$\ln P_1 \, \mathbf{X} \sim N\left(\begin{pmatrix} \eta \\ 0 \end{pmatrix}, \begin{bmatrix} \sigma^2 & 0 \\ 0 & \xi^2 \end{bmatrix}\right),$$

$$\ln P_2 \, \mathbf{X} \sim N\left(\begin{pmatrix} 0 \\ 0 \end{pmatrix}, \begin{bmatrix} 1 & 0 \\ 0 & 1 \end{bmatrix}\right),$$

(8.7)

where

$$B = \eta \Big/ \left(\frac{1+\sigma^2}{2}\right)^{\frac{1}{2}} = \frac{1}{4}, \frac{1}{2}, \frac{3}{4}, 1; \sigma^2 = \frac{1}{2}, 1, 2; \quad \text{and} \quad \xi^2 = \frac{1}{2}, 1, 2.$$

The last factor defines the nonlinear response surface $W(\mathbf{X})$. The $++$ notation for the eight levels of $W(\mathbf{X})$ refers to nonlinearity along the discriminant and non-linearity orthogonal to the discriminant. Specifically, we let

$$W(\mathbf{X}) = W(u, v) = \exp\left[a\left(\frac{2}{1+\sigma^2}\right)^{\frac{1}{2}}(u - \eta/2) + b\left(\frac{2}{1+\xi^2}\right)^{\frac{1}{2}} v\right] \quad (8.8)$$

where $(a, b) = (+\frac{1}{2}, +\frac{1}{2}), (+\frac{1}{2}, 0), (+\frac{1}{2}, -\frac{1}{2}), (+.1, +\frac{1}{2}), (-.1, -\frac{1}{2}), (-\frac{1}{2}, +\frac{1}{2}),$ $(-\frac{1}{2}, 0), (-\frac{1}{2}, -\frac{1}{2})$. Values of a were set to $\pm.1$ instead of 0 in order to avoid cases in which $\hat{\tau}_o$ is unbiased in random samples; η, σ^2, and ξ^2 appear in (8.8) so that the response surfaces (8.8) with the distributions of $\mathbf{X}$ given by (8.7) are equivalent to the response surfaces $W(u, v) = \exp(au + bv)$ with the distributions of $\mathbf{X}$ standardized so that $\mathbf{E}_1(\mathbf{X}) + \mathbf{E}_2(\mathbf{X}) = 0$ and $\frac{1}{2}[\mathbf{V}_1(\mathbf{X}) + \mathbf{V}_2(\mathbf{X})] = \mathbf{I}$.

The response surfaces given by (8.8) are moderately nonlinear for the distributions given by (8.7). In order to justify the use of the phrase *moderately nonlinear* to describe these response surfaces, we calculate the percentage of the variance of $W(\mathbf{X})$ that can be attributed to the linear regression on $\mathbf{X}$ in P_i:

$$R_i{}^2 = \mathbf{C}_i(\mathbf{X}, W(\mathbf{X}))^T \mathbf{V}_i(\mathbf{X})^{-1} \mathbf{C}_i(\mathbf{X}, W(\mathbf{X})) / \mathbf{V}_i(W(\mathbf{X})),$$

where $\mathbf{C}_i(\cdot, \cdot)$ is the covariance in P_i. Straightforward algebra using (8.7), (8.8), and the fact that if $t \sim N(\theta, \phi)$, then

$$E[\exp(\gamma t)] = \exp[\theta\gamma + \gamma^2\phi/2]$$

and

$$E[t \cdot \exp(\gamma t)] = (\theta + \gamma\phi)E[\exp(\gamma t)]$$

shows that

$$R_i{}^2 = A_i/[\exp(A_i) - 1]$$

where

$$A_i = 2a^2[1 + \sigma^{2(2i-3)}] + 2b^2[1 + \xi^{2(2i-3)}].$$

It follows that the percentage of variance of $W(\mathbf{X})$ that can be attributed to the linear regression on $\mathbf{X}$ varies between 70 and 92 percent across all conditions of this Monte Carlo study.

4. REGRESSION ADJUSTMENTS WITH RANDOM SAMPLES

In practice, it is not uncommon for researchers to conduct observational studies without any matching. Random samples from P_1 and P_2 are chosen and regression adjustments are used to control the $\mathbf{X}$ variables. We begin our study of the Monte Carlo results by considering estimators with $r = 1$.

When $r = 1$, $\hat{\tau}_o$ yields no reduction in squared bias with either matching procedure, and $\hat{\tau}_p$ is the same for both matching procedures because it is the usual analysis of covariance estimator with two groups and no blocking; therefore, when $r = 1$, only three of the six possible estimators defined by the first two factors are of interest. Although when $r = 1$ $\hat{\tau}_d$ metric matched and $\hat{\tau}_d$ discriminant matched use the same units they in general yield different percentage reductions in expected squared bias because they pair the units in different ways before performing the regression adjustment on matched pair differences.

Table 8.2 presents the Monte Carlo percentage reductions in expected squared bias for $\hat{\tau}_p$. Although $\hat{\tau}_p$ does quite well in many conditions, especially when $\sigma^2 = \xi^2 = 1$, in other conditions it does quite poorly, especially when $\sigma^2 = \xi^2 = 2$ and $\sigma^2 = \xi^2 = \frac{1}{2}$. The negative values in Table 8.2 indicate that the regression adjustment actually increases bias. These results show that $\hat{\tau}_p$ based on random samples cannot be counted on to control the X-variables when the response surfaces are nonlinear.

Some insight into the problem with $\hat{\tau}_p$ can be achieved by considering the large sample case. In large samples, $\hat{\tau}_p$ is approximately $E_1(W(\mathbf{X})) - E_2(W(\mathbf{X})) - \eta c$ where c is the pooled slope of $W(\mathbf{X})$ on the discriminant, that is, the first component of $[\mathbf{C}_1(\mathbf{X}, W(\mathbf{X})) + \mathbf{C}_2(\mathbf{X}, W(\mathbf{X}))]/(1 + \sigma^2)$. From (8.7) and (8.8), we can write $\eta c = a B[E_1(W(\mathbf{X}))(\sigma^2/(1 + \sigma^2)) + E_2(W(\mathbf{X}))/(1 + \sigma^2)]$; because $W(\mathbf{X})$ and B are positive, ηc has the same sign as a. If $\sigma^2 = \xi^2 = 1$, then the initial bias, $E_1(W(\mathbf{X})) - E_2(W(\mathbf{X}))$, has the same sign as a, implying that the adjustment $-\eta c$ is in the correct direction. However, if $\sigma_1^2 \neq 1$ and/or $\xi^2 \neq 1$, then the adjustment may be in the wrong direction and actually increase bias.

Somewhat surprising, estimating the regression coefficient from matched-pair differences in the random samples can result in better estimates. Table 8.3 presents Monte Carlo values for the percentage reduction in expected squared bias for $\hat{\tau}_d$ metric matched with $r = 1$; $\hat{\tau}_d$ is superior to $\hat{\tau}_p$ in 209 of 288 cases, and in only two cases ($\sigma^2 = 1, \xi^2 = \frac{1}{2}, B = \frac{1}{2}, ++$ and $\sigma^2 = 2, \xi^2 = \frac{1}{2}, B = \frac{1}{4}, --$) do the results favor $\hat{\tau}_p$ by more than 5 percent. The fact that $\hat{\tau}_d$ is usually better than $\hat{\tau}_p$ is consistent with Monte Carlo results and analytic considerations presented in Rubin (1973b) for the univariate case.

The results for $\hat{\tau}_d$ discriminant matched are similar but inferior to the results for $\hat{\tau}_d$ metric matched. The mean of the 288 metric minus discriminant differences

Table 8.2. *Percentage reduction in expected squared bias; Monte Carlo values for $\hat{\tau}_p$ in random samples of size 50 ($r = 1$)*

	Response surface	$\sigma^2 = \frac{1}{2}$				$\sigma^2 = 1$				$\sigma^2 = 2$			
		$B = \frac{1}{4}$	$\frac{1}{2}$	$\frac{3}{4}$	1	$\frac{1}{4}$	$\frac{1}{2}$	$\frac{3}{4}$	1	$\frac{1}{4}$	$\frac{1}{2}$	$\frac{3}{4}$	1
	++	−52	−55	17	54	43	65	80	87	87	93	95	96
	+0	40	81	93	97	96	99	99	100	87	96	98	99
	+−	−27	−13	47	75	51	74	87	92	87	94	97	97
$\xi^2 = \frac{1}{2}$	0+	37	22	10	05	40	28	18	16	43	33	26	25
	0−	65	72	78	82	66	73	78	82	67	73	78	82
	−+	74	87	93	96	83	91	95	97	87	93	96	97
	−0	88	96	99	99	96	99	99	100	38	81	94	97
	−−	75	88	94	96	83	92	95	97	82	91	95	96
	++	35	60	77	86	84	92	95	96	85	94	97	98
	+0	40	81	93	97	96	99	99	100	87	96	98	99
	+−	48	73	87	93	84	93	96	98	82	93	97	98
$\xi^2 = 1$	0+	82	84	85	87	83	85	87	89	84	87	89	90
	0−	86	88	90	91	86	87	89	91	85	87	89	91
	−+	86	94	97	98	89	95	97	98	58	79	90	95
	−0	88	96	99	99	96	99	99	100	39	81	94	97
	−−	85	93	97	97	83	92	95	97	39	68	85	92
	++	80	89	93	95	82	92	95	97	74	88	94	97
	+0	40	81	93	97	96	99	99	100	87	96	98	99
	+−	82	91	95	97	79	90	95	97	70	86	93	97
$\xi^2 = 2$	0+	64	71	77	81	63	71	77	81	63	71	77	81
	0−	43	34	29	30	41	30	23	24	38	26	18	17
	−+	91	96	97	97	60	79	89	93	−12	02	55	79
	−0	88	96	99	99	96	99	99	100	39	81	94	97
	−−	85	93	95	96	42	69	84	90	−43	−26	42	72

in percentage reduction in expected squared bias is 5.4, the minimum difference is −15.6, the maximum difference is 58.1, 219 differences are positive, and only three differences are less than −10 ($\sigma^2 = \frac{1}{2}, \xi^2 = \frac{1}{2}, B = \frac{3}{4}, ++; \sigma^2 = 1, \xi^2 = \frac{1}{2}, B = \frac{1}{2}, ++; \sigma^2 = 2, \xi^2 = \frac{1}{2}, B = \frac{1}{4}, --$).

Even though the results for $\hat{\tau}_d$ are somewhat better than for $\hat{\tau}_p$, in cases in which $\hat{\tau}_p$ does poorly, so does $\hat{\tau}_d$. Of course, with real data we could try fitting higher-order (e.g., quadratic) terms, although the nonlinearity might be difficult to detect because these response surfaces are only moderately nonlinear. Our study does not include quadratic terms in the regression adjustment but does include matched sampling. Hence, we turn to results with $r > 1$ to see if matched sampling improves the estimation of τ.

5. RESULTS WITH $r > 1$

We now consider the utility of matched sampling, alone and in combination with regression adjustment. Because the analyses of the 7-factor Monte Carlo study

Table 8.3. *Percentage reduction in expected squared bias for $\hat{\tau}_d$, metric matching in random sample of size 50 $(r = 1)$*

	Response surface	$\sigma^2 = \frac{1}{2}$				$\sigma^2 = 1$				$\sigma^2 = 2$			
		$B = \frac{1}{4}$	$\frac{1}{2}$	$\frac{3}{4}$	1	$\frac{1}{4}$	$\frac{1}{2}$	$\frac{3}{4}$	1	$\frac{1}{4}$	$\frac{1}{2}$	$\frac{3}{4}$	1
	++	−12	−43	24	66	54	49	87	86	91	92	93	93
	+0	53	87	96	99	97	99	99	99	92	98	99	98
	+−	−09	12	63	84	60	79	90	93	91	93	94	94
$\xi^2 = \frac{1}{2}$	0+	47	47	47	44	56	52	67	60	64	64	68	65
	0−	74	84	89	92	79	87	93	94	78	87	93	94
	−+	84	94	97	96	89	94	96	97	86	93	95	96
	−0	94	98	99	98	97	99	99	99	47	88	97	99
	−−	83	93	94	95	87	93	96	97	72	86	95	97
	++	47	70	77	86	88	91	95	95	91	96	94	95
	+0	60	89	96	98	96	99	99	99	92	98	99	98
	+−	60	80	92	95	86	95	97	97	91	97	98	97
$\xi^2 = 1$	0+	86	87	85	89	89	89	90	89	89	90	91	93
	0−	88	90	92	93	91	91	93	94	91	92	95	95
	−+	92	97	97	97	91	96	97	97	64	83	93	96
	−0	95	98	99	98	96	98	99	99	48	88	97	99
	−−	88	93	94	94	86	89	94	96	47	76	91	96
	++	87	89	88	93	88	93	96	93	86	94	94	94
	+0	57	89	96	98	96	98	99	99	92	98	99	98
	+−	84	93	96	97	83	94	97	97	81	94	97	97
$\xi^2 = 2$	0+	73	77	84	86	75	83	88	87	76	83	88	89
	0−	55	52	54	56	58	59	57	59	54	53	56	57
	−+	91	94	95	95	63	81	90	93	06	26	73	88
	−0	94	98	99	98	96	98	99	99	51	88	97	99
	−−	82	88	91	92	48	75	88	91	−19	07	65	87

are somewhat involved, we begin in Section 5.1 by presenting specific results for two procedures in order to convey the flavor of our conclusions. The remainder of Section 5 presents detailed analyses of the results of the Monte Carlo study with $r > 1$. Section 5.2 presents an analysis of variance (ANOVA) of the study. Section 5.3 shows that although the difference between metric and discriminant matching varies with the estimator, the distribution of **X**, and the response surface, metric matching is clearly superior to discriminant matching. Section 5.4 focuses on the results for metric matching and concludes that $\hat{\tau}_d$ is the best regression adjustment procedure that we have considered. Section 5.5 displays results for $\hat{\tau}_d$ metric matched that can be used to suggest ratios of sample sizes needed to remove nearly all of the bias for a variety of nonlinear response surfaces.

5.1. Two Specific Estimators

Tables 8.4 and 8.5 present percentage reductions in expected squared bias for metric matched samples with $r = 2$ for $\hat{\tau}_o$ and $\hat{\tau}_d$. By comparing Tables 8.3 and 8.5, we

Table 8.4. *Percentage reduction in expected squared bias for $\hat{\tau}_o$, metric matching with $r = 2$*

	Response surface	$\sigma^2 = \frac{1}{2}$				$\sigma^2 = 1$				$\sigma^2 = 2$			
		$B = \frac{1}{4}$	$\frac{1}{2}$	$\frac{3}{4}$	1	$\frac{1}{4}$	$\frac{1}{2}$	$\frac{3}{4}$	1	$\frac{1}{4}$	$\frac{1}{2}$	$\frac{3}{4}$	1
	++	95	88	88	82	83	77	73	67	50	49	49	46
	+0	95	96	94	87	89	85	79	71	65	61	57	51
	+−	95	91	92	87	87	82	77	70	60	54	52	48
$\xi^2 = \frac{1}{2}$	0+	98	94	90	80	96	93	88	78	92	87	78	67
	0−	98	98	95	91	98	97	93	77	95	94	89	82
	−+	99	99	99	98	99	98	98	96	93	96	93	90
	−0	99	99	99	98	98	98	97	95	85	93	90	87
	−−	99	99	99	97	99	98	97	95	91	96	93	90
	++	80	83	84	80	71	74	71	65	50	53	50	45
	+0	96	96	94	88	89	87	81	72	67	64	58	51
	+−	91	90	89	84	82	80	76	68	63	59	54	48
$\xi^2 = 1$	0+	83	83	79	73	79	76	73	68	70	67	62	57
	0−	90	91	93	92	90	92	91	90	88	89	87	83
	−+	99	99	99	98	95	98	98	96	81	93	93	90
	−0	99	99	99	98	98	98	97	95	85	93	91	88
	−−	98	99	99	98	93	98	98	96	77	92	91	89
	++	56	70	75	71	60	64	63	57	45	45	43	39
	+0	95	95	94	88	89	86	80	72	67	63	57	50
	+−	71	78	79	75	69	71	67	61	54	53	48	43
$\xi^2 = 2$	0+	56	60	58	57	56	56	55	51	51	48	47	42
	0−	55	50	51	55	60	56	56	63	60	63	64	66
	−+	89	96	98	99	72	93	97	97	54	82	91	91
	−0	99	99	98	97	97	97	97	95	84	93	90	87
	−−	89	96	98	98	75	92	96	96	51	80	90	88

immediately see advantages to matching even with $r = 2$. The estimator $\hat{\tau}_d$ metric matched with $r = 2$ usually removes most of the squared bias and is clearly better than $\hat{\tau}_d$ (or $\hat{\tau}_p$) with $r = 1$. Of the 288 differences between the percentage reductions in expected squared bias for $\hat{\tau}_d$ metric matched with $r = 2$ and $\hat{\tau}_d$ metric matched with $r = 1$, only two are negative; of the 288 differences between the percentage reductions in expected squared bias for $\hat{\tau}_d$ metric matched with $r = 2$ and $\hat{\tau}_p$ metric matched with $r = 1$, only seven are negative; and the most negative of these nine differences is only -1. In the next sections we will see that in those cases that are difficult for matching (e.g., $\sigma^2 = \xi^2 = 2$), larger ratios result in even better estimates.

By comparing Tables 8.4 and 8.5 we see that in all cases except with $\sigma^2 = \xi^2 = 2$, there is an advantage to using regression adjustment on matched samples. When $\sigma^2 = \xi^2 = 2$, $\hat{\tau}_o$ is superior to $\hat{\tau}_d$ in a few cases, but $\hat{\tau}_d$ is usually better. Without knowledge of the response surface, $\hat{\tau}_d$ is to be preferred to $\hat{\tau}_o$; with such knowledge, a more appropriate regression adjustment can be used. We will see that $\hat{\tau}_d$ is in this sense always preferable to $\hat{\tau}_o$ (or $\hat{\tau}_p$).

Table 8.5. *Percentage reduction in expected squared bias for $\hat{\tau}_d$, metric matching with $r = 2$*

	Response surface	$\sigma^2 = \frac{1}{2}$				$\sigma^2 = 1$				$\sigma^2 = 2$			
		$B = \frac{1}{4}$	$\frac{1}{2}$	$\frac{3}{4}$	1	$\frac{1}{4}$	$\frac{1}{2}$	$\frac{3}{4}$	1	$\frac{1}{4}$	$\frac{1}{2}$	$\frac{3}{4}$	1
	++	99	97	96	97	98	96	98	98	95	97	97	97
	+0	99	100	99	100	99	100	100	100	96	99	99	99
	+−	98	97	97	96	97	97	98	97	95	97	97	96
$\xi^2 = \frac{1}{2}$	0+	99	98	97	95	99	98	98	96	98	98	97	96
	0−	100	99	99	99	100	100	99	99	99	99	99	99
	−+	100	100	100	100	99	100	100	100	93	97	99	100
	−0	100	100	100	100	99	100	100	100	79	95	99	100
	−−	100	100	100	100	99	99	100	100	89	95	98	99
	++	98	98	98	97	98	98	99	99	95	98	96	98
	+0	99	100	100	100	99	100	100	100	96	99	99	99
	+−	98	99	99	98	98	99	99	98	95	98	97	97
$\xi^2 = 1$	0+	97	97	97	96	98	96	97	97	96	96	95	97
	0−	98	98	98	98	98	98	98	98	97	97	98	98
	−+	99	100	100	100	98	99	100	100	80	90	97	99
	−0	100	100	100	100	99	100	100	100	79	95	99	100
	−−	99	100	100	100	97	98	99	100	74	86	95	98
	++	93	96	98	97	93	97	98	98	92	97	96	98
	+0	98	99	100	100	99	99	100	100	96	99	99	99
	+−	92	96	98	98	93	97	98	98	91	96	98	98
$\xi^2 = 2$	0+	88	90	92	93	89	92	93	92	88	92	93	96
	0−	80	76	78	80	81	81	80	79	78	77	77	79
	−+	94	97	99	99	84	93	97	99	51	62	87	95
	−0	100	100	100	100	99	100	100	100	77	94	99	100
	−−	93	97	99	99	80	90	96	98	41	50	82	93

5.2. An ANOVA of the Results When $r > 1$

Table 8.6 presents an ANOVA of the 7-factor study where factor 3 has five levels $r = 2, 3, 4, 6, 9$. For simplicity of display, the response surface factor and the three distribution of **X** factors have been collapsed into one "distribution" factor with 288 levels. In fact, little information was lost by collapsing the distributional factors because of numerous large higher-order interactions among the distributional factors and because larger interactions between procedure factors and distributional factors tended to involve higher-order interactions among the distributional factors. The purpose of this ANOVA is simply to see which are the large sources of variation.

Table 8.7 summarizes the procedure factors by the average value of the percentage reduction in expected squared bias over the 288 distributional conditions. If there were no interactions between procedures and conditions, then Table 8.7 would be an adequate summary for our Monte Carlo study; that is, there would be good and bad procedures and easy and hard distributional conditions, but comparisons between procedures would be the same in each distributional condition.

Table 8.6. *Analysis of variance of 7-factor Monte Carlo study with $r > 1$*

Source	Degrees of freedom	Mean square
Matching Method (metric, discriminant)	1	36.615
Regression Adjustment ($\hat{\tau}_o, \hat{\tau}_p, \hat{\tau}_d$)	2	7.414
Ratio ($r = 2, 3, 4, 6, 9$)	4	.705
Distribution[a]	287	.442
M.M. × R.A.	2	.430
M.M. × Ratio	4	.214
M.M. × Distribution	287	.241
R.A. × Ratio	8	.114
R.A. × Distribution	574	.027
Ratio × Distribution	1148	.004
M.M. × R.A. × Ratio	8	.051
M.M. × R.A. × Distribution	514	.007
M.M. × Ratio × Distribution	1148	.002
R.A. × Ratio × Distribution	2296	.002
M.M. × R.A. × Ratio × Distribution	2296	.001

[a] Factors 4, 5, 6, and 7 defined in Section 3.

However, there are nontrivial interactions between procedures and distributions in the sense that if we fit the procedure-plus-distribution additive model to the 7-factor study, we are left with some large residuals to explain. Although Table 8.7 is not an entirely adequate summary for our study, we will see in Sections 5.3 through 5.5 that most trends displayed there are not misleading. The major trends in Table 8.7 are that

1. Metric matching is superior to discriminant matching.
2. $\hat{\tau}_d$ is superior to $\hat{\tau}_p$ especially with metric matched samples, and both $\hat{\tau}_d$ and $\hat{\tau}_p$ are superior to $\hat{\tau}_o$ especially for smaller r.
3. Larger ratios of sample sizes are better, although only modest benefits accrue when moving from $r = 2$ to larger ratios, the benefit being largest with $\hat{\tau}_o$ and smallest with $\hat{\tau}_d$.

Table 8.7. *Percentage reduction in expected squared bias averaging over distributional conditions*

| Ratio (r) | Discriminant matching | | | Metric matching | | |
	$\hat{\tau}_o$	$\hat{\tau}_p$	$\hat{\tau}_d$	$\hat{\tau}_o$	$\hat{\tau}_p$	$\hat{\tau}_d$
1	00[a]	78	78	00[a]	78	84
2	71	83	84	81	91	96
3	74	84	85	88	94	97
4	75	84	85	90	95	98
6	74	84	85	94	96	99
9	75	85	86	95	97	99

[a] Theoretical values.

Table 8.8. *Summary of differences in percentage reduction in expected squared bias for each estimator: metric matching minus discriminant matching*[a]

	Ratio $r =$	2	3	4	6	9
$\hat{\tau}_o$ (metric)	min	−19.9	−16.0	−13.0	−7.8	−6.1
$-\hat{\tau}_o$ (discrim)	max	124.5	120.2	137.6	139.3	175.6
	mean	9.9	13.3	15.4	19.2	20.6
	# > 0	194	219	232	245	254
$\hat{\tau}_p$ (metric)	min	−2.0	−.7	−.7	−.4	−.5
$-\hat{\tau}_p$ (discrim)	max	102.0	73.4	90.2	83.1	82.9
	mean	8.9	9.8	10.9	12.0	12.7
	# > 0	262	260	256	259	262
$\hat{\tau}_d$ (metric)	min	−1.2	−1.3	−.6	−.4	−.2
$-\hat{\tau}_d$ (discrim)	max	107.6	79.1	90.1	83.9	81.6
	mean	11.5	12.0	12.9	13.2	13.2
	# > 0	269	264	264	271	273

[a] 288 differences for each estimator, one for each distributional condition in Monte Carlo study.

Further analysis will show that the conclusion from 1. and 2., to the effect that τ_d metric matched is the best combination of matching method and regression adjustment, is correct. However, the conclusion from 3., to the effect that ratios larger than 2 are not needed, is not always true; with some combinations of distributions of **X** and response surfaces, using larger ratios for matching can result in substantial improvements.

5.3. Metric Matching vs. Discriminant Matching

Table 8.6 indicates that the matching method factor and its interactions with distributional factors make a large contribution to the variation in the Monte Carlo study. Table 8.7 suggests that metric matching is on the average superior to discriminant matching. But these tables do not tell us whether the interaction between matching method and distribution is the result of a varying superiority of metric matching over discriminant matching or an occasional superiority of discriminant matching. If discriminant matching were better than metric matching for only some distributions of **X**, then we should decide which matching method to use on the basis of the observed distribution of **X** in G_1 and G_2. Our results clearly show metric matching to be superior to discriminant matching for all distributions of **X** considered.

Focus on one estimator, that is, one adjustment ($\hat{\tau}_o$, $\hat{\tau}_d$, $\hat{\tau}_p$) and one ratio ($r = 2, 3, 4, 6, 9$); for each of the 288 distributional conditions, take the difference between the percentage reduction in expected squared bias obtained by metric matching and obtained by discriminant matching. If all 288 differences were positive, we would know that metric matching was superior to discriminant matching for that estimator (e.g., $\hat{\tau}_d$ with $r = 2$).

Table 8.8 summarizes the $3 \times 5 = 15$ metric minus discriminant sets of differences across the 288 distributional conditions. With $\hat{\tau}_d$ and $\hat{\tau}_p$ there are essentially no cases where discriminant matching is to be preferred to metric matching; the

Table 8.9. *Differences in percentage reduction in expected squared bias:* $\hat{\tau}_o$ *(metric matched) minus* $\hat{\tau}_o$ *(discriminant matched) when distributions of* X *are relatively favorable to discriminant matching* ($\sigma^2 = 2, \xi^2 = \frac{1}{2}$)

	$r = 2$				$r = 3$				$r = 4$			
$B = \frac{1}{4}$	$\frac{1}{2}$	$\frac{3}{4}$	1	$\frac{1}{4}$	$\frac{1}{2}$	$\frac{3}{4}$	1	$\frac{1}{4}$	$\frac{1}{2}$	$\frac{3}{4}$	1	
++	-08	-15	-18	-12	-03	-16	-15	-10	05	-05	-13	-12
+0	-01	-04	-03	-01	-03	-04	-03	-02	01	-03	-03	-02
+−	-04	-20	-13	-08	-01	-07	-09	-08	16	-01	-08	-06
0+	63	71	35	24	64	64	57	57	74	94	88	61
0−	52	49	30	20	55	73	36	30	61	45	31	29
−+	18	16	10	11	17	16	08	07	21	13	09	06
−0	11	-01	-01	-01	24	01	-01	-01	20	00	-01	-00
−−	32	26	10	06	27	26	11	06	29	15	09	06

most negative difference is only −2 percent. With $\hat{\tau}_o$, metric matching is usually superior to discriminant matching, but further study of these differences is enlightening. All cases where the differences are ≤ -10 percent occur when $\sigma^2 = 2$ and $\xi^2 = \frac{1}{2}$: five cases when $r = 2$, three cases when $r = 3$, and two cases when $r = 4$. Table 8.9 displays the metric minus discriminant differences for $\hat{\tau}_o$ with $r = 2, 3, 4$ for $\sigma^2 = 2$ and $\xi^2 = \frac{1}{2}$. Clearly, even for these distributions of X, metric matching is to be preferred unless exceptionally strong prior knowledge suggests a response surface that has curvature increasing as one moves from the P_2 range of X to the P_1 range of X, and even then little can be lost by metric matching especially if $r \geq 4$.

This conclusion holds for $\hat{\tau}_o$ even when the response surface is linear in the discriminant (i.e., $W(X) = X(1, 0)^T$). Of the $3(\sigma^2) \times 3(\xi^2) \times 4(B) \times 5(r) = 180$ differences in percentage reduction in expected squared bias for $\hat{\tau}_o$ (Mahalanobis metric matched) minus $\hat{\tau}_o$ (discriminant matched), only 26 were greater than 2 percent, and of these, 6 favored discriminant matching (five 3 percent differences, one 4 percent difference), whereas 20 favored Mahalanobis metric matching (four 3 percent differences, six 4 percent, two 5 percent, four 6 percent, three 7 percent, and one 9 percent difference).

Because there seems to be no reason to recommend discriminant matching over metric matching, we restrict further investigation of the Monte Carlo study to results obtained by metric matching.

5.4. Comparing Regression Adjustments

Table 8.10 compares the regression adjustments (based on metric matched samples) for each ratio across the 288 distributional conditions. The comparison of $\hat{\tau}_d$ with $\hat{\tau}_p$ shows that although there is usually not much difference between the adjustments, $\hat{\tau}_d$ is clearly superior to $\hat{\tau}_p$, the most negative difference in percentage reductions in expected squared bias being −2.2 percent when $r = 2$.

The comparison of $\hat{\tau}_d$ with $\hat{\tau}_o$ in Table 8.10 shows that although $\hat{\tau}_d$ is usually substantially better than $\hat{\tau}_o$, $\hat{\tau}_o$ is better than $\hat{\tau}_d$ in a few cases. All five cases having

Table 8.10. *Summary of differences in percentage reduction in expected squared bias for estimators based on metric matched samples*

	Ratio $r =$	2	3	4	6	9
$\hat{\tau}_d$ (metric)	min	−2.2	−.6	−.0	−.1	−.7
$-\hat{\tau}_p$ (metric)	max	34.1	26.3	20.9	16.2	12.6
	mean	4.2	3.4	3.0	2.2	1.8
	# > 0	275	281	285	284	283
$\hat{\tau}_d$ (metric)	min	−29.8	−16.1	−9.5	−9.3	−6.0
$-\hat{\tau}_o$ (metric)	max	59.7	45.0	37.2	26.9	21.6
	mean	14.9	9.2	7.5	5.0	3.8
	# > 0	270	259	270	271	270

differences in percentage reductions in expected squared bias favoring $\hat{\tau}_o$ by 10 percent or more occur when $\sigma^2 = \xi^2 = 2$, three with $r = 2$ and two with $r = 3$. Tables 8.4 and 8.5 provide the results for $\hat{\tau}_o$ and $\hat{\tau}_d$ when $r = 2$ and show that without strong prior knowledge of the response surface, τ_d is better than τ_o even when $\sigma^2 = \xi^2 = 2$; a researcher having rather specific knowledge of the response surface should be fitting a model relevant to that response surface and should not be using a linear approximation ($\hat{\tau}_d$ or $\hat{\tau}_p$) or no adjustments ($\hat{\tau}_o$).

The conclusion thus far is simple: Use $\hat{\tau}_d$ based on metric matched samples. It remains for us to summarize the efficacy of using different ratios for the matching.

5.5. The Effect of Using Different Ratios for Matching

Tables 8.2 through 8.5 give specific results for the estimators $\hat{\tau}_p$ ($r = 1$), $\hat{\tau}_d$ ($r = 1$, metric), $\hat{\tau}_o$ ($r = 2$, metric), and $\hat{\tau}_d$ ($r = 2$, metric). These tables show that in some cases any of these estimators can remove nearly all of the squared bias, whereas in other cases even the best of them, $\hat{\tau}_d$ with $r = 2$, removes less than 50 percent of squared bias. The results are sensitive to the distribution of **X** and the response surface, especially when $r = 1$. Table 8.5 shows that if $\sigma^2 \leq 1$ and $\xi^2 \leq 1$, $\hat{\tau}_d$ metric matched with $r = 2$ can be counted on to remove most of the bias, and in most cases with either $\sigma^2 > 1$ or $\xi^2 > 1$ usually removes more than 90 percent of the bias. The clearest need for improved estimation occurs when both $\sigma^2 > 1$ and $\xi^2 > 1$. Increasing r yields better estimates.

In order to indicate the advantages of increasing r, we present Table 8.11, which gives "pessimistic" results for $\hat{\tau}_d$ metric matched. Pessimistic means that for each ratio and each distribution of **X** defined by a value of (B, σ^2, ξ^2), we have produced the minimum percentage reduction in squared bias over the eight response surfaces. Hence, within the context of our Monte Carlo study, these represent the worst results that can be obtained by using $\hat{\tau}_d$ metric matched. We see that increases in r result in improved estimation. When $r = 4$, $\hat{\tau}_d$ metric matched removes 73 percent of the expected squared bias in the worst case and usually removes more than 90 percent even when $\sigma^2 = \xi^2 = 2$. The most difficult cases are those with small initial bias. Using a ratio equal to $2\sigma^2\xi^2$ for matching usually removes most of the bias.

Table 8.11. *Pessimistic percentage reductions in expected squared bias for $\hat{\tau}_d$, metric matched*

r	ξ^2	$\sigma^2 = \frac{1}{2}$				$\sigma^2 = 1$				$\sigma^2 = 2$			
		$B = \frac{1}{4}$	$\frac{1}{2}$	$\frac{3}{4}$	1	$\frac{1}{4}$	$\frac{1}{2}$	$\frac{3}{4}$	1	$\frac{1}{4}$	$\frac{1}{2}$	$\frac{3}{4}$	1
1	$\frac{1}{2}$	−12	−43	24	44	54	49	67	60	47	64	68	65
	1	47	70	77	86	86	89	90	89	47	76	91	93
	2	55	52	54	56	48	59	57	59	−19	07	56	57
2	$\frac{1}{2}$	98	97	96	95	97	96	98	96	79	95	97	96
	1	97	97	97	96	97	96	97	97	74	86	95	97
	2	80	76	78	80	80	81	80	79	41	50	77	79
3	$\frac{1}{2}$	99	99	99	98	97	98	98	98	85	96	98	97
	1	99	99	99	99	94	98	98	98	82	88	96	97
	2	88	84	83	83	87	85	85	85	61	65	82	85
4	$\frac{1}{2}$	100	99	100	99	98	98	99	99	92	97	98	98
	1	99	99	99	99	98	98	98	99	88	92	97	97
	2	92	89	90	88	90	92	91	89	73	77	89	89
6	$\frac{1}{2}$	99	100	100	100	99	99	100	100	94	98	99	98
	1	99	99	99	99	98	99	99	99	94	94	98	97
	2	94	94	94	94	94	94	94	92	80	81	92	92
9	$\frac{1}{2}$	100	100	99	100	100	99	100	100	97	99	99	99
	1	99	99	100	100	99	97	99	100	95	99	99	99
	2	96	96	96	96	95	96	96	96	85	85	95	95

6. DISCUSSION

The general conclusion from our analyses of the Monte Carlo study is that the best procedure we have considered is $\hat{\tau}_d$ (the regression-adjusted estimator based on matched pair differences) using large r (large ratio of size of reservoir to size of matched sample) and metric matched samples (specifically, using the Mahalanobis metric, (8.1)). Obtaining G_2 with large r can be expensive in practice, however, and we have seen that the improvements that accrue from using $r = 9$ rather than $r = 2$ are modest, except when the spread of the $\mathbf{X}$ distribution is larger in P_1 than P_2. Tentative advice would be to metric match using a ratio of $2\sigma^2\xi^2$ (i.e., perhaps twice the determinant of $\Sigma_1\Sigma_2^{-1}$ where Σ_i is the covariance of $\mathbf{X}$ in P_i) and perform regression adjustments on matched pair differences. Our results demonstrate quite clearly that matching can dramatically improve estimation.

Of course, a realistic criticism of this work is that we have not considered other procedures, ones that carefully try to search for nonlinear components in the response surfaces or try to perform sophisticated Bayesian or empirical Bayesian analyses that average over a variety of nonlinear models for the response surfaces. Our reaction to this criticism is that although we hope that in any real data analysis such techniques would be applied, considering all of them in a Monte Carlo study is impossible because good data-analytic techniques must be, by nature, conditional on the observed data. Presumably, such sophisticated estimators would do at least as well as $\hat{\tau}_d$, and so the results for $\hat{\tau}_d$ (metric matched) can be thought of as minimums for methods that try to estimate the response surface by more than a linear fit. We feel that the general benefits from obtaining matched samples would hold for more sophisticated estimators because with matched samples the sensitivity of the regression adjustment to model specification is reduced.

In conclusion, because we feel that results similar to ours would be obtained for more than two matching variables, for nonnormal matching variables, and other nonlinear response surfaces, we feel that an effective plan for controlling matching variables in observational studies is to perform regression adjustments on matched samples obtained by nearest available Mahalanobis metric pair matching. Of course this summary of advice assumes the sampling situation described in Section 2, with dependent variables recorded in the matched samples but not recorded in the initial random samples.

APPENDIX

Discussion of Monte Carlo Study

The Monte Carlo study is a 7-factor study as described in Section 3. The first three factors define 36 procedures, and the last four factors define 288 distributional conditions. For the moment, focus on one condition and one procedure: Over repeated matched samples we wish to know the expected value of BIAS2, expression (8.6).

Letting NSIM be the number of sampling replications in each condition, we could simply perform NSIM replications in each condition, drawing samples independently across the conditions and independently for each procedure. Such a sampling scheme would be more expensive than necessary for several reasons. First, it would generate many more random numbers than needed. Second, it would not provide efficient comparisons of procedures within conditions or a procedure across conditions because the independent sampling would not have created correlated estimates of BIAS2. Third, no attempt would have been made to increase the precision of the study by using simulation covariates, quantities that are defined for each procedure and condition and correlated with BIAS2, but whose expectations we know from analytic considerations, for example, the moments of $\mathbf{X}$ in the random samples G_1 and G_2.

First consider the third point, using simulation covariates to increase precision. In our study with normal random variables, we know the expectations of all sample moments in G_1 and G_2, and these should be related to the ease of obtaining well-matched samples and the utility of regression adjustment. For example, G_1 and G_2 samples with means farther apart than usual should imply that the resultant matched samples will have means farther apart than usual. We can let the data estimate these relationships between sample moments in G_1 and G_2 and BIAS2. Pilot studies indicated that using the first three moments in G_1 and G_2 as simulation covariates resulted in the most cost-effective plan. More covariates would have been more expensive because extra storage would have been required in core, and fewer covariates would have been more expensive because additional simulations would have been needed to obtain the same precision. Roughly speaking, our use of simulation covariates allowed us to reduce the number of simulations by a factor of 3. Typically, the squared multiple correlation between BIAS2 and the simulation covariates was about .6 and higher when the standard errors were higher; hence, without the covariance adjustment we would have needed roughly NSIM = 300 in order to obtain the precision that was obtained using NSIM = 100 and simulation covariates.

The issues of minimizing the number of random numbers generated in order to save cost and correlating the estimates across conditions to increase precision of comparisons are really handled in the same manner. First, focus on one distributional condition and consider a fixed ratio r and a fixed matching method. Then we want to compare the three adjustments $\hat{\tau}_o$, $\hat{\tau}_p$, $\hat{\tau}_d$ with respect to BIAS2. By calculating all adjustments on the same matched sample, we make the estimates of BIAS2 correlated across adjustments and hence make comparisons more precise. Now let us consider different matching methods with the same ratio; we cannot use the same matched samples from G_2 but we can and do use the same random samples (G_1, G_2) for matching. For example, BIAS2 for $\hat{\tau}_o$, $r = 2$, metric and $\hat{\tau}_o$, $r = 2$, discriminant are calculated in matched samples obtained from the same random samples. Using the same random samples correlates the metric and discriminant results, increasing the precision when comparing metric matching with discriminant matching for each estimator. Finally, consider different ratios for matching; we cannot use the same random samples G_2, but we can and do use the same random samples G_1 and overlapping random samples G_2 (i.e., the $r = 2$, G_2 sample includes the $r = 1$, G_2 sample, the $r = 3$, G_2 sample includes the $r = 2$, G_2 sample, etc.). Using overlapping random samples correlates the results for different ratios and hence increases the precision of comparisons between estimators using different ratios.

Furthermore, we can correlate results across response surfaces and distributions of **X**, thereby increasing precision of comparisons of procedures between distributional conditions and reducing computational costs by reducing the number of random deviates that must be generated. Correlating results across response surfaces is trivial because all nine response surfaces can be studied from the same matched sample. Correlating results across distributions of **X** is done by having the $4 \times 3 \times 3$ cases use the same $N(0, 1)$ deviates. Only the G_1 sample needs to be linearly transformed in accordance with equation (8.7).

The $N(0, 1)$ deviates were generated by Marsaglia's rectangle-wedge-tail method described in Knuth (1969).

9. Bias Reduction Using Mahalanobis-Metric Matching

Donald B. Rubin

Abstract: Monte Carlo methods are used to study the ability of nearest-available, Mahalanobis-metric matching to make the means of matching variables more similar in matched samples than in random samples.

1. INTRODUCTION

Matched sampling is a method for selecting treatment units (e.g., smokers) and control units (e.g., nonsmokers) with similar values of matching variables $\mathbf{X}$ (e.g., age, sex, family medical history). Rubin (1976b,c) presented analytical work on a class of matching methods for multivariate $\mathbf{X}$, which are called 'equal percent bias reducing' (EPBR) because they yield the same percent reduction in expected bias for each matching variable, and thus for each linear combination of the matching variables. If the matching method is not EPBR, then there are different percent reductions in bias for different matching variables; in fact, any non-EPBR matching method increases bias for some linear combinations of the matching variables.

This communication extends the previous work of Rubin (1976b,c) by presenting Monte Carlo values for the percent reduction in bias when $\mathbf{X}$ is bivariate normal, and nearest-available, Mahalanobis-metric, pair matching is used. Related theoretical work on this matching method appears in Carpenter (1977). We chose to study Mahalanobis-metric matching because it is: (i) EPBR under commonly assumed distributional conditions (see Rubin, 1976b, §2); (ii) a natural extension of univariate nearest-available-pair matching methods studied earlier (Rubin, 1973a,b; Cochran and Rubin, 1973), in the sense that it attempts to find pair matches close on all matching variables; (iii) easy to implement in practice by using commonly available computer programs for sorting and calculating sample covariance matrices; and (iv) superior to another EPBR pair-matching method for nonlinear response surfaces (Rubin, 1979b).

2. NOTATION

Let P_1 be the population of treatment units, let P_2 be the population of control units and let μ_i and Σ_i be the mean and covariance, respectively, of $\mathbf{X}$ in P_i, $i = 1, 2$.

Reprinted with permission from *Biometrics*, 1980, **36**, 293–298. © 1980 International Biometric Society.

Random samples G_1 and G_2 of sizes N and $rN > N$ are obtained from P_1 and P_2, and $\mathbf{X}$ is recorded for all units in G_1 and G_2. In the main study, the dependent variables $\mathbf{Y}$ will be measured on all of the G_1 units but only on N of the G_2 units: the recording of $\mathbf{Y}$ might require, for example, intensive medical examinations that are too expensive to conduct on all rN controls. Therefore, a subsample of G_2 of size N will be chosen on which to measure $\mathbf{Y}$. The values of $\mathbf{Y}$ in this subsample and in the G_1 sample will be used to estimate the effect of the P_1 vs P_2 treatment on $\mathbf{Y}$.

In order to control for differences due to $\mathbf{X}$, the subsample of G_2 will be chosen by a method that tries to match the $\mathbf{X}$ distributions in the final samples, namely by random-order, nearest-available, Mahalanobis-metric matching. This matching method is defined as follows. First, randomly order the G_1 units; then choose for each G_1 unit in turn the nearest-available (i.e., previously unmatched) control unit as a match, where nearest is defined by the sample Mahalanobis metric. That is, let $\mathbf{X}_1$ and $\mathbf{X}_2$ be the values of $\mathbf{X}$ for units from G_1 and G_2; then the distance between the units is $(\mathbf{X}_1 - \mathbf{X}_2)\mathbf{S}^{-1}(\mathbf{X}_1 - \mathbf{X}_2)^T$, where $\mathbf{S} = \{(N-1)\mathbf{S}_1 + (rN-1)\mathbf{S}_2\}/(N+rN-2)$ with $\mathbf{S}_i$ being the sample covariance of $\mathbf{X}$ in $G_i, i = 1, 2$. The resulting matched sample will have N pairs of $G_1 - G_2$ units 'close' with respect to the matching variables $\mathbf{X}$.

Let $\bar{\mathbf{X}}_i$ be the mean of $\mathbf{X}$ in $G_i, i = 1, 2$, and let $\bar{\mathbf{X}}_{2*}$ be the mean of $\mathbf{X}$ in the matched G_2 sample; also, let μ_{2*} be the expected value of $\bar{\mathbf{X}}_{2*}$. If the matching is EPBR then $\mu_1 - \mu_2 = k(\mu_1 - \mu_{2*})$, where $\theta = 100(1 - k)$ is the percent reduction in expected bias; when $\mathbf{Y}$ is linear in $\mathbf{X}$, θ is also the percent reduction in bias of $\mathbf{Y}$. For EPBR matching, θ is an obvious measure of how much closer the distributions of $\mathbf{X}$ are in matched samples than in random samples.

Nearest-available Mahalanobis-metric matching will be EPBR under two distributional conditions given in Theorem 2 of Rubin (1976b). The first condition is that $\mathbf{X}$ is ellipsoidally distributed; this will be the case if a linear transformation of $\mathbf{X}$ is spherically distributed (e.g., multivariate normal, multivariate t). The second condition is that, orthogonal to the discriminant, the covariances of $\mathbf{X}$ in P_1 and P_2 are proportional; that is,

$$\Sigma_1 \propto \Sigma_2 + c(\mu_1 - \mu_2)^T(\mu_1 - \mu_2), \tag{9.1}$$

where c is some scalar. Equation (9.1) implies that the best linear discriminant with respect to Σ_1 is the same as the best linear discriminant with respect to Σ_2.

3. MONTE CARLO PERCENT REDUCTIONS IN BIAS FOR BIVARIATE NORMAL X

In order to obtain Monte Carlo values of the percent reduction in bias for Mahalanobis-metric matching when it is EPBR, we choose the simple case with $\mathbf{X}$ bivariate normal in both P_1 and P_2; then, without loss of generality we can assume

$$\mathbf{X} \sim N\left\{\begin{bmatrix} \eta \\ 0 \end{bmatrix}, \begin{bmatrix} \sigma^2 & 0 \\ 0 & \xi^2 \end{bmatrix}\right\} \text{ in } P_1; \mathbf{X} \sim N\left\{\begin{bmatrix} 0 \\ 0 \end{bmatrix}, \begin{bmatrix} 1 & 0 \\ 0 & 1 \end{bmatrix}\right\} \text{ in } P_2. \tag{9.2}$$

This follows from the restriction (9.1) and the fact that any inner-product matching method is location and scale invariant. Note that η is the bias along the discriminant, σ^2 is the ratio of variances along the discriminant and ξ^2 is the ratio of variances

Table 9.1. *Percent reduction in bias of* **X***, Mahalanobis-metric matching,*
N = 50, **X** *normal*, Monte Carlo values*

r	ξ^2	$B=\frac{1}{4}$	$\frac{1}{2}$	$\frac{3}{4}$	1	$\frac{1}{4}$	$\frac{1}{2}$	$\frac{3}{4}$	1	$\frac{1}{4}$	$\frac{1}{2}$	$\frac{3}{4}$	1
			$\sigma^2=\frac{1}{2}$				$\sigma^2=1$				$\sigma^2=2$		
	$\frac{1}{2}$	92	90	85	77	84	78	71	63	61	56	52	47
2	1	92	90	86	79	84	80	73	65	63	58	53	48
	2	89	88	84	77	84	77	72	64	63	57	52	47
	$\frac{1}{2}$	96	96	94	90	90	88	85	79	73	70	66	61
3	1	94	95	94	91	90	89	86	80	75	72	67	61
	2	92	94	91	89	89	87	83	78	75	70	66	60
	$\frac{1}{2}$	97	97	97	95	93	92	90	85	80	77	73	68
4	1	96	96	96	94	92	92	90	86	81	78	73	69
	2	93	93	93	92	89	89	87	84	79	75	73	67
	$\frac{1}{2}$	100	98	98	97	95	95	94	91	85	83	80	75
6	1	96	98	97	97	95	94	93	91	86	83	80	76
	2	97	95	95	95	94	92	91	89	83	81	79	74
	$\frac{1}{2}$	99	99	99	98	98	97	96	94	89	87	85	81
9	1	98	98	98	98	96	96	96	94	90	87	85	81
	2	96	97	96	96	94	94	93	92	87	85	83	80

* See expression (9.2).

orthogonal to the discriminant. The number of standard deviations between the means of the discriminant is $B = \eta\{\frac{1}{2}(1+\sigma^2)\}^{-\frac{1}{2}}$.

Table 9.1 presents Monte Carlo results for Mahalanobis-metric matching when $N = 50$; $B = \frac{1}{4}, \frac{1}{2}, \frac{3}{4}, 1$; $\sigma^2 = \frac{1}{2}, 1, 2$; $\xi^2 = \frac{1}{2}, 1, 2$ and $r = 2, 3, 4, 6, 9$. These values of B, σ^2, ξ^2 and r are thought to be representative of values likely to occur in practice. Previous results suggest that values obtained for $N = 50$ are typical (on the conservative side) for $N > 50$, while $N < 50$ appears to be uncommon in practice. The standard errors of these values are typically less than 1%; details of the computational methods are found in the Appendix.

There are several trends in Table 9.1, which can be summarized as follows. (i) The value of ξ^2 has only a minor effect on θ. (ii) The value of σ^2 has a large effect on θ, $\sigma^2 = \frac{1}{2}$ being relatively favorable to matching and $\sigma^2 = 2$ being relatively unfavorable to matching. (iii) The value of B has a large effect on θ, $B = \frac{1}{4}$ being relatively favorable and $B = 1$ being relatively unfavorable. (iv) The value of r has a large effect on θ in cases unfavorable to matching, i.e., in cases with large B or σ^2, larger values of r being more favorable.

The fourth trend is qualitatively obvious: a larger pool of controls from which to match yields better matches. The other trends are also qualitatively obvious because, as σ^2 becomes smaller and B becomes larger, the P_1 distribution effectively becomes farther from the fixed P_2 distribution. The value of ξ^2 is fairly irrelevant although one might expect that $\xi^2 = 1$ is slightly more favorable to matching than $\xi^2 \neq 1$ because then the distributions of **X** orthogonal to the discriminant are the same in P_1 and P_2.

4. WHY NOT 100% BIAS REDUCTION?

In order to understand better the quantitative aspects of the results in Table 9.1, especially the inability to attain 100% bias reduction, let us consider how to obtain as close to 100% bias reduction as possible. Let $\Delta = (\mu_1 - \mu_2)\Sigma_2^{-1}$ be the coefficient of the best linear discriminant with respect to the P_2 inner product. For large N, when $\mathbf{X}$ is ellipsoidally distributed and restriction (9.1) holds, choosing for the matched G_2 sample the N G_2 units whose mean value of $\mathbf{X}\Delta^T$ is closest to $\bar{\mathbf{X}}_1\Delta^T$ (i.e., mean matching on the P_2-best linear discriminant) is EPBR and makes the means of $\mathbf{X}$ in the matched samples as similar as possible (see Rubin, 1976c, §3).

The resulting percent reduction in expected bias is given by min $(100, \theta_{\max})$, where $\theta_{\max}$ can be approximated using expressions from Rubin (1976c), reprinted as expressions (7.5) and (7.6) in Chapter 7 of this volume; thus:

$$\theta_{\max} \doteq 100 \times \frac{2^{\frac{\pi}{2}}}{\pi^{\frac{1}{2}}} \times \frac{r^{(1-\frac{\pi}{4})}(1-1/r)^{\frac{\pi}{4}}}{B(1+\sigma^2)^{\frac{1}{2}}}, \tag{9.3}$$

where $\sigma^2 = \sigma_1^2/\sigma_2^2$ is the ratio of the variances of the P_2-best linear discriminant in P_1 and P_2, $\sigma_i^2 = \Delta\Sigma_i^{-1}\Delta^T$ and $\mathbf{B} = (\eta_1 - \eta_2)/\{(\sigma_1^2 + \sigma_2^2)/2\}^{\frac{1}{2}}$ is the number of standard deviations between the means, $\eta_1 = \mu_1\Delta^T$ and $\eta_2 = \mu_2\Delta^T$, of the P_2-best linear discriminant in P_1 and P_2. This approximation for $\theta_{\max}$ appears to hold quite well for moderate r (say, <10), moderate N (say, >25) and for a variety of underlying distributions with both heavy and normal tails (see Rubin, 1976b, §4). Notice that the dimensionality of $\mathbf{X}$, that is, the number of matching variables, does not appear explicitly in (9.3) but enters only via σ^2 and B. The rows of Table 9.2 labelled MAX give the values of min $(100, \theta_{\max})$ for $B = \frac{1}{4}, \frac{1}{2}, \frac{3}{4}, 1; \sigma^2 = \frac{1}{2}, 1, 2$ and $r = 2, 3, 4$.

In order to study the loss in bias reduction due to using pair matching rather than mean matching, assume that $\mathbf{X}$ is univariate normal and matches are chosen by random-order, nearest-available, Mahalanobis-metric, pair matching (any metric matching method would yield the same results because $\mathbf{X}$ is univariate). The rows of Table 9.2 labelled MM-1 give Monte Carlo values of the percent reduction in expected bias when $N = 50$. The differences between the MAX values and the MM-1 values are smaller in easy cases (cases with small σ^2, small B, large r) than in hard cases. When $\mathbf{X}$ is multivariate the MM-1 values can also be interpreted as the percent reduction in bias obtained by pair matching on the P_2-best linear discriminant, that is, by calculating $\mathbf{X}\Delta^T$ for all units and pair matching on this scalar variable (this assumes Δ is known).

The rows of Table 9.2 labelled MM-2 reproduce values from Table 9.1 for Mahalanobis-metric matching with bivariate normal $\mathbf{X}$ and $N = 50$. The MM-2 values are always less than or equal to the MM-1 values indicating that with fixed B, σ^2 and r, pair matching on two variables results in less bias reduction than pair matching on one variable. Also, the differences between the MM-1 and MM-2 rows are relatively constant at around 5%, although there is a slight tendency for the differences to be larger in the easier distributional cases (i.e., smaller B and σ^2).

The results in Table 9.2, thus, suggest that with bivariate $\mathbf{X}$ and Mahalanobis-metric matching the inability to obtain the maximum bias reduction in easy cases is primarily due to the dimensionality of $\mathbf{X}$ (i.e., the attempt to have matches that

Table 9.2. *Percent reduction in bias of* **X** *(***X** *normal)*

r		$\sigma^2 = \frac{1}{2}$				$\sigma^2 = 1$				$\sigma^2 = 2$			
		$B = \frac{1}{4}$	$\frac{1}{2}$	$\frac{3}{4}$	1	$\frac{1}{4}$	$\frac{1}{2}$	$\frac{3}{4}$	1	$\frac{1}{4}$	$\frac{1}{2}$	$\frac{3}{4}$	1
	MAX	100	100	100	100	100	100	100	80	100	100	87	65
2	MM-1	99	98	93	84	92	87	78	69	66	59	53	51
	MM-2	92	90	86	79	84	80	73	65	63	58	53	48
	MAX	100	100	100	100	100	100	100	100	100	100	100	89
3	MM-1	100	99	99	97	96	95	91	84	79	75	69	63
	MM-2	94	95	94	91	90	89	86	80	75	72	67	61
	MAX	100	100	100	100	100	100	100	100	100	100	100	100
4	MM-1	100	100	100	99	98	97	94	89	86	81	75	71
	MM-2	96	96	96	94	92	92	90	86	81	78	73	69

are close orthogonal to the discriminant as well as along the discriminant), whereas the inability to obtain the maximum bias reduction in hard cases is primarily due to choosing to pair match rather than mean match.

These conclusions do not imply that in practice we should mean match on the discriminant in order to optimize the percent reduction in bias of **X**. Matching to optimize the percent reduction in bias makes most sense when we know dependent variables **Y** are linear in **X**, because then the percent reduction in bias of **Y** equals that of **X**. In practice, we must be concerned with the possibility of nonlinear relationships, and pair matching rather than mean matching will be more robust to nonlinearities. Results in Rubin (1979b) show that with nonlinear relationships, Mahalanobis-metric matching is to be preferred to matching on the discriminant whether or not regression adjustments are to be performed. The results in this communication indicate how much closer the means of the distributions can be made by nearest-available Mahalanobis-metric, pair matching, and consequently the results provide guidance for judging the ratios of sample sizes needed to obtain well-matched samples.

ACKNOWLEDGMENTS

I would like to thank D. T. Thayer for excellent and extensive programming support, and two referees for helpful editorial comments.

APPENDIX

1. *Simulation procedure: perform steps A, B, C below for* $i = 1, \ldots, 100$
    A. Generate raw data – 1000 normal (0, 1) deviates – using a method developed by Marsaglia, which is known as the 'rectangle-wedge-tail' method, Knuth (1969).
        (i) G_1 sample: $\mathbf{Z}_1 = 50 \times 2$.
        (ii) G_2 samples: $\mathbf{Z}_2 = 450 \times 2$ (first 100 rows represent G_2 when $r = 2$, first 150 rows represent G_2 when $r = 3, \ldots$, all 450 rows represent G_2 when $r = 9$).

B. Compute values of simulation covariates – first three moments of Z_1 in G_1 (two means, two variances, one covariance, two skewnesses, two crossed third moments) and first three moments of Z_2 in each of six G_2 samples.

   (i) $SC1(i, j), j = 1, \ldots 9$, gives the values of the nine simulation covariates in G_1.

   (ii) $SC2(i, j, k), j = 1, \ldots, 9$, gives the values of the nine simulation covariates in G_2 for the kth ratio, $k = 1, \ldots, 5$ ($r_1 = 2, r_2 = 3, r_3 = 4, r_4 = 6, r_5 = 9$).

C. Perform the matching and calculate BIAS – perform steps (i) and (ii) below for each of $4(B) \times 3(\sigma^2) \times 3(\xi^2) = 36$ distributions of X in P_1, and index these distributions of X by $l = 1, \ldots, 36$.

   (i) Calculate X values in G_1, X_1. Each row of X_1 is obtained by a linear transformation of each row of Z_1:

$$X_1 = (\eta, 0)^T + \begin{bmatrix} \sigma & 0 \\ 0 & \xi \end{bmatrix} Z_1. \qquad (A.1)$$

Let $B(0)$ be the mean of the first component of $X_1 = \eta + \sigma \times SC1(i, 1)$.

   (ii) For $k = 1, \ldots, 5$, select the matched sample from first $50 \times r_k$ rows of Z_2 and let $B(k)$ be the mean of the first component of Z_2 in the matched sample.

   (iii) Let BIAS $(i, k, l) = B(0) - B(k)$, where i indexes simulation replication, k indexes ratio and l indexes distribution of X.

2. *Notes on analysis of data generated by program*
For each ratio indexed by $k = 1, \ldots, 5$, there are 18 simulation covariates $[SC1(i, j) j = 1, \ldots, 9$ and $SC2(i, j, k) j = 1, \ldots 9]$ having known expectation $(0, 0, 1, 1, 0, 0, 0, 0, 0, 0, 0, 1, 1, 0, 0, 0, 0, 0)$ and 36 dependent variables corresponding to the 36 levels of l, BIAS(i, k, l). The covariance adjusted estimate of the expected value of each dependent variable can be found in the standard manner of sample surveys by regressing the dependent variable on the covariates where the units are the simulation replications indexed by i; see for example, Cochran (1963, Chapter 7).

PART IV: FUNDAMENTALS OF PROPENSITY
SCORE MATCHING

In 1978 I had a Guggenheim Fellowship to pursue work on causal inference. I had recently completed my *Annals of Statistics* article, Rubin (1978a), dealing with Bayesian inference for causal effects, and I was actively consulting with June Reinisch, Director of the Kinsey Institute, on matching problems with a large Danish cohort (used for illustrative purposes in several articles, including Rosenbaum and Rubin (1985a,b), both reprinted in this part, and Rubin and Thomas (1996, 2000), both reprinted in Part V. Some of the empirical context for this problem appears in Chapter 25 in Part VII.

Starting in the winter of 1978, I was able to take a one-semester leave from ETS to visit Harvard University, and taught a seminar on causal inference. There was an outstanding first-year PhD student who took my seminar and ended up writing his thesis largely under my supervision – Paul Rosenbaum. The combination of his brilliance and our real data problems led to a series of papers on propensity score methods that appear to have started an avalanche of very recent contributions addressing propensity score methods; Google, at the end of 2005, lists close to a million entries for "propensity scores" or "propensity score."

Chapter 10, Rosenbaum and Rubin (1983a), introduced the propensity score. It provided definitions of balancing scores, propensity scores, strongly ignorable treatment assignment (a stronger version of ignorable treatment assignment, defined in Rubin (1978a) but implied in Rubin (1976a)). Moreover, this article proved key theorems and corollaries about the propensity score's role in observational studies, for matching, for subclassification, and for model-based adjustment. Also, we tied this work to earlier work by Cochran and by me, referenced in the introductions to Parts I and II. In some ways, in hindsight I wish that we had more carefully distinguished between the use of matching and subclassification on the propensity for the design of observational studies, on one hand, and the use of model-based adjustment for the propensity score, on the other. We did not say anything technically wrong, but I do think that there has been some misguided use of the propensity score for model-based adjustment that possibly could have been avoided if this first paper on the topic had more clearly distinguished between the issues of design and analysis.

Also appearing that same year and reprinted here was Chapter 11, Rosenbaum and Rubin (1983b), on sensitivity analyses following propensity score subclassification for assessing the assumption of strongly ignorable assignment. This method has been used in recent medical studies (e.g., Conners et al., 1996), and is a direct descendant of the classic work by Jerome Cornfield et al. (1959), and later Bross (1966). These two classic references addressed the association between smoking and lung cancer that had been seen in many observational studies. The issue of whether cigarette smoking caused lung cancer was a matter of debate in the 1950s and 1960s, and the sensitivity of conclusions to relevant but unobserved covariates was a major concern.

The key idea of sensitivity analyses is to assume that treatment assignment would be strongly ignorable if a key unobserved covariate U were observed, and then to posit various plausible associations between U and the outcomes, and between U and treatment assignment. If adjustment for U based on plausible associations made the estimated relationship between treatment and outcome vanish, then the conclusion is that the causal effect of treatment might not exist. If, on the other hand, the estimated treatment effect remained strong after such adjustment, then there is evidence that the effect is real.

Two other papers with Paul, also reprinted here, applied the propensity score methods for design outlined in the seminal paper to real examples. Rosenbaum and Rubin (1984b) applied subclassification, or as it is sometimes called, stratified matching, to a real data set that Paul had worked on during the period of his thesis – part of a Duke databank dealing with coronary bypass surgery. This was a careful methodological paper, providing guidance in the context of a detailed example about how to estimate propensity scores and carry out diagnostic checks on the resulting balance, which is still an important topic that needs substantial work. This example had 74 covariates, and we showed how Cochran's (1968a) advice concerning the number of subclasses to use with a single X worked with many X's when using the propensity score to summarize them all. Also, we described, in theory, what it meant to create balance on X when there were missing values in some of the X variables for some of the units; the inability to implement this theory in most practical situations led to D'Agostino and Rubin (2000), reprinted here as Chapter 21.

Another paper, Rosenbaum and Rubin (1985a), reprinted here as Chapter 13, applied propensity score matching, including other optional refinements, such as Mahalanobis-metric matching on key covariates, to the aforementioned Kinsey data set. The conclusion was that Mahalanobis-metric matching on a few key continuous covariates within propensity score calipers (e.g., $\frac{1}{4}$ standard deviation of the propensity score) was generally a very effective matching method, more effective than simply matching on the propensity score, and far more effective than attempting to Mahalanobis-metric match on many covariates. Of course, this advice was in the context of the example, which had 15 covariates to be balanced (and their squares and cross products), and an unusually large control reservoir, roughly 20 times the size of the treated group. It is important to note in this example and in the previous Duke data set, that the estimation of the propensity score, diagnostics for it, re-estimation, et cetera, took place prior to any access to any outcome data. In this sense, these efforts comprise the design phase of an observational study and correspond to the design phase of a randomized experiment – no outcome variables are available. This issue has been mentioned earlier, is discussed in various articles reprinted in this volume, especially in Rubin (2001b), reprinted in Part VI, and will be mentioned again in the concluding advice.

The final paper with Paul reprinted here is Chapter 14, Rosenbaum and Rubin (1985b), which used the same Kinsey data to illustrate how an attempt to categorize covariates and find exact matches on the categorized variables is definitely an inferior technique. Not only does it typically lead to an unnecessarily large fraction of unmatched units, but also the resulting matches are inferior because of the coarseness of the matching created by the categorized covariates. Many of the sticky issues with such "old-fashioned" matching methods are avoided using propensity score methods.

A final comment here is that Paul Rosenbaum has gone on to make very substantial contributions to matching that are entirely independent of mine, including to the topics of "greedy" versus "optimal matching" and "full" matching, as well as to other areas of causal inference, including sensitivity analysis and randomization-based inference (e.g., in his book, Rosenbaum (2002)). Optimal matching attempts to obtain one–one matches that satisfy some criterion for optimality in the entire matched samples (e.g., by using network flow theory). In contrast, greedy matching orders the treated units in some way (I considered high–low, low–high, and random in my 1973 *Biometrics* papers), and finds the nearest available match from the control reservoir after setting aside previously selected controls. My early assessment was that there was not much payoff in trying to be optimal because essentially the same set of controls was chosen in any case; I still believe that this is often the case. Full matching refers to matched samples created by allowing both one treated unit to be matched to a variable number of controls, or having a variable number of treated units matched to one control unit. In some situations, another name for this procedure could be "maximal subclassification," because each subclass has exactly one treated or one control unit. I believe that there is a lot of merit in the associated ideas and techniques. A variety of Paul's publications, some joint with coauthors, have explored these interesting and important topics (Gu and Rosenbaum, 1993; Ming and Rosenbaum, 2000; Rosenbaum, 1989, 1991, 1995, 2002).

10. The Central Role of the Propensity Score in Observational Studies for Causal Effects

Paul R. Rosenbaum and Donald B. Rubin

Abstract: The propensity score is the conditional probability of assignment to a particular treatment given a vector of observed covariates. Both large and small sample theory show that adjustment for the scalar propensity score is sufficient to remove bias due to all observed covariates. Applications include: (i) matched sampling on the univariate propensity score, which is a generalization of discriminant matching, (ii) multivariate adjustment by subclassification on the propensity score where the same subclasses are used to estimate treatment effects for all outcome variables and in all subpopulations, and (iii) visual representation of multivariate covariance adjustment by a two-dimensional plot.

1. DEFINITIONS

1.1. The Structure of Studies for Causal Effects

Inferences about the effects of treatments involve speculations about the effect one treatment would have had on a unit which, in fact, received some other treatment. We consider the case of two treatments, numbered 1 and 0. In principle, the ith of the N units under study has both a response r_{1i} that would have resulted if it had received treatment 1, and a response r_{0i} that would have resulted if it had received treatment 0. In this formulation, causal effects are comparisons of r_{1i} and r_{0i}, for example $r_{1i} - r_{0i}$ or r_{1i}/r_{0i}. Since each unit receives only one treatment, either r_{1i} or r_{0i} is observed, but not both, so comparisons of r_{1i} and r_{0i} imply some degree of speculation. In a sense, estimating the causal effects of treatments is a missing data problem, since either r_{1i} or r_{0i} is missing.

This formulation is that used in the literature of experimental design, for example, in the classic books by Fisher (1935) and Kempthorne (1952), and follows the development by Rubin (1974, 1977a, 1978a, 1980a); Hamilton (1979) adopts a similar approach. The structure would not be adequate when, for example, the response of unit i to treatment t depends on the treatment given to unit j, as could happen if they compete for resources. The assumption that there is a unique value r_{ti} corresponding to unit i and treatment t has been called the stable unit-treatment value assumption (Rubin, 1980a), and will be made here. For discussion of some possible violations of this assumption, see Cox (1958, Chapter 2) or Rubin (1978a, §2.3).

Reprinted from *Biometrika*, 1983, **70**, 1, 41–55, by permission of Oxford University Press.

In this paper, the N units in the study are viewed as a simple random sample from some population, and the quantity to be estimated is the average treatment effect, defined as

$$E(r_1) - E(r_0), \tag{10.1}$$

where $E(.)$ denotes expectation in the population.

Let $z_i = 1$ if unit i is assigned to the experimental treatment, and $z_i = 0$ if unit i is assigned to the control treatment. Let x_i be a vector of observed pretreatment measurements or covariates for the ith unit; all of the measurements in x are made prior to treatment assignment, but x may not include all covariates used to make treatment assignments. It is assumed that the numbering of units is done at random, so that the index i contains no information; observed information about unit i is contained in x_i. Throughout, we ignore measure theoretic details.

1.2. Balancing Scores and the Propensity Score

In randomized experiments, the results in the two treatment groups may often be directly compared because their units are likely to be similar, whereas in nonrandomized experiments, such direct comparisons may be misleading because the units exposed to one treatment generally differ systematically from the units exposed to the other treatment. Balancing scores, defined here, can be used to group treated and control units so that direct comparisons are more meaningful.

A balancing score, $b(x)$, is a function of the observed covariates x such that the conditional distribution of x given $b(x)$ is the same for treated ($z = 1$) and control ($z = 0$) units; that is, in Dawid's (1979) notation,

$$x \perp\!\!\!\perp z \mid b(x).$$

The most trivial balancing score is $b(x) = x$. More interesting balancing scores are many-one functions of x. In Section 2 we identify all functions of x that are balancing scores and identify the coarsest function of x that is a balancing score, namely the propensity score. We also show that easily obtained estimates of balancing scores behave like balancing scores. Also, we show that if treatment assignment is strongly ignorable given x, as defined in Section 1.3, then the difference between treatment and control means at each value of a balancing score is an unbiased estimate of the treatment effect at that value, and consequently pair matching, subclassification and covariance adjustment on a balancing score can produce unbiased estimates of the average treatment effect (10.1). Moreover in Section 3 we see that common methods of multivariate adjustment in observational studies, including covariance adjustment for x and discriminant matching (Cochran & Rubin, 1973), implicitly adjust for an estimated scalar balancing score.

In order to motivate formally adjustment for a balancing score, we must consider the sampling distribution of treatment assignments. Let the conditional probability of assignment to treatment one, given the covariates, be denoted by

$$e(x) = \mathrm{pr}(z = 1 \mid x), \tag{10.2}$$

where we assume

$$\text{pr}(z_1, \ldots, z_n \mid x_1, \ldots, x_n) = \prod_{i=1}^{N} e(x_i)^{z_i} \{1 - e(x_i)\}^{1-z_i}.$$

Although this strict independence assumption is not essential, it simplifies notation and discussion. The function $e(x)$ is called the propensity score, that is, the propensity towards exposure to treatment 1 given the observed covariates x. In Section 2, $e(x)$ is shown to be the coarsest balancing score.

1.3. Strongly Ignorable Treatment Assignment

Randomized and nonrandomized trials differ in two distinct ways because in randomized experiments z_i has a distribution determined by a specified random mechanism. First, in a randomized trial, the propensity score is a known function so that there exists one accepted specification for $e(x)$. In a nonrandomized experiment, the propensity score function is almost always unknown so that there is not one accepted specification for $e(x)$; however, $e(x)$ may be estimated from observed data, perhaps using a model such as a logit model. To a Bayesian, estimates of these probabilities are posterior predictive probabilities of assignment to treatment 1 for a unit with vector x of covariates.

The second way randomized trials differ from nonrandomized trials is that, with properly collected data in a randomized trial, x is known to include all covariates that are both used to assign treatments and possibly related to the response (r_1, r_0). More formally, in a randomized trial, treatment assignment z and response (r_1, r_0) are known to be conditionally independent given x,

$$(r_1, r_0) \perp\!\!\!\perp z \mid x.$$

This condition is usually not known to hold in a nonrandomized experiment. Moreover, in a randomized experiment, every unit in the population has a chance of receiving each treatment. Generally, we shall say treatment assignment is strongly ignorable given a vector of covariates v if

$$(r_1, r_0) \perp\!\!\!\perp z \mid v, \quad 0 < \text{pr}(z = 1 \mid v) < 1 \tag{10.3}$$

for all v. For brevity, when treatment assignment is strongly ignorable given the observed covariates x, that is, when (10.3) holds with $v = x$, we shall say simply that treatment assignment is strongly ignorable. If treatment assignment is strongly ignorable, then it is ignorable in Rubin's (1978a) sense, but the converse is not true.

2. THEORY

2.1. Outline

Section 2 presents five theorems whose conclusions may be summarized as follows.

(i) The propensity score is a balancing score.
(ii) Any score that is 'finer' than the propensity score is a balancing score; moreover, x is the finest balancing score and the propensity score is the coarsest.

(iii) If treatment assignment is strongly ignorable given x, then it is strongly ignorable given any balancing score.
(iv) At any value of a balancing score, the difference between the treatment and control means is an unbiased estimate of the average treatment effect at that value of the balancing score if treatment assignment is strongly ignorable. Consequently, with strongly ignorable treatment assignment, pair matching on a balancing score, subclassification on a balancing score and covariance adjustment on a balancing score can all produce unbiased estimates of treatment effects.
(v) Using sample estimates of balancing scores can produce sample balance on x.

2.2. Large-Sample Theory

The results in this section treat $e(x)$ as known, and are therefore applicable to large samples.

Theorem 1. *Treatment assignment and the observed covariates are conditionally independent given the propensity score, that is*

$$x \perp\!\!\!\perp z \mid e(x).$$

The above theorem is a special case of Theorem 2. Cochran & Rubin (1973) proved Theorem 1 in the particular case of multivariate normal covariates x; the result holds regardless of the distribution of x.

Theorem 2. *Let $b(x)$ be a function of x. Then $b(x)$ is a balancing score, that is,*

$$x \perp\!\!\!\perp z \mid b(x), \tag{10.4}$$

if and only if $b(x)$ is finer than $e(x)$ in the sense that $e(x) = f\{b(x)\}$ for some function f.

Proof. First suppose $b(x)$ is finer than $e(x)$. Since $e(x) = \text{pr}(z = 1 \mid x)$, to show $b(x)$ is a balancing score it is sufficient to show

$$\text{pr}\{z = 1 \mid b(x)\} = e(x). \tag{10.5}$$

Now by the definition of $e(x)$,

$$\text{pr}\{z = 1 \mid b(x)\} = E\{e(x) \mid b(x)\}.$$

But since $b(x)$ is finer than $e(x)$,

$$E\{e(x) \mid b(x)\} = e(x),$$

as required, so that $b(x)$ is a balancing score.

Now, for the converse, suppose $b(x)$ is a balancing score, but that $b(x)$ is not finer than $e(x)$, so that there exists x_1 and x_2 such that $e(x_1) \neq e(x_2)$ but $b(x_1) = b(x_2)$. But then, by the definition of $e(\cdot)$, $\text{pr}(z = 1 \mid x_1) \neq \text{pr}(z = 1 \mid x_2)$, so that z and x are not conditionally independent given $b(x)$, and thus $b(x)$ is not a balancing score. Therefore, to be a balancing score, $b(x)$ must be finer than $e(x)$.

Theorem 1 implies that if a subclass of units or a matched treatment-control pair is homogeneous in $e(x)$, then the treated and control units in that subclass or matched pair will have the same distribution of x. Theorem 2 implies that if subclasses or matched treated-control pairs are homogeneous in both $e(x)$ and certain chosen components of x, it is still reasonable to expect balance on the other components of x within these refined subclasses or matched pairs. The practical importance of Theorem 2 beyond Theorem 1 arises because it is sometimes advantageous to subclassify or match not only for $e(x)$ but for other functions of x as well; in particular, such a refined procedure may be used to obtain estimates of the average treatment effect in subpopulations defined by components of x, for example males, females.

Theorem 3 is the key result for showing that if treatment assignment is strongly ignorable, then adjustment for a balancing score $b(x)$ is sufficient to produce unbiased estimates of the average treatment effect (10.1).

Theorem 3. *If treatment assignment is strongly ignorable given x, then it is strongly ignorable given any balancing score b(x); that is,*

$$(r_1, r_0) \perp\!\!\!\perp z \mid x$$

and

$$0 < pr(z = 1 \mid x) < 1$$

for all x imply

$$(r_1, r_0) \perp\!\!\!\perp z \mid b(x)$$

and

$$0 < pr\{z = 1 \mid b(x)\} < 1$$

for all b(x).

Proof. The inequality given $b(x)$ follows immediately from the inequality given x. Consequently, it is suffcient to show that

$$pr\{z = 1 \mid r_1, r_0, b(x)\} = pr\{z = 1 \mid b(x)\},$$

which by Theorem 2, equation (10.5), is equivalent to showing that

$$pr\{z = 1 \mid r_1, r_0, b(x)\} = e(x).$$

Now

$$pr\{z = 1 \mid r_1, r_0, b(x)\} = E\{pr(z = 1 \mid r_1, r_0, x) \mid r_1, r_0, b(x)\},$$

which by assumption equals $E\{pr(z = 1 \mid x) \mid r_1, r_0, b(x)\}$, which by definition equals $E\{e(x) \mid r_1, r_0, b(x)\}$, which, since $b(x)$ is finer than $e(x)$, equals $e(x)$ as required.

Theorem 3 also can be proved using Lemmas 4.2(i) and 4.3 of Dawid (1979).

We are now ready to relate balancing scores and ignorable treatment assignment to the estimation of treatment effects.

The response r_t to treatment t is observed only if the unit receives treatment t, that is if $z = t$. Thus, if a randomly selected treated unit, $z = 1$, is compared to a randomly selected control unit, $z = 0$, the expected difference in response is

$$E(r_1 \mid z = 1) - E(r_0 \mid z = 0). \qquad (10.6)$$

Expression (10.6) does not equal (10.1) in general because the available samples are not from the marginal distribution of r_t, but rather from the conditional distribution of r_t given $z = t$.

Suppose a specific value of the vector of covariates x is randomly sampled from the entire population of units, that is, both treated and control units together, and then a treated unit and a control unit are found both having this value for the vector of covariates. In this two-step sampling process, the expected difference in response is

$$E_x\{E(r_1 \mid x, z = 1) - E(r_0 \mid x, z = 0)\}, \tag{10.7}$$

where E_x denotes expectation with respect to the distribution of x in the entire population of units. If treatment assignment is strongly ignorable, that is if (10.3) holds with $v = x$, then (10.7) equals

$$E_x\{E(r_1 \mid x) - E(r_0 \mid x)\},$$

which does equal the average treatment effect (10.1).

Now suppose a value of a balancing score $b(x)$ is sampled from the entire population of units and then a treated unit and a control unit are sampled from all units having this value of $b(x)$, but perhaps different values of x. Given strongly ignorable treatment assignment, it follows from Theorem 3 that

$$E\{r_1 \mid b(x), z = 1\} - E\{r_0 \mid b(x), z = 0\} = E\{r_1 \mid b(x)\} - E\{r_0 \mid b(x)\}$$

from which it follows that

$$E_{b(x)}[E\{r_1 \mid b(x), z = 1\} - E\{r_0 \mid b(x), z = 0\}]$$
$$= E_{b(x)}[E\{r_1 \mid b(x)\} - E\{r_0 \mid b(x)\}] = E(r_1 - r_0), \tag{10.8}$$

where $E_{b(x)}$ denotes expectation with respect to the distribution of $b(x)$ in the entire population. In words, under strongly ignorable treatment assignment, units with the same value of the balancing score $b(x)$ but different treatments can act as controls for each other, in the sense that the expected difference in their responses equals the average treatment effect.

The above argument has established the following theorem and corollaries.

Theorem 4. *Suppose treatment assignment is strongly ignorable and $b(x)$ is a balancing score. Then the expected difference in observed responses to the two treatments at $b(x)$ is equal to the average treatment effect at $b(x)$, that is,*

$$E\{r_1 \mid b(x), z = 1\} - E\{r_0 \mid b(x), z = 0\} = E\{r_1 - r_0 \mid b(x)\}.$$

Corollary 4.1. *Pair matching on balancing scores. Suppose treatment assignment is strongly ignorable. Further suppose that a value of a balancing score $b(x)$ is randomly sampled from the population of units, and then one treated, $z = 1$, unit and one control, $z = 0$, unit are sampled with this value of $b(x)$. Then the expected difference in response to the two treatments for the units in the matched pair equals the average treatment effect at $b(x)$. Moreover, the mean of matched pair differences obtained by this two-step sampling process is unbiased for the average treatment effect (10.1).*

Corollary 4.2. *Subclassification on balancing scores. Suppose treatment assignment is strongly ignorable. Suppose further that a group of units is sampled using $b(x)$ such that: (i) $b(x)$ is constant for all units in the group, and (ii) at least one unit in the group received each treatment. Then, for these units, the expected difference in treatment means equals the average treatment effect at that value of $b(x)$. Moreover, the weighted average of such differences, that is, the directly adjusted difference, is* *unbiased for the treatment effect* (10.1), *when the weights equal the fraction of the population at* $b(x)$.

Corollary 4.3. *Covariance adjustment on balancing scores. Suppose treatment assignment is strongly ignorable, so that in particular,* $E\{r_t \mid z = t, b(x)\} = E\{r_t \mid b(x)\}$ *for balancing score* $b(x)$. *Further suppose that the conditional expectation of* r_t *given* $b(x)$ *is linear:*

$$E\{r_t \mid z = t, b(x)\} = \alpha_t + \beta_t b(x) \qquad (t = 0, 1).$$

Then the estimator

$$(\hat{\alpha}_1 - \hat{\alpha}_0) + (\hat{\beta}_1 - \hat{\beta}_0)\, b(x)$$

is conditionally unbiased given $b(x_i)$ $(i = 1, \ldots, n)$ *for the treatment effect at* $b(x)$, *namely* $E\{r_1 - r_0 \mid b(x)\}$, *if* $\hat{\alpha}_t$ *and* $\hat{\beta}_t$ *are conditionally unbiased estimators of* α_t *and* β_t, *such as least squares estimators. Moreover,*

$$(\hat{\alpha}_1 - \hat{\alpha}_0) + (\hat{\beta}_1 - \hat{\beta}_0)\bar{b},$$

where $\bar{b} = n^{-1}\Sigma b(x_i)$, *is unbiased for the average treatment effect* (10.1) *if the units in the study are a simple random sample from the population.*

2.3. Some Small-Sample Theory

Usually the propensity scores $e(x_i)$ must be estimated from available data, $(z_i, x_i)(i = 1, \ldots, N)$. Define the sample conditional proportion prop $(A \mid B)$ as the proportion of those vectors (z_i, x_i) satisfying condition B that also satisfy condition A, leaving prop $(A \mid B)$ undefined if no vector satisfies condition B. For example, prop $\{z = 1 \mid x = (1, 0)\}$ is the proportion of the N units with $z = 1$ among all units with $x = (1, 0)$. Estimate $e(x)$ by $\hat{e}(a) = $ prop $(z = 1 \mid x = a)$. If $\hat{e}(a) = 0$ or 1 then all units with $x = a$ received the same treatment. Theorem 5, which parallels Theorem 1, shows that at all intermediate values of $\hat{e}(a)$, that is for $0 < \hat{e}(a) < 1$, there is sample balance. Of course, intermediate values of $\hat{e}(x)$ will exist only when x takes on relatively few values.

Theorem 5. *Suppose* $0 < \hat{e}(a) < 1$. *Then*

$$\text{prop}\{z = 0, x = a \mid \hat{e}(x) = \hat{e}(a)\}$$
$$= \text{prop}\{z = 0 \mid \hat{e}(x) = \hat{e}(a)\}\text{prop}\{x = a \mid \hat{e}(z) = \hat{e}(a)\}. \qquad (10.9)$$

An analogous theorem about sample balance parallels Theorem 2, and the proofs parallel the corresponding proofs of Theorems 1 and 2 because proportions follow essentially the same axioms as probabilities.

Corollary 5.1. *Suppose the* N *units are a random sample from an infinite population, and suppose* x *takes on only finitely many values in the population and at each such value* $0 < e(x) < 1$. *Then with probability 1 as* $N \to \infty$, *subclassification on* $\hat{e}(x)$ *produces sample balance, that is,* (10.9) *holds.*

In practice, except when x takes on only a few values, $\hat{e}(a)$ will be either zero or one for most values of a. Consequently, in order to estimate propensity scores, some modelling will be required.

The propensity score can often be modelled using an appropriate logit model (Cox, 1970) or discriminant score.

Clearly,

$$e(x) = \text{pr}(z = 1 \mid x) = \frac{\text{pr}(z = 1)\,\text{pr}(x \mid z = 1)}{\text{pr}(z = 1)\,\text{pr}(x \mid z = 1) + \text{pr}(z = 0)\,\text{pr}(x \mid z = 0)}.$$

Elementary manipulations establish the following facts.

(i) If $\text{pr}\,(x \mid z = t) = N_p\,(\mu_t, \Omega)$ then $e(x)$ is a monotone function of the linear discriminant $x^T \Omega^{-1}(\mu_1 - \mu_2)$. Therefore, matching on $e(x)$ includes discriminant matching (Cochran & Rubin, 1973; Rubin 1976b,c; 1979b; 1980b) as a special case. Some related results appear in Section 3.2.

(ii) If $\text{pr}(x \mid z = t)$ is a polynomial exponential family distribution, i.e., if

$$\text{pr}(x \mid z = t) = h(x) \exp\{P_t(x)\},$$

where $P_t(x)$ is a polynomial in x of degree k, say, then $e(x)$ obeys a polynomial logit model

$$\log \frac{e(x)}{1 - e(x)} = \log \frac{\text{pr}(z = 1)}{1 - \text{pr}(z = 1)} + P_1(x) - P_0(x)$$

$$= \log \frac{\text{pr}(z = 1)}{1 - \text{pr}(z = 1)} + Q(x),$$

where $Q(x)$ is a degree k polynomial in x. This polynomial exponential family includes the linear exponential family resulting in a linear logit model for $e(x)$, the quadratic exponential family described by Dempster (1971), and the binary data model described by Cox (1972). Related discussion is given by Dawid (1976).

3. THREE APPLICATIONS OF PROPENSITY SCORES TO OBSERVATIONAL STUDIES

3.1. Techniques for Adjustment in Observational Studies

The general results we have presented suggest that, in practice, adjustment for the propensity score should be an important component of the analysis of observational studies because evidence of residual bias in the propensity score is evidence of potential bias in estimated treatment effects. We conclude with three examples of how propensity scores can be explicitly used to adjust for confounding variables in observational studies. The examples involve three standard techniques for adjustment in observational studies (Cochran, 1965; Rubin, 1984c), namely, matched sampling, subclassification, and covariance adjustment, that is, the three methods addressed by Corollaries 4.1, 4.2, and 4.3.

3.2. Use of Propensity Scores to Construct Matched Samples from Treatment Groups

Matching is a method of sampling from a large reservoir of potential controls to produce a control group of modest size in which the distribution of covariates is similar to the distribution in the treated group. Some sampling of a large control reservoir is often required to reduce costs associated with measuring the response, for example, costs associated with obtaining extensive follow-up data on patients in clinical studies (Rubin, 1973a; Cohn et al., 1981).

Although there exist model-based alternatives to matched sampling, e.g., covariance adjustment on random samples, there are several reasons why matching is appealing.

(I) Matched treated and control pairs allow relatively unsophisticated researchers to appreciate immediately the equivalence of treatment and control groups, and to perform simple matched pair analyses which adjust for confounding variables. This issue is discussed in greater detail below in Section 3.3 on balanced subclassification.

(II) Even if the model underlying a statistical adjustment is correct, the variance of the estimate of the average treatment effect (10.1) will be lower in matched samples than in random samples since the distributions of x in treated and control groups are more similar in matched than in random samples. To verify this reduced variance, inspect the formula for the variance of the covariance adjusted estimate (Snedecor & Cochran, 1980, p. 368, formula 18.2.3), and note that the variance decreases as the difference between treatment and control means on x decreases.

(III) Model-based adjustment on matched samples is usually more robust to departures from the assumed form of the underlying model than model-based adjustment on random samples (Rubin, 1973b, 1979b), primarily because of reduced reliance on the model's extrapolations.

(IV) In studies with limited resources but large control reservoirs and many confounding variables, the confounding variables can often be controlled by multivariate matching, but the small-sample sizes in the final groups do not allow control of all variables by model-based methods.

Ideally, treated and control units would be exactly matched on all covariates x, so that the sample distributions of x in the two groups would be identical. Theorem 2 shows that it is sufficient to match exactly on any balancing score $b(x)$ to obtain the same probability distributions of x for treated and control units in matched samples. Moreover, Corollary 4.1 shows that if treatment assignment is strongly ignorable, exact matching on a balancing score leads to an unbiased estimate of the average treatment effect. Unfortunately, exact matches even on a scalar balancing score are often impossible to obtain, so methods which seek approximate matches must be used. We now study properties of some matching methods based on the propensity score.

A multivariate matching method is said to be equal percent bias reducing if the bias in each coordinate of x is reduced by the same percentage (Rubin, 1976b,c). Matching methods which are not equal percent bias reducing have the potentially undesirable property that they increase the bias for some linear functions of x. If matched sampling is performed before the response (r_1, r_0) can be measured, and if all that is suspected about the relation between (r_1, r_0) and x is that it is approximately linear, then matching methods which are equal percent bias reducing are reasonable in that they lead to differences in mean response in matched samples that should be less biased than in random samples.

The initial bias in x is

$$B = E(x \mid z = 1) - E(x \mid z = 0). \qquad (10.10)$$

Let us suppose that we have a random sample of treated units and a large reservoir of randomly sampled control units, and suppose each treated unit is matched with a control unit from the reservoir. Then the expected bias in x in matched samples is

$$B_m = E(x \mid z = 1) - E_m(x \mid z = 0), \qquad (10.11)$$

where the subscript m indicates the distribution in matched samples. In general, from Theorem 2, B_m is a null vector if exact matches on a balancing score have been obtained. If $B_m = \gamma B$ for some scalar γ, with $0 < \gamma < 1$, then the matching method is equal percent bias reducing: the bias in each coordinate of x is reduced by $100(1 - \gamma)\%$. If the method is not equal percent bias reducing, then there exists a vector w such that $w B_m > w B$, so that matching has increased the bias for some linear function of x.

In Section 2.3 we observed that discriminant matching is equivalent to matching on the propensity score if the covariates x have a multivariate normal distribution. Assuming multivariate normality, Rubin (1976b) showed that matching on the population or sample discriminant is equal percent bias reducing. We now show that matching on the population propensity score is equal percent bias reducing under weaker distributional assumptions. It is assumed that the matching algorithm matches each treated, $z = 1$, unit with a control, $z = 0$, unit drawn from a reservoir of control units on the basis of a balancing score, for example, using nearest available matching on a scalar balancing score.

Theorem 6. *Let $b = b(x)$ be a balancing score. For any matching method that uses b alone to match each treated unit, $z = 1$, with a control unit, $z = 0$, the reduction in bias is*

$$B - B_m = \int E(x \mid b) \, \{\mathrm{pr}_m(b \mid z = 0) - \mathrm{pr}\,(b \mid z = 0)\} \, db, \qquad (10.12)$$

where $\mathrm{pr}_m(b \mid z = 0)$ denotes the distribution of b in matched samples from the control group.

Proof. From (10.10) and (10.11) we have

$$B - B_m = \int \{E_m(x \mid z = 0, b)\mathrm{pr}_m(b \mid z = 0) - E(x \mid z = 0, b)\mathrm{pr}(b \mid z = 0)\} \, db. \qquad (10.13)$$

For any matching method satisfying the condition of the theorem,

$$E_m(x \mid z = 0, b) = E(x \mid z = 0, b) \qquad (10.14)$$

because any matching method using b alone to match units alters the marginal distribution of b in the control group, $z = 0$, but does not alter the conditional distribution of x given b in the control group. However, by Theorem 2,

$$E(x \mid z = 0, b) = E(x \mid b). \tag{10.15}$$

Substitution of (10.14) and (10.15) into equation (10.13) yields the result (10.12).

Corollary 6.1. *If $E(x \mid b) = \alpha + \beta f(b)$ for some vectors α and β and some scalar-valued function $f(\cdot)$, then matching on b alone is equal percent bias reducing.*

Proof. The percent reduction in bias for the ith coordinate of x is, from (10.12)

$$100 \frac{\beta_i[E_m\{f(b) \mid z = 0\} - E\{f(b) \mid z = 0\}]}{\beta_i[E\{f(b) \mid z = 1\} - E\{f(b) \mid z = 0\}]},$$

which is independent of i, as required.

The following corollary shows that if subpopulations are defined using x so that some function $d(x)$ is constant within each subpopulation, then propensity matching within subpopulations is equal percent bias reducing in each subpopulation.

Corollary 6.2. *Let $d = d(x)$ be some function of x. If $E(x \mid b, d) = \alpha_d + \beta_d f_d$ (b) for vectors α_d, β_d, and some scalar-valued functions $f_d(\cdot)$, then matching on b alone at each value of d is equal percent bias reducing at each value of d, that is,*

$$E(x \mid d, z = 1) - E_m(x \mid d, z = 0) = \gamma_d\{E(x \mid d, z = 1) - E(x \mid d, z = 0)\}$$

for scalar γ_d.

Proof. Apply Theorem 6 and Corollary 6.1 within subpopulations.

Rubin's (1979b) simulation study examines the small-sample properties of discriminant matching in the case of normal covariates with possibly different covariances in the treatment groups. Thus, the study includes situations where the true propensity score is a quadratic function of x but the discriminant score is a linear function of x. Table 10.1 presents previously unpublished results from this study for situations in which the propensity score is a monotone function of the linear discriminant, so that propensity matching and discriminant matching are effectively the same. The covariates x are bivariate normal with common covariance matrix. In the simulation, 50 treated units are matched using nearest available matching (Cochran & Rubin, 1973) on the sample discriminant with 50 control units drawn from a reservoir of $50R$ potential control units, for $R = 2, 3, 4$; details are given by Rubin (1979b).

Assuming parallel linear response surfaces, Table 10.1 shows that even in the absence of additional adjustments, propensity, i.e., discriminant, matching alone can remove most of the initial bias if the reservoir is relatively large. Moreover, Table 10.1 shows that the population and sample propensity scores are approximately equally effective in removing bias, so that no substantial loss is incurred by having to estimate the propensity score. It should be noted that the conditions underlying Table 10.1 differ from the conditions underlying Theorem 1 because nearest available matching with imperfect matches provides only a partial adjustment for the propensity score.

Table 10.1. *Percent reduction in bias due to matched sampling based on the sample and population propensity scores*

Ratio of size control reservoir to size of treatment group	Type of score	Initial bias along standardized discriminant			
		0.25	0.50	0.75	1.00
2	Sample	92	85	77	67
	Population	92	87	78	69
3	Sample	101	96	91	83
	Population	96	95	91	84
4	Sample	97	98	95	90
	Population	98	97	94	89

Assuming bivariate normal covariates with common covariance matrix, parallel linear response surfaces, sample size of 50 in treated and control groups. Estimated percent reduction in bias from Rubin's (1979b) simulation study. The largest estimated standard error for this table is less than 0.03.

Propensity matching should prove especially effective relative to Mahalanobis metric matching (Cochran & Rubin, 1973; Rubin, 1976b,c; 1979b; 1980b) in situations where markedly nonspherically distributed x make the use of a quadratic metric unnatural as a measure of distance between treated and control units. For example, we have found in practice that if x contains one coordinate representing a rare binary event, then Mahalanobis metric matching may try too hard to match that coordinate exactly, thereby reducing the quality of matches on the other coordinates of x. Propensity matching can effectively balance rare binary variables for which it is not possible to match treated and control units adequately on an individual basis.

3.3. Subclassification on Propensity Scores

A second major method of adjustment for confounding variables is subclassification, whereby experimental and control units are divided on the basis of x into subclasses or strata (Cochran, 1965, 1968a; Cochran & Rubin, 1973). Direct adjustment with subclass total weights can be applied to the subclass differences in response to estimate the average treatment effect (10.1) whenever treatment assignment is strongly ignorable, without modelling assumptions such as parallel linear response surfaces; see Corollary 4.2.

As a method of multivariate adjustment, subclassification has the advantage that it involves direct comparisons of ostensibly comparable groups of units within each subclass and therefore can be both understandable and persuasive to an audience with limited statistical training. The comparability of units within subclasses can be verified by the simplest methods, such as bar charts of means.

A major problem with subclassification (Cochran, 1965) is that as the number of confounding variables increases, the number of subclasses grows dramatically, so that even with only two categories per variable, yielding 2^P subclasses for P variables, most subclasses will not contain both treated and control units. Subclassification on the propensity score is a natural way to avoid this problem.

Table 10.2. *Example of increased balance using subclassification on estimated propensity score as summarized by distributions of F statistics for 74 covariates*

	Minimum	Lower quartile	Median	Upper quartile	Maximum
Treatment main effect without subclassification	4.0	6.8	10.9	16.8	51.8
Treatment main effect with subclassification	0.0	0.1	0.2	0.6	3.6
Treatment by subclass interaction	0.0	0.4	0.8	1.2	2.9

We now use an estimate of the propensity score to subclassify patients in an actual observational study of therapies for coronary artery disease. The treatments are coronary artery bypass surgery, $z = 1$, and drug therapy, $z = 0$. The covariates x are clinical, haemodynamic, and demographic measurements on each patient made prior to treatment assignment. Even though the covariates have quite different distributions in the two treatment groups, within each of the five subclasses, the surgical and drug patients will be seen to have similar sample distributions of x.

The propensity score was estimated using a logit model for z given x. Covariates and interactions among covariates were selected for the model using a stepwise procedure. Based on Cochran's (1968a) observation that subclassification with five subclasses is sufficient to remove at least 90% of the bias for many continuous distributions, five subclasses of equal size were constructed at the quintiles of the sample distribution of the propensity score, each containing 303 patients. Beginning with the subclass with the highest propensity scores, the five subclasses contained 234 surgical patients, 164 surgical patients, 98 surgical patients, 68 surgical patients and 26 surgical patients, respectively.

For each of the 74 covariates, Table 10.2 summarizes the balance before and after subclassification. The first row describes the 74 F statistics, that is the squares of the usual two-sample t statistics, for comparing the surgical group and drug group means of each covariate prior to subclassification. The second and third rows describe F statistics for the main effect of treatment and for interaction in a 2×5, treatments by subclasses, analysis of variance, performed for each covariate. Although there is considerable imbalance prior to subclassification, within the constructed subclasses there is greater balance than would have been expected if treatments had been assigned at random within each subclass.

When subclasses are perfectly homogeneous in $b(x)$, Theorem 2 shows that x has the same distribution for treated, $z = 1$, and control, $z = 0$, units in each subclass. Moreover, by Corollary 4.2, if treatment assignment is strongly ignorable, then the directly adjusted estimate with population total weights is unbiased for the average treatment effect (10.1). However, in this example, and generally in practice, subclasses will not be exactly homogeneous in the balancing score $b(x)$ that was used in subclassification, so the directly adjusted estimate may contain some residual bias due to x.

The corollary to the following theorem shows that direct adjustment based on a balancing score $b = b(x)$ can be expected to reduce bias in each coordinate of x providing the adjustment reduces the bias in b.

Let I_s be the set of values of a balancing score which make up subclass s ($s = 1, \ldots, S$), so that $b(a) \in I_s$ implies that units with $x = a$ fall in subclass s. Suppose the weight applied to subclass s in direct adjustment is w_s.

Theorem 7. *The bias in x after direct adjustment for the subclasses $(I_s, s = 1, \ldots, S)$ is*

$$B_s = \sum_{s=1}^{S} w_s \int E(x \mid b)\{\mathrm{pr}(b \mid z = 1, b \in I_s) - \mathrm{pr}(b \mid z = 0, b \in I_s)\}db,$$

where $b = b(x)$.

Corollary 7.1. *If $E(x \mid b) = \alpha + \beta f(b)$ for some vectors α and β and some scalar valued function $f(\cdot)$ of b, and if the subclasses are formed using b, then the subclassification is equal percent bias reducing in the sense that the percent of bias in x remaining after adjustment is the same for each coordinate of x, namely, 100γ, where*

$$\gamma = \frac{\Sigma_s w_s \int f(b)\{\mathrm{pr}(b \mid z = 1, b \in I_s) - \mathrm{pr}(b \mid z = 0, b \in I_s)\}db}{\int f(b)\{\mathrm{pr}(b \mid z = 1) - \mathrm{pr}(b \mid z = 0)\}db},$$

where the sum is over $s = 1, \ldots, S$.

Proof. Apply Theorem 7 and follow the argument of Corollary 6.1.

In parallel with Corollary 6.2 direct adjustment based on a balancing score within subpopulations defined by x can be shown to be equal percent bias reducing within those subpopulations.

Subclassification on the propensity score is not the same as any of the several methods proposed by Miettinen (1976): the propensity score is not generally a 'confounder' score. For example, one of Miettinen's confounder scores is

$$\mathrm{pr}(z = 1 \mid r_z = 1, x) \neq \mathrm{pr}(z = 1 \mid x) = e(x).$$

Moreover, under strong ignorability,

$$e(x) = \mathrm{pr}(z = 1 \mid x) = \mathrm{pr}(z = 1 \mid r_1, r_0, x) \neq \mathrm{pr}(z = 1 \mid r_z = 1, x),$$

so strong ignorability does not convert a confounder score into the propensity score.

3.4. Propensity Scores and Covariance Adjustment

The third standard method of adjustment in observational studies is covariance adjustment. The point estimate of the treatment effect obtained from an analysis of covariance adjustment for multivariate x is, in fact, equal to the estimate obtained from univariate covariance adjustment for the sample linear discriminant based on x, whenever the same sample covariance matrix is used for both the covariance adjustment and the discriminant analysis. This fact is most easily demonstrated by linearly transforming x to the sample discriminant and components orthogonal to the sample discriminant which by construction have the same sample mean in both groups. Since covariance adjustment is effectively adjustment for the linear discriminant, plots of the responses r_{1i} and r_{0i} or residuals $r_{ki} - \hat{r}_{ki}$, where $\hat{r}_{ki}$ is the value of r_{ki} predicted from the regression model used in the covariance adjustment, versus the linear discriminant are useful in identifying nonlinear or nonparallel response surfaces, as well as extrapolations, which might distort the estimate of the average treament effect. Furthermore, such a plot is a bivariate display of multivariate adjustment, and as such might be useful for general presentation.

Generally, plots of responses and residuals from covariance analysis against the propensity score $e(x)$ are more appropriate than against the discriminant, unless of course the covariates are multivariate normal with common covariance matrix in which case the propensity score is a monotone function of the discriminant. The reason is that, by Corollary 4.3, if treatment assignment is strongly ignorable, then at each $e(x)$ the expected difference in response $E\{r_1 \mid z = 1, e(x)\} - E\{r_0 \mid z = 0, e(x)\}$ equals the average treatment effect at $e(x)$, namely $E\{r_1 \mid e(x)\} - E\{r_0 \mid e(x)\}$. This property holds for the propensity score $e(x)$ and for any balancing score $b(x)$, but does not generally hold for other functions of x; generally, plots against other functions of x are still confounded by x.

Cases where covariance adjustment has been seen to perform quite poorly are precisely those cases in which the linear discriminant is not a monotone function of the propensity score, so that covariance adjustment is implicitly adjusting for a poor approximation to the propensity score. In the case of univariate x, the linear discriminant is a linear function of x, whereas the propensity score may not be a monotone function of x if the variances of x in the treated and control groups are unequal. Intuitively, if the variance of x in the control group is much larger than the variance in the treated group, then individuals with the largest and smallest x values usually come from the control group. Rubin (1973b, Tables 4 and 6, with $r = 1$ and τ_p as the estimator) has shown that with nonlinear response surfaces, univariate covariance adjustment can either increase the bias or overcorrect for bias dramatically if the variances of x in the treated and control groups differ. Unequal variances of covariates are not uncommon in observational studies, since the subset of units which receives a new treatment is often more homogeneous than the general population. For example, in the observational half of the Salk vaccine trial, the parents of second graders who volunteered for vaccination had higher and therefore less variable educational achievement, x, than parents of control children who were parents of all first and third graders (Meier, 1978).

In the case of multivariate normal x, Rubin (1979b, Table 10.2) has shown that covariance adjustment can seriously increase the expected squared bias if the covariance matrices in treated and control groups are unequal, that is, if the discriminant is not a monotone function of the propensity score. In contrast, when the covariance matrices are equal, so that the discriminant is a monotone function of the propensity score, covariance adjustment removes most of the expected squared bias in the cases considered by Rubin (1979b, Table 10.2). In summary, covariance adjustment cannot be relied upon to perform well unless the linear discriminant is highly correlated with the propensity score.

ACKNOWLEDGMENTS

The authors acknowledge valuable discussions with Arthur P. Dempster, A. Philip Dawid, and Roderick J. A. Little on the subject of this paper.

11. Assessing Sensitivity to an Unobserved Binary Covariate in an Observational Study with Binary Outcome

Paul R. Rosenbaum and Donald B. Rubin

Abstract: This paper proposes a simple technique for assessing the range of plausible causal conclusions from observational studies with a binary outcome and an observed categorical covariate. The technique assesses the sensitivity of conclusions to assumptions about an unobserved binary covariate relevant to both treatment assignment and response. A medical study of coronary artery disease is used to illustrate the technique.

1. INTRODUCTION AND NOTATION

Inevitably, the results of clinical studies are subject to dispute. In observational studies, one basis for dispute is obvious: since patients were not assigned to treatments at random, patients at greater risk may be over-represented in some treatment groups. This paper proposes a method for assessing the sensitivity of causal conclusions to an unmeasured patient characteristic relevant to both treatment assignment and response. Despite their limitations, observational studies will continue to be a valuable source of information, and therefore it is prudent to develop appropriate methods of analysis for them.

Our sensitivity analysis consists of the estimation of the average effect of a treatment on a binary outcome variable after adjustment for observed categorical covariates and an unobserved binary covariate u, under several sets of assumptions about u. Both Cornfield et al. (1959) and Bross (1966) have proposed guidelines for determining whether an unmeasured binary covariate having specified properties could explain all of the apparent effect of a treatment, that is, whether the treatment effect, after adjustment for u could be zero. Our method has two advantages: first, Cornfield et al. (1959) and Bross (1966) adjust only for the unmeasured binary covariate u, whereas we adjust for measured covariates in addition to the unmeasured covariate u. Second, Cornfield et al. (1959) and Bross (1966, 1967) only judge whether the effect of the treatment could be zero having adjusted for u, where Cornfield et al. (1959) employ an implicit yet extreme assumption about u. In contrast, we provide actual estimates of the treatment effect adjusted for both u and the observed categorical covariates under any assumption about u.

Reprinted from *The Journal of the Royal Statistical Society*, **B, 45,** 2, 1983, 212–218.

In principle, the ith of the N patients under study has both a binary response r_{1i} that would have resulted if he had received the new treatment, and a binary response r_{0i} that would have resulted if he had received the control treatment. In this formulation, treatment effects are comparisons of r_{1i} and r_{0i}, such as $r_{1i} - r_{0i}$. Since each patient receives only one treatment, either r_{1i} or r_{0i} is observed, but not both, and therefore comparisons of r_{1i} and r_{0i} imply some degree of speculation. Treatment effects defined as comparisons of the two potential responses, r_{1i} and r_{0i}, of individual patients are implicit in Fisher's (1935) randomization test of the sharp null hypothesis that $r_{1i} = r_{0i}, i = 1, \ldots, N$. Such definitions are used explicitly by Kempthorne (1952) in his discussion of randomization-based inference in experiments, and by Rubin (1977a, 1978a), Hamilton (1979) and Rosenbaum and Rubin (1983a) in discussions of observational studies. The definition does contain some implicit assumptions, such as the assumption of non-interference between patients; see Cox (1958, Chapter 2) or Rubin (1978a, Section 2.3) for discussion. Here, the N patients in the study are viewed as a simple random sample from some population, and the average treatment effect is defined as

$$E(r_1) - E(r_0) = \text{pr}(r_1 = 1) - \text{pr}(r_0 = 1) = \tau_1 - \tau_0, \tag{11.1}$$

where $E(\cdot)$ and $\text{pr}(\cdot)$ denote expectation and probability, respectively, in the population.

For the ith patient of N patients in the study ($i = 1, \ldots, N$) let z_i be the indicator for treatment assignment, and let $z_i = 1$ if patient i is assigned to the new treatment, and $z_i = 0$ if patient i is assigned to the control treatment. Suppose that patients have been stratified or subclassified into one of J subclasses on the basis of an observed categorical covariate, and that patient i falls in subclass $s_i, i = 1, 2, \ldots, N$, where s_i is an integer between 1 and J. The population model for the subclassifying variable s is assumed to be a saturated multinomial.

If the study had been a randomized block experiment with blocks defined by subclasses, treatment assignment z and response (r_1, r_0) would be conditionally independent given s, or in Dawid's (1979) notation:

$$(r_1, r_0) \perp\!\!\!\perp z \mid s. \tag{11.2}$$

Moreover, in such a study, $1 > \text{pr}(z = 0 \mid s) > 0$ for all s. These conditions are not known to hold in an observational study. Generally, we shall say that treatment assignment is *strongly ignorable given some set of covariates* $\mathbf{v}$ if $(r_1, r_0) \perp\!\!\!\perp z \mid \mathbf{v}$ and $1 > \text{pr}(z = 0 \mid \mathbf{v}) > 0$ for all possible $\mathbf{v}$. For brevity, when treatment assignment is strongly ignorable given the observed covariates, we shall say simply that *treatment assignment is strongly ignorable*. If treatment assignment is strongly ignorable, then it is ignorable in Rubin's (1978a) sense, but the converse is not true.

We develop and apply a method to aid in judging the sensitivity of conclusions to certain plausible variations in assumptions about an unobserved binary covariate u. In particular, we assume that treatment assignment is not strongly ignorable given s, but is strongly ignorable given s and u; that is, we will assume that (11.2) may be false but that

$$(r_1, r_0) \perp\!\!\!\perp z \mid s, u \tag{11.3}$$

is true, where $1 > \text{pr}(z = 0 \mid s, u) > 0$ for all s, u. If conclusions are insensitive over a range of plausible assumptions about u, the number of interpretations of the data is reduced, and causal conclusions are more defensible.

An alternative approach is briefly mentioned in the review by Schlesselman (1978). In that approach, however, the parameter used to measure the effect of the treatment, namely,

$$\frac{\text{pr}(r_z = 0 \mid z = 0, u = 1, s)}{\text{pr}(r_z = 0 \mid z = 1, u = 1, s)} \tag{11.4}$$

(or r_A in Schlesselman's notation), refers only to a subpopulation of patients defined by the unobserved variable u, that is, the subpopulation with $u = 1$; as a result, as assumptions about u are varied, the meaning of the parameter changes. Unless additional assumptions are made about the absence of certain interaction effects (in particular, that expression (11.4) equals the corresponding expression with $u = 0$), this parameter is not suitable for sensitivity analyses that consider various assumptions about u. Moreover, if the problem is formulated in terms of the scalar observable response variable r_z, then it is impossible to state the crucial assumption (11.3) under which the sensitivity analysis is correct.

2. THE STRUCTURE OF THE SENSITIVITY ANALYSIS

Because, by (11.3), (r_1, r_0) and z are conditionally independent given u and s, we can write the joint distribution of (r_t, z, u, s) for $t = 0, 1$ as

$$\text{pr}(r_t \mid u, s)\,\text{pr}(z \mid u, s)\,\text{pr}(u \mid s)\,\text{pr}(s),$$

where without loss of generality

$$\text{pr}(s = j) = \phi_j; \quad \sum_{j=1}^{J} \phi_j = 1, \quad \text{pr}(u = 0 \mid s) = \pi_s, \quad s = 1, \ldots, J, \tag{11.5}$$

$$\text{pr}(z = 0 \mid u, s) = [1 + \exp(\gamma_s + u\alpha_s)]^{-1}, \quad s = 1, \ldots, J; \quad u = 0, 1; \tag{11.6}$$

and

$$\text{pr}(r_t = 0 \mid u, s) = [1 + \exp(\beta_{st} + u\delta_{st})]^{-1}, \quad s = 1, \ldots, J; \quad u = 0, 1; \quad t = 0, 1. \tag{11.7}$$

The parameters $\phi_1, \ldots, \phi_J$ give the probabilities in each of the J subclasses; π_s gives the probability that $u = 0$ in subclass s; γ_s gives the log odds of assignment to treatment 0 in subclass s when $u = 0$, and $\gamma_s + \delta_s$ gives the corresponding log odds when $u = 1$; and finally, β_{st} gives the log odds that $r_t = 0$ in subclass s when $u = 0$, while $\beta_{st} + \delta_{st}$ gives the corresponding log odds when $u = 1$. A formally similar model with an unobserved binary covariate appears in Rubin (1978a, Section 4.2).

Only some of the parameters in (11.5)–(11.7) can be estimated from observed data because u is never observed and r_t is only observed when $t = z$. Obviously, the ϕ_s can be estimated from the observed proportions of patients in each subclass. As we show explicitly in Section 4, for fixed values of sensitivity parameters π_s, α_s, δ_{st}

Table 11.1. *Proportion of patients improved at 6 months in each subclass*

Subclass[†]	Treatment	Number of patients	Proportion improved	Standard error
1	Surgical	26	0.54	0.10
	Medical	277	0.35	0.03
2	Surgical	68	0.70	0.06
	Medical	235	0.40	0.03
3	Surgical	98	0.70	0.05
	Medical	205	0.35	0.03
4	Surgical	164	0.71	0.04
	Medical	139	0.30	0.04
5	Surgical	234	0.70	0.03
	Medical	69	0.39	0.06

† The subclasses were constructed by the method of Rosenbaum and Rubin (1983a) in which the conditional probability of surgical treatment given the observed covariates is estimated, and patients with similar estimated probabilities are placed in the same subclass. The subclasses are predictive of treatment assignments; they are not prognostic subclasses. This method balances observed covariates within each subclass.

there exist unique maximum likelihood estimates of the remaining parameters, γ_s and β_{st}. Since the average treatment effect (11.1) is $\tau_1 - \tau_0$ where

$$\tau_t = \sum_{s=1}^{J} \phi_s \left[(1 - \pi_s) \frac{\exp{(\beta_{st} + \delta_{st})}}{1 + \exp{(\beta_{st} + \delta_{st})}} + \pi_s \frac{\exp{(\beta_{st})}}{1 + \exp{(\beta_{st})}} \right] \qquad (11.8)$$

the maximum likelihood estimate of the treatment effect can be calculated from the maximum likelihood estimates of ϕ_s and β_{st} corresponding to fixed values of the sensitivity parameters π_s, α_s and δ_{st}.

3. AN EXAMPLE USING DATA ON CORONARY ARTERY DISEASE

Before discussing the technical details underlying the sensitivity analysis, we present the results of applying it to an example concerning symptomatic relief from coronary artery disease[1]. Treatment 1 is coronary artery bypass surgery; treatment 0 is medical therapy. The response (r_1, r_0) is functional improvement 6 months after cardiac catheterization, with $r_1 = 1$ or $r_1 = 0$ indicating improvement or no improvement, respectively, under surgical treatment, and $r_0 = 1$ or $r_0 = 0$ indicating improvement or no improvement, respectively, under medical treatment.

Here, patients are stratified by the multivariate method described in Rosenbaum and Rubin (1983a) that balances 74 observed covariates. The observed proportion improved within each subclass is displayed in Table 11.1. In examining this table, it must be remembered that there exists evidence for a placebo effect of bypass surgery (Benson and McCalie, 1979).

When treatment assignment is strongly ignorable given the subclasses, then direct adjustment with subclass total weights yields the maximum likelihood estimate of the average treatment effect (11.1) under the saturated multinomial

---

[1] The data are used to illustrate methodology, and do not constitute a study of coronary artery disease.

Table 11.2. *Effects of an unobserved two-category covariate u on the probability of substantial symptomatic improvement at 6 months for medical (M) and surgical (S) patients*

Effect of $u = 1$ vs $u = 0$ on treatment assignment z	Effect of $u = 1$ vs $u = 0$ on response under M	Effect of $u = 1$ vs $u = 0$ on response under S	Fraction of patients with $u = 0$: π		
			0.1	0.5	0.9
Doubles the odds of surgery $\exp(\alpha) = 2$	Halves the odds of improvement $\exp(\delta_0) = \frac{1}{2}$	Halves the odds of improvement $\exp(\delta_1) = \frac{1}{2}$	S 0.67 M 0.36	S 0.68 M 0.35	S 0.68 M 0.36
		Doubles the odds of improvement $\exp(\delta_1) = 2$	S 0.66 M 0.36	S 0.65 M 0.35	S 0.66 M 0.36
	Doubles the odds of improvement $\exp(\delta_0) = 2$	Halves the odds of improvement $\exp(\delta_1) = \frac{1}{2}$	S 0.67 M 0.36	S 0.68 M 0.37	S 0.68 M 0.36
		Doubles the odds of improvement $\exp(\delta_1) = 2$	S 0.66 M 0.36	S 0.65 M 0.37	S 0.66 M 0.36
Triples the odds of surgery $\exp(\alpha) = 3$	Reduces by $\frac{2}{3}$ the odds of improvement $\exp(\delta_0) = \frac{1}{3}$	Reduces by $\frac{2}{3}$ the odds of improvement $\exp(\delta_1) = \frac{1}{3}$	S 0.68 M 0.35	S 0.70 M 0.34	S 0.69 M 0.35
		Triples the odds of improvement $\exp(\delta_1) = 3$	S 0.66 M 0.35	S 0.63 M 0.34	S 0.65 M 0.35
	Triples the odds of improvement $\exp(\delta_0) = 3$	Reduces by $\frac{2}{3}$ the odds of improvement $\exp(\delta_1) = \frac{1}{3}$	S 0.68 M 0.37	S 0.70 M 0.38	S 0.69 M 0.37
		Triples the odds of improvement $\exp(\delta_1) = 3$	S 0.66 M 0.37	S 0.63 M 0.38	S 0.65 M 0.37

model for s. The directly adjusted proportions improved are 0.36 for medicine and 0.67 for surgery, with standard errors 0.04 and 0.06 respectively (calculated following Mosteller and Tukey, 1977, Chapter 11c).

In order to study the sensitivity of estimates to the assumption of strongly ignorable assignment, we now assume that treatment assignment is not strongly ignorable, but rather that treatment assignment is strongly ignorable given s and unobserved binary covariate u. That is, we assume (11.2) is false but (11.3) is true. Table 11.2 displays the sensitivity of the estimate of the average treatment effect to 24 sets of assumptions about u, where α is the increase in the log odds of surgery associated with $u = 1$ rather than $u = 0$, δ_t is the increase in the log odds of improvement under treatment t associated with $u = 1$, and $\pi = \mathrm{pr}(u = 0)$. In order to limit the size of the sensitivity analysis, $(\alpha, \delta_1, \delta_0, \pi)$ is assumed in Table 11.2 to be the same across all subclasses. The general method presented in Sections 2 and 4 does not require this restriction; moreover, the Appendix displays subclass specific results which do not depend on this assumption.

In Table 11.2, the estimates of the proportion improved vary from 0.34 to 0.38 for medicine and from 0.63 to 0.70 for surgery; these ranges of values of $\hat{\tau}_0$ and $\hat{\tau}_1$ are about the same as the standard error of the directly adjusted proportions (i.e., 0.04 for medicine, 0.06 for surgery). Consequently, we see that this hypothetical, unobserved covariate u, which has defied the cardiologists' attempt to record all variables used in assigning treatments, would have to more than triple the odds of surgery and more than triple the odds of improvement, before altering the conclusion that the proportion improved under surgery far exceeds the proportion improved under medicine. Although this difference may reflect a placebo effect of surgery, the difference does not seem to be easily explained as the result of an imbalance due to the non-randomized nature of the study.

Another way of describing this analysis is to say that we have explored the extent to which the data might be an example of Simpson's paradox. For discussion from this perspective, see Lindley and Novick (1981).

4. CALCULATING THE MAXIMUM LIKELIHOOD ESTIMATES OF τ_1 AND τ_0

Since r_{ti} is observed only if $t = z_i$, the likelihood of the parameters given the observed data $(r_{z_i i}, z_i, s_i)$, $i = 1, \ldots, N$ is

$$\prod_{i=1}^{N} \mathrm{pr}\,(r_{t_i} \mid t = z_i, s_i)\,\mathrm{pr}\,(z_i \mid s_i)\,\mathrm{pr}\,(s_i), \tag{11.9}$$

where, by definition,

$$\mathrm{pr}\,(s = j) = \phi_j, \tag{11.10}$$

and, by (11.5) and (11.6)

$$\mathrm{pr}\,(z = 0 \mid s) = \frac{\pi_s}{1 + \exp\,[\gamma_s]} + \frac{(1 - \pi_s)}{1 + \exp\,[\gamma_s + \alpha_s]}, \tag{11.11}$$

and finally, by (11.3) and (11.7),

$$\begin{aligned}
\mathrm{pr}\,(r_t = 0 \mid z = t, s) &= \mathrm{pr}\,(r_t = 0 \mid s, u = 0)\,\mathrm{pr}\,(u = 0 \mid z = t, s) \\
&\quad + \mathrm{pr}\,(r_t = 0 \mid s, u = 1)\,\mathrm{pr}\,(u = 1 \mid z = t, s) \\
&= \frac{w_{sz}}{1 + e^{\beta_{st}}} + \frac{1 - w_{sz}}{1 + e^{\beta_{st} + \delta_{st}}},
\end{aligned} \tag{11.12}$$

where by (11.5), (11.6) and (11.11)

$$\begin{aligned}
w_{sz} &= \mathrm{pr}\,(u = 0 \mid z, s) \\
&= \frac{\pi_s \exp\,[z\gamma_s]}{1 + \exp\,[\gamma_s]} \left\{ \frac{\pi_s \exp\,[z\gamma_s]}{1 + \exp\,[\gamma_s]} + \frac{(1 - \pi_s)\,\exp\,[z(\gamma_s + \alpha_s)]}{1 + \exp\,[\gamma_s + \alpha_s]} \right\}^{-1}, \\
&= \pi_s \left\{ \pi_s + (1 - \pi_s)\,\frac{\exp\,(z\alpha_s)\,[1 + \exp\,(\gamma_s)]}{1 + \exp\,[\gamma_s + \alpha_s]} \right\}^{-1}.
\end{aligned} \tag{11.13}$$

Note that the w_{sz}'s define the conditional distribution of the unobserved covariate u given the observed treatment assignment.

Since the implied model for the $2 \times 2 \times J$ table of counts for the observed data, (r_z, z, s), is saturated, the maximum likelihood estimate of pr $(s = j) = \phi_j$ is the proportion $\hat{\text{pr}}\,(s = j)$ of patients in subclass j, the maximum likelihood estimate of pr $(z = 0|s)$ is the observed proportion $\hat{\text{pr}}\,(z = 0\,|\,s)$ of patients receiving treatment 0 in subclass s, and for $z = 0$ and 1, the maximum likelihood estimate of pr $(r_t = 0\,|\,t = z, s)$ is the observed proportion $\hat{\text{pr}}\,(r_t = 0\,|\,t = z, s)$ of patients with response $r_z = 0$ among patients who received treatment z in subclass s.

For each subclass s, equation (11.11) with $\hat{\text{pr}}\,(z = 0|s)$ substituted for pr $(z = 0\,|\,s)$ may be solved for the maximum likelihood estimate of γ_s; having found $\hat{\gamma}_s$ and calculated $\hat{w}_{st}$, equation (11.12) with $\hat{\text{pr}}\,(r_t = 0\,|\,t = z, s)$ substituted for pr $(r_t = 0\,|\,t = z, s)$ may be solved for $\hat{\beta}_{st}$. The maximum likelihood estimates of ϕ_s and β_{st} can be substituted into (11.8) to find the maximum likelihood estimate of τ_t.

Equations (11.11) and (11.12) are both of the form

$$p = \frac{a}{1 + \exp(\theta)} + \frac{1 - a}{1 + \exp(\theta + b)}. \qquad (11.14)$$

To find θ for fixed a, b and $p \neq 0$, we note that the right-hand side of (11.14) is monotone in $\omega = \exp(\theta)$, taking values between 0 and 1, and thus (11.14) has one positive solution in ω. To find this solution, we note that (11.14) implies a quadratic equation in ω:

$$[p \exp(b)]\,\omega^2 + [(p - a)\exp(b) + p - 1 + a]\,\omega + (p - 1) = 0. \qquad (11.15)$$

To calculate $(\hat{\gamma}_s, \hat{\beta}_{s0}, \hat{\beta}_{s1})$ for fixed $(\pi_s, \alpha_s, \delta_{s1}, \delta_{s0})$:

(a) Solve (11.11) for $\hat{\gamma}_s$ using (11.15) with $p = \hat{\text{pr}}\,(z = 0\,|\,s)$, $a = \pi_s$, $b = \alpha_s$.
(b) For $t = 0, 1$, find $\hat{w}_{st}$, using (11.13).
(c) For $t = 0, 1$, solve (11.12) for $\hat{\beta}_{st}$ using (11.15) with $p = \hat{\text{pr}}\,(r_t = 0\,|\,t = z, s)$, $a = \hat{w}_{st}, b = \delta_{st}$.

Steps (a), (b) and (c) are repeated for $s = 1, \ldots, J$ and the results combined using (11.8). The Appendix presents some specific results for each subclass of the example from Section 2.

ACKNOWLEDGMENTS

The authors wish to acknowledge valuable discussions with Arthur P. Dempster and Roderick J. A. Little on the subject of this paper, and thank the editor and referees for helpful comments on an earlier draft of the paper.

APPENDIX

Subclass Specific Results

Table 11.2 is based on the assumption that the sensitivity parameters $(\alpha_s, \delta_{1s}, \delta_{0s}, \pi_s)$ do not depend on the subclass s. This assumption is not crucial to the conclusions

Table 11.3. *Effects of an unobserved two-category covariate u on the probability of improvement of 6 months for medical ($t = 0$) and surgical ($t = 1$) patients within each subclass; association with treatment assignment* $\exp(\alpha)$, *equal to 3*

Association with improvement $\exp(\delta_t)$	$\pi = \mathrm{pr}(u = 0)$		Subclass				
			1	2	3	4	5
$\frac{1}{3}$	0.1	S	0.555	0.710	0.710	0.718	0.705
		M	0.348	0.396	0.343	0.287	0.366
	0.5	S	0.600	0.741	0.737	0.735	0.713
		M	0.345	0.385	0.330	0.270	0.342
	0.9	S	0.573	0.723	0.718	0.720	0.704
		M	0.348	0.394	0.343	0.291	0.377
3	0.1	S	0.524	0.685	0.686	0.699	0.693
		M	0.351	0.404	0.355	0.310	0.411
	0.5	S	0.478	0.651	0.657	0.682	0.686
		M	0.355	0.415	0.371	0.334	0.443
	0.9	S	0.508	0.682	0.686	0.703	0.697
		M	0.353	0.407	0.359	0.313	0.405

in Section 3, although it does simplify presentation. This Appendix documents the stability of the estimates of the subclass specific average response to treatment t,

$$\mathrm{pr}(r_t = 0 \mid s) = \frac{\pi_s}{1 + \exp(\beta_{st})} + \frac{1 - \pi_s}{1 + \exp(\beta_{st} + \delta_{st})} = 1 - \tau_{st};$$

Table 11.3 displays values of τ_{st} when $\alpha = 3$.

As was noted in Section 4, for $s = 1, 2, \ldots, s$, $\hat{\tau}_{st}$ depends only on the observed data and π_s, α_s, δ_{st}. Therefore, we may examine the effects of allowing the sensitivity parameters to vary from subclass to subclass by simply combining subclass specific estimates $\hat{\tau}_{st}$. For example, from Table 11.3 when $\alpha = 3$ for $\pi_1 = 0.1$, $\delta_{10} = \frac{1}{3}$, and $\delta_{11} = 3$, we have $\hat{\tau}_{10} = 0.348$ and $\hat{\tau}_{11} = 0.524$. If, in addition, we let $\pi_s = 0.9$, $\delta_{s0} = \delta_{s1} = \frac{1}{3}$ for $s = 2, 3, 4, 5$ we have

$$\hat{\tau}_0 = (0.348 + 0.394 + 0.343 + 0.291 + 0.377)/5 = 0.351$$

and

$$\hat{\tau}_1 = (0.524 + 0.723 + 0.718 + 0.720 + 0.704)/5 = 0.678,$$

since every subclass contains the same number of patients. By similar calculations it is seen that the estimate of the average treatment effect is greater than 0.25 for all combinations of $(\alpha_s, \delta_{s1}, \delta_{s0}, \pi_s)$ in Table 11.3.

12. Reducing Bias in Observational Studies Using Subclassification on the Propensity Score

Paul R. Rosenbaum and Donald B. Rubin

Abstract: The propensity score is the conditional probability of assignment to a particular treatment given a vector of observed covariates. Previous theoretical arguments have shown that subclassification on the propensity score will balance all observed covariates. Subclassification on an estimated propensity score is illustrated, using observational data on treatments for coronary artery disease. Five subclasses defined by the estimated propensity score are constructed that balance 74 covariates, and thereby provide estimates of treatment effects using direct adjustment. These subclasses are applied within subpopulations, and model-based adjustments are then used to provide estimates of treatment effects within these subpopulations. Two appendixes address theoretical issues related to the application: the effectiveness of subclassification on the propensity score in removing bias, and balancing properties of propensity scores with incomplete data.

1. INTRODUCTION: SUBCLASSIFICATION AND THE PROPENSITY SCORE

1.1. Adjustment by Subclassification in Observational Studies

In observational studies for causal effects, treatments are assigned to experimental units without the benefits of randomization. As a result, treatment groups may differ systematically with respect to relevant characteristics and, therefore, may not be directly comparable. One commonly used method of controlling for systematic differences involves grouping units into subclasses based on observed characteristics, and then directly comparing only treated and control units who fall in the same subclass. Obviously such a procedure can only control the bias due to imbalances in *observed* covariates.

Cochran (1968a) presents an example in which the mortality rates of cigarette smokers, cigar/pipe smokers, and nonsmokers are compared after subclassification on the covariate age. The age-adjusted estimates of the average mortality for each type of smoking were found by direct adjustment – that is, by combining the subclass-specific mortality rates, using weights equal to the proportions of the population within the subclasses. Cochran (1968a) shows that five subclasses are often sufficient to remove over 90% of the bias due to the subclassifying variable or covariate. However, as noted in Cochran (1965), as the number of covariates increases, the number of subclasses grows exponentially; so even with only two categories per covariate, there are 2^p subclasses for p covariates. If p is moderately large, some subclasses will contain no units, and many subclasses will contain either treated or control units but not both, making it impossible to form directly adjusted estimates for the entire population.

Fortunately, however, there exists a scalar function of the covariates, namely the propensity score, that summarizes the information required to balance the distribution of the covariates. Specifically, subclasses formed from the scalar propensity score will balance all p covariates. In fact, often five subclasses constructed from the propensity score will suffice to remove over 90% of the bias due to each of the covariates.

1.2. The Propensity Score in Observational Studies

Consider a study comparing two treatments, labeled 1 and 0, where z indicates the treatment assignment. The propensity score is the conditional probability that a unit with vector $\mathbf{x}$ of *observed* covariates will be assigned to treatment 1, $e(\mathbf{x}) = \Pr(z = 1 \mid \mathbf{x})$. Rosenbaum and Rubin (1983a, Theorem 1) show that subclassification on the population propensity score will balance $\mathbf{x}$, in the sense that within subclasses that are homogeneous in $e(\mathbf{x})$, the distribution of $\mathbf{x}$ is the same for treated and control units; formally, $\mathbf{x}$ and z are conditionally independent given $e = e(\mathbf{x})$,

$$\Pr(\mathbf{x}, z \mid e) = \Pr(\mathbf{x} \mid e) \Pr(z \mid e). \tag{12.1}$$

The proof is straightforward. Generally, $\Pr(\mathbf{x}, z \mid e) = \Pr(\mathbf{x} \mid e) \Pr(z \mid \mathbf{x}, e)$. But since e is a function of $\mathbf{x}$, $\Pr(z \mid \mathbf{x}, e) = \Pr(z \mid \mathbf{x})$. To prove (12.1), it is thus sufficient to show that $\Pr(z = 1 \mid \mathbf{x}) = \Pr(z = 1 \mid e)$. Now $\Pr(z = 1 \mid \mathbf{x}) = e$ by definition, and $\Pr(z = 1 \mid e) = E(z \mid e) = E\{E(z \mid \mathbf{x}) \mid e\} = E(e \mid e) = e$, proving (12.1).

Expression (12.1) suggests that to produce subclasses in which $\mathbf{x}$ has the same distribution for treated and control units, distinct subclasses should be created for each distinct value of the known propensity score. In common practice, $e(x)$ is not known, and it is not feasible to form subclasses that are exactly homogeneous in $e(x)$ and contain both a treated and control unit. In Section 2 we examine the balance obtained from subclassification on an estimated propensity score. Appendix A considers the consequences of coarse or inexact subclassification on $e(\mathbf{x})$. Of course, although we expect subclassification on an estimated $e(\mathbf{x})$ to produce balanced distributions of $\mathbf{x}$, it cannot, like randomization, balance unobserved covariates, except to the extent that they are correlated with $\mathbf{x}$.

Cochran and Rubin (1973) and Rubin (1970, 1976b,c) proposed and studied discriminant matching as a method for controlling bias in observational studies. As noted by Rosenbaum and Rubin (1983a, Sec. 2.3 (i)), with multivariate normal $\mathbf{x}$ distributions having common covariance in both treatment groups, the propensity score is a monotone function of the discriminant score. Consequently, subclassification on the propensity score is a strict generalization of this work to cases with arbitrary distributions of $\mathbf{x}$.

Subclassification on the propensity score is not, however, the same as any of the several methods proposed later by Miettinen (1976); as Rosenbaum and Rubin (1983a, Sec. 3.3) state formally, the propensity score is not generally a "confounder" score. First, the propensity score depends only on the joint distribution of $\mathbf{x}$ and z, whereas a confounder score depends additionally on the conditional distribution of a discrete outcome variable given $\mathbf{x}$ and z, and is not defined for continuous outcome variables. Second, by Theorem 2 of Rosenbaum and Rubin (1983a), the propensity score is the coarsest function of $\mathbf{x}$ that has balancing property (12.1), so unless a confounder score is finer than the propensity score, it will not have this balancing property.

2. FITTING THE PROPENSITY SCORE AND ASSESSING THE BALANCE WITHIN SUBCLASSES

2.1. The First Fit and Subclassification

We illustrate subclassification based on the propensity score with observational data on two treatments for coronary artery disease: 590 patients with coronary artery bypass surgery ($z = 1$), and 925 patients with medical therapy ($z = 0$). The vector of covariates, $\mathbf{x}$, contains 74 hemodynamic, angiographic, laboratory, and exercise test results.[1] The propensity score was estimated using a logit model (Cox 1970) for z,

$$\log[e(\mathbf{x})/(1 - e(\mathbf{x}))] = \alpha + \beta^T \mathbf{f}(\mathbf{x}), \tag{12.2}$$

where α and β are parameters and $f(\cdot)$ is a specified function.

Not all of the 74 covariates and their interactions were included in the logit model for the 1,515 patients in the study. Main effects of variables were selected for inclusion in the first logit model using an inexpensive stepwise discriminant analysis. A second stepwise discriminant analysis added cross-products or interactions of those variables whose main effects were selected by the first stepwise procedure. Using these selected variables and interactions, the propensity score was then estimated by maximum likelihood logistic regression (using the SAS system). The result was the first logit model. (Alternatively, stepwise logit regression could have been used to select variables, e.g., Dixon et al. 1981.)

Based on Cochran's (1968a) results and a new result in Appendix A of this article, we may expect approximately a 90% reduction in bias for each of the 74 variables when we subclassify at the quintiles of the distribution of the *population* propensity score. Consequently we subclassified at the quintiles of the distribution

---

[1] The data analysis that follows is intended to illustrate statistical techniques, and does not by itself constitute a study of coronary bypass surgery. The literature on the efficiency of coronary bypass surgery is quite extensive with many subtleties addressed and controversies exhibited. Furthermore, the data being used were considered "preliminary and unverified" for this application.

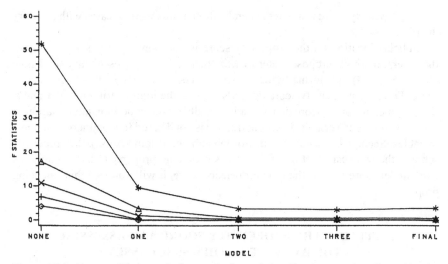

Figure 12.1. *F* tests of balance before and after subclassifications: main effects (5-point summary). (Minimum ◇; lower quartile +; median ×; upper quartile △; maximum *.)

of the *estimated* propensity score based on this initial analysis, which we term the first model.

We now examine the balance achieved by this first subclassification. Each of the 74 covariates was subjected to a two-way (2 (treatments) × 5 (subclasses)) analysis of variance. Above the word *none*, Figures 12.1 and 12.2 display a five-number summary (i.e., minimum, lower quartile, median, upper quartile, maximum) of the 74 *F* ratios prior to subclassification, that is, the squares of the usual two-sample *t* statistics for comparing the medical and surgical group means for each covariate prior to subclassification. Above the word *one*, *F* ratios are displayed for the main effect

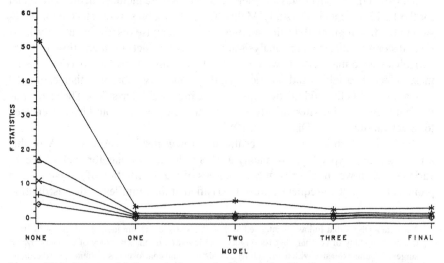

Figure 12.2. *F* tests of balance before and after subclassifications: interactions (5-point summary). (Minimum ◇; lower quartile +; median ×; upper quartile △; maximum *.)

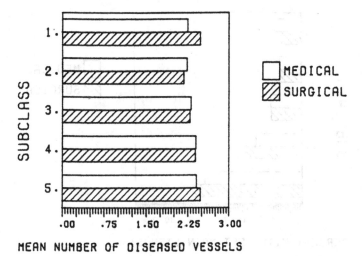

MEAN NUMBER OF DISEASED VESSELS

Figure 12.3. Balance within subclasses: baseline number of diseased vessels.

of the treatment (Figure 12.1) and the treatment × subclass interaction (Figure 12.2) in the two-way analysis of variance. Although there has been a substantial reduction in most F ratios, several are still quite large, possibly indicating that the propensity score is poorly estimated by the first model. Indeed, as a consequence of Theorem 1 of Rosenbaum and Rubin (1983a), each such F test is an approximate test of the adequacy of the model for the propensity score; the test is only approximate primarily because the subclasses are not exactly homogeneous in the fitted propensity score.

2.2. Refinement of the Fitted Propensity Score and the Balance Obtained in the Final Subclassification

Figures 12.1 and 12.2 display summaries of F ratios from a sequence of models constructed by a gradual refinement of the first model. At each step, variables with large F ratios that had previously been excluded from the model were added. All logistic models were fitted by maximum likelihood. If a variable produced a large F ratio even after inclusion in the model, then the square of the variable and cross-products with other clinically important variables were tried. In the final model, β and $\mathbf{f}(\mathbf{x})$ in (12.2) were of dimension 45, including 7 interaction degrees of freedom and 1 quadratic term. There is considerably greater balance on the observed covariates $\mathbf{x}$ within these final subclasses than would have been expected from randomized assignment to treatment within subclasses.

Figures 12.3–12.5 display the balance within subclasses for three important covariates. Although the procedure used to form the subclasses may not be accessible to some nonstatisticians, the comparability of patients within subclasses can be examined with the simplest methods, such as the bar charts used here. For example, Figure 12.5 indicates some residual imbalance on the percentage of patients with poor left ventrical (LV) contraction, at least for patients in subclass 1 – that is, in the subclass with the lowest estimated probabilities of surgery. This imbalance is less than would be expected from randomization within subclasses; the main-effect F

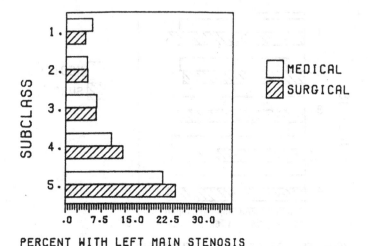

PERCENT WITH LEFT MAIN STENOSIS

Figure 12.4. Balance within subclasses: baseline left ventricular contraction.

ratio is .4 and the interaction F ratio is .9. Nonetheless, we would possibly want to adjust for this residual imbalance, perhaps using methods described in Section 3.3.

2.3. The Fitted Propensity Score: Overlap of Treated and Control Groups

Figure 12.6 contains boxplots (Tukey 1977) of the final fitted propensity scores. By construction, most surgical patients have higher propensity scores – that is, higher estimated probabilities of surgery – than most medical patients. There are a few surgical patients with higher estimated probabilities of surgery than any medical patient, indicating a combination of covariate values not appearing in the medical group. For almost every medical patient, however, there is a surgical patient who is comparable in the sense of having a similar estimated probability of surgery.

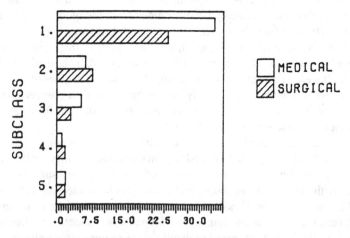

PERCENT WITH POOR LV CONTRACTION

Figure 12.5. Balance within subclasses: baseline left main stenosis.

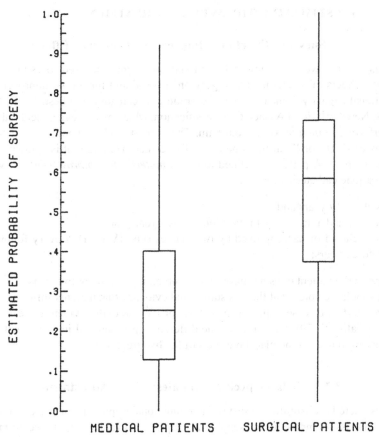

Figure 12.6. Boxplots of the estimated propensity score.

2.4. Incomplete Covariate Information

Five variables – four related to exercise tests and one quantitative measure of left ventrical function – were not measured during the early years of the study, so many patients are missing these covariate values. If the propensity score is defined as the conditional probability of assignment to treatment 1 given the observed covariate information *and* the pattern of missing data, then Appendix B shows that subclassification on the propensity score will balance both the observed data and the pattern of missing data. Essentially, we estimated the probabilities of surgical treatment separately for early and late patients, and then used these estimated probabilities as propensity scores. Subclassification on the corresponding population propensity scores can be expected to balance, within subclasses, each of the following: (a) the distribution of those covariates that are measured for both early and late patients, (b) the proportions of early and late patients, and (c) the distribution of all covariates for the late patients. (For proof, see Corollary B.1 of Appendix B.) The observed values of these five covariates were indeed balanced by our procedure: the main-effect F ratios were 2.1, .1, .3, .2, and .0; the interaction F ratios were .4, 1.4, .1, .6, and .3.

3. ESTIMATING THE AVERAGE TREATMENT EFFECT

3.1. Survival; Functional Improvement; Placebo Effects

In this section, we show how balanced subclasses may be used to estimate the average effects of medicine and surgery on survival and functional improvement. Functional capacity is measured by the crude four-category (I = best, II, III, IV = worst) New York Heart Association classification, which measures a patient's ability to perform common tasks without pain. The current study is confined to patients in classes II, III, or IV at the time of cardiac catheterization, that is, patients who could improve. A patient is defined to have *uninterrupted improvement* to t years after cardiac catheterization if he:

1. is alive at t years and
2. has not had a myocardial infarction before t years and
3. is in class I or has improved by two classes (i.e., IV to II) at every follow-up before t years;

otherwise the patient does not have uninterrupted improvement to t years.

It should be noted that there is substantial evidence that patients suffering from coronary artery disease respond to placebos; for a review of this evidence, see Benson and McCallie (1979). Part or all of the difference in functional improvement may reflect differences in the placebo effects of the two treatments.

3.2. Subclass-Specific Estimates; Direct Adjustment

The estimated probabilities of survival and functional improvement at six months in each subclass for medicine and surgery are displayed in Table 12.1. (These estimates take censoring into account by using the Kaplan–Meier (1958) procedure.) In each subclass the proportion improved under surgery exceeds the proportion improved under medical therapy; the proportion surviving to six months is higher following medical treatment, although the standard errors are quite large.

Each subclass contains 303 patients. Therefore, for medical therapy and surgery, the directly adjusted proportions, with subclass total weights, are simply the averages of the five subclass-specific proportions. These adjusted proportions are displayed in Table 12.2 for $t = 6$ months, 1 year, and 3 years. Note that for $t = 6$ months, 1 year, and 3 years, the medical versus surgical differences in survival are small compared to their standard errors, but consistently higher probabilities of improvement are estimated for surgical treatment. As noted previously, improvement may be affected by differential placebo effects of surgery (Benson and McCallie 1979).

If the subclasses were perfectly homogeneous in the propensity score and the sample sizes were large, the distributions of **x** for surgical and medical patients would be identical within each subclass. Consequently, if this were the case, the difference between surgical and medical adjusted proportions would have no bias due to **x**, or using terminology in Cochran (1968a), subclassification on the propensity score (followed by direct adjustment) would remove all of the initial bias due to **x**. Of course, initial bias due to unmeasured covariates will be removed only to the extent that they are correlated with **x**.

Table 12.1. *Subclass specific results at six months*

Subclass[a]	Treatment group	No. of patients	Survival to 6 months		Substantial improvement at 6 months	
			Estimate	Standard error	Estimate	Standard error
1	Medical	277	.892	(.019)	.351	(.030)
	Surgical	26	.846	(.071)	.538	(.098)
2	Medical	235	.953	(.014)	.402	(.032)
	Surgical	68	.926	(.032)	.705	(.056)
3	Medical	205	.922	(.019)	.351	(.034)
	Surgical	98	.898	(.031)	.699	(.047)
4	Medical	139	.941	(.020)	.303	(.042)
	Surgical	164	.933	(.020)	.706	(.036)
5	Medical	69	.924	(.033)	.390	(.063)
	Surgical	234	.914	(.018)	.696	(.030)
Directly Adjusted Across Subclasses	Medical	–	.926	(.022[b])	.359	(.042[b])
	Surgical	–	.903	(.039[b])	.669	(.059[b])

[a] Based on estimated propensity score.
[b] Standard errors for the adjusted proportions were calculated following Mosteller and Tukey (1977, Chap. 11c).

In our example, however, the five subclasses are not perfectly homogeneous in **x**. Results in Cochran (1968a) show that for many examples of univariate x, five subclasses will remove approximately 90% of the initial bias in x; Cochran did not consider multivariate **x**. Nevertheless, his results can be applied, using a theorem presented in Appendix A, to suggest that adjustment with five subclasses based on the propensity score will remove approximately 90% of the initial bias in each coordinate of multivariate **x**.

Table 12.2. *Directly adjusted probabilities of survival and uninterrupted improvement (and standard errors[*])*

	6 months		1 year		3 years	
	Pr	SE	Pr	SE	Pr	SE
Survival						
Medical	.926	(.022)	.902	(.025)	.790	(.040)
Surgical	.903	(.039)	.891	(.040)	.846	(.049)
Uninterrupted Improvement						
Medical	.359	(.042)	.226	(.040)	.126	(.036)
Surgical	.669	(.059)	.452	(.060)	.298	(.057)

[*]Standard errors (SE) for the adjusted proportions were calculated following Mosteller and Tukey (1977, Chapter 11c).

Table 12.3. *Directly adjusted estimated probabilities*
of substantial improvement

No. of diseased vessels	Initial functional class		
	II	III	IV
1			
Medical Therapy	.469	.277	.487
Surgery	.708	.629	.635
2			
Medical Therapy	.404	.221	.413
Surgery	.780	.706	.714
3			
Medical Therapy	.248	.133	.278
Surgery	.709	.649	.657

3.3. Adjustment and Estimation Within Subpopulations Defined by x

It is often of interest to estimate average treatment effects within subpopulations. This section shows how balanced subclassification may be combined with model-based adjustment to obtain estimates of the average effect of the treatment within subpopulations defined by x. Specifically, we estimate the probabilities of uninterrupted improvement to six months for subpopulations of patients defined by the number of diseased vessels (N) and the New York Heart Association functional class at the time of cardiac catheterization (F). To avoid an excessive number of subpopulations, the small but clinically important subset of patients with significant left main stenosis has been excluded.

Patients were cross-classified according to the number of diseased vessels (N), initial functional class (F), treatment (Z), subclass based on the estimated propensity score (S), and condition at six months (I; improved = substantial treatment as defined in Sec. 3.1). A log-linear model, which fixed the IZN, IZF, ISN, SZ, SF, and FN margins, provided a good fit to this table (likelihood ratio $\chi^2 = 122.5$ on 120 degrees of freedom). (Here IZN denotes the marginal table formed by summing the entries in the table over initial functional class F and subclass S, leaving a three-way table.)

The directly adjusted estimates in Table 12.3 were calculated from the fitted counts, using the NFS marginal table for weights; in other words, within each subpopulation defined by the number of diseased vessels (N) and the initial functional class (F), estimates of the probabilities of improvement were adjusted using subclass (S) total weights. In all six subpopulations, the estimated probabilities of substantial improvement at six months are higher following surgery than following medical treatment (between 30% and 387% higher). The estimated probabilities differ least for one-vessel disease, functional class IV, and differ most for three-vessel disease, functional class III. The definition of substantial improvement has resulted in lower estimated probabilities of improvement for class III patients than for class II and IV patients. The estimated probabilities of improvement under surgery vary less than the estimated probabilities of improvement under medicine.

4. SENSITIVITY OF ESTIMATES TO THE ASSUMPTION OF STRONGLY IGNORABLE TREATMENT ASSIGNMENT

The estimates presented in Section 3 are approximately unbiased under the assumption that all variables related to *both* outcomes and treatment assignment are included in x. This condition is called strongly ignorable treatment assignment by Rosenbaum and Rubin (1983a), and in fact Corollary 4.2 of that paper asserts that if (a) treatment assignment is strongly ignorable, (b) samples are large, and (c) subclasses are perfectly homogeneous in the population propensity score, then direct adjustment will produce unbiased estimates of the average treatment effect. In randomized experiments, x is constructed to include all covariates used to make treatment assignments (e.g., block indicators) with the consequence that treatment assignment is strongly ignorable.

Of course with most observational data, such as the data presented here, we cannot be sure that treatment assignment is strongly ignorable given the observed covariates because there may remain unmeasured covariates that affect both outcomes and treatment assignment. It is then prudent to investigate the sensitivity of estimates to this critical assumption.

Rosenbaum and Rubin (1983b) develop and apply to the current example a method for assessing the sensitivity of these estimates to a particular violation of strong ignorability. They assume that treatment assignment is not strongly ignorable given the observed covariates x, but is strongly ignorable given (x, u), where u is an unobserved binary covariate. The estimate of the average treatment effect was recomputed under various assumptions about u. A related Bayesian approach was developed by Rubin (1978a).

5. CONCLUSIONS: THE PROPENSITY SCORE AND MULTIVARIATE SUBCLASSIFICATION

With just five subclasses formed from an estimated scalar propensity score, we have substantially reduced the bias in 74 covariates simultaneously. Although the process of estimating the propensity score for use in balanced subclassification does require some care, the comparability of treated and control patients within each of the final subclasses can be verified using the simplest statistical methods, and therefore results based on balanced subclassification can be persuasive even to audiences with limited statistical training. The same subclasses can also be used to estimate treatment effects within subpopulations defined by the covariates x. Moreover, balanced subclassification may be combined with model-based adjustments to provide improved estimates of treatment effects within subpopulations.

APPENDIX A

The Effectiveness of Subclassification on the Propensity Score in Removing Bias

Cochran (1968a) studies the effectiveness of univariate subclassification in removing bias in observational studies. In this Appendix, we show how Cochran's results are related to subclassification on the propensity score.

Let $f = f(\mathbf{x})$ be any scalar valued function of $\mathbf{x}$. The initial bias in f is $B_I = E(f \mid z = 1) - E(f \mid z = 0)$. The (asymptotic) bias in f after subclassification on the propensity score and direct adjustment with subclass total weights is

$$B_S = \sum_{j=1}^{J} \{E(f \mid z = 1, e \in I_j) - E(f \mid z = 0, e \in I_j)\} \Pr(e \in I_j),$$

where there are J subclasses, and I_j is the fixed set of values of e that define jth subclass. The percent reduction in bias in f due to subclassification on the propensity scores is $100\,(1 - B_S/B_I)$.

Cochran's (1968a) results do not directly apply to subclassification on the propensity score, since his work is concerned with the percent reduction in bias in f after subclassification on f, rather than the percent reduction in bias in f after subclassification on e. Nonetheless, as the following theorem shows, Cochran's results are applicable providing (a) the conditional expectation of f given e, that is $E(f \mid e) = \bar{f}$, is a monotone function of e, and (b) $\bar{f}$ has one of the distributions studied by Cochran. In particular, under these conditions, subclassification at the quintiles of the distribution of the propensity score, e, will produce approximately a 90% reduction in the bias of f. Note that in the following theorem, Cochran's (1968a) results apply directly to the problem of determining the percent reduction in bias in $\bar{f}$ after subclassification on $\bar{f}$.

Theorem A.1. *The percent reduction in the bias,* $100(1 - B_S/B_I)$, *in f following subclassification at specified quantiles of the distribution of the propensity score, e, equals the percent reduction in the bias in $\bar{f}$ after subclassification at the same quantiles of the distribution of $\bar{f}$, providing $\bar{f}$ is a strictly monotone function of e.*

Proof. First, we show that within a subclass defined by $e \in S$, the bias in f equals the bias in $\bar{f}$; that is, we show that

$$E(f \mid e \in S, z = 1) - E(f \mid e \in S, z = 0)$$
$$= E(\bar{f} \mid e \in S, z = 1) - E(\bar{f} \mid e \in S, z = 0). \qquad \text{(A.1)}$$

To show this it is sufficient to observe that for $t = 0, 1$,

$$E(f \mid e \in S, z = t) = E\{E(f \mid e, e \in S, z = t) \mid e \in S, z = t\}$$
$$= E\{E(f \mid e) \mid e \in S, z = t\}$$
$$= E(\bar{f} \mid e \in S, z = t).$$

where the second equality follows from the fact that e is the propensity score (i.e., from Equation (12.1)).

From (A.1) with $S = [0, 1]$, it follows that the initial bias in f equals the initial bias in $\bar{f}$. To complete the proof, we need to show that the bias in f after subclassification on e equals the bias in $\bar{f}$ after subclassification on $\bar{f}$. Since by assumption $\bar{f}$ is a strictly monotone function of e, subclasses defined at specified quantiles of the distribution of e contain exactly the same units as subclasses defined at the same quantiles of the distribution of $\bar{f}$. It follows from this observation and (A.1) that the bias in f within each subclass defined by e equals the bias in $\bar{f}$ within each subclass defined by $\bar{f}$. Since (a) the initial biases in f and $\bar{f}$ are equal, (b) the subclasses formed from e contain the same units as the subclasses formed from $\bar{f}$, and

(c) within each subclass, the bias in f equals the bias in $\bar{f}$, it follows that the percent reduction in bias in f after subclassification on e equals the percent reduction in bias in $\bar{f}$ after subclassification on $\bar{f}$.

APPENDIX B

Balancing Properties of the Propensity Score with Incomplete Data

In Section 2.4, we noted that several covariates were missing for a large number of patients. Let $\mathbf{x}^*$ be a p-coordinate vector, where the jth coordinate of $\mathbf{x}^*$ is a covariate value if the jth covariate was observed, and is an asterisk if the jth covariate is missing. (Formally, $\mathbf{x}^*$ is an element of $\{R,^*\}^p$.) Then $e^* = \Pr(z = 1 \mid \mathbf{x}^*)$ is a generalized propensity score. The following theorem and corollary show that e^* has balancing properties that are similar to the balancing properties of the propensity score e. The notation $a \perp\!\!\!\perp b \mid c$ means that a is conditionally independent of b given c (see Dawid 1979).

Theorem B.1. $\mathbf{x}^* \perp\!\!\!\perp z \mid e^*$

Proof. The proof of Theorem B.1 is identical to the proof of Theorem 1 of Rosenbaum and Rubin (1983a), with $\mathbf{x}^*$ in place of $\mathbf{x}$ and e^* in place of e.

Theorem B.1 implies that subclassification on the generalized propensity score e^* balances the observed covariate information and the pattern of missing covariates. Note that Theorem B.1 does *not* generally imply that subclassification on e^* balances the unobserved coordinates of $\mathbf{x}$; that is, it does not generally imply

$$\mathbf{x} \perp\!\!\!\perp z \mid e^*.$$

The consequences of Theorem B.1 are clearest when there are only two patterns of missing data, with $\mathbf{x} = (\mathbf{x}_1, \mathbf{x}_2)$, where $\mathbf{x}_1$ is always observed and $\mathbf{x}_2$ is sometimes missing. Let $c = 1$ when $\mathbf{x}_2$ is observed, and let $c = 0$ when $\mathbf{x}_2$ is missing. Then $e^* = \Pr(z = 1 \mid \mathbf{x}_1, \mathbf{x}_2, c = 1)$ for units with $\mathbf{x}_2$ observed, and $e^* = \Pr(z = 1 \mid \mathbf{x}_1, c = 0)$ for units with $\mathbf{x}_2$ missing. Subclasses of units may be formed using e^*, ignoring the pattern of missing data.

Corollary B.1. (a) For units with $\mathbf{x}_2$ missing, there is balance on $\mathbf{x}_1$ at each value of e^*; that is,

$$\mathbf{x}_1 \perp\!\!\!\perp z \mid e^*, c = 0.$$

(b) For units with $\mathbf{x}_2$ observed, there is balance on $(\mathbf{x}_1, \mathbf{x}_2)$ at each value of e^*; that is,

$$(\mathbf{x}_1, \mathbf{x}_2) \perp\!\!\!\perp z \mid e^*, c = 1.$$

(c) There is balance on $\mathbf{x}_1$ at each value of e^*; that is,

$$\mathbf{x}_1 \perp\!\!\!\perp z \mid e^*.$$

(d) The frequency of missing data is balanced at each value of e^*; that is,

$$c \perp\!\!\!\perp z \mid e^*.$$

Proof. Parts a and b follow immediately from Theorem 1 of Rosenbaum and Rubin (1983a), and Parts c and d follow immediately from Theorem B.1.

In practice, we may estimate e^* in several ways. In a large study with only a few patterns of missing data, we may use a separate logit model for each pattern of missing data. In general, however, there are 2^p potential patterns of missing data with p covariates. If the covariates are discrete, then we may estimate e^* by treating the * as an additional category for each of the p covariates, and we may apply standard methods for discrete cross-classifications (Bishop, Fienberg, and Holland 1975).

ACKNOWLEDGMENTS

The authors acknowledge Arthur Dempster for valuable conversations on the subject of this paper and Bruce Kaplan for assistance with Figures 12.1 and 12.2.

13. Constructing a Control Group Using Multivariate Matched Sampling Methods That Incorporate the Propensity Score

Paul R. Rosenbaum and Donald B. Rubin

Abstract: Matched sampling is a method for selecting units from a large reservoir of potential controls to produce a control group of modest size that is similar to a treated group with respect to the distribution of observed covariates. We illustrate the use of multivariate matching methods in an observational study of the effects of prenatal exposure to barbiturates on subsequent psychological development. A key idea is the use of the propensity score as a distinct matching variable.

1. INTRODUCTION: BACKGROUND; WHY MATCH?

Matched Sampling in Observational Studies. In many observational studies, there is a relatively small group of subjects exposed to a treatment and a much larger group of control subjects not exposed. When the costs associated with obtaining outcome or response data from subjects are high, some sampling of the control reservoir is often necessary. Matched sampling attempts to choose the controls for further study so that they are similar to the treated subjects with respect to background variables measured on all subjects.

The Danish Cohort. We examine multivariate matched sampling using initial data from a proposed study of the effects on psychological development of prenatal exposure to barbiturates. The analyses presented are preliminary and intended only to explore methodological options; none of the matched samples are the actual ones to be used for study of in utero exposure to barbiturates. The children under study were born between 1959 and 1961 and have been the object of other studies (e.g., Mednick et al. 1971; Zachau-Christiansen and Ross 1975). Prenatal and perinatal information is available for 221 barbiturate-exposed children and 7,027 unexposed children. A battery of measures of subsequent psychological development are to be obtained from all 221 exposed children, but cost considerations require sampling of the unexposed children to create a control group in which the measures will be obtained. The cost of the study will be approximately linear in the number of children studied – with basic costs that are largely independent of the number of children, and other costs associated with locating and examining the children that are approximately proportional to the number of children studied.

Approximate Efficiency Considerations. To obtain a rough idea of the loss in efficiency involved in not including all unexposed children as controls, suppose for the moment that there is no concern with biases between exposed and unexposed groups, in the sense that the mean difference between the groups can be regarded as an unbiased estimate of the effect of prenatal exposure to barbiturates. If control children are randomly sampled from among the 7,027 unexposed children, and if the variance, σ^2, of a particular psychological response is the same in the treated and control groups, then the standard error of the treated versus control difference in means will be $\sigma(1/221 + 1/N_c)^{1/2}$, where N_c is the number of control children studied. For $N_c = 100, 221, 442 \ (= 2 \times 221), 663 \ (= 3 \times 221), 884 \ (= 4 \times 221)$, $2{,}210 \ (= 10 \times 221)$, and $7{,}027 (= 31.8 \times 221)$, the multipliers $(1/221 + 1/N_c)^{1/2}$ are, respectively, .121, .095, .082, .078, .075, .071, and .068. The cost of studying all 7,027 control children would be substantially greater than the cost of a modest sample, and the gain in precision would not be commensurate with the increase in cost. Cost considerations in this study led to a sample of 221 matched controls.

Distributions of Background Variables Before Matching. In fact, forming a control group by random sampling of the unexposed children is not a good idea. Many of the unexposed children may not be good controls because they are quite different from all exposed children with respect to background variables. Consequently, controls will not be selected by random sampling, but rather by matched sampling on the basis of the covariates listed in Table 13.1. From the *t* statistics and standardized differences in Table 13.1, we see that the exposed and unexposed children differ considerably. The hope is that matched sampling will produce a control group that is similar to the treated group with respect to these covariates.

For Nontechnical Audiences, Matched Sampling Is Often a Persuasive Method of Adjustment. One virtue, not the least important, of matched sampling is that nontechnical audiences often find that matching, when successful, is a persuasive method of adjusting for imbalances in observed covariates. Although matching algorithms can be complex, the simplest methods, such as comparisons of sample moments, often suffice to indicate whether treated and matched control groups can be directly compared without bias due to observed covariates.

The Limitations of Incomplete Categorical Matching. Perhaps the most obvious matching method involves categorizing each of the 20 variables in Table 13.1 and considering a treated child and control child as a suitable matched pair only if they fall in the same category on each variable. Unfortunately, even if each of the 20 variables is divided into just two categories, the 7,027 potential controls will be distributed among $2^{20} \doteq 1$ million matching categories, so exact matches may be hard to find for some treated children. If a version of categorical matching described by Rosenbaum and Rubin (1984c) is applied to the current data, only 126 of the 221 treated children have exact matches, so 95 (43%) of the treated children are discarded as unmatchable. Discarding treated children in this way can lead to serious biases, since the unmatched treated children differ systematically from the matched treated children. We consider only methods that match all 221 exposed children.

Table 13.1. *Covariate imbalance prior to matching*

Covariate	Description of original covariate	As used for estimating the propensity score	Differences in covariate means prior to matching	
			Two-sample t statistic	Standardized difference in %*
Child characteristics				
Sex	Female/male	0, 1	−1.02	−7
Twin	Single/multiple birth	0, 1	−1.28	−10
SIBPOS	Oldest child (no, yes)	0, 1	−2.33	−16
C-age	Age at start of study	Months	.46	3
Mother characteristics				
SES	Socioeconomic status (9 ordered categories)	Integers 1–9	3.66	26
Education	Mother's education (4 ordered categories)	Integers 1–4	2.09	15
Single	Unmarried (no, yes)	0, 1	−5.70	−43
M-age	Age (years)	Years	8.99	59
Height	Mother's height (5 ordered categories)	Integers 1–5	2.55	18
Characteristics of the pregnancy				
WGTHGT3	(Weight gain)/height³ (30 values based on category midpoints)	30 values	−.00	−0
PBC415	Pregnancy complications (an index)	Index value and its square	2.61	17
PRECLAM	Preeclamsia (no, yes)	0, 1	1.82	9
RESPILL	Respiratory illness (no, yes)	0, 1	1.73	10
LENGEST	Length of gestation (10 ordered categories)	$(10 - i)^{1/2}$ and i for $i = 1, 2, \ldots, 10$	.72	6
Cigarette	Cigarette consumption, last trimester (0 = none, plus 4 ordered categories)	Integers 0–4 and their squares	−.48	−3
Other Drugs				
Antihistamine	No. of exposures to antihistamines (0–6)	Integers 0–6 and their squares	1.76	10
Hormone	No. of exposures to hormones (0–6)	Integers 0–6 and their squares	8.41	28
HRMG1	Exposed to hormone type 1 (no, yes)	0, 1	2.67	15
HRMG2	Exposed to hormone type 2 (no, yes)	0, 1	3.75	19
HRMG3	Exposed to hormone type 3 (no, yes)	0, 1	3.46	18

* The standardized difference in percent is the mean difference as a percentage of the average standard deviation: $100(\bar{x}_1 - \bar{x}_{0R})/[(s_1^2 + s_{0R}^2)/2]^{1/2}$, where for each covariate, $\bar{x}_1$ and $\bar{x}_{0R}$ are the sample means in the treated group and the control reservoir and s_1^2 and s_{0R}^2 are the corresponding sample variances.

2. THEORY RELEVANT TO THE CHOICE OF A MATCHING METHOD

2.1. The Most Important Scalar Matching Variable:
The Propensity Score

Let **x** denote the vector of covariates for a particular child, and let the binary variable z indicate whether the child was exposed ($z = 1$) or unexposed ($z = 0$). The propensity score, $e(\mathbf{x})$, is the conditional probability of exposure given the covariates; that is, $e(\mathbf{x}) = \Pr(z = 1|\mathbf{x})$. Treated children and control children selected to have the same value of $e(\mathbf{x})$ will have the same distributions of **x**; formally, z and **x** are conditionally independent given $e(\mathbf{x})$. Exact matching on $e(\mathbf{x})$ will, therefore, tend to balance the **x** distributions in the treated and control groups. Moreover, matching on $e(\mathbf{x})$ and any function of **x**, such as selected coordinates of **x**, will also balance **x**. (For proofs of these balancing properties of propensity scores, see Rosenbaum and Rubin 1983a Theorems 1 and 2. For related discussions of propensity scores, see Rubin 1983, 1984; Rosenbaum 1984a; and Rosenbaum and Rubin 1984b.) The propensity score is a potential matching variable because it does not depend on response information that will be collected after matching. Since exact adjustment for a known propensity score will, on average, remove all of the bias in **x**, the propensity score $e(\mathbf{x})$ is in a sense the most important scalar matching variable.

Matching on $e(\mathbf{x})$ balances the *observed* covariates **x**; however, unlike randomization, matching on $e(\mathbf{x})$ does not balance *unobserved* covariates except to the extent that they are correlated with **x**. For discussion of methods for addressing the possible effects of unobserved covariates in observational studies, see Rosenbaum and Rubin (1983b) and Rosenbaum (1984b and 1984c).

In practice, several issues need to be addressed before the propensity score can be used as a matching variable. First, the functional form of $e(\mathbf{x})$ is rarely if ever known, at least in observational studies such as the one we describe, and therefore $e(\mathbf{x})$ must be estimated from the available data. Second, exact matches will rarely be available, and so issues of closeness on $e(\mathbf{x})$ must be addressed. Third, adjustment for $e(\mathbf{x})$ balances **x** only in expectation, that is, averaging over repeated studies. In any particular study, further adjustment for **x** may be required to control chance imbalances in **x**. Such adjustments, for example by covariance analysis, are often used in randomized experiments to control chance imbalances in observed covariates.

As noted by Rosenbaum and Rubin (1983a, sec. 2.3), matching on the propensity score is generalization to arbitrary **x** distributions of discriminant matching for multivariate normal **x** as proposed by Rubin (1970) and discussed by Cochran and Rubin (1973) and Rubin (1976b,c; 1979b; 1980b). Propensity matching is not, however, the same as any of the several procedures proposed by Miettinen (1976): the propensity score is not generally a confounder score (see Rosenbaum and Rubin 1983a, sec. 3.3, for discussion).

2.2. Estimating the Propensity Score

We estimated the propensity score in the Danish cohort using a logit model (Cox 1970):

$$q(\mathbf{x}) \equiv \log[(1 - e(\mathbf{x}))/e(\mathbf{x})] = \alpha + \boldsymbol{\beta}^T \mathbf{f}(\mathbf{x}),$$

where α and β are parameters to be estimated, $q(\mathbf{x})$ is the log odds against exposure, and $\mathbf{f}(\mathbf{x})$ is a specified function, which in this instance included quadratic terms and transforms (see Table 13.1 for details). A logit model for $e(\mathbf{x})$ can be formally derived from $\Pr(\mathbf{x}|z = t)$ if the latter has any of a variety of exponential family distributions, such as the multivariate normal $N(\mathbf{\mu}_t, \mathbf{\Sigma})$, the multivariate logit model for binary data in Cox (1972), the quadratic exponential family of Dempster (1971), or the multinomial/multivariate normal distribution of Dempster (1973); see Rosenbaum and Rubin [1983a, sec. 2.3(ii)] for details.

The sample means of the maximum likelihood estimates $\hat{q}(\mathbf{x})$ of $q(\mathbf{x})$ are 3.06 and 3.76 in the treated and control groups, respectively. The sample variance of $\hat{q}(\mathbf{x})$ is 2.3 times greater in the treated group than in the control group. The standardized difference for $\hat{q}(\mathbf{x})$ is .77 (calculated as in the footnote to Table 13.1), which, as one would expect, is larger than the standardized difference for any single variable in Table 13.1. In the treated and control groups, respectively, the minimum values of $\hat{q}(\mathbf{x})$ are -3.9 and $-.6$; the lower quartiles, 2.5 and 3.2; the medians, 3.1 and 3.7; the upper quartiles, 3.8 and 4.2; and the maximums, 5.1 and 7.5. Three treated children have $\hat{q}(\mathbf{x})$ values lower than any control child: their $\hat{q}(\mathbf{x})$ values are -3.9, -1.3, and -1.2.

There is, then, a substantial difference along the propensity score. The larger variance in the treated groups suggests that finding appropriate matches will be relatively more difficult than if the variances were equal (Cochran and Rubin 1973, Table 2.3.1; Rubin 1973a, Table 5.1; Rubin 1980b, Table 1). The reason for the difficulty is the concentration of the $\hat{q}(\mathbf{x})$'s around 3.76 in the control group and the wider dispersion of the $\hat{q}(\mathbf{x})$'s around 3.06 in the treated group: for matching, controls are required with low values of $\hat{q}(\mathbf{x})$, which are relatively uncommon in the control group.

2.3. Matching Methods That Are Equal-Percent Bias Reducing

The mean bias or expected difference in $\mathbf{x}$ prior to matching is $E(\mathbf{x}|z = 1) - E(\mathbf{x}|z = 0)$, whereas the mean bias in $\mathbf{x}$ after matching is $E(\mathbf{x}|z = 1) - \mathbf{\mu}_{0M}$, where $\mathbf{\mu}_{0M}$ is the expected value of $\mathbf{x}$ in the matched control group. Generally, $\mathbf{\mu}_{0M}$ depends on the matching method used, whereas $E(\mathbf{x}|z = 1)$ and $E(\mathbf{x}|z = 0)$ depend only on population characteristics. As defined by Rubin (1976b,c), a matching method is equal-percent bias reducing (EPBR) if the reduction in bias is the same for each coordinate of $\mathbf{x}$, that is, if

$$E(\mathbf{x}|z = 1) - \mathbf{\mu}_{0M} = \gamma\{E(\mathbf{x}|z = 1) - E(\mathbf{x}|z = 0)\}$$

for some scalar $0 \leq \gamma \leq 1$. If a matching method is not EPBR, then matching actually increases the bias for some linear functions of $\mathbf{x}$. If little is known about the relationship between $\mathbf{x}$ and the response variables that will be collected after matching, then EPBR matching methods are attractive, since they are the only methods that reduce bias in all variables having linear regression on $\mathbf{x}$. Rosenbaum and Rubin (1983a, sec. 3.2) show that matching on the population propensity score alone is EPBR whenever $\mathbf{x}$ has a linear regression on some scalar function of e; that is, whenever $E(\mathbf{x}|e) = \alpha + \mathbf{\gamma}^T \mathbf{g}(e)$ for some scalar function $g(\cdot)$.

3. CONSTRUCTING A MATCHED SAMPLE: AN EMPIRICAL COMPARISON OF THREE MULTIVARIATE METHODS

3.1. Overview: How Much Importance Should Be Given to the Propensity Score?

Matched samples were constructed by using three different methods that matched every treated child to one control child. By design, all three methods required exact matches on sex. The three methods differed in the importance given to the estimated propensity score relative to the other variables in **x**.

3.2. Nearest Available Matching on the Estimated Propensity Score

With nearest available propensity score matching, (a) treated and control children are randomly ordered; (b) the first treated child is matched with the control child of the same sex having the nearest $\hat{q}(\mathbf{x})$, and both children are removed from the lists of treated and control children; (c) step (b) is repeated for the remaining unmatched treated children. The decision to define distance in terms of $\hat{q}(\mathbf{x})$ rather than $\hat{e}(\mathbf{x})$ was somewhat but not entirely arbitrary and probably had negligible effect. It did, however, avoid the compression of the $\hat{e}(\mathbf{x})$ scale near 0 and 1, and moreover, $\hat{q}(\mathbf{x})$ was more nearly normally distributed, which is relevant in the context of Section 3.4. Nearest available matching on a scalar covariate x was studied by Rubin (1970), reviewed by Cochran and Rubin (1973), and extended by Rubin (1973a,b); its application to matching on linear discriminant scores with bivariate normal **x** was studied, using Monte Carlo, by Rubin (1979b, 1980b). The effects of random ordering in step (a) rather than ordering by $q(\hat{\mathbf{x}})$ are discussed by Rubin (1973a).

In Tables 13.2 and 13.3, column 1 describes the balance obtained in the samples matched by nearest available propensity matching. Note that the standardized differences in Table 13.2 have the same denominator as the standardized differences in Table 13.1, whereas the t statistics indicated by the footnotes to Table 13.2 have denominators that are affected by the matching and so are not directly comparable. (For discussion of the relationship between t statistics on covariate means and the coverage of confidence intervals for treatment effects formed by ignoring the variable, see, e.g., Cochran 1965, sec. 3.1.) The *two-sample t* statistics (values indicated by pluses in Table 13.2) are relevant for comparing the distributions of the covariates in the treated and matched control groups. The *paired t* statistics (values indicated by asterisks) are relevant for assessing the effects of residual biases in the covariates in analyses of outcome variables based on matched pair differences.

In Table 13.3, the sample percent reduction in bias for a covariate is $100(1 - b_M|b_I)$, where b_I and b_M are the treated versus control differences in covariate means initially and after matching, respectively. When the initial mean sample bias, b_I, is small, the sample percent reduction in bias, $100(1 - b_M|b_I)$, is quite unstable; therefore, we report percent reductions only for variables with large initial biases (i.e., standardized differences above 20% in Table 13.1).

Tables 13.2 and 13.3 both suggest that nearest available matching on the propensity score has removed almost all of the mean difference along the propensity score – arguably the most important variable – and that there has been substantial reduction

Table 13.2. *Covariate imbalance in matched samples: Standardized differences (%)*

Factor	Nearest available matching on the propensity score	Mahalanobis metric matching	
		Including the propensity score	Within propensity score calipers
Child characteristics			
Sex	0	0	0
Twin	−3	0	0
SIBPOS	−5	5*	0
C-age	7	6	−6
Mother characteristics			
SES	−10*	5	−1
Education	−17+	3	−7*
Single	−7	−3*	−2
M-age	−8*	5	−1
Height	−8	3	−9+**
Characteristics of the pregnancy			
WGTHGT3	−0	−3	1
PBC415	−14+*	6*	1
PRECLAM	0	0	0
RESPILL	−7	0	0
LENGEST	−12*	−3*	−4
Cigarette	0	10+***	9***
Drugs			
Antihistamine	−3	4	9*
Hormone	8*	6***	6*
HRMG1	−2	−7**	−6
HRMG2	−2	−5**	−9*
HRMG3	−3	−8**	−11*
$\hat{q}(\mathbf{x})$	−3**	−20++****	−3**

Note: The standardized difference in percent is $100(\bar{x}_1 - \bar{x}_{0M})/[(s_1^2 + s_{0R}^2)/2]^{1/2}$, where for each covariate, $\bar{x}_1$ and $\bar{x}_{0M}$ are the sample means in the treated group and matched control group and s_1^2 and s_{0R}^2 are the sample variances in the treated group and control reservoir. Note that the denominator of the standardized difference is the same for all three matching methods. The values of paired (*) and two-sample (+) *t*-statistics are indicated as follows: * and +, between 1.0 and 1.5 in absolute value; ** and ++, between 1.5 and 2.0 in absolute value; ***, between 2 and 3 in absolute value; ****, above 3 in absolute value.

in the standardized differences for most variables. Still, the residual differences on several variables (Education, PBC415, LENGEST) are bothersome; further analytical adjustments for these variables might be required, for example, using analysis of covariance on matched-pair differences (Rubin 1973b, 1979b).

3.3. Mahalanobis Metric Matching Including the Propensity Score

Mahalanobis metric matching has been described by Cochran and Rubin (1973) and Rubin (1976b) and studied in detail by Carpenter (1977) and Rubin (1979b, 1980b). With nearest available Mahalanobis metric matching, treated and control children

Table 13.3. *Covariate imbalance in matched samples: percent reductions in bias for variables with substantial initial bias (standardized absolute bias of 20% or greater)*

	Nearest available matching on the propensity score	Mahalanobis metric matching	
		Including the propensity score	Within propensity score calipers
Single	84	93	95
Hormone	71	79	79
SES	140	81	105
M-age	114	91	102
$\hat{q}(\mathbf{x})$	96	74	96

are randomly ordered. The first treated child is matched with the closest control child of the same sex, where distance is defined by the Mahalanobis distance:

$$d(\mathbf{u}, \mathbf{v}) = (\mathbf{u} - \mathbf{v})^T \mathbf{C}_{0R}^{-1}(\mathbf{u} - \mathbf{v}), \tag{13.1}$$

where $\mathbf{u}$ and $\mathbf{v}$ are values of $\{\mathbf{x}^T, \hat{q}(\mathbf{x})\}^T$ and $\mathbf{C}_{0R}$ is the sample covariance matrix of $\{\mathbf{x}^T, \hat{q}(\mathbf{x})\}$ in the control reservoir. The two matched children are then removed from the treated and control lists, and the process is repeated. In the case of multivariate normal covariates with common covariance matrix in treated and control groups, Rubin (1976b, Theorem 2) has shown that Mahalanobis metric matching is EPBR.

The results of applying Mahalanobis metric matching are given in column 2 of Tables 13.2 and 13.3. As one might expect, Mahalanobis metric matching is somewhat more successful than propensity matching in reducing the standardized differences for individual coordinates of $\mathbf{x}$, but it is far less successful in reducing the standardized difference along the propensity score. The standardized difference of 20% and two-sample t statistic of -1.99 for $\hat{q}(\mathbf{x})$ are disturbing.

Mahalanobis metric matching produces several large matched-pair t statistics, as indicated in the footnotes to Table 13.2. The standard deviations of the within-pair differences in covariate values are smaller than under nearest available propensity score matching, so the matched-pair t statistics are larger. As noted previously (Sec. 3.2), the t statistics for different methods are not directly comparable. Nevertheless, the large values of the matched-pair t statistics indicate that analyses based on matched-pair differences can be misleading unless analysis of covariance is used to control within-pair differences due to $\mathbf{x}$.

3.4. Nearest Available Mahalanobis Metric Matching Within Calipers Defined by the Propensity Score

In an effort to obtain the best features of both previous methods, we now consider a hybrid matching method that first defines a subset of potential controls who are close to each treated child on the propensity score (i.e., within "calipers," Althauser and Rubin 1971) and then selects the control child from this subset by using nearest available Mahalanobis metric matching (for variables $\{\mathbf{x}, \hat{q}(\mathbf{x})\}$). The details of the procedure are given in Figure 13.1. With multivariate normal covariates having common covariance matrices in treated and control groups, and with $\hat{q}(\mathbf{x})$ replaced

1. Randomly order the treated children.
2. *Caliper Matching on the Propensity Score*: For the first treated child, find all available untreated children of the same sex with $\hat{q}(\mathbf{x})$ values that differ from the $\hat{q}(\mathbf{x})$ value for the treated child by less than a specified constant c. If there is no such untreated child, match the treated child to the untreated child of the same sex with the nearest value of $\hat{q}(\mathbf{x})$.
3. *Nearest Available Mahalanobis Metric Matching Within Calipers*: From the subset of children defined in step 2, select as a match the untreated child of the same sex who is closest in the sense of the Mahalanobis distance for the variables $\{\mathbf{x}, \hat{q}(\mathbf{x})\}$.
4. Remove the treated child and the matched control child from the lists of treated and untreated children. Go to step 2 for the next treated child.

Figure 13.1. Nearest available Mahalanobis metric matching within calipers defined by the propensity score.

by its population value (the linear discriminant), this matching method would be EPBR, since each of the two stages would reduce bias in $\mathbf{x}$ by a constant percentage. A computational advantage of this method is a substantial reduction in the number of Mahalanobis distances that need to be computed. The method in Section 3.3 required the computation of about 1.5 million Mahalanobis distances.

The caliper width, c, used in step 2 of Figure 13.1 was determined by using results from Cochran and Rubin (1973) concerning the performance of caliper matching. Write σ_1^2 and σ_{0R}^2 for the variances of $\hat{q}(\mathbf{x})$ in the treated and untreated groups, and let $\sigma = [(\sigma_1^2 + \sigma_{0R}^2)/2]^{1/2}$. Table 2.3.1 of Cochran and Rubin (1973) suggests that when $\sigma_1^2/\sigma_{0R}^2 = 2$, a caliper width of $c = .2\sigma$ would remove 98% of the bias in a normally distributed covariate, that $c = .4\sigma$ would remove 93%, and that $c = .6\sigma$ would remove 86%. That table also suggests that narrower caliper widths (i.e., smaller values of c) are required as σ_1^2/σ_{0R}^2 increases. The point estimate of σ_1^2/σ_{0R}^2 for $\hat{q}(\mathbf{x})$ is $s_1^2/s_{0R}^2 = 2.3$. Therefore, in the hope of removing at least 90% of the bias along $\hat{q}(\mathbf{x})$ by caliper matching, we took $c = .25s = (.25)(.930) = .232$, where $s = [(s_1^2 + s_{0R}^2)/2]^{1/2}$. (For further discussion of caliper matching, see Cochran 1972 and Raynor 1983.) There were four treated children who had no available matches within the calipers; following step 2 of Figure 13.1, they were matched with the nearest available control on $\hat{q}(\mathbf{x})$.

Some results of this matching appear in column 3 of Tables 13.2 and 13.3. Mahalanobis metric matching within calipers defined by the propensity score appears superior to the two other methods: it is better than matching on the propensity score in that it yields fewer standardized differences above 10% in absolute value, and it is better than Mahalanobis metric matching in controlling the difference along the propensity score.

3.5. Nonlinear Response Surfaces

Tables 13.1–13.3 compare the three matching methods in terms of the means of $\mathbf{x}$ in the treated and matched control groups. If, however, the response has a nonlinear

Table 13.4. *Summarized standardized differences (in %) for covariates, squares of covariates, and cross products of covariates*

	Root mean square[*]	Maximum absolute
Prior to matching	24	78
Nearest available matching on the		
propensity score	9	49
Mahalanobis metric matching		
Including the propensity score	9	76
Using propensity score calipers	7	27

[*] $100(\bar{b}^2 + s_b^2)^{1/2}$, where $\bar{b}$ and s_b^2 are the sample mean and variance of standardized differences for the $\binom{20}{2} + 2\binom{20}{1} = 230$ variables.

regression on $\mathbf{x}$ in the treated and control groups, then equal $\mathbf{x}$ means in matched samples do not necessarily indicate the absence of bias due to $\mathbf{x}$. For example, if the regressions on $\mathbf{x}$ are quadratic, then the means on $\mathbf{x}$ and $\mathbf{xx}^T$ are both relevant. Table 13.4 summarizes the standardized biases for $230 = \binom{20}{2} + 2 \times 20$ variables: the 20 coordinates of $(\mathbf{x}, \hat{q}(\mathbf{x}))$ with sex excluded because exact matches for sex were obtained, the squares of these 20 variables, and the cross products of pairs of these variables. The third matching method – Mahalanobis metric matching within propensity score calipers – appears clearly superior.

4. SUMMARY

When combined with covariance adjustments of matched-pair differences, multivariate matched sampling is known to be one of the most robust methods for reducing bias due to imbalances in observed covariates (Rubin 1973b, 1979b). Three methods for multivariate matched sampling have been illustrated and compared on data concerning the effects of prenatal barbiturate exposure. The first method was nearest available matching on the estimated propensity score; this method required less computation than the others and was fairly successful in reducing bias. The second method was nearest available Mahalanobis metric matching using all variables and the estimated propensity score; this method produced smaller standardized differences for individual variables but left a substantial difference along the propensity score. The third method – Mahalanobis metric matching within calipers defined by the estimated propensity score – appeared superior to the others with respect to balancing the covariates, their squares, and their cross products. Since the current study has examined just three possible methods on a single set of data, additional work on multivariate matching is needed in several areas: (a) theory concerning the effects on the best choice of matching method of (i) reservoir size, (ii) the magnitude of initial biases, (iii) dimensionality of $\mathbf{x}$, and (iv) covariate distributions; (b) theory concerning multivariate measures of the quality of matched samples involving non-Gaussian covariates; and (c) further empirical studies of multivariate matching methods.

ACKNOWLEDGMENT

The authors are grateful to Robert T. Patrick for extensive assistance in computing.

14. The Bias Due to Incomplete Matching

Paul R. Rosenbaum and Donald B. Rubin

Abstract: Observational studies comparing groups of treated and control units are often used to estimate the effects caused by treatments. Matching is a method for sampling a large reservoir of potential controls to produce a control group of modest size that is ostensibly similar to the treated group. In practice, there is a trade-off between the desires to find matches for all treated units and to obtain matched treated–control pairs that are extremely similar to each other. We derive expressions for the bias in the average matched pair difference due to (i) the failure to match all treated units – incomplete matching, and (ii) the failure to obtain exact matches – inexact matching. A practical example shows that the bias due to incomplete matching can be severe, and moreover, can be avoided entirely by using an appropriate multivariate nearest available matching algorithm, which, in the example, leaves only a small residual bias due to inexact matching.

1. INTRODUCTION

1.1. The Effects Caused by Treatments

A treatment is an intervention that can, in principle, be given to or withheld from any experimental unit under study. With an experimental treatment and a control treatment, each unit has two potential responses: a response r_1 that would be observed if the unit received the experimental treatment, and a response r_0 that would be observed if the unit received the control treatment. Causal effects of treatments are, by definition, comparisons of r_1 and r_0, such as $r_1 - r_0$ or r_1/r_0. Since each unit receives only one treatment, either r_1 or r_0 is observed but not both; the estimation of causal effects, therefore, requires an inference about the unobserved response that would have been observed under the treatment the unit did not actually receive (Kempthorne, 1952, §8.2; Rubin, 1974, 1977a, 1978a; Hamilton, 1979).

Let the binary variable z indicate the treatment assigned to a particular unit in an infinite population of units. We wish to estimate the average effect of the experimental treatment for those units in this population that received the experimental treatment, that is,

$$\tau_1 = E(r_1 - r_0 \mid z = 1), \tag{14.1}$$

where $E(\cdot \mid z = 1)$ refers to the average value in the subpopulation of units receiving treatment 1. When the subpopulation receiving the treatment is a small and special portion of the entire population, estimating τ_1 can be an easier task than estimating the average treatment effect for the entire population. Furthermore, τ_1 is often more relevant than the average effect for the entire population when, as is frequently the case, the treated group is representative of those that may be exposed to the treatment in the future.

1.2. Matching as a Method for Subsampling a Large Reservoir of Potential Controls

We assume the existence of a large random sample of N_1 units from the population of treated ($z = 1$) units and a still larger random sample of N_0 ($> N_1$) control ($z = 0$) units. Formally, each unit in the population has an associated binary random variable s indicating whether the unit was sampled ($s = 1$) or not ($s = 0$). A vector of covariates $\mathbf{x}$ has been measured on all sampled units, but responses have not yet been measured, at least on the control units. (The assumed structure – an infinite population together with large but finite random samples – permits us to discuss the issues that arise in finite samples where exact matches on $\mathbf{x}$ may be unobtainable.)

In order to reduce costs associated with measuring the response for the entire control group, it may be reasonable to use only a comparatively small sample of the N_0 control units. Matching is a method for selecting control units so that the sampled control group is similar to the treated group with respect to the distribution of the observed covariates $\mathbf{x}$. For an early reference on matched sampling, see Cochran (1953a); for review papers see Cochran (1965), Cochran and Rubin (1973), and Rubin (1984c); and for applications see Althauser and Rubin (1970) and Cohn et al. (1981).

Faced with cost constraints that limit the size of the control group, the alternative to matched sampling of controls is some other form of sampling, such as simple random sampling. Whether or not matched sampling is used, further analytical adjustments, such as regression adjustment or subclassification, may be desirable to control any residual bias and to increase efficiency. Rubin (1973b, 1979b) finds that regression adjustment of matched-pair differences is a robust technique.

There is a practical trade-off when obtaining matched samples between the desires to (i) find matches for all treated units and (ii) use only matched treated–control pairs that are extremely similar to each other. Here we compare the biases in the average matched-pair difference due to incomplete matching (i.e., treated units without matches) and to inexact matching (i.e., treated–control matched pairs with different values of $\mathbf{x}$). We provide theoretical expressions for both components of bias, and furthermore illustrate the effects of different matching methods with a practical example. The results from this example suggest that the bias due to incomplete matching can be severe and can be avoided entirely by using an appropriate multivariate nearest available matching algorithm, which, in the example, leaves only a small residual bias due to inexact matching.

In order to define the components of bias in matched samples, we need to define several concepts: effect surfaces, response surfaces, strongly ignorable treatment assignment, and matching methods that match on the basis of **x** alone. For simplicity, we assume that vectors $(r_1, r_0, z, \mathbf{x})$ are independent across units.

1.3. Effect Surfaces and Response Surfaces

The *effect surface* is the population regression of the treatment effect, $r_1 - r_0$, on **x** (Rosenbaum and Rubin, 1984a); that is,

$$\tau(\mathbf{x}) = E(r_1 - r_0 \mid \mathbf{x}).$$

The quantity $\tau(x)$ is the average treatment effect at **x**. In Rubin (1977a), the average treatment effect at **x** is defined in the same manner as $\tau(\mathbf{x}) = \mu_1(\mathbf{x}) - \mu_0(\mathbf{x})$, where $\mu_t(\mathbf{x}) = E(r_t \mid \mathbf{x})$, $t = 0, 1$. We cannot estimate $\tau(\mathbf{x})$ directly since r_1 and r_0 are never observed jointly on the same unit.

Response surfaces, in contrast to effect surfaces, can be estimated directly. The *response surface* among treated units in the population, $\rho_1(\mathbf{x})$, is the regression of r_1 on **x** among units receiving the treatment, that is, $\rho_1(\mathbf{x}) = E(r_1 \mid z = 1, \mathbf{x})$; analogously, the response surface among control units in the population is $\rho_0(\mathbf{x}) = E(r_0 \mid z = 0, \mathbf{x})$. The estimation of response surfaces is straightforward since, for sampled (i.e., $s = 1$) units, r_t is observed when $z = t$, for $t = 0, 1$, and moreover, since the sampling is simple random sampling, $E(r_t \mid z = t, \mathbf{x}) = E(r_t \mid z = t, s = 1, \mathbf{x})$ for $t = 0, 1$. The difference in response surfaces, $\rho_1(\mathbf{x}) - \rho_0(\mathbf{x})$, does not generally equal the effect surface $\tau(\mathbf{x})$ because of the conditioning on z in the response surface; that is, $\rho_t(\mathbf{x}) = E(r_t \mid z = t, \mathbf{x})$ whereas $\mu_t(\mathbf{x}) = E(r_t \mid \mathbf{x})$, $t = 0, 1$. However, the effect surface does equal the difference in response surfaces when treatment assignment is strongly ignorable given **x**, as it is in randomized experiments.

1.4. Strongly Ignorable Treatment Assignment

Since the matching methods that we examine adjust for the observed covariates **x**, it is important to define a simple and useful condition sufficient to ensure that appropriate adjustment for **x** yields unbiased estimates. As defined in Rosenbaum and Rubin (1983a), treatment assignment is *strongly ignorable* given **x** if (i) the responses (r_1, r_0) are conditionally independent of the treatment z given **x**, and (ii) at each value of **x**, there is a positive probability of receiving each treatment; that is, if

$$\Pr(r_1, r_0, z \mid \mathbf{x}) = \Pr(r_1, r_0 \mid \mathbf{x})\Pr(z \mid \mathbf{x})$$

and

$$0 < \Pr(z = 1 \mid \mathbf{x}) < 1 \quad \text{for all possible } \mathbf{x}.$$

For example, treatment assignment is strongly ignorable if the treatment assignment is randomized – that is, if z is determined by the flip of a fair coin. More generally, treatment assignment is strongly ignorable if treatments are randomly assigned with probabilities that depend on **x** alone (e.g., Rubin, 1977a). A weaker condition appropriate for Bayesian inference – ignorable treatment assignment – is defined by Rubin (1978a).

When treatment assignment is strongly ignorable given the observed covariates $\mathbf{x}$,

$$\tau(\mathbf{x}) = \rho_1(\mathbf{x}) - \rho_0(\mathbf{x}) \tag{14.2}$$

and

$$\tau_1 = \mathrm{E}[\tau(\mathbf{x}) \mid z = 1] . \tag{14.3}$$

Thus, when treatment assignment is strongly ignorable given the observed $\mathbf{x}$, we can estimate τ_1, the average treatment effect among treated units in the population, by (i) estimating the difference in population response surfaces, $\rho_1(\mathbf{x}) - \rho_0(\mathbf{x})$, which equals $\tau(\mathbf{x})$ by (14.2), and (ii) averaging this estimate of $\tau(\mathbf{x})$ over an estimate of the distribution of $\mathbf{x}$ in the treated subpopulation, $\Pr(\mathbf{x} \mid z = 1)$, as called for in (14.3).

1.5. Matching on the Basis of x Alone

We shall say that we are *matching on the basis of* $\mathbf{x}$ *alone* if the probability that a sampled treated or control unit is included in the matched subsample depends on the covariates $\mathbf{x}$ and on the treatment group z, but not on the response (r_1, r_0). Formally, write $\mathbf{R}$ for the $(N_0 + N_1) \times 2$ matrix of response pairs (r_1, r_0); $\mathbf{X}$ for the matrix with $N_0 + N_1$ rows containing the $\mathbf{x}$'s; $\mathbf{Z}$ for the vector of $N_0 + N_1$ treatment assignments; and $\mathbf{M}$ for the vector of $N_0 + N_1$ subsampling indicators, where the ith coordinate of $\mathbf{M}$, m_i, is 1 if the ith sampled unit is included in the final matched subsample, and is 0 if the unit is excluded. [Formally, let $m = -1$ for units in the population who are not sampled ($s = 0$), and therefore not eligible for matching.] Then we are matching on the basis of $\mathbf{x}$ alone if $\mathbf{M}$ is conditionally independent of $\mathbf{R}$ given $\mathbf{X}$, and $\mathbf{Z}$; i.e., if

$$\Pr(\mathbf{M}, \mathbf{R} \mid \mathbf{Z}, \mathbf{X}) = \Pr(\mathbf{M} \mid \mathbf{Z}, \mathbf{X})\Pr(\mathbf{R} \mid \mathbf{Z}, \mathbf{X}) \tag{14.4}$$

Most commonly discussed matching methods involve matching on $\mathbf{x}$ alone. These include all those described by Cochran (1953a, 1965, 1968a, 1972), Billewicz (1965), Althauser and Rubin (1970), Cochran and Rubin (1973), Rubin (1973a, 1976b, 1979b, 1980b), McKinlay (1975, 1977), Carpenter (1977), Raynor and Kupper (1981), Anderson et al. (1980, Chap. 6), Raynor (1983), and Rosenbaum and Rubin (1983a). Matching methods using "confounder scores" (Miettinen, 1976), which have been estimated using the observed response data, are not on the basis of $\mathbf{x}$ alone.

When the matching has been on the basis of $\mathbf{x}$ alone, the conditional expectations of r_t given $z = t$ and $\mathbf{x}$ in matched samples – that is, $\rho_{Mt}(\mathbf{x}) = \mathrm{E}(r_t \mid z = t, m = 1, \mathbf{x})$ – equal the corresponding response surfaces; i.e.,

$$\rho_{Mt}(\mathbf{x}) = \rho_t(\mathbf{x}) \quad \text{for} \quad t = 0, 1, \quad \text{and for all } \mathbf{x}. \tag{14.5}$$

Equation (14.5) follows immediately from (14.4), since

$$\Pr(\mathbf{R} \mid \mathbf{M}, \mathbf{Z}, \mathbf{X}) = \Pr(\mathbf{R} \mid \mathbf{Z}, \mathbf{X}) = \prod_{i=1}^{N_1+N_0} \Pr(r_{1i}, r_{0i} \mid z_i, \mathbf{x}_i) .$$

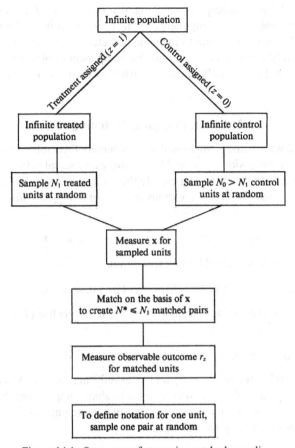

Figure 14.1. Summary of stages in matched sampling.

Throughout, we assume matching has been on the basis of **x** alone, so that (14.5) holds. Let $N^* \leqslant N_1$ denote the number of matched pairs resulting from application of the matching method, where we assume $N^* \geqslant 1$.

2. THE BIAS RESULTING FROM VARIOUS MATCHING PROCEDURES

2.1. Overview: The Three Components of the Bias

Consider a single matched pair, randomly selected from the N^* matched pairs, as obtained by the process summarized in Figure 14.1. The expected difference in the responses of the treated and control units in this pair is denoted by

$$\Delta_M = E(r_1 \mid z = 1, m = 1) - E(r_0 \mid z = 0, m = 1). \tag{14.6}$$

In practice, we would, of course, average the N^* matched-pair differences to obtain an estimate of τ_1. The bias, $\Delta_M - \tau_1$, can be written as the sum of three components: a component (Ψ) due to departures from strongly ignorable treatment assignment;

a component (ι) due to incomplete matching, that is, the discarding of some treated units as unmatchable; and a component (γ) due to coarse or inexact matching on $\mathbf{x}$. In order to address the bias due to incomplete matching, expressions for all three sources have to be derived. A fourth possible component of bias, which we are not considering, is the bias due to failure to obtain representative samples from the target populations.

2.2. Exact, Complete Matching on x

In this section, we assume a situation that is extremely favorable to matching (e.g., discrete $\mathbf{x}$ of low dimension, large N_0/N_1), so that every sampled treated unit can be matched with a control unit having exactly the same value of $\mathbf{x}$. Since all sampled treated units are matched, $z = 1$ implies $m = 1$ for sampled units. Hence, from (14.6),

$$\Delta_M = E(r_1 \mid z = 1) - E(r_0 \mid z = 0, m = 1). \tag{14.7}$$

By familiar properties of conditional expectation,

$$E(r_0 \mid z = 0, m = 1) = E[\rho_{M0}(\mathbf{x}) \mid z = 0, m = 1]. \tag{14.8}$$

Since matching is on the basis of $\mathbf{x}$ alone, by (14.5) it follows that (14.8) is equivalent to

$$E(r_0 \mid z = 0, m = 1) = E[\rho_0(\mathbf{x}) \mid z = 0, m = 1]. \tag{14.9}$$

Since each sampled treated ($z = 1$) unit is exactly matched on $\mathbf{x}$ with a control ($z = 0$) unit, and since the treated units are a random sample from the population of treated units, it follows that the distribution of $\mathbf{x}$ among matched controls equals the distribution of $\mathbf{x}$ among treated units; i.e.,

$$\Pr(\mathbf{x} \mid z = 0, m = 1) = \Pr(\mathbf{x} \mid z = 1, m = 1) = \Pr(\mathbf{x} \mid z = 1).$$

Hence, from (14.9), we see that in the case of exact complete matching,

$$\Delta_M = E(r_1 \mid z = 1) - E[\rho_0(\mathbf{x}) \mid z = 1]. \tag{14.10}$$

Recalling that $\tau_1 = E(r_1 \mid z = 1) - E(r_0 \mid z = 1)$, we have from (14.10) that the bias $\Delta_M - \tau_1$ equals

$$E(r_0 \mid z = 1) - E[\rho_0(\mathbf{x}) \mid z = 1] = \Psi, \quad \text{say.} \tag{14.11}$$

If treatment assignment is strongly ignorable given $\mathbf{x}$, then $\rho_0(\mathbf{x}) = E(r_0 \mid \mathbf{x}) = E(r_0 \mid z = 1, \mathbf{x})$, and it follows that $\Psi = 0$. Hence, with strongly ignorable treatment assignment, exact complete matching implies $\Delta_M = \tau_1$; Ψ given by (14.11) is the bias due to departure from strongly ignorable treatment assignment.

2.3. Exact, Incomplete Matching on x

If some treated and control units are exactly matched for $\mathbf{x}$, but other treated units are discarded because exactly matching controls cannot be found, then by (14.5),

$$\Delta_M = E(r_1 \mid z = 1, m = 1) - E[\rho_0(\mathbf{x}) \mid z = 0, m = 1]. \tag{14.12}$$

But (14.12) equals

$$E(r_1 \mid z = 1, m = 1) - E[\rho_0(\mathbf{x}) \mid z = 1, m = 1] \qquad (14.13)$$

since the exact matching on $\mathbf{x}$ forces the distribution of $\mathbf{x}$ among matched control units to exactly equal the distribution of $\mathbf{x}$ among matched treated units; i.e.,

$$\Pr(\mathbf{x} \mid z = 0, m = 1) = \Pr(\mathbf{x} \mid z = 1, m = 1).$$

Using (14.13) and (14.11), we have for exact incomplete matching on $\mathbf{x}$,

$$\Delta_{\mathrm{M}} = \tau_1 + \Psi + \iota \qquad (14.14)$$

where

$$\iota = E[\rho_1(\mathbf{x}) - \rho_0(\mathbf{x}) \mid z = 1, m = 1] - E[\rho_1(\mathbf{x}) - \rho_0(\mathbf{x}) \mid z = 1] \qquad (14.15)$$

is the bias due to the incomplete matching of treated units. This component of bias, ι, can be 0 for several reasons. Clearly, $\iota = 0$ if all treated units are matched. More generally, ι will be 0 when $\mathbf{x}$ has the same distribution among matched and unmatched sampled treated units – that is, when $\Pr(\mathbf{x} \mid z = 1, m = 1) = \Pr(\mathbf{x} \mid z = 1, m = 0) = \Pr(\mathbf{x} \mid z = 1, s = 1) = \Pr(\mathbf{x} \mid z = 1)$ – because expectations given $(z = 1, m = 1)$ are then equivalent to expectations given $(z = 1)$. Another way ι can equal 0 is if $\rho_1(\mathbf{x}) - \rho_0(\mathbf{x})$ is constant for all $\mathbf{x}$, that is, if the response surfaces are parallel.

Consider, as an example, the case of linear response surfaces:

$$\rho_t(\mathbf{x}) = \alpha_t + \beta_t^{\mathrm{T}} \mathbf{x} \quad \text{for} \quad t = 0, 1. \qquad (14.16)$$

Then

$$\iota = (\beta_1 - \beta_0)^{\mathrm{T}} [E(\mathbf{x} \mid z = 1, m = 1) - E(\mathbf{x} \mid z = 1)]. \qquad (14.17)$$

Since, by random sampling,

$$
\begin{aligned}
E(\mathbf{x} \mid z = 1) &= E(\mathbf{x} \mid z = 1, s = 1) \\
&= [1 - \Pr(m = 0 \mid z = 1, s = 1)]E(\mathbf{x} \mid z = 1, m = 1) \\
&\quad + \Pr(m = 0 \mid z = 1, s = 1)E(\mathbf{x} \mid z = 1, m = 0),
\end{aligned}
$$

we may rewrite (14.17) as

$$
\begin{aligned}
\iota = (\beta_1 - \beta_0)^{\mathrm{T}} &[E(\mathbf{x} \mid z = 1, m = 1) - E(\mathbf{x} \mid z = 1, m = 0)] \\
&\times \Pr(m = 0 \mid z = 1, s = 1).
\end{aligned}
\qquad (14.18)
$$

Thus, with linear response surfaces, ι is a linear function of (i) the difference between the expectation of $\mathbf{x}$ in matched and unmatched treated samples, and (ii) the probability that a sampled treated unit will not be matched; of course, when $\beta_1 = \beta_0$, we have $\iota = 0$ from (14.18) or the fact that $\rho_1(\mathbf{x}) - \rho_0(\mathbf{x})$ is constant.

Exact, incomplete matching has been criticized by Billewicz (1965) and McKinlay (1975, 1977) for the loss of efficiency associated with discarding unmatched treated units. In Section 3, we show in a practical example that incomplete matching can introduce potentially serious bias as well.

2.4. Approximate, Incomplete Matching on x

Suppose that treated and control units are matched on $\mathbf{x}$ by a method that does not produce exact matches on $\mathbf{x}$. Two such methods are caliper matching (Althauser and Rubin, 1970; Cochran and Rubin, 1973; Rubin, 1973a; Raynor, 1983) and stratified matching – i.e., exact matching on a discretized or categorized summary of $\mathbf{x}$ (Cochran, 1968a; 1972, §6; Raynor and Kupper, 1981). Other such methods are described in Rubin (1973a, 1976b, 1980b). Then, from (14.6),

$$\Delta_M = \mathrm{E}[\rho_1(\mathbf{x}) \mid z = 1, m = 1] - \mathrm{E}[\rho_0(\mathbf{x}) \mid z = 0, m = 1]. \qquad (14.19)$$

If exact matches on $\mathbf{x}$ are not obtained, then the distribution of $\mathbf{x}$ among matched treated units does not generally equal the distribution among matched control units; that is, $\Pr(\mathbf{x} \mid z = 1, m = 1)$ does not generally equal $\Pr(\mathbf{x} \mid z = 0, m = 1)$. Therefore, in (14.19), the response surfaces are being averaged over different $\mathbf{x}$ distributions. Now (14.19) can be rewritten as

$$\Delta_M = \tau_1 + \Psi + \iota + \gamma \qquad (14.20)$$

where

$$\gamma = \mathrm{E}[\rho_0(\mathbf{x}) \mid z = 1, m = 1] - \mathrm{E}[\rho_0(\mathbf{x}) \mid z = 0, m = 1] \qquad (14.21)$$

is the bias due to coarse or inexact matching on $\mathbf{x}$. Clearly, $\gamma = 0$ when exact matches on $\mathbf{x}$ are obtained, since exact matching implies $\Pr(\mathbf{x} \mid z = 1, m = 1) = \Pr(\mathbf{x} \mid z = 0, m = 1)$. In the case of a linear response surface for r_0 given by (14.16), we have

$$\gamma = \beta_0^{\mathrm{T}}[\mathrm{E}(\mathbf{x} \mid z = 1, m = 1) - \mathrm{E}(\mathbf{x} \mid z = 0, m = 1)] , \qquad (14.22)$$

so that γ is a linear function of the difference between the expectation of $\mathbf{x}$ in matched treated and matched control samples.

In the special case of approximate but complete matching on $\mathbf{x}$, we have $\iota = 0$ in (14.20) and $\mathrm{E}[\rho_0(\mathbf{x}) \mid z = 1, m = 1] = \mathrm{E}[\rho_0(\mathbf{x}) \mid z = 1]$ in (14.21).

2.5. Addressing the Three Components of Bias

The bias Ψ due to departures from strongly ignorable treatment assignment cannot be directly estimated in a single observational study and, moreover, this component of bias is unaffected by supplementary regression adjustments of the kind described by Rubin (1973b, 1979b). These facts follow immediately from (14.11) since $\mathrm{E}(r_0 \mid z = 1)$ cannot be directly estimated. We can, however, examine the possible magnitude of Ψ under a range of assumptions about unobserved covariates; for example, see Rosenbaum and Rubin (1983b). Moreover, it is often possible to check whether the assumption of strongly ignorable treatment assignment is consistent with both the observed data and the causal mechanism through which the treatment is thought to produce its effects. The absence of such consistency would suggest Ψ cannot be assumed to be 0; see Rosenbaum (1984b) for discussion.

As is clear from (14.11), the component Ψ does not depend on the matching method; however, from (14.15) and (14.21), both ι and γ do depend on the matching method. A matching method that requires close matches on $\mathbf{x}$ may produce a small value of γ but a larger value of ι since close matches may be hard to find, resulting in a large number of unmatched treated units. Conversely, a matching method that requires only approximate matches on $\mathbf{x}$ may produce a small value of ι and a large value of γ.

One approach is to eliminate the ι component of bias entirely by matching *every* treated unit, using an algorithm such as nearest available matching (Cochran and Rubin, 1973; Rubin, 1973a), and then to reduce the size of the γ component of bias by supplementing the matched sampling with analytical adjustments for residual imbalances in $\mathbf{x}$; Rubin (1973b, 1979b) describes and evaluates a least squares regression procedure, and Rosenbaum (1984a) discusses a less model-dependent method for obtaining tests and confidence intervals. An alternative approach is to use an incomplete matching algorithm and then attempt to control both ι and γ components of bias through supplemental regression adjustments; however, this approach will often be unattractive, because the bias in $\mathbf{x}$ created by incomplete matching can exceed the initial bias in $\mathbf{x}$, as we illustrate in a practical example in Section 3.

2.6. An Artificial Example

To fix ideas, Figure 14.2 displays a simple artificial example in which x is a single binary covariate. We assume N_1 is large and ignore sampling fluctuations in the 2×2 tables in Fig. 14.2. Prior to matching, half of the treated sample has $x = 1$, whereas only 10% of the control reservoir has $x = 1$.

When the control reservoir is at least five times larger than the treated sample – i.e., when $R = N_0/N_1 \geqslant 5$ – exact matching is feasible. The resulting x distributions for the treated sample, the matched treated subsample, and the matched control subsample are identical; each contains $.5N_1$ units with $x = 1$ and $.5N_1$ units with $x = 0$.

When the control reservoir is between 1 and 5 times larger than the treated sample – i.e., when $1 < R < 5$ – exact matching is infeasible; there are too few control units with $x = 1$ – i.e., $.1RN_1 < .5N_1$ in Fig. 14.2. Incomplete matching distorts the x distribution in the matched treated subsample by discarding treated units, but it retains exact treatment-vs.-control balance on x. Inexact matching does not distort the x distribution in the matched treated subsample and does not discard any treated units, but it sacrifices exact treatment-vs.-control balance on x.

If adjustment by subclassification (Cochran, 1968a) is applied in the matched samples using population weights (i.e., .5 for $x = 1$ and .5 for $x = 0$), then the ι and γ components of bias will be eliminated. Inexact matching followed by adjustment by subclassification will be more efficient than incomplete matching followed by adjustment, since it is based on a larger sample.

3. AN EXAMPLE ILLUSTRATING THE INADEQUACIES OF INCOMPLETE CATEGORICAL MATCHING

3.1. Introduction: Purpose; Background

As noted in Section 2.3, the bias due to incomplete matching, ι, is due to the difference between the distributions of the covariates among matched and unmatched

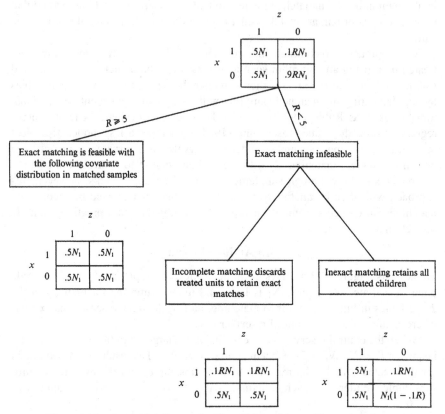

Figure 14.2. An artificial example of matching with one binary covariate x.

treated subsamples; that is, the difference between $\Pr(\mathbf{x} \mid z = 1, m = 1)$ and $\Pr(\mathbf{x} \mid z = 1, m = 0)$. As noted in Section 2.4, the bias γ due to coarse or approximate matching arises from the difference between the distribution of covariates among matched treated and matched control units, that is, the difference between $\Pr(\mathbf{x} \mid z = 1, m = 1)$ and $\Pr(\mathbf{x} \mid z = 0, m = 1)$. The example in this section illustrates the difference in these covariate distributions that can arise when incomplete categorical matching is used. The data analyses presented here are preliminary and exploratory, and none of the resultant matched samples being compared are the matched samples to be used for actual study.

The example is from a proposed study of the effects of prenatal exposure to barbiturates on subsequent psychological attributes. The children under study were born between 1959 and 1961, and have been the object of other studies (e.g., Mednick et al., 1971; Zachau-Christiansen and Ross, 1975). The data included 221 children with prenatal exposures to barbiturates and a reservoir of 7027 potential control children without such exposure. Twenty covariates were available for each child, including sex (male, female), twin (yes, no), mother unmarried (yes, no), an index of socioeconomic status, mother's age, mother's cigarette consumption during pregnancy, several variables describing other drugs taken, and various measures of the

Table 14.1. *Description of covariates and matching categories*

Name	Description of original covariate	As used for categorical matching	As used for complete propensity matching
SEX	Male/Female	0, 1	0, 1
TWIN	Single/Multiple birth	0, 1	0, 1
PRECLAM	Preeclampsia (yes, no)	0, 1	0, 1
HRMG1	Exposed to hormone type 1 (yes, no)	0, 1	0, 1
HRMG2	Exposed to hormone type 2 (yes, no)	0, 1	0, 1
HRMG3	Exposed to hormone type 3 (yes, no)	0, 1	0, 1
SINGLE	Mother unmarried (yes, no)	0, 1	0, 1
SIBPOS	Oldest child (yes, no)	0, 1	0, 1
RESPILL	Respiratory illness (yes, no)	0, 1	0, 1
HORMONE	Number of exposures to hormones (0–6)	0 vs. 1–6	Integers 0–6 and their squares
ANTIHIST	Number of exposures to antihistamines (0–6)	0 vs. 1–6	Integers 0–6 and their squares
SES	Socioeconomic status (9 ordered categories)	1–3, 4–6, 7–9	Integers 1 to 9
EDUC	Mother's education (4 ordered categories)	1–2, 3–4	Integers 1 to 4
PBC415	Pregnancy complications (an index)	0–3000, 3001–6000	Index value and its square
HEIGHT	Mother's height (5 ordered categories)	1–3, 4–5	Integers 1 to 5
M-AGE	Mother's age (years)	$\leqslant 22$, $\geqslant 23$	Years
WGTHGT3	(Weight gain)/height^3 (30 values based on category midpoints)	Low, High	30 values
CIGARET	Cigarette consumption, last trimester (0 = none, plus 4 ordered categories)	0, 1–4	Integers 0 to 4 and their squares
LENGEST	Length of gestation (10 ordered categories)	1–6, 7–8, 9–10	$(10 - i)^{1/2}$ and i for $i = 1, \ldots, 10$
C-AGE	Child's age at start of study	Divided at median	Months

health of the mother. The covariates are listed in Table 14.1. Outcome variables were to be obtained for all 221 treated children, but cost considerations required sampling of the 7027 potential controls.

We note that not all twenty matching variables are truly proper covariates, in the sense that some of the variables may possibly have been affected by exposure to barbiturate treatment. For example, perhaps exposure to barbiturates caused some mothers to smoke more or less during pregnancy. Matching or adjustment for such variables must be carefully considered. Cox (1958, §4.6) discusses the consequences in randomized experiments of adjustment for a posttreatment concomitant variable that may have been affected by the treatment; such adjustments can introduce a bias. In observational studies, adjustment for a posttreatment variable can increase bias if the variable is affected by the treatment, or it can decrease bias if the variable is, in a certain sense, a surrogate for relevant unobserved covariates, or it can do both to some extent (Rosenbaum, 1984c).

3.2. Methods: Incomplete Categorical Matching; The Standardized Difference in Covariate Means as a Measure of Covariate Imbalance

Each of the twenty covariates was coded into two or more categories, as described in Table 14.1. Each treated child was matched with one control child who was in the same category on all twenty covariates; if no such control child existed, the treated child was left unmatched.

To investigate the ι and γ components of bias, we will examine imbalances in the distribution of $\mathbf{x}$ before and after matching. The measures of imbalance we use are most appropriate when the response surfaces are linear in $\mathbf{x}$. In particular, we estimate $E(\mathbf{x} \mid z = 1, m = 1) - E(\mathbf{x} \mid z = 1, m = 0)$ in (14.18) by $\bar{\mathbf{x}}_{M1} - \bar{\mathbf{x}}_{U1}$ where $\bar{\mathbf{x}}_{M1}$ is the sample mean of $\mathbf{x}$ among matched treated children and $\bar{\mathbf{x}}_{U1}$ is the mean of $\mathbf{x}$ among unmatched treated children; $\Pr(m = 0 \mid z = 1)$ is estimated by the proportion of treated children that are unmatched. Moreover, we estimate $E(\mathbf{x} \mid z = 1, m = 1) - E(\mathbf{x} \mid z = 0, m = 1)$ in (14.22) by $\bar{\mathbf{x}}_{M1} - \bar{\mathbf{x}}_{M0}$ where $\bar{\mathbf{x}}_{M0}$ is the sample mean of $\mathbf{x}$ among matched control children. The impact of a specific difference in $\mathbf{x}$ means on the ι and γ components of bias depends, of course, on the values of β_0 and β_1 in (14.18) and (14.22), which cannot be estimated until response data are obtained.

Table 14.2 reports three measures of imbalance for each covariate $i = 1, \ldots, 20$: B_{0i}, a measure of covariate imbalance prior to matching; B_{Ci}, a measure of covariate imbalance due to coarse or inexact matching; and B_{Ii}, a measure of covariate imbalance due to incomplete matching. The measures are defined using the same denominator, namely

$$\bar{s}_i = \left(\frac{s_{1i}^2 + s_{0i}^2}{2} \right)^{1/2}$$

where s_{ti} is the standard deviation of covariate i in treatment group t, for $t = 0, 1$, prior to matching. For covariates $i = 1, \ldots, 20$.

$$B_{0i} = \frac{\bar{x}_{1i} - \bar{x}_{0i}}{\bar{s}_i}$$

where $\bar{x}_{ti}$ is the mean of covariate i in treatment group t prior to matching,

$$B_{Ci} = \frac{\bar{x}_{M1i} - \bar{x}_{M0i}}{\bar{s}_i},$$

and

$$B_{Ii} = \frac{\bar{x}_{M1i} - \bar{x}_{U1i}}{\bar{s}_i}.$$

Table 14.2. *Standardized differences in covariate means for twenty covariates before and after matching*

	All treated $(n = 221)$ vs. all control $(n = 7027)$ children (B_0)	Incomplete categorical matching		Complete propensity matching
		Matched treated $(n = 126)$ vs. matched control $(n = 126)$ children (B_C)	Matched treated $(n = 126)$ vs. unmatched treated $(n = 95)$ children (B_I)	Matched treated $(n = 221)$ vs. matched control $(n = 221)$ children
Variable				
SEX	−.07	.00	.01	.00
TWIN	−.10	.00	−1.09	−.03
PRECLAM	.09	.00	−.11	−.00
HRMG1	.15[a]	.00	−.49[b]	−.02
HRMG2	.19[b]	.00	−.45[b]	−.02
HRMG3	.18[b]	.00	−.49[b]	−.03
SINGLE	−.43[b]	.00	.30[a]	−.07
SIBPOS	−.16[b]	.00	−.03	−.05
RESPILL	.10	.00	−.36[a]	−.07
HORMONE	.28[b]	.00	−.56[b]	.08
ANTIHIST	.10	.08	−.36[a]	−.03
SES	.26[b]	.05	.10	−.10
EDUC	.15[a]	.00	.21	−.17
PBC415	.17[a]	.23	.14	−.14
HEIGHT	.18[a]	.14	−.04	−.08
M-AGE	.59[b]	.21	−.00	−.08
WGTHGT3	−.00	.26	.13	−.00
CIGARET	−.03	.06	.08	.00
LENGEST	.06	.05	.19	−.12
C-AGE	.03	.07	.03	.07

[a] Indicates $2 \leqslant$ | two-sample t statistic | < 3.
[b] Indicates $3 \leqslant$ | two-sample t statistic |.

The footnotes to Table 14.2 indicate the size of the usual two-sample t statistics. Unlike the standardized difference measures – B_0, B_I, B_C – the t statistics have denominators that are affected by the matching, which alters the sample sizes and generally the standard deviations in the control group.

3.3. Results: Substantial Differences Between Matched and Unmatched Treated Children

Matches were obtained for 126 of the 221 treated children. The measures of covariate imbalance are listed in Table 14.2. The absolute difference before matching, $|B_{0i}|$, is above .2 for four variables, and is .1 or above for fourteen variables. The comparison of the matched treated and matched control children yielded absolute differences due to coarse matching, $|B_{Ci}|$, that are above .2 for three covariates, and above .1 for four covariates; of course, $B_{Ci} = 0$ if variable i was originally dichotomous, since exact matches were obtained for such variables. Note that categorical matching substantially increased the difference in means for WGTHGT3 (i.e., mother's weight gain during pregnancy divided by the cube of her height).

The largest standardized differences, however, are most often found in the third column, which compares the 126 matched and the 95 unmatched treated children. For nine covariates, these absolute standardized differences due to incomplete matching, $|B_{Ii}|$, are above 0.2, and for fourteen covariates are 0.1 or above. In a sense, *the matched and unmatched treated children are less comparable than the treated and control children were prior to matching.* As noted in Section 2.4, differences between the covariate distributions among matched and unmatched treated units can cause the ι component of bias to be large. This potential source of bias can often be avoided by using a nearest available matching algorithm (Rubin, 1973a, 1973b, 1976b, 1979b, 1980b; Rosenbaum and Rubin, 1983a, §3.2). An appropriate procedure is illustrated in the next section.

3.4. Complete Matching Using the Propensity Score

The propensity score, $e(\mathbf{x})$, is in this context the conditional probability of exposure to barbiturates for a child with a specified covariate vector $\mathbf{x}$; that is, $e(\mathbf{x}) = \Pr(z = 1 \mid \mathbf{x})$. Rosenbaum and Rubin (1983a) have shown that matching on the propensity score will tend to balance the observed covariates $\mathbf{x}$. Here, we estimate the propensity score using a logit model (Cox, 1970), and then match every treated child to a control child using nearest available matching (Rubin, 1973a) on the estimated propensity score. Recall that with the categorical matching method, only 126 of the 221 treated children were matched. However, with the current procedure, all 221 treated children were matched, so the ι component of bias is 0. In fact, all but four of the treated children had matches within a quarter of a standard deviation of the log odds of the estimated $e(\mathbf{x})$'s. The values of B_{Ci} for nearest available propensity score matching are given in the last column in Table 14.2. Even though all 221 treated children have been matched, the γ component of bias has not been increased for most linear response surfaces. In fact, the mean of the absolute values of the standardized differences, $|B_{Ci}|$, is .058 for both columns 2 and 4, and the standard deviation of the absolute standardized differences is .085 for column 2 and .049 for column 4; propensity matching has produced absolute standardized differences due to coarse matching which are, on average, equal to the standardized differences of incomplete categorical matching, but which tend to be more uniform in size.

In short, in this instance it was possible to match all treated children, thereby eliminating the ι component of bias and permitting more complete use of available data, while not compromising the comparability of the final matched groups.

We are not necessarily advocating the particular nearest available matching method used here, but we are advocating the use of complete matching methods, like nearest available propensity matching, in examples such as this one, in order to avoid the potential bias due to incomplete matching. Current work suggests that more complex nearest available matching methods produce even better matched samples for these data. Some of this work is reported in Rosenbaum and Rubin (1985a) in which two more complex nearest available matching methods are compared with the propensity matching described here. The first other method is Mahalanobis metric matching separately by sex using the remaining nineteen variables (linear terms only) and the estimated propensity score to measure distance. The second other method uses the same Mahalanobis metric matching but within propensity score calipers, and it is generally superior to the other two methods. The methods actually used to match subjects are still more complex and reflect concerns with importance of variables and quality of back-up matches to be used in the event of nonresponse.

ACKNOWLEDGMENT

The authors acknowledge Robert Patrick for assistance in computing.

Before moving to Harvard University in the summer of 1983, I spent two wonderful years as a Professor at the University of Chicago, 75% in the Department of Statistics and 25% in the Department of Education. During that brief period of time, I supervised several great PhD students, one of whom was Neal Thomas, who, despite a strong mathematics background, had a deep appreciation for applied statistics. The issue of the subasymptotic properties of propensity score matching was of real interest to me at the time, and remains an important one today. His PhD thesis addressed this topic, but he and I made more progress after he was out and working at RAND, and then ETS. The result was a sequence of four articles, which started with quite general theoretical results and then moved to apply that theory.

The first of these, Chapter 15, Rubin and Thomas (1992a), was published in the *Annals of Statistics* and was built on a substantial extension of EPBR matching methods studied in Chapter 6, Rubin (1976b). Affinely invariant matching methods are ones that are closed under any wide-sense linear transformation of the data, such as discriminant matching (using either the estimated or the population discriminant), propensity score matching (again, using either the population or estimated, e.g., by logistic regression), as well as Mahalanobis-metric matching, and many other methods. An ellipsoidally symmetric distribution is defined by a center, an inner product (which together define the affine transformation to sphericity), and the distribution on the radius of concentric hyperspheres on each of whose surfaces there is equal probability density. Rubin and Thomas (1992a) considered what happens when any affinely invariant matching method is applied to distributions that are ellipsoidally symmetric in the treatment and control groups (with proportional inner products). It showed that the result can be decomposed into what happens in one direction, along the best linear discriminant, and what happens in any direction orthogonal to the discriminant. Not only does this decomposition hold for results concerning the expectation of the centers (e.g., means) of the distributions in the matched samples (whence EPBR), but the analogous decomposition also holds for the inner products (e.g., variance–covariance matrices) and for the sampling distribution of the centers in the matched samples. Actually, the results proved were somewhat more general in that they were extended to methods that were conditionally affinely invariant and distributions that were conditionally ellipsoidally symmetric. The results, presented in a series of theorems, corollaries, and a lemma, are the key to understanding what happens subasymptotically with estimated propensity scores rather than population propensity scores. This connection was made in subsequent articles.

The next most theoretical article in Part V, Chapter 16, Rubin and Thomas (1992b), appeared in *Biometrika* and applied the abstract results of the previous chapter to the theoretical situation of normally distributed matching variables. The normal distribution is the special case of ellipsoidal distributions where orthogonality implies independence. Exact results were stated in a sequence of theorems and corollaries, and derivations were also provided for a variety of useful approximations, some of which could be viewed as extensions of the approximations for the maximum percentage bias reduction given in Chapter 7 (Rubin, 1976c). In particular, these results showed that in many cases where matching can remove most of the initial bias (i.e., the mean difference between the treatment and control groups), matching on the estimated propensity score will be superior to matching on the population propensity score because the former will have better variance reduction properties. Neither of these first articles had any data, either real or simulated.

The third in this sequence, Chapter 17, Rubin and Thomas (1996), appeared in *Biometrics* and had both simulated and real data. This article finally bridged the earlier results to simulated ellipsoidally symmetric, but nonnormal, data and to dramatically nonnormal real data, in particular, the Kinsey data set referenced and used in Part IV. Specifically, the approximations based on normality, derived in Chapter 16 (Rubin and Thomas, 1992b), were applied even though they were not really appropriate. Remarkably, these approximations appeared to work quite well, despite the violations of normality. One reason for this is that some of the results hold under more general conditions than stated in Chapter 15 (Rubin and Thomas, 1992a), as shown in the very recent Rubin and Stuart (2006). Another reason may have to do with the approximate truth of the results for a much larger class of distributions, which could possibly be expressed as an expansion about the special case with ellipsoidal symmetry – a topic for future work.

The fourth and final article in this sequence, Chapter 18, Rubin and Thomas (2000), appeared in the *Journal of the American Statistical Association*. This article extended and completed the bridge to practice by conducting extensive simulations with real and artificial data, and presented evaluations of different kinds of matching methods, including in some situations with more than one control matched to each treated, and different forms of regression adjustment. The article in some ways followed the general simulation design of Chapter 8 (Rubin, 1979b) but in higher dimensions and with more flexible functional forms. It also included some theoretical results, proved in appendices, but the focus was on providing practical advice.

The concluding advice is relatively straightforward: use propensity score matching on all covariates, supplemented with Mahalanobis-metric matching on prognostically important covariates within propensity calipers, rather than using only the prognostically important covariates; this matching is important even when regression adjustment is to be used later; the combination of matching with regression adjustment is generally better than either alone. There were some benefits to many–one matching in some cases, but those were relatively limited.

Subsequent work with Neal on matching in the context of a particular application appears as Chapter 20 in Part VI.

15. Affinely Invariant Matching Methods with Ellipsoidal Distributions

Donald B. Rubin and Neal Thomas

Abstract: Matched sampling is a common technique used for controlling bias in observational studies. We present a general theoretical framework for studying the performance of such matching methods. Specifically, results are obtained concerning the performance of affinely invariant matching methods with ellipsoidal distributions, which extend previous results on equal percent bias reducing methods. Additional extensions cover conditionally affinely invariant matching methods for covariates with conditionally ellipsoidal distributions. These results decompose the effects of matching into one subspace containing the best linear discriminant, and the subspace of variables uncorrelated with the discriminant. This characterization of the effects of matching provides a theoretical foundation for understanding the performance of specific methods such as matched sampling using estimated propensity scores. Calculations for such methods are given in subsequent articles.

1. BACKGROUND

Matched sampling is a popular and important technique for controlling bias in observational studies. It has received increasing attention in the statistical literature in recent years [Cochran (1968a); Cochran and Rubin (1973); Rubin (1973a,b), (1976b,c), (1979b); Carpenter (1977); and Rosenbaum and Rubin (1983a, 1985a)]. The basic situation has two populations of units, treated (e.g., smokers) and control (e.g., nonsmokers), and a set of observed matching variables $\mathbf{X} = (X_1, \ldots, X_p)$ (e.g., age, gender, weight). The objective is to compare the distributions of the outcome variables having adjusted for differences in the distributions of $\mathbf{X}$ in the two populations. Matched sampling is a way of adjusting for $\mathbf{X}$ through data collection.

Suppose there exist random samples from the treated and control populations of sizes N_t and N_c, respectively, where $\mathbf{X}$ is recorded on all $N_t + N_c$ units; typically, due to cost considerations, outcomes and additional covariates can only be recorded on subsamples of the initial samples. Instead of randomly choosing subsamples, often matched subsamples of sizes $n_t \leq N_t$ and $n_c \leq N_c$ are chosen in such a way that the distributions of $\mathbf{X}$ among the n_t and n_c matched units are more similar than they would be in random subsamples. Commonly, $N_c \gg N_t$ and $n_c = n_t = N_t$, so that only the controls are subsampled, as when the treated group has been exposed to an unusual occurrence (e.g., radiation) and the controls are all those who were not exposed.

Reprinted from *The Annals of Statistics*, 1992, **20**, 2, 1079–1093.

The standard matched-sample estimator of the treatment's effect on an outcome Y is the difference in the means of Y between the n_t and n_c matched treated and control units, $\overline{Y}_{mt} - \overline{Y}_{mc}$. If the distributions of Y differ in the treated and control populations primarily because of the treatment effect and differences in $\mathbf{X}$, then $\overline{Y}_{mt} - \overline{Y}_{mc}$ should be closer to the treatment effect than the difference of Y means in random subsamples of size n_t and n_c, $\overline{Y}_{rt} - \overline{Y}_{rc}$, thereby reflecting bias reduction arising from the matched sampling.

Here we present theoretical results on the performance of affinely invariant matching methods, defined in Section 2, with ellipsoidal distributions for $\mathbf{X}$, also defined in Section 2. The multivariate normal and the multivariate t are special cases of commonly referenced ellipsoidal distributions, and many practical matching methods are affinely invariant, such as discriminant matching, Mahalanobis metric matching, matching based on propensity scores estimated by logistic regression, and combinations of these considered in the statistics literature.

Results in Section 3 extend and generalize those in Rubin (1976b) on bias reduction in theoretically and practically important ways. In particular, we provide general theoretical results not only on bias reduction for any linear combination of $\mathbf{X}$, $Y = \alpha'\mathbf{X}$, as in Rubin (1976b), but also on the variance of $\overline{Y}_{mt} - \overline{Y}_{mc}$ relative to that of $\overline{Y}_{rt} - \overline{Y}_{rc}$, and on the expectation of the second moments of Y in the matched samples. These results are of importance to practice because they include cases in which the matching methods are based on estimated discriminant or estimated propensity scores, as well as on estimated metrics such as the Mahalanobis metric, and therefore lay the foundation for obtaining valid standard errors in samples matched using estimated propensity scores.

In Section 4, extensions are presented for matching methods that are only conditionally affinely invariant because they use a subset of the matching variables in a special way. These extensions only require conditionally ellipsoidal distributions, a generalization of the normal general location model [e.g., Olkin and Tate (1961)], and are relevant to the important practice of forcing better matching with respect to key covariates [e.g., Mahalanobis metric matching on a subset of variables within calipers of the estimated propensity scores, Rosenbaum and Rubin (1985a)].

Explicit analytic expressions based on the results of Sections 3 and 4 under multivariate normality using discriminant matching are given in Rubin and Thomas (1992b). Qualitative descriptions are also given indicating how these results for matching with estimated scores change for different types of nonnormal distributions; the implications of these results for practice are studied using simulation techniques and real data in Rubin and Thomas (1996, 2000).

2. DEFINITIONS AND NOTATION

2.1. Ellipsoidal Distributions

Ellipsoidal distributions are characterized by the fact that there exists a linear transformation of the variables that results in a spherically symmetric distribution for the transformed variables. An ellipsoidal distribution is fully specified by (i) its center of symmetry, (ii) its inner product defined by the linear transformation to sphericity and (iii) the distribution on the radii of concentric hyperspheres on which there is uniform probability density [Dempster (1969)]. Such multivariate distributions play an important role in the theory of matching methods because the symmetry allows general results to be obtained.

In the general matching situation with ellipsoidal distributions, let μ_t and μ_c be the centers of $\mathbf{X}$, and Σ_t and Σ_c be the inner products of $\mathbf{X}$ in the treated and control populations, respectively. Although some of our results hold with weaker restrictions, we assume that $\Sigma_t \propto \Sigma_c$, so that one common linear transformation can reduce $\mathbf{X}$ to sphericity in both treated and control populations; however, the two ellipsoidal distributions may differ (e.g., normal in the control group and t_3 in the treated group).

When $\Sigma_t \propto \Sigma_c$, the ellipsoidal distributions will be called proportional. For simplicity of description, we assume finite first and second moments of $\mathbf{X}$ so that μ_t and μ_c are the expectations (means) of $\mathbf{X}$ and Σ_t and Σ_c are its covariances. Little generality of practical relevance is lost by making these latter restrictions.

2.2. Canonical Form for Proportional Ellipsoidal Distributions

As noted by Cochran and Rubin (1973), Rubin (1976b), Efron (1975) and other authors, with proportional ellipsoidal distributions, there exists an affine transformation of $\mathbf{X}$ to the following canonical form,

$$\mu_t = \delta \mathbf{1}, \tag{15.1}$$

where δ is a positive scalar constant and $\mathbf{1}$ is the unit vector,

$$\mu_c = \mathbf{0}, \tag{15.2}$$

where $\mathbf{0}$ is the zero vector,

$$\Sigma_t = \sigma^2 \mathbf{I}, \tag{15.3}$$

where σ^2 is a positive scalar and $\mathbf{I}$ is the identity matrix, and

$$\Sigma_c = \mathbf{I}. \tag{15.4}$$

This canonical form for the distributions of $\mathbf{X}$ is very useful because the resulting distributions are fully exchangeable (symmetric) in the coordinates of $\mathbf{X}$, so that many results can be obtained by simple symmetry arguments. For these symmetry arguments to apply with matched samples, the matching method must also possess the corresponding symmetry, and so we define affinely invariant matching methods to have this property.

2.3. Affinely Invariant Matching Methods

Let $(\mathscr{X}_t, \mathscr{X}_c)$ be a pair of (units by variables) data matrices, where $\mathscr{X}_t$ is the N_t by p matrix with elements (X_{tij}) and $\mathscr{X}_c$ is the N_c by p matrix with elements (X_{cij}). A general matching method is a mapping from $(\mathscr{X}_t, \mathscr{X}_c)$ to a pair of sets of indices, (T, C), of those units chosen in the matched samples, where T has n_t elements from $(1, \ldots, N_t)$ and C has n_c elements from $(1, \ldots, N_c)$. (A more restrictive definition of general matching would require $n_t = n_c$, with a $1:1$ correspondence between the elements of T and C). An affinely invariant matching method is one such that the matching output is the same following any affine transformation, A, of $\mathbf{X}$:

$$(\mathscr{X}_t, \mathscr{X}_c) \mapsto (T, C),$$

implies

$$(A(\mathscr{X}_t), A(\mathscr{X}_c)) \mapsto (T, C).$$

Matching methods based on population or sample inner products, such as discriminant matching or Mahalanobis metric matching, are affinely invariant, as are common methods using propensity scores based on linear logistic regression estimators, which are affinely invariant as noted in Efron (1975). Methods that are not affinely invariant include those where one coordinate of $\mathbf{X}$ is treated differently from the other components (e.g., weighted in the Mahalanobis metric to reflect greater importance) or where nonlinear estimators (e.g., of the discriminant) are used, such as trimmed moment estimators.

When an affinely invariant matching method is used with proportional ellipsoidal distributions, the canonical form for the distributions of $\mathbf{X}$ given in (15.1)–(15.4) can be assumed. This canonical form will be used throughout the remainder of Section 2 and in Section 3.

2.4. The Best Linear Discriminant

A particularly important linear combination of $\mathbf{X}$ is the best linear discriminant, $(\boldsymbol{\mu}_t, -\boldsymbol{\mu}_c)'\Sigma_c^{-1}\mathbf{X}$, which is proportional to $\mathbf{1}'\mathbf{X}$ in the canonical form of the distributions defined by (15.1)–(15.4). The standardized discriminant, Z, is defined as the discriminant with unit variance in the control population,

$$Z = \mathbf{1}'\mathbf{X}/\sqrt{p}. \tag{15.5}$$

The standardized discriminant is a log-likelihood ratio statistic when $\mathbf{X}_t$ and $\mathbf{X}_c$ are multivariate normal with proportional covariance matrices, and it has the maximum difference in population means among all standardized linear combinations in the more general setting.

An arbitrary linear combination $\mathbf{Y} = \alpha'\mathbf{X}$ can be expressed as a sum of components along and orthogonal to the standardized discriminant; for notational simplicity, let Y be standardized, $\alpha'\alpha = 1$. Let W be the standardized linear combination of $\mathbf{X}$ orthogonal to Z,

$$W = \gamma'\mathbf{X}, \quad \gamma'\mathbf{1} = 0, \quad \gamma'\gamma = 1, \tag{15.6}$$

with γ chosen so that

$$Y = \rho Z + \sqrt{1 - \rho^2}\, W, \tag{15.7}$$

where ρ is the correlation of Y with the standardized discriminant Z. By their construction, Z and W have several properties useful in the derivations. In the control population,

$$E(\overline{Z}_{rc}) = E(\overline{W}_{rc}) = E(\overline{Y}_{rc}) = 0,$$
$$\text{var}(\overline{Z}_{rc}) = \text{var}(\overline{W}_{rc}) = \text{var}(\overline{Y}_{rc}) = \frac{1}{n_c}, \tag{15.8}$$

where the subscript rc refers to a randomly chosen sample of n_c control units. Likewise, in the treated population,

$$E(\overline{Z}_{rt}) = \delta\sqrt{p}, \quad E(\overline{W}_{rt}) = 0, \quad E(\overline{Y}_{rt}) = \rho\delta\sqrt{p},$$
$$\text{var}(\overline{Z}_{rt}) = \text{var}(\overline{W}_{rt}) = \text{var}(\overline{Y}_{rt}) = \frac{\sigma^2}{n_t}, \tag{15.9}$$

where the subscript rt refers to a randomly chosen sample of n_t treated units.

Corollaries in Section 3 decompose the effects of matching into two pieces: the effect on the best linear discriminant Z and the effect on any covariate W orthogonal to Z.

3. DECOMPOSING THE MATCHING EFFECTS

3.1. Results Based on Symmetry

Theorem 3.1. *Suppose an affinely invariant matching method is applied with fixed sample sizes (N_t, N_c, n_t, n_c) and proportional ellipsoidal distributions, which are represented in canonical form. Then,*

$$E(\overline{\mathbf{X}}_{mt}) \propto \mathbf{1}, \tag{15.10}$$
$$E(\overline{\mathbf{X}}_{mc}) \propto \mathbf{1} \tag{15.11}$$

and

$$\text{var}(\overline{\mathbf{X}}_{mt} - \overline{\mathbf{X}}_{mc}) \propto \mathbf{I} + c\mathbf{1}\mathbf{1}', \quad c > -1/p, \tag{15.12}$$

where $\overline{\mathbf{X}}_{mt}$ and $\overline{\mathbf{X}}_{mc}$ are the mean vectors in the matched treated and control samples, and $E(\cdot)$ and $\text{var}(\cdot)$ are the expectation and variance over repeated random draws from the initial populations of treated and control units. Furthermore,

$$E(v_{mt}(\mathbf{X})) \propto \mathbf{I} + c_t\mathbf{1}\mathbf{1}', \quad c_t \geq -1/p, \tag{15.13}$$
$$E(v_{mc}(\mathbf{X})) \propto \mathbf{I} + c_c\mathbf{1}\mathbf{1}', \quad c_c \geq -1/p, \tag{15.14}$$

where $v_{mt}(\mathbf{X})$ and $v_{mc}(\mathbf{X})$ are the sample covariance matrices of $\mathbf{X}$ in the matched treated and control samples with divisors $(n_t - 1)$ and $(n_c - 1)$. If $\delta = 0$, then

$$E(\overline{\mathbf{X}}_{mt}) = E(\overline{\mathbf{X}}_{mc}) = 0,$$
$$\text{var}(\overline{\mathbf{X}}_{mt} - \overline{\mathbf{X}}_{mc}) \propto \mathbf{I},$$
$$E(v_{mt}(\mathbf{X})) \propto \mathbf{I},$$
$$E(v_{mc}(\mathbf{X})) \propto \mathbf{I}.$$

Proof of Theorem 3.1. The proof of Theorem 3.1 follows from symmetry with little derivation. The expectations of the sample means of each coordinate of $\mathbf{X}$ must be the same and thus proportional to $\mathbf{1}$ in both treated and matched control samples. The covariance matrices must be exchangeable in the treated and matched control samples. The general form for the covariance matrix of exchangeable variables is proportional to

$$\mathbf{I} + c\mathbf{1}\mathbf{1}', \quad c \geq -1/p.$$

When $\delta = 0$, the complete rotational symmetry implies the final set of claims.   □

The symmetry results of Theorem 3.1 imply that any W orthogonal to the discriminant Z has the same distribution, as summarized in the following corollary.

Corollary 3.1. *The quantities* $\text{var}(\overline{W}_{mt} - \overline{W}_{mc})$, $E(v_{mt}(W))$ *and* $E(v_{mc}(W))$ *take the same three values for all standardized W orthogonal to Z. The analogous three results apply for statistics in random subsamples indexed by rt and rc. Since Z is the discriminant, defined without regard to the choice of Y, the analogous quantities for Z are also the same for all Y.*

3.2. Corollaries That Decompose the Effect of Matching

Although Theorem 3.1 and Corollary 3.1 follow almost immediately from the symmetry of ellipsoidal distributions and restrictions placed on the matching algorithms, their consequences are not as apparent. The corollaries that follow from it, stated here and proved in Section 3.3, show that under the conditions of Theorem 3.1, the moments of any Y in the matched samples are determined by the moments in the matched samples of the discriminant and any single covariate uncorrelated with the discriminant, and furthermore, that those moments of Y involve Y only through ρ.

Suppose proportional ellipsoidal distributions with $\mu_t \neq \mu_c$, and affinely invariant matching methods, and let Y have correlation ρ with the best linear discriminant Z and correlation $\sqrt{1 - \rho^2}$ with W uncorrelated with Z. Then, the following four corollaries hold.

Corollary 3.2. *Matching is equal percent bias reducing, EPBR (Rubin, 1976a),*

$$\frac{E(\overline{Y}_{mt} - \overline{Y}_{mc})}{E(\overline{Y}_{rt} - \overline{Y}_{rc})} = \frac{E(\overline{Z}_{mt} - \overline{Z}_{mc})}{E(\overline{Z}_{rt} - \overline{Z}_{rc})}, \qquad (15.15)$$

where the subscript rt refers to a randomly chosen sample of n_t treated units, and the subscript rc refers to a randomly chosen sample of n_c control units. Equation (15.15) implies that the percent reduction in bias is the same for any linear combination of **X**, *because* $E(\overline{Z}_{mt} - \overline{Z}_{mc})/E(\overline{Z}_{rt} - \overline{Z}_{rc})$ *takes the same value for all Y.*

Corollary 3.3. *The matching is ρ^2-proportionate modifying of the variance of the difference in matched sample means,*

$$\frac{\text{var}(\overline{Y}_{mt} - \overline{Y}_{mc})}{\text{var}(\overline{Y}_{rt} - \overline{Y}_{rc})} = \rho^2 \frac{\text{var}(\overline{Z}_{mt} - \overline{Z}_{mc})}{\text{var}(\overline{Z}_{rt} - \overline{Z}_{rc})} + (1 - \rho^2)\frac{\text{var}(\overline{W}_{mt} - \overline{W}_{mc})}{\text{var}(\overline{W}_{rt} - \overline{W}_{rc})}, \qquad (15.16)$$

where the ratios

$$\frac{\text{var}(\overline{Z}_{mt} - \overline{Z}_{mc})}{\text{var}(\overline{Z}_{rt} - \overline{Z}_{rc})}, \qquad \frac{\text{var}(\overline{W}_{mt} - \overline{W}_{me})}{\text{var}(\overline{W}_{rt} - \overline{W}_{rc})},$$

take the same two values for all Y.

Corollary 3.4. The matching is ρ^2-proportionate modifying of the expectations of the sample variances

$$\frac{E(v_{mt}(Y))}{E(v_{rt}(Y))} = \rho^2 \frac{E(v_{mt}(Z))}{E(v_{rt}(Z))} + (1 - \rho^2)\frac{E(v_{mt}(W))}{E(v_{rt}(W))} \tag{15.17}$$

and

$$\frac{E(v_{mc}(Y))}{E(v_{rc}(Y))} = \rho^2 \frac{E(v_{mc}(Z))}{E(v_{rc}(Z))} + (1 - \rho^2)\frac{E(v_{mc}(W))}{E(v_{rc}(W))} , \tag{15.18}$$

where $v_{rt}(\cdot)$ is the sample variance of n_t randomly chosen treated units (computed using $n_t - 1$ in the denominator) and likewise for $v_{rc}(\cdot)$ and n_c randomly chosen control units (using $n_c - 1$). The ratios

$$\frac{E(v_{mt}(Z))}{E(v_{rt}(Z))}, \quad \frac{E(v_{mt}(W))}{E(v_{rt}(W))}, \quad \frac{E(v_{mc}(Z))}{E(v_{rc}(Z))}, \quad \frac{E(v_{mc}(W))}{E(v_{rc}(W))},$$

take the same four values for all Y.

Corollary 3.5. When $\mu_t = \mu_c$,

$$E(\overline{Y}_{mt} - \overline{Y}_{mc}) = E(\overline{Y}_{rt} - \overline{Y}_{rc}) = 0,$$

and the ratios

$$\frac{\mathrm{var}(\overline{Y}_{mt} - \overline{Y}_{mc})}{\mathrm{var}(\overline{V}_{rt} - \overline{Y}_{rc})}, \quad \frac{E(v_{mt}(Y))}{E(v_{rt}(Y))}, \quad \frac{E(v_{mc}(Y))}{E(v_{rc}(Y))},$$

take the same three values for all Y.

3.3. Proofs of the Corollaries

Proof of Corollary 3.2. After matching, from (15.7),

$$E(\overline{Y}_{mt} - \overline{Y}_{mc}) = \rho E(\overline{Z}_{mt} - \overline{Z}_{mc}) + (\sqrt{1 - \rho^2})E(\overline{W}_{mt} - \overline{W}_{mc}),$$

where by (15.6),

$$E(\overline{W}_{mt} - \overline{W}_{mc}) = \gamma' E(\overline{\mathbf{X}}_{mt} - \overline{\mathbf{X}}_{mc}) .$$

But by Theorem 3.1, $E(\overline{\mathbf{X}}_{mt} - \overline{\mathbf{X}}_{mc}) \propto \mathbf{1}$, and from (15.6), $\gamma'\mathbf{1} = 0$. Hence,

$$E(\overline{Y}_{mt} - \overline{Y}_{mc}) = \rho E(\overline{Z}_{mt} - \overline{Z}_{mc}), \tag{15.19}$$

and Corollary 3.2 follows by noting from (15.7) that $E(\overline{Y}_{rt} - \overline{Y}_{rt}) = \rho E(\overline{Z}_{rt} - \overline{Z}_{rc})$. $\qquad \square$

Proof of Corollary 3.3. After matching, from (15.7),

$$\mathrm{var}(\overline{Y}_{mt} - \overline{Y}_{mc}) = \rho^2 \mathrm{var}(\overline{Z}_{mt} - \overline{Z}_{mc}) + (1 - \rho^2)\mathrm{var}(\overline{W}_{mt} - \overline{W}_{mc}),$$

because from (15.5) and (15.6),

$$\mathrm{cov}(\overline{Z}_{mt} - \overline{Z}_{mc}, \overline{W}_{mt} - \overline{W}_{mc}) = \frac{1}{\sqrt{p}}\mathbf{1}'\mathrm{var}(\overline{\mathbf{X}}_{mt} - \overline{\mathbf{X}}_{mc})\gamma,$$

which from (15.12), is proportional to

$$1'(I + c11')\gamma = 1'\gamma + cp1'\gamma = 0.$$

Equation (15.16) follows because in random subsamples, the treated and control subsamples are independent with

$$\text{var}(\overline{Y}_{rt} - \overline{Y}_{rc}) = \text{var}(\overline{Z}_{rt} - \overline{Z}_{rc}) = \text{var}(\overline{W}_{rt} - \overline{W}_{rc}) = \frac{\sigma^2}{n_t} + \frac{1}{n_c}.$$

The final claim of Corollary 3.3 follows from Corollary 3.1.   □

Proof of Corollary 3.4. After matching, in the treated sample,

$$E(v_{mt}(Y)) = \rho^2 E(v_{mt}(Z)) + (1 - \rho^2)E(v_{mt}(W))$$

because the expected matched treated sample covariance of Z and W is

$$\frac{1}{\sqrt{p}} E(1' v_{mt}(X)\gamma) \propto 1'(1 + c_t p 11')\gamma = 0.$$

Noting that $E(v_{rt}(Y)) = E(v_{rt}(Z)) = E(v_{rt}(W)) = 1$, implies equation (15.17), and an analogous derivation for the control sample establishes (15.18). Corollary 3.1 completes the proof of Corollary 3.4.   □

Corollary 3.5 follows immediately from symmetry considerations.

4. EXTENSIONS INVOLVING A SPECIAL SET OF COVARIATES

4.1. Conditionally Affinely Invariant Matching Methods

An important class of extended results covers methods that treat a subset of the covariates, denoted by $\mathbf{X}^{(s)} = (X_1, \ldots, X_s)$, differently from the remaining covariates, $\mathbf{X}^{(r)} = (X_{s+1}, \ldots, X_p)$. For example, in a study of the effects of smoking on health in human populations, a match for each smoker might be selected according to the criteria (a) the closest nonsmoker with respect to the discriminant (computed using numerous personal characteristics) who exactly matches the smoker's gender and is within ± 5 years of the smoker's age or (b) the closest nonsmoker with respect to the Mahalanobis metric on age and an index measuring environmental exposure to carcinogens who is also within $\pm 1/4$ standard deviations of the smoker on the discriminant, as in Rosenbaum and Rubin (1985a). In these examples, the special matching variables, $\mathbf{X}^{(s)}$ denote (a) gender and age and (b) age and environmental exposure.

Using a construction similar to the definition of affinely invariant matching in Section 2.3, these matching methods are called *conditionally affinely invariant* because they satisfy the condition that their matching output is the same following an affine transformation A of $\mathbf{X}^{(r)}$:

$$\left(\left(\mathcal{X}_t^{(s)}, \mathcal{X}_t^{(r)} \right), \left(\mathcal{X}_c^{(s)}, \mathcal{X}_c^{(r)} \right) \right) \mapsto (T, C)$$

implies

$$\left(\left(\mathcal{X}_t^{(s)}, A\mathcal{X}_t^{(r)} \right), \left(\mathcal{X}_c^{(s)}, A\mathcal{X}_c^{(r)} \right) \right) \mapsto (T, C).$$

Conditionally affinely invariant matching methods include affinely invariant matching methods as a subclass. When using conditionally affinely invariant matching methods, similar, but weaker results are obtained under correspondingly weaker distributional assumptions that remove the requirement of rotational invariance for the special variables $\mathbf{X}^{(s)}$.

4.2. Conditionally Ellipsoidal Distributions

Distributions satisfying this weaker set of restrictions are called *conditionally ellipsoidal distributions:* The conditional distribution of $\mathbf{X}^{(r)}$ given $\mathbf{X}^{(s)}$ is ellipsoidal with the conditional mean a linear function of $\mathbf{X}^{(s)}$ and constant conditional covariance matrix. A special case of conditionally ellipsoidal distributions is the normal general location model widely discussed in the statistics literature, starting with Olkin and Tate (1961).

Denote the covariance matrices and means for the two subsets of covariates by $\Sigma_t^{(s)}, \Sigma_t^{(r)}, \mu_t^{(s)}$ and $\mu_t^{(r)}$ in the treated population, and $\Sigma_c^{(s)}, \Sigma_c^{(r)}, \mu_c^{(s)}$ and $\mu_c^{(r)}$ in the control population, and the conditional covariance matrices and means of $\mathbf{X}^{(r)}$ given $\mathbf{X}^{(s)}$ by $\Sigma_t^{(r|s)}, \Sigma_c^{(r|s)}\mu_t^{(r|s)}$ and $\mu_c^{(r|s)}$. The distributions are called proportional conditionally ellipsoidal distributions if the conditional distributions $\mathbf{X}^{(r)}|\mathbf{X}^{(s)}$ are ellipsoidal in both the treated and control populations with proportional conditional covariance matrices, $\Sigma_t^{(r|s)} \propto \Sigma_c^{(r|s)}$ and common linear regressions of the $(p - s)$ covariates in $\mathbf{X}^{(r)}$ on the s variables in $\mathbf{X}^{(s)}$, $B = (\beta_{s+1}, \dots, \beta_p)$. The elements in the kth column of the s by $(p - s)$ matrix, $\mathbf{B}$, are the multiple regression coefficients of X_k on $(X_1, \dots, X_s)$ for $k = (s + 1), \dots, p$.

The special case of proportional conditionally ellipsoidal distributions with $\mathbf{X}^{(s)}$ binomial or multinomial and $\mathbf{X}^{(r)}$ multivariate normal has been studied extensively in the discrimination literature [e.g., Krzanowski (1975), (1980); Daudin (1986)]. This model gives rise to a linear logistic regression model for predicting population membership based on the covariates, $\mathbf{X}$, and thus is relevant to the applied practice that estimates linear propensity scores using logistic regression [Rosenbaum and Rubin (1985a)].

4.3. Canonical Form for Conditionally Ellipsoidal Distributions

Using conditionally affinely invariant matching methods with proportional conditionally ellipsoidal distributions, a canonical form for the distributions can be assumed corresponding to the canonical form used in Section 2.2,

$$\mu_t^{(r)} \propto \mathbf{1}, \tag{15.20}$$

$$\mu_c^{(r)} = \mathbf{0}, \tag{15.21}$$

$$\Sigma_t = \begin{bmatrix} \Sigma_t^{(s)} & \mathbf{0} \\ \mathbf{0} & \sigma^2\mathbf{I} \end{bmatrix} \tag{15.22}$$

and

$$\Sigma_c = \begin{bmatrix} \Sigma_c^{(s)} & \mathbf{0} \\ \mathbf{0} & \mathbf{I} \end{bmatrix}. \tag{15.23}$$

This form is obtained by leaving the $\mathbf{X}^{(s)}$ unchanged and letting the canonical $\mathbf{X}^{(r)}$ be defined as the components of $\mathbf{X}^{(r)}$ uncorrelated with $\mathbf{X}^{(s)} : \mathbf{X}^{(r)} - \mathbf{B}\mathbf{X}^{(s)}$. When the covariates are transformed to this canonical form, the conditional distributions of $\mathbf{X}^{(r)}|\mathbf{X}^{(s)}$ satisfy

$$\mu_t^{(r|s)} = \mu_t^{(r)} \propto 1, \tag{15.24}$$

$$\mu_c^{(r|s)} = \mu_c^{(r)} = 0 \tag{15.25}$$

and

$$\Sigma_t^{(r|s)} = \Sigma_t^{(r)} \propto \mathbf{I}, \tag{15.26}$$

$$\Sigma_c^{(r|s)} = \Sigma_c^{(r)} \propto \mathbf{I}, \tag{15.27}$$

so that the distributions of $(\mathbf{X}^{(s)}, \mathbf{X}^{(r)})$ are exchangeable under permutations of components of $\mathbf{X}^{(r)}$ conditional on $\mathbf{X}^{(s)}$ in both populations, and thus the unconditional distributions of $(\mathbf{X}^{(s)}, \mathbf{X}^{(r)})$ are also exchangeable under permutations of components of $\mathbf{X}^{(r)}$.

4.4. Representation for a Linear Combination Y

An arbitrary standardized linear combination Y can be represented in a simple form analogous to (15.7),

$$Y = \rho \mathfrak{Q} + (\sqrt{1 - \rho^2})\mathcal{W}, \tag{15.28}$$

where the vectors $\mathfrak{Q}$ and $\mathcal{W}$ are the standardized projections of Y along and orthogonal to the subspace $\{\mathbf{X}^{(s)}, Z\}$, where Z requires definition. It is the standardized discriminant of the covariates uncorrelated with $\mathbf{X}^{(s)}$,

$$\left[\left(\Sigma_c^{(r|s)} \right)^{-1} \left(\mu_t^{(r)} - \mu_c^{(r)} - \mathbf{B}' \left(\mu_t^{(s)} - \mu_c^{(s)} \right) \right) \right]' \mathbf{X}^{(r)}, \tag{15.29}$$

or in canonical form,

$$Z = \mathbf{1}'\mathbf{X}^{(r)}/\sqrt{p - s}, \tag{15.30}$$

except when $\mu_t^{(r)} = \mu_c^{(r)}$, in which case Z is defined to be 0. Writing $\mathfrak{Q}$ and $\mathcal{W}$ as

$$\mathfrak{Q} = \psi'\mathbf{X} = \left(\psi^{(s)'}, \psi^{(r)'} \right) \begin{pmatrix} \mathbf{X}^{(s)} \\ \mathbf{X}^{(r)} \end{pmatrix}, \tag{15.31}$$

$$\mathcal{W} = \gamma'\mathbf{X} = \left(\gamma^{(s)'}, \gamma^{(r)'} \right) \begin{pmatrix} \mathbf{X}^{(s)} \\ \mathbf{X}^{(r)} \end{pmatrix}, \tag{15.32}$$

a characterization of ψ and γ, in canonical form, paralleling (15.6), is summarized in the following lemma.

Lemma 4.1. *The coefficients γ and ψ must satisfy*

$$\gamma^{(s)} = 0 \tag{15.33}$$

and

$$\gamma^{(r)'}\psi^{(r)} = 0. \tag{15.34}$$

When $\mu_t^{(r)} \neq \mu_c^{(r)}$,

$$\psi^{(r)} \propto 1, \tag{15.35}$$

$$1'\gamma^{(r)} = 0. \tag{15.36}$$

When $\mu_t^{(r)} = \mu_c^{(r)}$,

$$\psi^{(r)} = 0. \tag{15.37}$$

Proof of Lemma 4.1. First, (15.33) is true because $\mathcal{W}$ is a linear combination of the X_i uncorrelated with $\{X^{(s)}, Z\}$ and thus with $\{X^{(s)}\}$; (15.34) follows from (15.33) and the fact that $\mathcal{Q}$ is uncorrelated with $\mathcal{W}$; (15.35) and (15.37) follow from the definition of Z in (15.30), and (15.34) and (15.35) imply (15.36). □

4.5. Extensions of Results in Section 3.3

Theorem 4.1. *Suppose a conditionally affinely invariant matching method is applied with fixed sample sizes* (N_t, N_c, n_t, n_c) *and proportional conditionally ellipsoidal distributions, which are represented in canonical form. Then,*

$$E(\overline{\mathbf{X}}_{mt}) = \begin{bmatrix} E\left(\overline{\mathbf{X}}_{mt}^{(s)}\right) \\ \mu_{mt}^{(r)} 1 \end{bmatrix}, \tag{15.38}$$

$$E(\overline{\mathbf{X}}_{mc}) = \begin{bmatrix} E\left(\overline{\mathbf{X}}_{mc}^{(s)}\right) \\ \mu_{mc}^{(r)} 1 \end{bmatrix}, \tag{15.39}$$

where $\mu_{mt}^{(r)}$ *is the common mean of each component of* $\overline{\mathbf{X}}_{mt}^{(r)}$, *and likewise for* $\mu_{mc}^{(r)}$ *and* $\overline{\mathbf{X}}_{mc}^{(r)}$. *The variance of the difference in matched samples means is given by*

$$\mathrm{var}(\overline{\mathbf{X}}_{mt} - \overline{\mathbf{X}}_{mc}) = \begin{bmatrix} \mathrm{var}\left(\overline{\mathbf{X}}_{mt}^{(s)} - \overline{\mathbf{X}}_{mc}^{(s)}\right) & \mathbf{C}1' \\ 1\mathbf{C}' & k(\mathbf{I} + c_0 11') \end{bmatrix}, \tag{15.40}$$

where $k \geq 0$, $c_0 \geq -1/(p - s)$, *and* $\mathbf{C}' = (c_1, \ldots, c_2)$. *Furthermore,*

$$E(v_{mt}(\mathbf{X})) = \begin{bmatrix} E(v_{mt}(\mathbf{X}^{(s)})) & \mathbf{C}_t 1' \\ 1\mathbf{C}_t' & k_t(\mathbf{I} + c_{t0} 11') \end{bmatrix}, \tag{15.41}$$

$$E(v_{mc}(\mathbf{X})) = \begin{bmatrix} E(v_{mc})(\mathbf{X}^{(s)}) & \mathbf{C}_c 1' \\ 1\mathbf{C}_c' & k_c(\mathbf{I} + c_{c0} 11') \end{bmatrix}, \tag{15.42}$$

where $k_t \geq 0, c_{t0} \geq -1/(p - s), \mathbf{C}_t' = (c_{t1}, \ldots, c_{ts})$, *and analogously for the matched control sample covariance matrix. When* $\mu_t^{(r)} = \mu_c^{(r)}$,

$$E(\overline{\mathbf{X}}_{mt}) \quad = \begin{bmatrix} E\left(\overline{\mathbf{X}}_{mt}^{(s)}\right) \\ 0 \end{bmatrix}, \tag{15.43}$$

$$E(\overline{\mathbf{X}}_{mc}) \quad = \begin{bmatrix} E\left(\overline{\mathbf{X}}_{mc}^{(s)}\right) \\ 0 \end{bmatrix}, \tag{15.44}$$

$$\text{var}\left(\overline{\mathbf{X}}_{mt}^{(r)} - \overline{\mathbf{X}}_{mc}^{(r)}\right) = \begin{bmatrix} \text{var}\left(\overline{\mathbf{X}}_{mt}^{(s)} - \overline{\mathbf{X}}_{mc}^{(s)}\right) & 0 \\ 0 & k\mathbf{I} \end{bmatrix}, \tag{15.45}$$

$$E(v_{mt}(\mathbf{X}^{(r)})) = \begin{bmatrix} E(v_{mt}(\mathbf{X}^{(s)})) & 0 \\ 0 & k_t\mathbf{I} \end{bmatrix}, \tag{15.46}$$

$$E(v_{mc}(\mathbf{X}^{(r)})) = \begin{bmatrix} E(v_{mc}(\mathbf{X}^{(s)})) & 0 \\ 0 & k_c\mathbf{I} \end{bmatrix}. \tag{15.47}$$

The proof of Theorem 4.1 is nearly identical to the proof of Theorem 3.1. The only difference is that covariances between components in $\mathbf{X}^{(s)}$ and $\mathbf{X}^{(r)}$ appear in Theorem 4.1, which have no analog in Theorem 3.1; these covariances are exchangeable in the coordinates of $\mathbf{X}^{(r)}$. Under the additional condition $\mu_t^{(r)} = \mu_c^{(r)}$, the rotational invariance of $\mathbf{X}^{(r)}$ implies the further simplifications in (15.43)–(15.47). The symmetry results of Theorem 4.1 imply that any $\mathcal{W}$ orthogonal to $\mathcal{Q}$ has the same distribution, as is summarized in the following corollary.

Corollary 4.1. *The quantities* $\text{var}(\overline{\mathcal{W}}_{mt} - \overline{\mathcal{W}}_{mc})$, $E(v_{mt}(\mathcal{W}))$ *and* $E(v_{mc}(\mathcal{W}))$ *take the same three values for all standardized Y. The analogous results apply for statistics of random subsamples indexed by rt and rc. However, the corresponding expressions involving $\mathcal{Q}$ do depend on the choice of Y.*

Under the conditions of Theorem 4.1, using the linear combination, Y, defined in (15.28), the following three corollaries hold.

Corollary 4.2. *The percent bias reduction of Y equals the percent bias reduction of Y in the subspace* $\{\mathbf{X}^{(s)}, Z\}$,

$$\frac{E(\overline{Y}_{mt} - \overline{Y}_{mc})}{E(\overline{Y}_{rt} - \overline{Y}_{rc})} = \frac{E(\overline{\mathcal{Q}}_{mt} - \overline{\mathcal{Q}}_{mc})}{E(\overline{\mathcal{Q}}_{rt} - \overline{\mathcal{Q}}_{rc})}. \tag{15.48}$$

Corollary 4.3. *The matching is ρ^2-proportionate modifying of the variance of the difference in matched sample means,*

$$\frac{\text{var}(\overline{Y}_{mt} - \overline{Y}_{mc})}{\text{var}(\overline{Y}_{rt} - \overline{Y}_{rc})} = \rho^2 \frac{\text{var}(\overline{\mathcal{Q}}_{mt} - \overline{\mathcal{Q}}_{mc})}{\text{var}(\overline{\mathcal{Q}}_{rt} - \overline{\mathcal{Q}}_{rc})} + (1 - \rho^2)\frac{\text{var}(\overline{\mathcal{W}}_{mt} - \overline{\mathcal{W}}_{mc})}{\text{var}(\overline{\mathcal{W}}_{rt} - \overline{\mathcal{W}}_{rc})}, \tag{15.49}$$

where the ratio $\text{var}(\overline{\mathcal{W}}_{mt} - \overline{\mathcal{W}}_{mc})/\text{var}(\overline{\mathcal{W}}_{rt} - \overline{\mathcal{W}}_{rc})$ *takes the same value for all Y.*

Corollary 4.4. *The matching is ρ^2-proportionate modifying of the expectations of the sample variances,*

$$\frac{E(v_{mt}(Y))}{E(v_{rt}(Y))} = \rho^2 \frac{E(v_{mt}(\mathcal{Q}))}{E(v_{rt}(\mathcal{Q}))} + (1 - \rho^2)\frac{E(v_{mt}(\mathcal{W}))}{E(v_{rt}(\mathcal{W}))} \tag{15.50}$$

and

$$\frac{E(v_{mc}(Y))}{E(v_{rc}(Y))} = \rho^2 \frac{E(v_{mc}(\mathcal{Q}))}{E(v_{rc}(\mathcal{Q}))} + (1 - \rho^2)\frac{E(v_{mc}(\mathcal{W}))}{E(v_{rc}(\mathcal{W}))}, \tag{15.51}$$

where the ratios $E(v_{mt}(W))/E(v_{rt}(W))$ and $E(v_{mc}(W))/E(v_{rc}(W))$ take the same two values for all Y.

A result corresponding to Corollary 3.5 with $\mu_t = \mu_c$ can be obtained but the simplification that occurs in the case when there are no special covariates is not present for the extended results.

Proof of Corollary 4.2. The proof is similar to that of Corollary 3.2 after noting from (15.32) and (15.33) that $W = \gamma^{(r)'} X^{(r)}$ and from (15.36), (15.38) and (15.33) that $\gamma^{(r)'} E(\overline{X}_{mt}^{(r)} - \overline{X}_{mc}^{(r)}) = 0$. $\qquad\square$

Proof of Corollary 4.3. The proof is analogous to that of Corollary 3.3; from Theorem 4.1 and (15.33),

$$\mathrm{cov}(\overline{\mathcal{Q}}_{mt} - \overline{\mathcal{Q}}_{mc}, \overline{W}_{mt} - \overline{W}_{mc}) = \psi' \left[\begin{array}{cc} \mathrm{var}\left(\overline{X}_{mt}^{(s)} - \overline{X}_{mc}^{(s)}\right) & \mathbf{C1}' \\ \mathbf{1C}' & k(\mathbf{I} + c_0 \mathbf{11}') \end{array} \right] \left[\begin{array}{c} \mathbf{0} \\ \gamma^{(r)} \end{array} \right].$$

Using V_s to represent a vector with s components, the covariance can be written as

$$\mathrm{cov}(\overline{\mathcal{Q}}_{mt} - \overline{\mathcal{Q}}_{mc}, \overline{W}_{mt} - \overline{W}_{mc})$$

$$= [V_s', \psi^{(s)'} \mathbf{C1}' + k\psi^{(r)'} + kc_0\psi^{(r)'}\mathbf{11}'] \left[\begin{array}{c} \mathbf{0} \\ \gamma^{(r)} \end{array} \right]$$

$$= \psi^{(s)'} \mathbf{C1}'\gamma^{(r)} + k\psi^{(r)'}\gamma^{(r)} + kc_0\psi^{(r)'}\mathbf{11}'\gamma^{(r)}$$

$$= \mathbf{0},$$

from (15.34) and (15.35) when $\mu_t^{(r)} \neq \mu_c^{(r)}$. When $\mu_t^{(r)} = \mu_c^{(r)}$, the result follows from (15.37) and (15.45), which implies $\mathbf{C} = \mathbf{0}$ and $c_0 = 0$. $\qquad\square$

The proof of Corollary 4.4 is analogous to the proof of Corollary 3.4 with a modification like that in Corollary 4.3, and is not presented in detail.

5. CONCLUSIONS

Our theoretical framework has established three general results (Corollaries 3.2–3.4) concerning the performance of affinely invariant matching methods, and three general results (Corollaries 4.2–4.4) concerning the performance of conditionally affinely invariant matching methods. The first collection of results can be regarded as a special case of the latter results. They exhibit four ratios involving Z and three ratios involving W, all of which are free of dependence on the particular outcome variable Y, but do depend on the matching setting (i.e., distributional forms, sample sizes and the particular matching method employed). The latter collection of results exhibit four ratios involving $\mathcal{Q}$, which depend on Y as well as the matching setting, and three ratios involving $\mathcal{W}$, which are free of Y. Thus, the corollaries show that only a small number of quantities are needed to evaluate the sampling properties of complex matching procedures for a large class of theoretically important distributions.

Under normality, simple approximations for all ratios involving W and $\mathcal{W}$ can be obtained analytically for matching methods that use population or estimated discriminants. Furthermore, under these conditions, bounds can be obtained for the ratios involving Z and $\mathcal{D}$. These results are presented in Rubin and Thomas (1992b), and of particular importance, they describe the difference between using estimated and population discriminants. Subsequent work [Rubin and Thomas (1992c)] using simulated and real data supports the relevance of these results to practice when distributions do not satisfy underlying assumptions.

16. Characterizing the Effect of Matching Using Linear Propensity Score Methods with Normal Distributions

Donald B. Rubin and Neal Thomas

Abstract: Matched sampling is a standard technique for controlling bias in observational studies due to specific covariates. Since Rosenbaum & Rubin (1983a), multivariate matching methods based on estimated propensity scores have been used with increasing frequency in medical, educational, and sociological applications. We obtain analytic expressions for the effect of matching using linear propensity score methods with normal distributions. These expressions cover cases where the propensity score is either known, or estimated using either discriminant analysis or logistic regression, as is typically done in current practice. The results show that matching using estimated propensity scores not only reduces bias along the population propensity score, but also controls variation of components orthogonal to it. Matching on estimated rather than population propensity scores can therefore lead to relatively large variance reduction, as much as a factor of two in common matching settings where close matches are possible. Approximations are given for the magnitude of this variance reduction, which can be computed using estimates obtained from the matching pools. Related expressions for bias reduction are also presented which suggest that, in difficult matching situations, the use of population scores leads to greater bias reduction than the use of estimated scores.

1. INTRODUCTION

1.1. Matched Sampling

Matched sampling refers to an important class of applied statistical procedures designed to control bias in observational studies (Cochran & Rubin, 1973; Rubin, 1973a,b; McKinlay, 1977; Carpenter, 1977; Rosenbaum & Rubin, 1985a; Kane, 1991). Recent theoretical work by Rubin & Thomas (1992a) extends earlier work of Rubin (1976b,c) to a general class of matching methods called affinely invariant. In particular, Rubin & Thomas show that, when these matching methods are applied to the class of ellipsoidal distributions, their performance can be decomposed into effects in two orthogonal subspaces defined by the population best linear discriminant and its complement. Here we provide explicit analytic expressions under normality for the effects on the subspaces when matching on an estimated linear propensity score. Of particular importance, the results show that matching using

Reprinted from *Biometrika*, 1992, **79**, 4, 797–809, by permission of Oxford University Press.

estimated propensity scores is a very effective algorithm for selecting samples with similar means, typically more effective than matching using population scores.

1.2. Proportional Normally Distributed Matching Variables

Consider random samples of sizes N_t and N_c from the 'treated' and control populations respectively, for example, smokers and nonsmokers. In each sample there are p normally distributed matching variables recorded, $X = (X_1, \ldots, X_p)$, with means μ_t, and μ_c, and variance-covariance matrices Ψ_t and Ψ_c respectively in the treated and control populations. We shall assume the special case of proportional normally distributed matching variables, which have $\Psi_t = \sigma^2 \Psi_c$, for a positive constant σ. Matching methods will be used to select a subset of n_t treated units and a subset of n_c control units whose distributions of X are more similar to each other than in the original samples, where the matching ratios in each group are defined by $N_t = R_t n_t$ and $N_c = R_c n_c$.

1.3. Linear Propensity Score Matching Methods

All of the matching methods considered here are linear propensity score methods, which match on a scalar linear summary of X designed to be a monotone function of the probability that a unit receives the treatment. The coefficients of the linear combination are typically estimated and so are functions of X, resulting in conditionally linear functions of X. The two most common examples of such conditionally linear functions of X in practice are the sample linear discriminant,

$$\hat{\beta}_p' X = \left\{ S_p^{-1}(\bar{X}_t - \bar{X}_c) \right\}' X, \tag{16.1}$$

where S_p^{-1} is the pooled sample covariance matrix, and $\bar{X}_t$ and $\bar{X}_c$ are the means of the treated and control samples, and the logistic regression estimator,

$$\hat{\beta}_L' X, \tag{16.2}$$

where $\hat{\beta}_L$ is the maximum likelihood estimate of β from a logistic regression with the treated units assigned the value '1' and the control units assigned the value '0'. When only the controls are subsampled, another common choice is obtained by replacing S_p^{-1} in (16.1) by S_c^{-1}, the covariance matrix in the control sample:

$$\hat{\beta}_c' X = \left\{ S_c^{-1}(\bar{X}_t - \bar{X}_c) \right\}' X. \tag{16.3}$$

Theoretical results for the estimates in (16.1), (16.2) and (16.3) are derived in Sections 3 and 4 by obtaining exact and asymptotic results for a closely related quantity, the semi-estimated discriminant, which is computed using the sample means and the population covariance matrix to calculate the coefficient of X, $\hat{\beta}_\Psi = \Psi_c^{-1}(\bar{X}_t - \bar{X}_c)$. For notational convenience, the semi-estimated discriminant is expressed in the standardized form

$$Z^* = \left\{ \Psi_c^{-1}(\bar{X}_t - \bar{X}_c) \right\}' X \left\{ (\bar{X}_t - \bar{X}_c)' \Psi_c^{-1}(\bar{X}_t - \bar{X}_c) \right\}^{-\frac{1}{2}}. \tag{16.4}$$

Asymptotic calculations and simulation study show that matching methods using the semi-estimated discriminant (16.4) produce very similar matched samples to matching methods using estimators (16.1), (16.2) or (16.3) over the range of settings typically encountered in practice. As a consequence, our approximations for matching based on 'estimated' linear propensity scores do not distinguish between the methods used to estimate the propensity scores.

We assume throughout that the matching on the estimated discriminant has resulted in some reduction in its initial bias. More precisely, letting the subscripts *mt* and *mc* refer to the matched treated and matched control groups of sizes n_t and n_c, for matching on the estimated discriminant, we assume that

$$(\bar{Z}^*_{mt} - \bar{Z}^*_{mc})^2 < (\bar{Z}^*_t - \bar{Z}^*_c)^2, \tag{16.5}$$

which is essentially verifiable from the data. We also exclude unreasonable matching methods that inflate the variances of the matched samples along the estimated discriminant to arbitrarily large sizes. In particular, we use the weak restriction

$$\frac{v_{mt}(Z^*)}{v_t(Z^*)} \leqslant c, \qquad \frac{v_{mc}(Z^*)}{v_c(Z^*)} \leqslant c, \tag{16.6}$$

for some nonnegative constant c, where $v_{mt}(\cdot)$ is the sample variance of the n_t matched treated units and $v_t(\cdot)$ is the sample variance of the full treated sample with N_t units, computed using $n_t - 1$ and $N_t - 1$ in the denominators respectively, and likewise for $v_{mc}(\cdot)$ and $v_c(\cdot)$ with control samples of size n_c and N_c. The only other restriction on the matching method imposed here is that the matching depends only on the estimated propensity scores; our results can be extended to the situation where other functions of the matching variables are also used in the matching in a manner analogous to the extension by Rubin & Thomas (1992a) to 'conditionally affinely invariant matching'.

Results for matching on the population best linear discriminant, which is expressed in standardized form as

$$Z = \left\{ \Psi_c^{-1}(\mu_t - \mu_c) \right\}' X \{ (\mu_t - \mu_c)' \Psi_c^{-1}(\mu_t - \mu_c) \}^{-\frac{1}{2}}, \tag{16.7}$$

are also obtained for comparison with the matching based on the estimated discriminant. When discussing the effects of matching on Z, we assume restrictions analogous to (16.5) and (16.6), that is, the same expressions with Z in place of Z^*.

1.4. Application of Matching Results from Rubin & Thomas (1992a)

Let Y be a linear combination of X, Z be the population best linear discriminant defined by (16.7), W be any fixed linear combination of X uncorrelated with this discriminant, and ρ be the correlation of Y and Z. When $\mu_t = \mu_c$, ρ is defined to be zero and Z is defined to be an arbitrarily chosen fixed linear combination of X with unit variance in the control population. In our context of proportional normal matching variables, Z and W are independent covariates in both the treated and control populations.

Because the normal distribution is a special case of ellipsoidal distributions and linear propensity score matching methods using population or estimated discriminants are a special case of affinely invariant matching methods, the general results of Rubin & Thomas (1992a) imply the following characterization.

(i) The matching is equal percent bias reducing, EPBR of Rubin (1976b),

$$\frac{E(\bar{Y}_{mt} - \bar{Y}_{mc})}{E(\bar{Y}_{rt} - \bar{Y}_{rc})} = \frac{E(\bar{Z}_{mt} - \bar{Z}_{mc})}{E(\bar{Z}_{rt} - \bar{Z}_{rc})} \quad (\rho^2 > 0), \quad E(\bar{Y}_{mt} - \bar{Y}_{mc}) = 0 \quad (\rho^2 = 0),$$

(16.8)

where the subscript rt refers to a randomly chosen sample of n_t treated units, and the subscript rc refers to a randomly chosen sample of n_c control units.

(ii) The matching is ρ^2-proportionate modifying of the variance of the difference in matched sample means (Rubin & Thomas, 1992a),

$$\frac{\text{var}(\bar{Y}_{mt} - \bar{Y}_{mc})}{\text{var}(\bar{Y}_{rt} - \bar{Y}_{rc})} = \rho^2 \frac{\text{var}(\bar{Z}_{mt} - \bar{Z}_{mc})}{\text{var}(\bar{Z}_{rt} - \bar{Z}_{rc})} + (1 - \rho^2)\frac{\text{var}(\bar{W}_{mt} - \bar{W}_{mc})}{\text{var}(\bar{W}_{rt} - \bar{W}_{rc})}. \quad (16.9)$$

(iii) The matching is ρ^2-proportionate modifying of the expectations of the sample variances (Rubin & Thomas, 1992a),

$$\frac{E\{v_{mt}(Y)\}}{E\{v_{rt}(Y)\}} = \rho^2 \frac{E\{v_{mt}(Z)\}}{E\{v_{rt}(Z)\}} + (1 - \rho^2)\frac{E\{v_{mt}(W)\}}{E\{v_{rt}(W)\}}, \quad (16.10)$$

$$\frac{E\{v_{mc}(Y)\}}{E\{v_{rc}(Y)\}} = \rho^2 \frac{E\{v_{mc}(Z)\}}{E\{v_{rc}(Z)\}} + (1 - \rho^2)\frac{E\{v_{mc}(W)\}}{E\{v_{rc}(W)\}}, \quad (16.11)$$

where $v_{rt}(\cdot)$ is the sample variance of n_t randomly chosen treated units computed using $n_t - 1$ in the denominator, and likewise for $v_{rc}(\cdot)$ with n_c randomly chosen control units computed using $n_c - 1$.

Because Z is the discriminant and W is any uncorrelated covariate, the seven ratios on the right-hand sides of (16.8)–(16.11), which are constants when the sample sizes, population moments and matching methods are fixed, describe the results for the first two moments of matching for all linear combinations of X. Here we obtain analytic approximations for the dominant ratios when matching on population propensity scores and when matching on estimated propensity scores.

2. OVERVIEW OF RESULTS

2.1. Matching on the Population Discriminant, Z

Discriminant matching, that is, matching on Z or a sample estimate of Z, has been studied in several papers (Cochran & Rubin, 1973; Rubin, 1976b,c; 1979b, 1980b). With proportional normal matching variables, matching on the population best linear discriminant, Z, does not affect the distribution of any uncorrelated component, W, because Z and W are independent. The results in the following theorem are then immediate, except for the approximation in (16.16), which was obtained by Rubin (1976c) with an appropriate change in notation.

Theorem 2.1. *For proportional normal matching variables, matching on the population discriminant that is bias reducing on the population discriminant implies that*

$$\frac{E(\bar{Z}_{mt} - \bar{Z}_{mc})}{E(\bar{Z}_{rt} - \bar{Z}_{rc})} = g_0, \qquad (16.12)$$

where $g_0 \leqslant 1$ with equality if and only if $\mu_t = \mu_c$. Also, for any covariate W uncorrelated with Z,

$$\text{var}(\bar{W}_{mt} - \bar{W}_{mc}) = \text{var}(\bar{W}_{rt} - \bar{W}_{rc}) = \text{var}(\bar{W}_{rt}) + \text{var}(\bar{W}_{rc}), \qquad (16.13)$$

$$E\{v_{mt}(W)\} = E\{v_{rt}(W)\} = \sigma^2 E\{v_{rc}(W)\}, \qquad (16.14)$$

$$E\{v_{mc}(W)\} = E\{v_{rc}(W)\}. \qquad (16.15)$$

An approximation for g_0 is given by

$$g_0 \simeq \tilde{g}_0 = \left[1 - \frac{\pi\{\sigma \log(R_t) + \log(R_c)\}}{4B} \right]_+, \qquad (16.16)$$

where

$$B = \{(\mu_t - \mu_c)' \Psi_c^{-1} (\mu_t - \mu_c)\}^{\frac{1}{2}}$$

is the number of control group standard deviations between the means of X in the treated and control groups.

2.2. Matching on Estimated Linear Propensity Scores

The following results and approximations are derived in Sections 3 and 4.

Theorem 2.2. *For proportional normal matching variables, matching on the semiestimated discriminant Z^* that is bias reducing in the sense of (16.5) implies that*

$$\frac{E(\bar{Z}_{mt} - \bar{Z}_{mc})}{E(\bar{Z}_{rt} - \bar{Z}_{rc})} = g, \qquad (16.17)$$

where $g \leqslant 1$ with equality if and only if $\mu_t = \mu_c$. Also, for any covariate W uncorrelated with Z,

$$\begin{aligned}
\text{var}(\bar{W}_{mt} - \bar{W}_{mc}) &= \text{var}(\bar{W}_{rt})\left\{ 1 - \frac{1 - k^2}{R_t} - \left(1 - \frac{1}{R_t}\right)h^2 \right\} \\
&\quad + \text{var}(\bar{W}_{rc})\left\{ 1 - \frac{1 - k^2}{R_c} - \left(1 - \frac{1}{R_c}\right)h^2 \right\} \\
&< \text{var}(\bar{W}_{rt}) + \text{var}(\bar{W}_{rc}), \qquad (16.18)
\end{aligned}$$

where $k^2 < 1$ and $h^2 \leqslant 1/p$. Furthermore, if the matching also has bounded variance in the sense of (16.6), then

$$E\{v_{mt}(W)\} = E\{v_{rt}(W)\} + \varepsilon_t, \qquad (16.19)$$

$$E\{v_{mc}(W)\} = E\{v_{rc}(W)\} + \varepsilon_c, \tag{16.20}$$

with $\varepsilon_c, \varepsilon_t \to 0$ as $N_t \to \infty$ or $N_c \to \infty$ provided that $\mu_t \neq \mu_c$.

Approximations for g, h and k are given by

$$g \simeq \tilde{g} = \left[1 - \frac{\pi\{\sigma \log{(R_t)} + \log{(R_c)}\}}{4\{p(\sigma^2/N_t + 1/N_c) + B^2\}^{\frac{1}{2}}}\right]_+ \geq \tilde{g}_0, \tag{16.21}$$

$$h \simeq \tilde{h} = \left\{\left(\frac{1}{p}\right) \frac{\sigma^2/N_t + 1/N_c}{\sigma^2/N_t + 1/N_c + B^2/p}\right\}^{\frac{1}{2}}, \tag{16.22}$$

$$k \simeq \tilde{k} = \tilde{g}. \tag{16.23}$$

Section 4 shows that when $\tilde{g}$ is zero, close mean matching on Z^* is possible, but not otherwise. Note that $\tilde{g}_0$, $\tilde{g}$, $\tilde{h}$ and $\tilde{k}$ can be estimated directly from the full treated and control samples using the estimated discriminant in place of Z to estimate B and σ^2.

2.3. Summarized Results

From (16.12) and (16.17), matching on either the population or the estimated linear propensity score results in bias reduction in the population discriminant, Z, and thus for all linear combinations of X by (16.8). When close matches on the estimated scores are not available, matching on the population propensity scores is expected to remove more bias, as shown by the inequality (16.21). From (16.8), there is no bias in covariates uncorrelated with the population discriminant in matched samples formed using the population or estimated discriminant.

Equations (16.14), (16.15), (16.19) and (16.20) show that matching on either the estimated or population linear propensity scores results in the same expected second moments in the matched samples as in random samples for all covariates uncorrelated with the population discriminant.

Equation (16.13) shows that matching on the population linear propensity score results in the same variance of the difference in the sample means in the matched samples as in random samples for all covariates uncorrelated with the discriminant. In contrast, (16.18) shows that when matching on an estimated linear propensity score, the variance of the difference in the means of the matched samples is less than the variance of the difference in the means of random samples for all covariates uncorrelated with the population discriminant. To see how much less this variance typically is, consider the common situation with no subsampling of the treated units, $R_t = 1$, one matched control unit selected for each treated unit, $n_t = n_c$, approximately equal variances in the treated and control populations, $\sigma^2 \simeq 1$, and close mean matching on the estimated discriminant, $k_2 \simeq 0$. Then the ratio of the variances of the difference in the matched sample means based on estimated and population propensity scores reduces to $\frac{1}{2}(1 - 1/R_c)$ for all covariates uncorrelated with the population discriminant. Thus, with large pools of controls, matching using estimated linear propensity scores results in approximately half the variance for the difference in the matched sample means as in corresponding random samples for all covariates uncorrelated with the population discriminant, and thus, with large p, the superior performance of the estimated score applies to most linear combinations of X. This phenomenon has also been discussed by Rosenbaum (1984a, 1987) in the context of randomization-based inference.

3. DERIVATION OF EXACT RESULTS WHEN MATCHING ON Z^*

3.1. Reduction to Canonical Form

Following Rubin & Thomas (1992a), because we are assuming affinely invariant matching methods and proportional covariance matrices, we can without loss of generality assume the following canonical form for the matching variables:

$\mu_t = \delta 1$, where δ is a positive scalar constant and 1 is the unit vector,
$\mu_c = 0$, where 0 is the zero vector, and
$\Psi_c = I$, where I is the identity matrix.

Similar canonical forms have been used in a variety of related problems; see, for example, Efron (1975), Rubin (1976b,c), and Ashikaga & Chang (1981).

3.2. The Semi-Estimated Discriminant, Z^*, and an Uncorrelated Component, W^*

In order to derive the results in Section 2.2, we need to have precise notation for the semi-estimated discriminant introduced in Section 1.3, and for covariates conditionally uncorrelated with it. All of the results that follow use the canonical form for the matching variables.

Henceforth, let

$$Z^* = \{(\bar{X}_t - \bar{X}_c)'(\bar{X}_t - \bar{X}_c)\}^{-\frac{1}{2}}(\bar{X}_t - \bar{X}_c)'X \tag{16.24}$$

be the standardized semi-estimated discriminant. Then form a standardized linear combination, W^*, which is uncorrelated with Z^* given $(\bar{X}_t, \bar{X}_c)$. With the matching variables represented in canonical form, W^* satisfies

$$W^* = \gamma'_*X, \quad \gamma'_*(\bar{X}_t - \bar{X}_c) = 0 = \bar{W}_t^* - \bar{W}_c^*, \quad \gamma'_*\gamma_* = 1. \tag{16.25}$$

Note that γ_* in this construction is a function of $(\bar{X}_t, \bar{X}_c)$.

3.3. The Distribution of W^*

Theorem 3.1. *For proportional normal matching variables, matching on the semi-estimated discriminant Z^* implies*

$$E\{\bar{W}_{mt}^* - \bar{W}_{mc}^* \mid (\bar{X}_t, \bar{X}_c)\} = 0,$$

$$\text{var}\{\bar{W}_{mt}^* - \bar{W}_{mc}^* \mid (\bar{X}_t, \bar{X}_c)\} = \frac{\sigma^2}{n_t}\left(1 - \frac{1}{R_t}\right) + \frac{1}{n_c}\left(1 - \frac{1}{R_c}\right),$$

$$E\{v_{mt}(W^*) \mid (\bar{X}_t, \bar{X}_c)\} \quad = \sigma^2,$$
$$E\{v_{mc}(W^*) \mid (\bar{X}_t, \bar{X}_c)\} \quad = 1,$$
$$E\{\text{cov}_{mc}(Z^*, W^*) \mid \bar{X}_t, \bar{X}_c)\} = 0.$$

Furthermore, conditional on $(\bar{X}_t, \bar{X}_c)$, the sample moments $v_{mt}(W^*)$ and $v_{mc}(W^*)$ are independent, and $\{\bar{W}_{mc}^*, v_{mc}(W^*), \bar{W}_{mt}^*, v_{mt}(W^*)\}$ are independent of $\{\bar{Z}_{mc}^*, v_{mc}(Z^*), Z_{mt}^*, v_{mt}(Z^*)\}$. Finally, each equation holds without conditioning on $(\bar{X}_t, \bar{X}_c)$.

Proof of Theorem 3.1. In canonical form, all of the matching variable components, $(X_{t,ij}, i = 1, \ldots, N_t, j = 1, \ldots, p)$ and $(X_{c,ij}, i = 1, \ldots, N_c, j = 1, \ldots, p)$ are mutually independent. Much but not all of this independence is retained after conditioning on $(\bar{X}_t, \bar{X}_c)$. Specifically, after conditioning on $(\bar{X}_t, \bar{X}_c)$: (a) the treated and control samples are independent; (b) different matching variables from the same unit remain independent, that is, $(X_{t,i1}, \ldots, X_{t,ip})$ are mutually independent as are $(X_{c,i1}, \ldots, X_{c,ip})$; and (c) different matching variables from different units also remain independent, although $(X_{t,1j}, \ldots, X_{t,N_t j})$ are correlated given $\bar{X}_t$, as are $(X_{c,1j}, \ldots, X_{c,N_c j})$ given $\bar{X}_c$.

Since we are conditioning on a linear combination of the X, $(X_1, \ldots, X_p)$ is still multivariate normal in both the treated and control populations, so that the linear combinations W^* and Z^* are also normally distributed. Since W^* and Z^* are constructed to be uncorrelated given $(\bar{X}_t, \bar{X}_c)$, they are conditionally independent. As a result, matching on Z^*, or on any function of it, has no effect on W^* other than selecting a random subset of W^* values. This demonstrates that the matched treated values of W^* are independent of the matched control values of W^*, thereby establishing the independence of all of the matched quantities involving Z^* with matched quantities involving W^*, and the independence of $v_{mt}(W^*)$ and $v_{mc}(W^*)$.

The distribution of W^* in the matched control sample is equivalent to the distribution of a random subsample of size n_c from a larger sample of N_c independent normal random variables conditional on the average of the N_c normal variables. The moments of this distribution can be derived using standard calculations from finite sampling theory (Cochran, 1977) and the fact that with normal data, the sample mean and sample variance are independent. For the control units, we have

$$E\{\bar{W}_{mc}^* \mid (\bar{X}_t, \bar{X}_c)\} = \bar{W}_c^*,$$
$$\text{var}\{\bar{W}_{mc}^* \mid (\bar{X}_t, \bar{X}_c)\} = \left(\frac{N_c - n_c}{N_c}\right)\frac{1}{n_c} = \frac{1}{n_c}\left(1 - \frac{1}{R_c}\right),$$
$$E\{v_{mc}(W^*) \mid (\bar{X}_t, \bar{X}_c)\} = 1.$$

Likewise for the matched treated units,

$$E\{\bar{W}_{mt}^* \mid (\bar{X}_t, \bar{X}_c)\} = \bar{W}_t^*,$$
$$\text{var}\{\bar{W}_{mt}^* \mid (\bar{X}_t, \bar{X}_c)\} = \left(\frac{N_t - n_t}{N_t}\right)\frac{\sigma^2}{n_t} = \frac{\sigma^2}{n_t}\left(1 - \frac{1}{R_t}\right),$$
$$E\{v_{mt}(W^*) \mid (\bar{X}_t, \bar{X}_c)\} = \sigma^2.$$

Since $\bar{W}_t^*$ equals $\bar{W}_c^*$, and $\bar{W}_{mt}^*$ and $\bar{W}_{mc}^*$ are independent, the formulae for the conditional mean and variance of $\bar{W}_{mt}^* - \bar{W}_{mc}^*$ follow. The final statement about unconditional results is immediate since the conditional expressions are free of $(\bar{X}_t, \bar{X}_c)$. $\qquad\square$

Theorem 3.1 gives results for the first and second moments only, but can be easily extended to derive the complete distribution of W^* in the matched samples. The first two moments are the most important for covariate adjustment in observational studies, however, and the results for these moments generalize, at least approximately, to nonnormal data.

3.4. A Corollary for W

Consider a fixed linear combination uncorrelated with the linear discriminant, $W = \gamma'X$, which for notational convenience is scaled to have variance one in the control population, $\gamma'\gamma = 1$, $\gamma'1 = 0$ with the canonical covariates. Then, a decomposition corresponding to Z and W can be formed using Z^* and W^*,

$$W = r_* Z^* + (1 - r_*^2)^{\frac{1}{2}} W^*,$$

where W^* is defined by (16.25), and r_* is the correlation of W and Z^* conditional on $(\bar{X}_t, \bar{X}_c)$ given by

$$r_* = \gamma'(\bar{X}_t - \bar{X}_c)\{(\bar{X}_t - \bar{X}_c)'(\bar{X}_t - \bar{X}_c)\}^{-\frac{1}{2}}, \qquad (16.26)$$

in both the treated and control populations, which follows from (16.24).

Corollary 3.1. *With proportional normal matching variables, matching on the semi-estimated discriminant Z^* implies*

$$E(\bar{W}_{mt} - \bar{W}_{mc}) = E\{r_*(\bar{Z}_{mt}^* - \bar{Z}_{mc}^*)\} = 0, \qquad (16.27)$$

$$\mathrm{var}\,(\bar{W}_{mt} - \bar{W}_{mc}) = E\{r_*^2(\bar{Z}_{mt}^* - \bar{Z}_{mc}^*)^2\}$$
$$+ \{1 - E(r_*^2)\}\left\{\frac{\sigma^2}{n_t}\left(1 - \frac{1}{R_t}\right) + \frac{1}{n_c}\left(1 - \frac{1}{R_c}\right)\right\}, \qquad (16.28)$$

$$E\{v_{mt}(W)\} = E\{r_*^2 v_{mt}(Z^*)\} + \sigma^2\{1 - E(r_*^2)\}, \qquad (16.29)$$

$$E\{v_{mc}(W)\} = E\{r_*^2 v_{mc}(Z^*)\} + \{1 - E(r_*^2)\}. \qquad (16.30)$$

Proof of Corollary 3.1. To establish (16.28), decompose $\bar{W}_{mt} - \bar{W}_{mc}$ as

$$\bar{W}_{mt} - \bar{W}_{mc} = r_*(\bar{Z}_{mt}^* - \bar{Z}_{mc}^*) + (1 - r_*^2)^{\frac{1}{2}}(\bar{W}_{mt}^* - \bar{W}_{mc}^*),$$

and compute (16.27) and (16.28) by first conditioning on $(\bar{X}_t, \bar{X}_c)$ using Theorem 3.1. The right-hand side of (16.27) is immediate from (16.8). To demonstrate (16.30), decompose $v_{mc}(W)$ as

$$v_{mc}(W) = r_*^2 v_{mc}(Z^*) + (1 - r_*^2) v_{mc}(W^*) + 2r_*(1 - r_*^2)^{\frac{1}{2}} \mathrm{cov}_{mc}(Z^*, W^*),$$

and compute conditionally on $(\bar{X}_t, \bar{X}_c)$. Analogous calculations establish (16.29) for $v_{mt}(W)$.   □

3.5. The Distribution of r_*^2

From (16.26), and the construction of Z^* and W^*, the distribution of r_*^2 can be represented as

$$r_*^2 = \frac{e_i^2}{\sum_j \left\{ e_j + \delta(\sigma^2/N_t + 1/N_c)^{-\frac{1}{2}} \right\}^2}, \tag{16.31}$$

where the e_j ($j = 1, \ldots, p$) are independent standard normal deviates and e_i is any of the e_j.

Letting r_{*0}^2 be (16.31) with δ set to zero, it is immediate that r_*^2 is stochastically smaller than r_{*0}^2, which has a beta distribution with parameters $\frac{1}{2}$ and $(p-1)/2$, and therefore has mean $1/p$, variance $(1 - 1/p)(p + p^2/2)$, and is J-shaped with infinite density at zero and zero density at one. The expectation of r_*^2 is thus bounded above:

$$E(r_*^2) \leqslant \frac{1}{p}. \tag{16.32}$$

From the representation in (16.31), it also follows that $E(r_*^2) \to 0$ as $N_t \to \infty$ or $N_c \to \infty$ except when $\delta = 0$. For values of δ and σ that occur in practice, for example, $\delta > 0.1$, $\sigma^2 \simeq 1$, r_*^2 is typically very small because sample sizes are usually at least 25 to 50.

3.6. Proof of Theorem 2.2

Using the constructions of Sections 3.2 and 3.4 for Z^* and W^*, decompose Z as

$$Z = a_* Z^* + (1 - a_*^2)^{\frac{1}{2}} W^*,$$

where a_* is the correlation of Z and Z^* conditional on $(\bar{X}_t, \bar{X}_c)$, and Z^* and W^* are functions of $(\bar{X}_t, \bar{X}_c)$. Then,

$$\bar{Z}_t - \bar{Z}_c = a_*(\bar{Z}_t^* - \bar{Z}_c^*), \tag{16.33}$$

because, by the construction of W^*, $\bar{W}_t^* - \bar{W}_c^* = 0$. Also,

$$E\{\bar{Z}_{mt} - \bar{Z}_{mc} \mid (\bar{X}_t, \bar{X}_c)\} = a_* E\{\bar{Z}_{mt}^* - \bar{Z}_{mc}^* \mid (\bar{X}_t, \bar{X}_c)\}, \tag{16.34}$$

because, by Theorem 3.1, $E\{\bar{W}_{mt}^* - \bar{W}_{mc}^* \mid (\bar{X}_t, \bar{X}_c)\} = 0$. From the bias reducing assumption in (16.5), and the fact that by construction $\bar{Z}_t^* \geqslant \bar{Z}_c^*$,

$$-(\bar{Z}_t^* - \bar{Z}_c^*) < (\bar{Z}_{mt}^* - \bar{Z}_{mc}^*) < (\bar{Z}_t^* - \bar{Z}_c^*),$$

so by (16.33) and (16.34) and the fact that $a_*^2 > 0$ except when $\bar{X}_t = \bar{X}_c$, which has probability zero,

$$-E\{\bar{Z}_t - \bar{Z}_c \mid (\bar{X}_t, \bar{X}_c)\} < E\{\bar{Z}_{mt} - \bar{Z}_{mc} \mid (\bar{X}_t, \bar{X}_c)\} < E\{\bar{Z}_t - \bar{Z}_c \mid (\bar{X}_t, \bar{X}_c)\},$$

thereby establishing (16.17).

The equation and inequality in (16.18) follow immediately from (16.28) of Corollary 3.1 and the inequality in (16.32) with the substitutions

$$h^2 = E(r_*^2), \quad k^2 = \frac{E\{r_*^2(\bar{Z}_{mt}^* - \bar{Z}_{mc}^*)^2\}}{\text{var}\,(\bar{Z}_t - \bar{Z}_c)}, \tag{16.35}$$

where $\text{var}\,(\bar{Z}_t - \bar{Z}_c) = \text{var}\,(\bar{W}_t) + \text{var}\,(\bar{W}_c)$. The inequality $k^2 < 1$ follows from the bias reducing assumption in (16.5), which from (16.35) implies

$$
\begin{aligned}
k^2 \text{var}\,(\bar{Z}_t - \bar{Z}_c) &< E\{r_*^2(\bar{Z}_t^* - \bar{Z}_c^*)^2\} \\
&= E[\{\gamma'(\bar{X}_t - \bar{X}_c)\}^2\{(\bar{X}_t - \bar{X}_c)'(\bar{X}_t - \bar{X}_c)\}^{-\frac{1}{2}}(\bar{Z}_t^* - \bar{Z}_c^*)^2] \\
&= E\{\gamma'(\bar{X}_t - \bar{X}_c)\}^2 \\
&= \text{var}\,(\bar{Z}_t - \bar{Z}_c).
\end{aligned}
$$

The inequality, $h^2 \leqslant 1/p$, follows immediately from (16.32) in Section 3.5.

Equations (16.20) and (16.19) follow immediately from (16.30) and (16.29) of Corollary 3.1 with

$$\varepsilon_c = E\{r_*^2 v_{mc}(Z^*)\} - E(r_*^2), \quad \varepsilon_t = E\{r_*^2 v_{mt}(Z^*)\} - \sigma^2 E(r_*^2).$$

Using the bounded variance assumption in (16.6), it can be shown that $v_{mc}(Z^*)$ and $v_{mt}(Z^*)$ are uniformly integrable, and since $0 \leqslant r_*^2 \leqslant 1$ and $E(r_*^2) \to 0$ as $N_t \to \infty$ or $N_c \to \infty$ when $\delta \neq 0$, it follows that $\varepsilon_c, \varepsilon_t \to 0$ as $N_t \to \infty$ or $N_c \to \infty$. □

4. APPROXIMATIONS FOR h, g AND k

4.1. Derivation of Approximation for h

The approximation for h in (16.22) is obtained from (16.31) by noting that since (16.31) holds with each e_i in the numerator,

$$E(r_*^2) = \frac{1}{p} E\left[\frac{\sum e_j^2}{\sum \left\{ e_j + \delta(\sigma^2/N_t + 1/N_c)^{-\frac{1}{2}} \right\}^2} \right],$$

where the summations are over the range $j = 1, \ldots p$, which for large p gives

$$E(r_*^2) \simeq \frac{1}{p} \left\{ \frac{1}{1 + \delta^2(\sigma^2 N_t + 1/N_c)^{-1}} \right\},$$

producing the formula for $\tilde{h}^2 \simeq E(r_*^2)$ in (16.22).

4.2. Derivation of Approximation for g

Consider the following matching rule, used to obtain results on maxima of bias reduction by Rubin (1976c), that is impractical for real applications but is useful for approximating g and k.

Choose for the matched treated sample the n_t of N_t treated units with the smallest values of Z^*, and choose for the matched control sample the n_c of N_c control units with the largest values of Z^*.

Then, for this matching method,

$$E\{\bar{Z}^*_{mt} \mid (\bar{X}_t, \bar{X}_c)\} = \bar{Z}^*_t - \sigma\Omega(N_t, n_t), \qquad (16.36)$$

$$E\{\bar{Z}^*_{mc} \mid (\bar{X}_t, \bar{X}_c)\} = \bar{Z}^*_c + \Omega(N_c, n_c), \qquad (16.37)$$

where $\Omega(N, n)$ is the average of the expectations of the n largest of N randomly sampled standard normal variables. Equations (16.36) and (16.37) can be derived from two observations. First, for a generic sample of independent identically distributed normal random variables, $e_i \sim N(0, 1) \, (i = 1, \ldots, N_t)$, the deviations about the sample mean, $e_1 - \bar{e}, \ldots, e_{N_t} - \bar{e}$, are independent of $\bar{e}$, from which it follows that the expected value of the kth order statistic, $E(e_{(k)})$, is equal to $E\{(e - \bar{e})_{(k)} \mid \bar{e}\}$. The second observation is that the joint conditional distribution of

$$(Z^*_{t,1} - \bar{Z}^*_t), \ldots, (Z^*_{t,N_t} - \bar{Z}^*_t) \mid (\bar{X}_t, \bar{X}_c)$$

is the same as the joint conditional distribution of $\sigma(e_1 - \bar{e}), \ldots, \sigma(e_{N_t} - \bar{e}) \mid \bar{e}$, and likewise for $\bar{Z}^*_c - \bar{Z}^*_c$. This follows because the distribution involving Z^* is a rank $(N_t - 1)$ exchangeable multivariate normal distribution with component means 0 and component variances $\sigma^2(1 - 1/N_t)$; these quantities specify the distribution.

It follows immediately from (16.36) and (16.37) that

$$E\{\bar{Z}^*_{mt} - \bar{Z}^*_{mc} \mid (\bar{X}_t, \bar{X}_c)\} = \bar{Z}^*_t - \bar{Z}^*_c - \sigma\Omega(N_t, n_t) - \Omega(N_c, n_c), \qquad (16.38)$$

from which it follows that

$$\frac{E(\bar{Z}^*_{mt} - \bar{Z}^*_{mc})}{E(\bar{Z}^*_t - \bar{Z}^*_c)} = 1 - \frac{\sigma\Omega(N_t, n_t) + \Omega(N_c, n_c)}{E(\bar{Z}^*_t - \bar{Z}^*_c)}, \qquad (16.39)$$

so the bias can be eliminated if the right-hand side of (16.39) is less than or equal to zero. Note from (16.33) and (16.34) that

$$\frac{E(\bar{Z}^*_{mt} - \bar{Z}^*_{mc})}{E(\bar{Z}^*_t - \bar{Z}^*_c)} = E\left(\frac{\bar{Z}_{mt} - \bar{Z}_{mc}}{a_*}\right) \Big/ E\left(\frac{\bar{Z}_t - \bar{Z}_c}{a_*}\right)$$

$$\simeq E(\bar{Z}_{mt} - \bar{Z}_{mc})/E(\bar{Z}_t - \bar{Z}_c), \qquad (16.40)$$

because in large samples $a_*^2 \to 1$ except when $\mu_t = \mu_c$, and then $a_*^2 \to 1/p$. Then, from definition (16.24),

$$\bar{Z}^*_t - \bar{Z}^*_c = \{(\bar{X}_t - \bar{X}_c)'(\bar{X}_t - \bar{X}_c)\}^{\frac{1}{2}},$$

which can be represented in a manner similar to (16.31) as

$$
\left[\sum_{j=1}^{p} \left\{ e_j \left(\frac{\sigma^2}{N_t} + \frac{1}{N_c} \right)^{\frac{1}{2}} + \delta \right\}^2 \right]^{\frac{1}{2}},
\tag{16.41}
$$

so that for large p

$$
E(\bar{Z}_t^* - \bar{Z}_c^*) \simeq \left\{ p\delta^2 + p \left(\frac{\sigma^2}{N_t} + \frac{1}{N_c} \right) \right\}^{\frac{1}{2}}.
\tag{16.42}
$$

Using (16.39) and the approximations in (16.40) and (16.42) gives

$$
\begin{aligned}
g &= E(\bar{Z}_{mt} - \bar{Z}_{mc})/E(\bar{Z}_t - \bar{Z}_c) \\
&\simeq \tilde{g} = \left[1 - \frac{\sigma \Omega(N_t, n_t) + \Omega(N_c, n_c)}{\{ p\delta^2 + p(\sigma^2/N_t + 1/N_c) \}^{\frac{1}{2}}} \right]_+,
\end{aligned}
\tag{16.43}
$$

where the positive part is used in (16.43) to reflect the fact that when bias can be eliminated, matching will no longer select only the extreme values of Z^*.

A simple approximation for $\Omega(N, n)$ given by Rubin (1976c) is $\frac{1}{4}\pi \log(N/n)$. The approximation works well for common ratios of (N/n), in fact, conservatively for ratios less than 10, and produces the approximation in (16.21) for g. The inequality in (16.21) follows because the denominator in the right-most term of $\tilde{g}$ is larger than the denominator of the right-most term in $\tilde{g}_0$.

4.3. Derivation of Approximation for k

Several steps are required to derive the approximations for k in (16.23), which is a lower bound for k. First, conditioning on $(\bar{X}_t, \bar{X}_c)$,

$$
\begin{aligned}
E\{ r_*^2 (\bar{Z}_{mt}^* - \bar{Z}_{mc}^*)^2 \mid (\bar{X}_t, \bar{X}_c) \} &= r_*^2 E\{ (\bar{Z}_{mt}^* - \bar{Z}_{mc}^*)^2 \mid (\bar{X}_t, \bar{X}_c) \} \\
&\geqslant r_*^2 [E\{ (\bar{Z}_{mt}^* - \bar{Z}_{mc}^*) \mid (\bar{X}_t, \bar{X}_c) \}]^2
\end{aligned}
\tag{16.44}
$$

$$
\geqslant r_*^2 [E\{ (\bar{Z}_{mt}^* - \bar{Z}_{mc}^*) \mid (\bar{X}_t, \bar{X}_c) \}]_+^2.
\tag{16.45}
$$

The inequalities in (16.44) and (16.45) provide tight bounds except for matching settings near the boundary in (16.39), where close mean matching is nearly achievable on average, $E(\bar{Z}_{mt}^* - \bar{Z}_{mc}^*) \simeq 0$. In these cases, $\bar{Z}_{mt}^* - \bar{Z}_{mc}^* < 0$ or $(\bar{Z}_{mt}^* - \bar{Z}_{mc}^*)^2$ is relatively large compared to $E(\bar{Z}_{mt}^* - \bar{Z}_{mc}^*) \simeq 0$ for a substantial proportion of all samples, so that in these settings, the lower bound in (16.45) can be substantially less than

$$
E\{ r_*^2 (\bar{Z}_{mt}^* - \bar{Z}_{mc}^*)^2 \mid (\bar{X}_t, \bar{X}_c) \}.
$$

Substituting (16.38) into (16.45) gives

$$
E\{ r_*^2 (\bar{Z}_{mt}^* - \bar{Z}_{mc}^*)^2 \mid (\bar{X}_t, \bar{X}_c) \} \geqslant r_*^2 [\bar{Z}_t^* - \bar{Z}_c^* - \{ \sigma \Omega(N_t, n_t) + \Omega(N_c, n_c) \}]_+^2,
$$

so that

$$k^2 \operatorname{var}(\bar{Z}_t - \bar{Z}_c) \geqslant E\left(r_*^2[\bar{Z}_t^* - \bar{Z}_c^* - \{\sigma\Omega(N_t, n_t) + \Omega(N_c, n_c)\}]_+^2\right).$$

Using the representations for r_*^2 in (16.31) and $\bar{Z}_t^* - \bar{Z}_c^*$ in (16.41), and noting that we can average across each choice of e_i in the numerator of (16.31), gives

$$k^2 \operatorname{var}(\bar{Z}_t - \bar{Z}_c)$$

$$\geqslant E\left\{\left\{\frac{(\sigma^2/N_t + 1/N_c)\sum_i e_i^2}{p}\right\}\left(1 - \frac{\sigma\Omega(N_t, n_t) + \Omega(N_c, n_c)}{\left[\sum_j\{e_j(\sigma^2/N_t + 1/N_c)^{\frac{1}{2}} + \delta\}^2\right]^{\frac{1}{2}}}\right)_+^2\right\},$$

where the summations are over the ranges $i = 1, \dots, p$ and $j = 1, \dots, p$ respectively. For large p, this gives

$$\tilde{k}^2 = \left[1 - \frac{\sigma\Omega(N_t, n_t) + \Omega(N_c, n_c)}{\{p\delta^2 + p(\sigma^2/N_t + 1/N_c)\}^{\frac{1}{2}}}\right]_+^2 = \tilde{g}^2,$$

with the latter equality following from (16.43).

4.4. Comments on the Approximations

The approximations for g_0, g, h and k given by $\tilde{g}_0, \tilde{g}, \tilde{h}$ and $\tilde{k}$ can be improved. For instance, a more accurate approximation for $\Omega(N, n)$, which is given by Rubin (1976c), leads to more accurate versions of $\tilde{g}_0, \tilde{g}$ and $\tilde{k}$. Also, a small p approximation for $E(r_*^2)$ can be derived from the exact normal theory expression in Section 4.1, and analogous methods can be applied to the exact normal theory bound for k in Section 4.3. Simulation results support the utility of the approximations, $\tilde{g}_0, \tilde{g}, \tilde{h}$ and $\tilde{k}$, even for small p, but do indicate that an improved approximation for $\Omega(N, n)$ can be helpful in some cases.

17. Matching Using Estimated Propensity Scores: Relating Theory to Practice

Donald B. Rubin and Neal Thomas

Abstract: Matched sampling is a standard technique in the evaluation of treatments in observational studies. Matching on estimated propensity scores comprises an important class of procedures when there are numerous matching variables. Recent theoretical work (Rubin, D. B., and Thomas, N., 1992a, reprinted in this volume as Chapter 15) on affinely invariant matching methods with ellipsoidal distributions provides a general framework for evaluating the operating characteristics of such methods. Moreover, Rubin and Thomas (1992b, reprinted in this volume as Chapter 16) uses this framework to derive several analytic approximations under normality for the distribution of the first two moments of the matching variables in samples obtained by matching on estimated linear propensity scores. Here we provide a bridge between these theoretical approximations and actual practice. First, we complete and refine the nomal-based analytic approximations, thereby making it possible to apply these results to practice. Second, we perform Monte Carlo evaluations of the analytic results under normal and nonnormal ellipsoidal distributions, which confirm the accuracy of the analytic approximations, and demonstrate the predictable ways in which the approximations deviate from simulation results when normal assumptions are violated within the ellipsoidal family. Third, we apply the analytic approximations to real data with clearly nonellipsoidal distributions, and show that the thoretical expressions, although derived under artificial distributional conditions, produce useful guidance for practice. Our results delineate the wide range of settings in which matching on estimated linear propensity scores performs well, thereby providing useful information for the design of matching studies. When matching with a particular data set, our theoretical approximations provide benchmarks for expected performance under favorable conditions, thereby identifying matching variables requiring special treatment. After matching is complete and data analysis is at hand, our results provide the variances required to compute valid standard errors for common estimators.

1. INTRODUCTION

Matched sampling is a methodology for reducing bias due to observed covariates in observational studies for causal effects. The basic situation involves one sample of N_t treated subjects and a larger sample of N_c control subjects, where the p matching variables $X = (X_1, \ldots, X_p)$ have density f_t among the treated subjects with mean μ_t and variance–covariance matrix Σ_t, and density f_c among the control subjects with mean μ_c and variance–covariance matrix Σ_c. The covariates X are potential confounding or disturbing variables (Cochran, 1965) that are fully observed for

Reprinted with permission from *Biometrics*, 1996, **52**, 249–264. © 1996 International Biometric Society.

all $N_t + N_c$ subjects. In typical examples, N_t is between 25 and 250, the ratio $R = N_c/N_t$ is between 2:1 and 20:1, and the number of matching variables, p, is between 5 and 50, although in some examples N_t may be 1000 or more, R a hundred or more, and p may be a hundred or more.

Matched sampling is most useful when there are many potential outcome variables, but none has yet been observed. For example, in a study of the effect of prenatal exposure to phenobarbital on intellectual development (Reinisch et al., 1995; Rosenbaum and Rubin, 1985a,b), X consists of copious prenatal records, $N_t \approx 100$, $N_c \approx 8{,}000$, and Y consists of dozens of quantities to be obtained via examinations and interviews of the subjects as young adults. In such a case where it is prohibitively expensive to collect Y on all potential controls, it is desirable to select a subset of matched control subjects, typically of size N_t, "most like" the N_t treated subjects with respect to X, and then collect Y on the treated and matched control subjects. If the ratio R is large and it is financially feasible to collect information on more control subjects, m-to-one matching with $R \geq m \geq 1$ can be used to reduce the standard errors of comparisons, although the gain in precison achieved by increasing the matched control sample size is typically modest (Rosenbaum and Rubin, 1985b). Previous research has established that preliminary matching reduces the bias and increases the precision of subsequent regression adjustment (Rubin, 1973b, 1979b).

In typical matching settings such as the phenobarbital exposure example, p is so large that it is virtually impossible to find matched pairs that are close to each other on all components of X, as demonstrated in Rosenbaum and Rubin (1985a). Moreover, matching on all of the components using a distance measure, which effectively regards all interactions among the p covariates as equally important, does not work particularly well (e.g., see Rubin, 1980b; Gu and Rosenbaum, 1993; Rosenbaum and Rubin, 1985a, for results in the phenobarbital example). A better method in such cases calculates a scalar summary of X, an "estimated propensity score" (Rosenbaum and Rubin, 1983a) and uses this as the primary matching variable (e.g., Rosenbaum and Rubin, 1985a,b; Rosenbaum, 1986; Eastwood and Fisher, 1988; Kane et al., 1991; Aiken, Smith, and Lake, 1994).

To introduce essential ideas, consider *population* propensity score methods. Such methods first calculate for each subject a scalar value that is a function of X, the propensity score, $e(X) = P(I = 1|X)$, where $I = 1$ indicates treated and $I = 0$ indicates control. Propensity score matching then uses $e(X)$, or a monotone function of it, to select control subjects, for example, choosing in turn for each of the N_t treated subjects the closest not yet chosen control subject (nearest remaining neighbor – or nearest available – matching). Rosenbaum and Rubin (1983a) show that exact matching on $e(X)$ implies that the resulting matched control and treated subjects have the same distribution of multivariate X, and thus the bias due to X has been controlled; they call this the balancing property of propensity scores. One important monotone function of $e(X)$ is

$$\frac{f_t(X)}{f_c(X)} \propto \frac{e(X)}{1 - e(X)},$$

the likelihood ratio statistic for judging whether a subject is a treated or control subject. From this perspective, the balancing property of propensity scores, given in Theorem 1 of Rosenbaum and Rubin (1983a), is equivalent to the sufficiency of this likelihood ratio statistic. Their Theorem 2 describing the propensity score's role as the coarsest summary with this balancing property is equivalent to the minimal sufficiency of the likelihood ratio statistic, described in Cox and Hinkley (1974). Estimated propensity score methods simply replace $e(X)$ with an estimate.

In common analytic examples, as discussed by Rosenbaum and Rubin (1983a), the logistic transformation of the propensity score is linear in X. In particular, the logistic linear model applies when f_t and f_c are p-variate normal with a common variance–covariance matrix, and more generally, it is true or approximately true for broad classes of distributions, for instance, the general location model (Olkin and Tate, 1961). The fact that some monotone function of $e(X)$ is often approximately linear in X supports the practice of matching using linear functions of X with estimated coefficients, typically found either using discriminant analysis, motivated by normal distribution theory (Cochran and Rubin, 1973; Rubin, 1976b,c, 1980b), or by logistic regression, motivated by the likelihood ratio perspective.

There has been only limited statistical work evaluating this practice beyond the theoretical work in Rubin and Thomas (1992a,b) and some simulations in Drake (1993) and Gu and Rosenbaum (1993). Here we show that the effect of this practice on the distributions of X in the resultant matched samples can be predicted using closed-form approximations that are functions of quantities known in the samples (N_t, R, p), or easily estimated from the samples (B, σ^2), where B is the number of standard deviations between the means of f_t and f_c with respect to Σ_c,

$$B^2 = (\mu_t - \mu_c)' \Sigma_c^{-1} (\mu_t - \mu_c),$$

and σ^2 is the ratio of the variance of the best linear discriminant (with respect to Σ_c) in f_t and f_c,

$$\sigma^2 = \frac{(\mu_t - \mu_c)' \Sigma_c^{-1} \Sigma_t \Sigma_c^{-1} (\mu_t - \mu_c)}{(\mu_t - \mu_c)' \Sigma_c^{-1} (\mu_t - \mu_c)}.$$

In particular, we provide analytic expressions in Section 2 for the distribution of the first two moments of X in the matched samples, and then discuss some of the practical implications of the results. The approximations are assessed by simulations using ellipsoidal data in Section 3 and real data in Section 4, where they are shown to be accurate across a wide range of practically relevant situations.

Our results demonstrate that matching on estimated linear propensity scores is an effective way to match the first moments of the treated and control distributions in many settings, thereby reducing bias due to linear trends in the covariates. Also, the results can be used to predict, in a particular setting with fixed distributions, the likely quality of resultant matched samples as a function of the size of the control pool, a critical aspect in the design of matched samples. With a particular data set, the theoretical results can be used to identify covariates requiring special consideration. Moreover, our results provide the basis for calculating valid standard errors of estimates of treatment effect based on the matched samples; these uses are discussed in Sections 2.3 and 5.

2. ANALYTIC RESULTS UNDER NORMALITY WITH
PROPORTIONAL VARIANCE–COVARIANCE MATRICES

Ellipsoidal distributions (e.g., the multivariate normal and t) and affinely invariant matching methods are important theoretical tools for the study of matching methods (Rubin, 1976b,c, 1979b). With an affinely invariant matching method, the identical set of matches will be obtained after every affine transformation of X. Many practical matching methods are affinely invariant (e.g., discriminant and Mahalanobis metric matching), notable exceptions being extensions that either include some product terms or nonlinear transformations of X or require closer matches with respect to some specially designated covariates. Of particular relevance to this paper, Rubin and Thomas (1992a) extended earlier results with multivariate ellipsoidal distributions to show that when $\Sigma_t = \sigma^2 \Sigma_c$, the effect of using affinely invariant matching methods can be stated in terms of scalar results for two components: first, the population discriminant, $Z = \beta' X$, where $\beta = \Sigma_c^{-1}(\mu_t - \mu_c)$; and second, an arbitrary covariate uncorrelated with the discriminant, $W = \gamma' X$, where $\gamma' \Sigma_c \beta = 0$. Assuming normality and matched sampling on the estimated linear propensity scores, $\hat{\beta}' X$, Rubin and Thomas (1992b) derived approximations for most of the constants appearing in the theoretical decomposition into Z and W. Here we summarize and extend these results to obtain approximations for the remaining constants under the additional assumption of close mean and variance matching on the estimated propensity scores, thereby allowing us to compute all of the needed approximations.

2.1. Overview of the Effects of Matching

When specialized to our setting with (i) no subsampling of the treated sample, (ii) m-to-one matching using propensity scores estimated by linear discriminant analysis or logistic regression, and (iii) proportional normal distributions, our analytic results can be stated in terms of an arbitrary linear combination of the covariates, $Y = \alpha' X$, which has correlation ρ with the discriminant, Z; Y can be decomposed into its regression (i.e., projection) on Z and its component, W, uncorrelated with Z, where the means of W in f_t and f_c are the same.

The three primary results concerning matching in the estimated Z direction are given explicitly in equations (17.1)–(17.11), but we begin by giving a more intuitive summary of them. All of the results compare expected sample moments in matched control samples to the same moments in randomly selected control samples of the same size. First, matching in the estimated Z direction does not create bias in the W direction, implying that the reduction in the bias of any Y due to matching is proportional to the reduction in the bias of Z due to matching, as expressed by equation (17.1); the fractional remaining bias can be approximated analytically by equations (17.2)–(17.5). Second, the fractional reduction in the variance of the difference between the treated and matched control sample means of Y due to matching can be written as a weighted sum of (i) the fractional reduction of the variance of the difference between the means of Z due to matching, with weight ρ^2, and (ii) the fractional reduction of the variance of the difference between the means of W due to matching, with weight $1 - \rho^2$, as expressed by equation (17.6); the fractional reduction in the variance of W can be approximated by equation (17.7), and the fractional reduction in the variance of Z can be approximated by equation (17.8) when the matching produces the same means for the estimated propensity scores in the matched samples. The entire differential effect of the matching on different Y variables is determined by ρ^2, the effect of matching being the same for all W.

Third, the relative change in the control sample variance of Y due to matching can be analogously written as a weighted sum of (i) the relative change in the control sample variance of Z due to matching, with weight, ρ^2, and (ii) the relative change in the control sample variance of W with weight $1 - \rho^2$, as expressed by equation (17.9); when the matching produces the same sample variances of the estimated propensity scores in the matched samples, the relative changes in the variances of W and Z are approximated by equations (17.10) and (17.11), respectively. Again, the entire differential effect of matching on different Y variables is determined by ρ^2, the squared correlation of Y with Z.

2.2. Analytic Approximations for the Matched Sample Moments

1. The matching is equal percent bias reducing (EPBR, Rubin, 1976a),

$$\frac{E(\overline{Y}_t - \overline{Y}_{mc})}{E(\overline{Y}_t - \overline{Y}_{rc})} = \frac{E(\overline{Z}_t - \overline{Z}_{mc})}{E(\overline{Z}_t - \overline{Z}_{rc})}, \tag{17.1}$$

where the overbar indicates sample means, the subscript t refers to the statistics in the treated sample of size N_t, the subscripts rc and mc refer to the statistics in randomly chosen control samples and matched control samples of size mN_t, respectively, and moreover,

$$\frac{E(\overline{Z}_t - \overline{Z}_{mc})}{E(\overline{Z}_t - \overline{Z}_{rc})} \approx \tilde{g} = (1 - \theta^*_{\max})_+, \tag{17.2}$$

$$\theta^*_{\max} = \Omega(R/m)\tilde{h}V^{-1/2} = \frac{\Omega(R/m)}{\left\{p\left(\sigma + \frac{1}{R}\right)N_t^{-1} + B^2\right\}^{1/2}}, \tag{17.3}$$

where $\theta^*_{\max}$ is the maximum possible fractional bias reduction, $\Omega(R/m)$ is the expectation in the upper m/R tail of a normal distribution, and

$$V = \frac{1}{N_t}\left(\sigma^2 + \frac{1}{R}\right), \quad \tilde{h} = (p + B^2/V)^{-1/2}. \tag{17.4}$$

Rubin (1976a) derived the approximation

$$\Omega(R/m) \approx 2^{\pi/2}(2\pi)^{-1/2}(R/m)^{(1-\pi/4)}\left(1 - \frac{m}{R}\right)^{\pi/4}, \tag{17.5}$$

which we will use when numerically evaluating $\theta^*_{\max}$. The representation in (17.1)–(17.3) is a special case of (4.8) in Rubin and Thomas (1992b). With proportional ellipsoidal variables and either no initial bias or Y such that $\rho^2 = 0$ (i.e., $Y = W$), then $E(\overline{Y}_t - \overline{Y}_{mc}) = 0$; there is no bias in variables uncorrelated with the discriminant before matching, and the matching cannot create bias.

2. The matching is ρ^2-proportionate modifying of the variance of the difference in matched sample means of Y,

$$\frac{\text{var}(\overline{Y}_t - \overline{Y}_{mc})}{\text{var}(\overline{Y}_t - \overline{Y}_{rc})} = \rho^2 \frac{\text{var}(\overline{Z}_t - \overline{Z}_{mc})}{\text{var}(\overline{Z}_t - \overline{Z}_{rc})} + (1 - \rho^2)\frac{\text{var}(\overline{W}_t - \overline{W}_{mc})}{\text{var}(\overline{W}_t - \overline{W}_{rc})}, \qquad (17.6)$$

where

$$\frac{\text{var}(\overline{W}_t - \overline{W}_{mc})}{\text{var}(\overline{W}_t - \overline{W}_{rc})} \approx \left\{ \sigma^2 \tilde{g}^2 + \left(\frac{1 - \tilde{h}^2}{m}\right) - \left(\frac{1 - \tilde{h}^2 - \tilde{g}^2}{R}\right) \right\} (\sigma^2 + 1/m)^{-1}. \qquad (17.7)$$

When matching is close in the sense that the means of the estimated propensity scores in the treated and matched control samples are equal, then

$$\frac{\text{var}(\overline{Z}_t - \overline{Z}_{mc})}{\text{var}(\overline{Z}_t - \overline{Z}_{rc})} \approx \frac{p - 1}{p + B^2/V}\left(\frac{1/m - 1/R}{\sigma^2 + 1/m}\right); \qquad (17.8)$$

approximation (17.7) is a special case of (17.7) in Rubin and Thomas (1992b), and (17.8) is derived in the Appendix.

3. The matching is ρ^2-proportionate modifying of the expectation of the matched control sample variance of Y,

$$\frac{E\{v_{mc}(Y)\}}{E\{v_{rc}(Y)\}} = \rho^2 \frac{E\{v_{mc}(Z)\}}{E\{v_{rc}(Z)\}} + (1 - \rho^2)\frac{E\{v_{mc}(W)\}}{E\{v_{rc}(W)\}}, \qquad (17.9)$$

where v_{rc} and v_{mc} refer to the sample variances in random and matched control samples of size mN_t, respectively. When matching is close in the sense that the variances of the estimated propensity scores in the treated and matched control samples are equal, then

$$\frac{E\{v_{mc}(W)\}}{E\{v_{rc}(W)\}} \approx 1 + \frac{\sigma^2 - 1}{p + B^2/V}, \qquad (17.10)$$

and

$$\frac{E\{v_{mc}(Z)\}}{E\{v_{rc}(Z)\}} \approx \sigma^2 \left(\frac{1 + B^2/V}{p + B^2/V}\right) + \frac{p - 1}{p + B^2/V}; \qquad (17.11)$$

approximations (17.10) and (17.11) are derived in the Appendix. As B approaches zero, (17.10) and (17.11) converge to the common value $\sigma^2/p + (p - 1)/p$, which is exact when $B = 0$ under close variance matching for any linear combination of X, as proven by a simple symmetry argument.

2.3. Discussion of Practical Recommendations

The formulas in Section 2.2 can be evaluated using estimates based on easily computed moments in the full treated and control samples. The approximate bias in (17.1) gives a quick screening method for determining the potential utility of a data set for estimating treatment effects, because if substantial differences between the treated and control sample means will remain after matching, generally undesirable extrapolations will be needed to eliminate bias; Section 5 provides additional discussion.

Several practical recommendations follow from (17.1)–(17.3) and their derivation. Except when the treated sample is very small, the most important determinant of the comparability of the subsequent matched samples is the initial bias as represented by B^2 in the denominator of $\theta^*_{\max}$. The magnitude of B^2 is basically beyond the control of the researcher, but when it is too large ($B^2 > 1.5$, except when R is large), identification of variables contributing heavily to B^2 can be useful in guiding the search for alternative sources of control subjects or the reconsideration of whether each matching variable really is an appropriate covariate, rather than, for instance, an outcome variable affected by the treatment, or a variable almost certainly unrelated to the outcome of interest, both of which have occurred in our practical experience with matching. Applying (17.3) when N_t is moderately large, so that $\theta^*_{\max} \approx \Omega(R/m)B^{-1}$, matching ratios, R/m, of at least 2, 3, and 6 are required to eliminate differences in the sample means associated with initial B^2 values of 0.5, 1.0, and 1.5, respectively; for smaller values of N_t, somewhat larger values of R/m are required as suggested by (17.3). Using two-to-one instead of one-to-one matching doubles the number of control subjects needed to achieve the same bias reduction. For larger values of B^2, $\theta^*_{\max}$ depends on R primarily through $\Omega(R/m)$, so it is easy to assess the effect of increasing the pool of control subjects. Although the matching ratios indicated here are sufficient to eliminate differences in the sample means, somewhat larger matching ratios, R/m, are desirable in order to minimize differences in the sample variances of the estimated propensity scores.

Expressions (17.6)–(17.8) show that matching on the estimated linear propensity scores can be an effective method for reducing differences between the treated and control sample means. Suppose that close one-to-one matching on the estimated scores is achieved so that $\tilde{g} = 0$; then approximation (17.7) reduces to $\left\{1 - (p + B^2/V)^{-1}\right\}(1 - 1/R)(1 + \sigma^2)^{-1}$, the variance ratio involving Z is less than the ratio for W, and both ratios are less than $(1 - 1/R)(1 + \sigma^2)^{-1}$. Hence, when the treated and control variance matrices are approximately equal so that $\sigma^2 = 1$, then the variability in $\overline{Y}_t - \overline{Y}_{mc}$ is no larger than half that of a corresponding randomized design.

An important issue whenever there are many matching variables is how many of the variables should be included in the propensity score model, and whether to remove variables from the propensity score model to achieve a more parsimonious model. Unless a variable can be excluded because there is consensus that it is unrelated to the outcome variables or not a proper covariate, it is advisable to include it in the propensity score model even if it is not statistically significant. Excluding potentially relevant variables should be done only when the resultant matched samples are closely balanced with respect to these variables as will typically occur when the treated and full control sample means of the excluded variables are exceptionally close or when the excluded variables are highly correlated with variables already in the propensity score model. Including variables whose treated and control sample means differ only because of sampling variability and not because of substantial differences in their population means contributes only modestly to the difficulty of finding adequate matches on the estimated propensity scores. Examination of the formula for the maximum possible bias reduction, (17.3), shows that the decrease in achievable bias reduction depends weakly on the number of covariates provided $p(\sigma^2 + \frac{1}{R})N_t^{-1}$ remains small.

Although matching on the estimated linear propensity scores is an effective method for creating treated and matched control samples with similar means, it produces only small changes in the expected value of the sample variances for variables not highly correlated with Z. In particular, if the treated and control variances for a Y are different, the treated and matched control sample variances are also likely to be different. More explicitly, first consider the situation when (i) close variance matching obtains for the estimated scores, (ii) there is at least a moderate difference in the population treated and control means, and (iii) N_t is large enough so that B^2/V is large. Then (17.10) is approximately one, that is, the expected value of the matched control sample variance of each W variable is unchanged, whereas (17.11) approaches σ^2, so that the expected value of the variance of Z in the matched control sample is adjusted to equal the corresponding variance in the treated sample. The original X variables, however, are typically in the $(p - 1)$ dimensional space of W variables, and thus their sample variances are largely unchanged. When B^2/V is small, the effect of the matching is still a mixture of no change, and correction toward the treated value, with weights of $(p - 1)/p$ and $1/p$ respectively, but there is no simple allocation of the change to the Z and W variables. The preceding description still applies to the sample variances of W when close matching on the sample variances of the estimated propensity scores is not achieved. Close matching of the estimated propensity scores typically fails to occur when there are large differences in the treated and control means so that the matched sample is selected almost exclusively from the tail of the control distribution. In this case, the expected value of the variance of Z is reduced in the matched control sample regardless of the variance of Z in the treated sample. This phenomenon is discussed and documented with simulated data in Section 3.

An important feature of matching on the estimated linear propensity scores is that it does not require close pairwise matches. The effective balancing of a large number of sample means, as quantified in (17.1)–(17.8), is achieved by any propensity score matching procedure that produces good agreement between the treated and matched control sample means of the estimated linear propensity scores. Cochran and Rubin (1973) provide guidance for determining acceptable 'calipers' for individual matches to ensure close mean matching on a scalar variable, in our case, the estimated propensity scores. These calipers are generally wide, calipers of even one half of a standard deviation being tight enough to remove ninety percent of the initial bias; somewhat tighter calipers are often desirable to ensure that the sample variances are also well matched.

Propensity score matching shares with randomization the feature that it makes no distinction between covariates highly predictive of the outcome variables and ones that are only weakly predictive of the outcome variables. Because only fairly coarse matching on the estimated scores is required to achieve the practical benefits of propensity score matching, additional matching on a small number of substantively important variables will often be possible. Rosenbaum and Rubin (1985a) includes an example with additional matching.

3. SIMULATION STUDY WITH ELLIPSOIDAL DATA

A simulation study was performed using several ellipsoidal distributions to investigate the accuracy of the approximations in Section 2 for multivariate normal data and

Table 17.1. *Simulation settings*

Types of factor	Simulation factor	Levels
Distributional	Shape (v)	5, 10, ∞
factors – estimable	Difference in population means B	0.0, 0.25, 0.50, 0.75, 1.0, 1.5
but not known	Variance σ^2	0.75, 1.0, 1.25
to investigator		
Design factors –	Matching ratio, $R(N_c = RN_t)$	2, 5, 10
known to	Dimension of covariate space, p	5, 10
investigator	Treated sample size, N_t	25, 50
	Discriminant estimator	$\beta_*, \hat{\beta}_c, \hat{\beta}_p, \hat{\beta}_r$

the sensitivity of the approximations to nonnormal data. Ellipsoidal distributions are like the multivariate normal distribution in that the conditional expectation (regression) of any linear combination of variables on any other linear combination is linear, so that deviations from normality due to nonlinear regressions cannot be studied using ellipsoidal distributions. The conditional variances of ellipsoidal distributions, however, are not constant in general, but a function of the Mahalanobis distance of the conditioning variable from its center of symmetry (Fang, Kotz, and Ng, 1990). General ellipsoidal distributions therefore allow the study of this deviation from normality. In particular, we use the multivariate t_v distribution, $t_v = U/\sqrt{\chi_v^2/v}$, where U is p-variate standard normal and χ_v^2 is a chi square random variable with v degrees of freedom; the conditional variance is $(v + D^2)/(v + d^* - 2)$, where D is the Mahalanobis distance of the conditioning variable from its mean, and d^* is its dimensionality (Fang et al., 1990). Although not required for our general theoretical results to hold, we set v to be the same in f_t and f_c.

Ellipsoidal distributions with proportional variance–covariance matrices are especially useful for a simulation study like this one using affinely invariant matching methods because the resultant symmetry implies that a few quantities summarize the entire simulation. Following Ashikaga and Chang (1981), Efron (1975), Rubin (1976b,c), Rubin and Thomas (1992a,b), and many other authors, without loss of generality, we assume the following canonical form: $\mu_c = (0, 0, \ldots, 0)'$, $\mu_t = B(1, 1, \ldots, 1)'/\sqrt{p}$, Σ_c equals the p dimensional identity matrix, and $\Sigma_t = \sigma^2 \Sigma_c$. The matches produced by affinely invariant methods are the same for the original data and the data transformed to the canonical form. Thus, the results for all distributions in an ellipsoidal family of distributions can be characterized by results for the distributions in canonical form.

3.1. Design

A seven-way factorial simulation study was performed as summarized in Table 17.1. The degrees of freedom were chosen to represent moderate (10 df) and large (5 df) deviations from the constant conditional variance of the normal distribution (∞ df). Four estimates of the coefficients of the linear propensity scores were considered: the semi-estimate, $\beta_* = \Sigma_c^{-1}(\overline{X}_t - \overline{X}_c)$, which was used to derive the analytic approximations; the sample best linear discriminant using S_c, the control sample variance–covariance matrix, $\hat{\beta}_c = S_c^{-1}(\overline{X}_t - \overline{X}_c)$; the sample best linear discriminant using S_p, the pooled sample variance–covariance matrix, $\hat{\beta}_p = S_p^{-1}(\overline{X}_t - \overline{X}_c)$; and the logistic regression maximum likelihood estimate, $\hat{\beta}_r$. Each of these estimates is affinely invariant because the estimate of the linear propensity score is the same after any affine transformation of X as before it.

Simulation data were generated so that settings involving lower dimensional covariates and smaller sample sizes were obtained by subsetting the simulation settings with the largest sample sizes and dimensions. Each of the four estimators of β was applied to each simulated data set to obtain the estimated propensity scores. Treated and control subjects were matched using a one-to-one (i.e., $m = 1$) nearest remaining neighbor matching algorithm. The nesting of data sets and estimation methods reduced the computational time and increased the precision of the comparisons between settings by introducing positive correlations among the simulation estimates. Five hundred replications were performed for each setting on a SUN workstation.

Estimates of the moments, $E(\overline{Z}_t - \overline{Z}_{mc})$, $\text{var}(\overline{Z}_t - \overline{Z}_{mc})$, $E\{v_{mc}(Z)\}$, $\text{var}(\overline{W}_t - \overline{W}_{mc})$, and $E\{v_{mc}(W)\}$, were obtained from the means and variances of the 500 replications of $\overline{Z}_t - \overline{Z}_{mc}$, $v_{mc}(Z)$, $\overline{W}_t - \overline{W}_{mc}$, and $v_{mc}(W)$ at each setting for each discriminant estimator. There are $(p - 1)$ uncorrelated components, W, in the population, which are approximately uncorrelated in the matched samples. Because each provides an unbiased estimate of the W quantities in (17.7) and (17.10), the results from the $p - 1$ uncorrelated components were averaged to provide a single, more accurate simulation estimate.

We present results for the $\hat{\beta}_r$ estimator. The results for the other estimators were very similar, diverging slightly in the settings where the treated and control distributions were most different.

3.2. Results When Close Matching Is Possible

Table 17.2 summarizes the results for the $\hat{\beta}_r$ estimator from the settings where close mean matching is possible, defined as settings where $\theta^*_{\max}$ is greater than one, implying that 100% bias reduction is theoretically possible. The 288 settings summarized in Table 17.2 include all settings except those with $R = 2$ or with $B = 1.0, 1.5$. The five ratios involving Z and W in (17.1), (17.6), and (17.9) were estimated directly from the 500 simulation replications at each setting, and then compared to the theoretical approximations for them given in (17.2), (17.7), (17.8), (17.10), and (17.11). Simulation standard errors, based on 500 replications for each of the 288 settings, were small relative to the variation across conditions displayed in Table 17.2.

The analytic approximations perform very well, even in those settings where the data were generated from a t_5 distribution with its markedly nonconstant variance. The only evidence of inadequacy occurs with the approximation (17.8) for the ratio $\text{var}(\overline{Z}_t - \overline{Z}_{mc})/\text{var}(\overline{Z}_t - \overline{Z}_{rc})$, which is based on the assumption of exact mean matching of the estimated propensity scores; (17.8) performed well for the settings with $R = 10$ or small differences in f_t and f_c, but (17.8) began to fail in some of the more difficult settings summarized in Table 17.2, that is, cases with $\theta^*_{\max}$ close to one.

Table 17.2. *Summary of the results across the 288 settings with close matching using* $\hat{\beta}_r$: *simulation ratio divided by theoretical ratio*

	Min	Lower quartile	Median	Upper quartile	Max
$\left\{1 - \dfrac{E(\overline{Z}_t - \overline{Z}_{mc})}{E(\overline{Z}_t - \overline{Z}_{rc})}\right\}/(1 - \text{eqn}\,(17.2))$	0.80	0.90	0.93	0.96	1.02
$\left\{\dfrac{\text{var}(\overline{Z}_t - \overline{Z}_{mc})}{\text{var}(\overline{Z}_t - \overline{Z}_{rc})}\right\}/\text{eqn}\,(17.8)$	0.89	1.10	1.25	1.43	2.60
$\left\{\dfrac{E\{v_{mc}(Z)\}}{E\{v_{rc}(Z)\}}\right\}/\text{eqn}\,(17.11)$	0.82	0.95	0.99	1.00	1.16
$\left\{\dfrac{\text{var}(\overline{W}_t - \overline{W}_{mc})}{\text{var}(\overline{W}_t - \overline{W}_{rc})}\right\}/\text{eqn}\,(17.7)$	0.90	1.03	1.08	1.14	1.35
$\left\{\dfrac{E\{v_{mc}(W)\}}{E\{v_{rc}(W)\}}\right\}/\text{eqn}\,(17.10)$	0.93	0.99	1.00	1.04	1.19

Note: The table summarizes the 288 simulation settings where close matching is possible in the sense that the theoretical predictions for $1 - E(\overline{Z}_t - \overline{Z}_{mc})/E(\overline{Z}_t - \overline{Z}_{rc})$ are all one, i.e., $\theta^*_{\max} \geq 1$; simulation estimates of $1 - E(\overline{Z}_t - \overline{Z}_{mc})/E(\overline{Z}_t - \overline{Z}_{rc})$ are excluded when $B = 0$ because the theoretical value is undefined $(1 - 0 \div 0)$.

3.3. Trends for Difficult Matching Settings

We describe several trends that occur in difficult matching settings, defined by $\theta^*_{\max} < 1$, implying that 100% bias reduction is impossible. Substantial bias remains after matching in the more difficult settings, thereby diminishing the practical utility of the data for comparing treated and control subjects. The most difficult settings are therefore included here primarily to magnify the trends that occur in the analytic approximations as close mean and variance matching becomes more difficult to attain; thus we include all of the settings described in Table 17.1 in the summaries in this section.

Figure 17.1 displays the simulation estimate of the proportional bias reduction due to matching for each simulation setting versus $\theta^*_{\max}$. When the bias remaining is substantial, it is well approximated by $\theta^*_{\max}$. The bias is eliminated, as predicted, when $\theta^*_{\max}$ is much greater than one. The approximation in (17.2) fails to predict the moderate biases that remain when $\theta^*_{\max}$ is near one, however, because it is based on the assumption that the matches on the estimated propensity scores will be taken exclusively from the tail of the estimated propensity score distribution, which is not true in the simulation. More practical matching algorithms, such as the nearest remaining neighbor matching algorithm that was employed in the simulations, typically do not make such extreme choices for matches because they attempt to balance other characteristics of the matched distributions. In the most difficult matching settings with $\theta^*_{\max} < 1$, such as $R = 2$ and $B = 1.5$, the nearest available neighbor matching algorithm also selects primarily from the tail of the estimated propensity score distribution, so that (17.2) predicts the remaining bias well. In these difficult settings, however, the other two approximations for the Z component do poorly because they are based on the assumption of close mean and

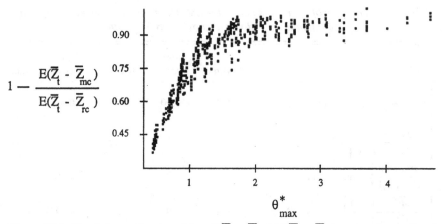

Figure 17.1. Simulation estimates of $1 - E(\overline{Z}_t - \overline{Z}_{mc})/E(\overline{Z}_t - \overline{Z}_{rc})$ versus the predicted maximum possible fractional bias reduction, θ^*_{max}. Matching based on the logistic regression estimator $\hat{\beta}_r$.

variance matching, which is impossible to attain. The approximation for the ratio $\mathrm{var}(\overline{Z}_t - \overline{Z}_{mc})/\mathrm{var}(\overline{Z}_t - \overline{Z}_{rc})$ in (17.8), which is based on exact mean matching of the estimated propensity scores, can be low by a factor of five to ten in cases where θ^*_{max} is much less than one. These poor estimates can cause substantial errors in the approximation of $\mathrm{var}(\overline{Y}_t - \overline{Y}_{mc})/\mathrm{var}(\overline{Y}_t - \overline{Y}_{rc})$ for covariates highly correlated with the discriminant. This problem does not strongly affect the approximation of $\mathrm{var}(\overline{Y}_t - \overline{Y}_{mc})/\mathrm{var}(\overline{Y}_t - \overline{Y}_{rc})$ for most covariates, however, because most covariates have relatively low correlations with the discriminant. Also in these difficult cases, the matching produces $E\{v_{mc}(Z)\}/E\{v_{rc}(Z)\}$ ratios as small as one half the value predicted in (17.11), because (17.11) is based on the assumption that the sample variances of the estimated propensity score are the same in the treated and matched control samples, whereas in these difficult cases, the matched controls are all being drawn from the tail of the control distribution.

The dominant trend for the moments of the uncorrelated variables W in the settings where close mean matching is not possible is the increase in variability over that predicted for normal matching variables, which is a consequence of the increase in the conditional variance of W as a function of the magnitude of the estimated propensity scores with t_v variables. Figure 17.2 displays the simulation estimates of $\mathrm{var}(\overline{W}_t - \overline{W}_{mc})/\mathrm{var}\ t_v(\overline{W}_t - \overline{W}_{rc})$ divided by its theoretical approximation (17.7) plotted against θ^*_{max}. The settings with data generated from the t_5 distribution are highlighted; they show the resultant inflation in the actual variance ratio relative to the theoretical value that occurs as the matching selects control subjects with more extreme propensity scores. The ratio $E\{v_{mc}(W)\}/E\{v_{rc}(W)\}$ exhibits a similar trend.

4. SIMULATION STUDY USING REAL DATA

The approximations in Section 2 are derived under the stringent assumptions of multivariate normality and proportional variance–covariance matrices, and for some approximations, the additional assumption of close mean or variance matching on the estimated propensity scores. The simulation study summarized in Section 3 supports

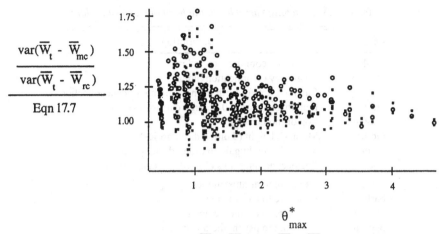

$$\frac{\text{var}(\overline{W}_t - \overline{W}_{mc})}{\text{var}(\overline{W}_t - \overline{W}_{rc})}$$

Eqn 17.7

$$\theta^*_{max}$$

Figure 17.2. Simulation estimates of $\text{var}(\overline{W}_t - \overline{W}_{mc})/\text{var}(\overline{W}_t - \overline{W}_{rc})$ divided by its theoretical approximation in (17.7) versus the predicted maximum possible fractional bias reduction, θ^*_{max}. Matching based on the logistic regression estimator $\hat{\beta}_r$. Simulation settings with data generated from the t_5 distribution are highlighted with a circle.

the utility of these approximations in practical settings when the conditional variance is quadratic rather than constant and the propensity score matching is not exact. In this section, a simulation study using real data shows that the approximations in Section 2 can provide useful information about the matching performance for multivariate data that are far from proportional or ellipsoidal.

4.1. Simulation Design

The simulations were generated from data consisting of matching variables collected on a treated sample of 97 children whose mothers were exposed to a hormone during pregnancy and 7848 potential control children. The matching variables, described in Table 17.3, are a mix of dichotomous, ordinal, and continuous. Rosenbaum and Rubin (1985a) discuss propensity score matching on a related data set, and Reinisch et al. (1995) contains substantive analyses of these data.

The data set was used to create 32 samples to generate replications of propensity score matching. First, four treated samples of size $N_t = 24$ were randomly selected without replacement from the 97 treated children. Second, for each treated sample, eight control samples of size $N_c = 240$ ($R = 10$) were randomly selected without replacement from the 7848 control children, thereby yielding a nested one-way design with 32 treatment–control sample pairs. Third, for each of the 32 treatment–control pairs of samples, $\hat{\beta}_r$ was computed, then the estimated linear propensity scores were computed in the treatment and control samples using $\hat{\beta}_r$, and nearest remaining neighbor matching on the estimated scores was applied to form 24 treated–matched control sample pairs of subjects. Fourth, for each of the matching variables, denoted generically by Y, the 32 replications of $\overline{Y}_t - \overline{Y}_{mc}$ and $v_{mc}(Y)$ were averaged to estimate $E(\overline{Y}_t - \overline{Y}_{mc})$ and $E\{v_{mc}(Y)\}$. The within and between variance estimates of the nested values of $\overline{Y}_t - \overline{Y}_{mc}$ were summed to estimate $\text{var}(\overline{Y}_t - \overline{Y}_{mc})$. Finally, the whole process was repeated: the subsampling of the

Table 17.3. *Matching variables in the hormone exposure data*

Label	Description
pbc420	Pregnancy complication index
pbc415	Pregnancy complication index
cage	Calendar time of birth from beginning of data collection
lgest	Length gestation (10 ordered categories)
lmotage	Log of mother's age
motht	Mother's height (5 ordered categories)
motwt	Mother's weight gain (6 ordered categories)
ses	Socioeconomic status (10 ordered categories)
antih	Exposure to antihistamines
barb	Exposure to barbiturates
chemo	Exposure to chemotherapy agents
psydrug	Exposure to psychotherapy drugs
sex	Male indicated by a 1, female by a 0
cigar	Mother smoked cigarettes
sib	First child indicated by a 1, other children by a 0

Note: The variables in the upper portion of the table are ordinal or continuous.
The variables in the lower portion of the table are dichotomous.

full treated and control samples and the subsequent matching and estimation of the moments, $E(\overline{Y}_t - \overline{Y}_{mc})$, $E\{v_{mc}(Y)\}$, and $\text{var}(\overline{Y}_t - \overline{Y}_{mc})$, were replicated 400 times, and the resulting estimates were averaged. Each replication of the subsampling provided an unbiased estimate of the matching performance, although these estimates are correlated across replications because they all use the common full samples of 97 treated and 7848 control subjects.

The population values μ_t, μ_c, Σ_t, and Σ_c required for the approximations in Section 2 were estimated using the sample moments $\overline{Y}_t$, $\overline{Y}_c$, S_t, and S_c from the full treated and control samples. The proportionality constant σ^2 was approximated by averaging the eigenvalues of $S_c^{-1}S_t$ (i.e., trace $(S_c^{-1}S_t)/p$); the ratio of the sample variances of the estimated discriminant gave essentially the same result in this data set. The standardized bias B, the discriminant, and the resulting propensity scores were also approximated using the full sample moments, and, ρ, the correlation of each matching variable with the propensity score was similarly estimated. This matching setting is similar to the more difficult settings described in Section 3.1; here $B = 1.28$, $\sigma^2 = 1.11$, $R = 10$, $p = 15$, $N_t = 24$, and $\hat{\theta}^*_{\max} = 1.3$.

4.2. Results

As anticipated from the relatively large value of $\hat{\theta}^*_{\max}$, the propensity score matching performed very well, removing almost all of the expected difference in the sample means of each matching variable, even those with large initial differences; summaries of the matching performance are given in Table 17.4. As anticipated from the normal-based theory and the t-based simulations, matching on the estimated linear propensity scores produced little change in the sample variances of most variables, and only modest changes in the variances of the variables most correlated with

Table 17.4. *Performance of the propensity score matching with real data set*

Variable	Correlation with discriminant ρ	Full unmatched data set		Matching performance		Simulation experiment[*] Simulation ratio divided by theoretical ratio		
		$\dfrac{(\bar{Y}_t - \bar{Y}_c)}{\sqrt{v_c(Y)}}$	$\dfrac{v_t(Y)}{v_c(Y)}$	$\dfrac{E(\bar{Y}_t - \bar{Y}_{mc})}{\sqrt{v_c(Y)}}$	$\dfrac{E\{v_t(Y)\}}{E\{v_{mc}(Y)\}}$	$1 - \dfrac{E(\bar{Y}_t - \bar{Y}_{mc})}{E(\bar{Y}_t - \bar{Y}_{rc})}$ Eq.17.2	$\dfrac{\mathrm{var}(\bar{Y}_t - \bar{Y}_{mc})}{\mathrm{var}(\bar{Y}_t - \bar{Y}_{rc})}$ Eq.17.6	$\dfrac{E\{v_{mc}(Y)\}}{E\{v_{rc}(Y)\}}$ Eq.17.9
pbc420	0.53	0.73	0.58	0.04	0.90	0.95	0.90	0.61
pbc415	0.20	0.27	1.23	0.04	1.22	0.86	0.83	1.00
cage	−0.05	−0.06	1.20	−0.05	1.24	0.22[b]	0.83	0.97
lgest	−0.05	−0.07	0.81	0.00	0.96	0.96[b]	0.85	0.84
lmotage	0.50	0.68	0.55	0.01	0.55	0.98	1.21	0.95
motht	0.20	0.28	0.58	0.01	0.72	0.98	0.90	0.80
motwt	0.01	0.01	1.28	−0.03	1.17	3.46[b]	0.87	1.09
ses	0.60	0.81	1.17	0.06	1.13	0.92	1.05	0.96
antih[a]	0.46	0.19[a]		0.02[a]		0.87	1.20	
barb[a]	0.26	0.10[a]		0.01[a]		0.87	1.08	
chemo[a]	0.10	0.04[a]		0.00[a]		0.94	1.00	
psydrug[a]	0.34	0.13[a]		0.03[a]		0.77	1.16	
sex[a]	−0.02	−0.01[a]		0.01[a]		1.53[b]	0.85	
cigar[a]	−0.25	−0.17[a]		0.00[a]		1.00	0.86	
sib[a]	0.02	0.01[a]		0.02[a]		−0.71[b]	0.86	

[a] Dichotomous variables – not divided by $\sqrt{v_c(Y)}$.

[b] Note small initial bias.

[*] Most simulation errors are less than five percent of the reported values; $E(\cdot)$ and $\mathrm{var}(\cdot)$ refer to simulation averages and variances.

the propensity scores; it is somewhat surprising that the variance ratio was notably improved for the variable *pbc420*, but unchanged for the variable *lmotage*.

To check how well the simulation results were predicted by our analytic approximations, the simulation estimates of $1 - E(\overline{Y}_t - \overline{Y}_{mc})/E(\overline{Y}_t - \overline{Y}_{rc})$, $E\{v_{mc}(Y)\}/E\{v_{rc}(Y)\}$, and $\mathrm{var}(\overline{Y}_t - \overline{Y}_{mc})/\mathrm{var}(\overline{Y}_t - \overline{Y}_{rc})$, divided by their theoretical approximations, are displayed in the three rightmost columns of Table 17.4. The approximations perform very well despite the sharp deviations from the idealized model underlying them. The values of $1 - E(\overline{Y}_t - \overline{Y}_{mc})/E(\overline{Y}_t - \overline{Y}_{rc})$ for the variables *sex, sib, cage,* and *motwt,* which differ substantially from one, are a consequence of the small initial differences in the full samples for these variables. The standardized biases of these variables are still very small in the treated and matched control samples.

These simulations constructed from real matching data demonstrate that our theoretical approximations can provide useful design information for markedly non-normal data despite the stringent assumptions employed in the derivations of the approximations. The relative robustness of the analytic approximations is due to the fact that the derivations formally require only the conditional first and second moment structure of the multivariate normal distribution. Although this moment structure is specific to the normal distribution, it is approximately shared by many multivariate distributions (Diaconis and Freedman, 1984; Li, 1991). The approximations deteriorate when there are large initial differences between f_t and f_c.

5. DISCUSSION OF BIAS REDUCTION AND PRECISION

The approximations in Section 2 can be used to predict the distributions of the first two sample moments of the original matching variables. These moments are particularly important summaries because they determine the bias for linear and quadratic models relating outcome variables to the covariates, which are motivated by first and second order Taylor series approximations. Although the bias for a specific linear model (conditional on the matching variables) can be computed directly from the matching data after it has been collected, standard error calculations require the approximations in Section 2.

To briefly illustrate the types of analyses that are possible from our results, consider the simple linear model with an additive treatment effect, Δ, and a regression coefficient, α on X, for an outcome variable $\mathcal{Y}$,

$$\mathcal{Y} = \alpha_0 + \Delta I + \alpha'X + \epsilon, \tag{17.12}$$

where $I = 1$ indicates treated and $I = 0$ indicates control as in Section 1, ϵ has mean zero and variance τ_I^2 conditional on X and I, and $Y = \alpha'X$ as in Section 2. From (17.1), the bias in the estimator $\overline{\mathcal{Y}}_t - \overline{\mathcal{Y}}_{mc}$ is approximately

$$\alpha'E(\overline{X}_t - \overline{X}_{rc})\tilde{g}. \tag{17.13}$$

Denoting the right side of approximations (17.7) and (17.8) by C_w and C_z, respectively, and applying (17.6), the variance of $\overline{\mathcal{Y}}_t - \overline{\mathcal{Y}}_{mc}$ is

$$\mathrm{var}(\overline{\mathcal{Y}}_t - \overline{\mathcal{Y}}_{mc}) \approx \{\mathrm{var}(\overline{Y}_t) + \mathrm{var}(\overline{Y}_c)\}\{\rho^2 C_z + (1 - \rho^2)C_w\} + \tau_1^2/N_t + \tau_0^2/(mN_t). \tag{17.14}$$

At the time of the matching, the quantities $E(\overline{X}_t - \overline{X}_{rc})$, $\tilde{g}$, C_z, and C_w in (17.13) and (17.14) are either known or can be estimated directly by substituting the moments from the unmatched samples as was done in Section 4. For any specified value of α, estimates of $\text{var}(\overline{Y}_t)$, $\text{var}(\overline{Y}_c)$, and ρ can be obtained from the sample moments of X. Similarly, the values of τ_0^2 and τ_1^2 can be obtained from the moments of X for any specified correlation between $\mathcal{Y}$ and Y in f_t and f_c. Once the outcome variables are collected, estimates of α and the multiple correlation between y and $\mathcal{Y}$ in f_t and f_c can be estimated by least squares regression. If data on $\mathcal{Y}$ are not available during the design stage, then estimates of the potential precision can be obtained by trying several values of both α and the correlation between Y and $\mathcal{Y}$.

When close mean matching is attained, $\tilde{g} \approx 0$, the bias with respect to model (17.12) is eliminated, so that matched sampling on estimated linear propensity scores effectively eliminates bias arising from linear trends in numerous matching variables. A qualitative description of the precision of $\overline{\mathcal{Y}}_t - \overline{\mathcal{Y}}_{mc}$ with close matching can be derived by recalling from Section 2.2 that C_z and C_w are no larger than $(1/m - 1/R)(\sigma^2 + 1/m)^{-1}$, so the variability in $\overline{\mathcal{Y}}_t - \overline{\mathcal{Y}}_{mc}$ due to X is typically less than half that of a corresponding completely randomized design, and the variability in $\overline{\mathcal{Y}}_t - \overline{\mathcal{Y}}_{mc}$ uncorrelated with X is the same as from a completely randomized design.

If we expand model (17.12) to include quadratic dependence on the covariates,

$$\mathcal{Y} = \alpha_0 + \Delta I + \alpha'X + X'AX + \epsilon, \qquad (17.15)$$

where $A = (a_{jk})$ is a symmetric matrix, then the bias in $\overline{\mathcal{Y}}_t - \overline{\mathcal{Y}}_{mc}$ is the expectation of

$$\alpha'(\overline{X}_t - \overline{X}_{rc}) + N_t^{-1}\sum_{i=1}^{N_t}(X_i^t)'AX_i^t - (mN_t)^{-1}\sum_{j=1}^{mN_t}(X_j^{mc})'AX_j^{mc}, \qquad (17.16)$$

where $X_i^t, i = 1, \ldots, N_t$, and $X_j^{mc}, j = 1, \ldots, mN_t$, denote the covariate values of the treated and matched control subjects respectively, with $(X_i^t)' = (X_{i1}^t, \ldots, X_{ip}^t)$, $(X_j^{mc})' = (X_{j1}^{mc}, \ldots, X_{jp}^{mc})$. The expectation of the first term in (17.16) can be approximated by (17.13). The contribution to the bias of the (kl)th quadratic term is

$$a_{kl}\left\{E\left(N_t^{-1}\sum_{i=1}^{N_t}X_{ik}^t X_{il}^t\right) - E\left((mN_t)^{-1}\sum_{j=1}^{mN_t}X_{jk}^{mc} X_{jl}^{mc}\right)\right\}$$

$$\approx a_{kl}\left[\{E(\text{cov}_t(X_k, X_l)) - E(\text{cov}_{mc}(X_k, X_l))\}\right.$$

$$+ \left.\left\{E\left(\overline{X}_k^t\right)E\left(\overline{X}_l^t\right) - E\left(\overline{X}_k^{mc}\right)E\left(\overline{X}_l^{mc}\right)\right\}\right],$$

where cov_t and cov_{mc} are the treated and matched control sample covariances. The terms involving the sample means and the treated sample covariances can be estimated from the sample moments and (17.13). The expectations of the matched control sample variances can be obtained directly from (17.9), and the expectations of the matched control sample covariances can be obtained by applying (17.9) to each covariate individually, and then to their sum. Computation of standard errors under model (17.15) requires the distribution of the fourth moments of the covariates in the matched control sample. Approximations for higher moments can be obtained using analytic methods similar to those used to derive the approximations in Section 2, but we expect that such approximations would be sensitive to the assumption of multivariate normality.

A key feature of propensity score matching is the potential to combine it with other statistical methods, during both the matching and the analysis, which utilize substantive knowledge about the relationships between the covariates and the outcome variables, while retaining the desirable properties of the propensity score matching documented in Sections 3 and 4. Extensions of the approximations given in Section 2 can be used to calculate the bias and precision of methods that combine propensity score matching with extra matching on key covariates and regression adjustments.

ACKNOWLEDGMENTS

We thank Raj Bahadur for very helpful comments about the role of likelihood ratio statistics and two editorial reviewers for exceptionally helpful comments, which clarified the presentation.

APPENDIX

Derivation of the Approximations

We derive the new approximations in Section 2.2 for the semi-estimated propensity scores, $\beta'_* X$, with $\beta_* = \Sigma_c^{-1}(\overline{X}_t - \overline{X}_c)$, where $\overline{X}_t$ and $\overline{X}_c$ are the means of the treated and control samples. We assume close mean matching on the semi-estimated scores for (17.8), and close variance matching for (17.10) and (17.11). The simulation results in Sections 3 and 4 demonstrate that the performance of estimated propensity scores based on the sample linear discriminant and logistic regression estimators is predicted very well by the performance of the semi-estimated scores. Without loss of generality, we assume a canonical form for f_t and f_c; specifically, we assume f_t is multivariate normal with mean zero and the identity variance–covariance matrix, and f_c is multivariate normal with mean proportional to $(1, \ldots, 1)'$ and variance–covariance matrix proportional to the identity. This canonical form is used and explained in the simulations in Section 3 as well as in the theoretical work of Rubin and Thomas (1992b).

We begin by computing an approximate expectation used in the derivation of (17.8), (17.10), and (17.11). Let a_* be the conditional (given $(\overline{X}_t, \overline{X}_c)$) correlation between the population propensity scores, Z, and the semi-estimated propensity scores, $Z^* = \beta'_* X$. Standardizing Z and Z^* to have variance one in f_c, in canonical form, $Z = (1/p)^{1/2} 1' X$ and $Z^* = \{(\overline{X}_t - \overline{X}_c)'(\overline{X}_t - \overline{X}_c)\}^{-1/2}(\overline{X}_t - \overline{X}_c)' X$, so that

$$a_*^2 = \left\{ \frac{1'(\overline{X}_t - \overline{X}_c)}{p} \right\}^2 \left\{ \frac{(\overline{X}_t - \overline{X}_c)'(\overline{X}_t - \overline{X}_c)}{p} \right\}^{-1}, \qquad (A.1)$$

where the p components of $(\overline{X}_t - \overline{X}_c)$ are independent and identically distributed normal random variables with mean $Bp^{-1/2}$, and variance V. Approximating the expectation of a_*^2 by the ratio of the expectations of its numerator and denominator in (A.1) gives

$$E(a_*^2) \approx \frac{B^2/p + V/p}{B^2/p + V} = \frac{1 + B^2/V}{p + B^2/V}. \tag{A.2}$$

Approximations more refined than (A.2) are possible, but (A.2) has performed well with real and simulated data.

Approximation (17.8) can be obtained using the canonical representation for the matching variables and applying Corollary 3.1 of Rubin and Thomas (1992b) to decompose Z given $(\overline{X}_t, \overline{X}_c)$ into a component along Z^* and a component uncorrelated to Z^*. Corollary 3.1, although stated for a component W uncorrelated to Z, can be applied to Z with the substitution of $\text{var}\left\{r_*\left(\overline{Z}_{mt}^* - \overline{Z}_{mc}^*\right)\right\}$ for $E\left\{r_*^2\left(\overline{Z}_{mt}^* - \overline{Z}_{mc}^*\right)^2\right\}$ in equation (3.5)[1] of Corollary 3.1, and the replacement of each occurrence of r_* by a_*. The approximation in (17.8) then follows by setting $\overline{Z}_{mt}^* = \overline{Z}_t^*$, $R_t = 1$, $R_c = R/m$, $n_c = mN_t$, using (A.2) for $E(a_*^2)$, and approximating $\text{var}\left\{a_*\left(\overline{Z}_t^* - \overline{Z}_{mc}^*\right)\right\}$ to be zero, which follows from the assumption of close mean matching.

Approximation (17.11) is also based on Corollary 3.1 of Rubin and Thomas (1992b) with Z in place of W, v_t in place of v_{mt}, and a_* in place of r_*. It is obtained by substituting equation (3.6)[2] into equation (3.7)[3] of Corollary 3.1, assuming that $v_t(Z^*) = v_{mc}(Z^*)$, to yield

$$E\{v_{mc}(Z) - v_t(Z)\} = (1 - \sigma^2)\{1 - E(a_*^2)\},$$

and then substituting σ^2 for $E\{v_t(Z)\}$ and (A.2) for $E(a_*^2)$. Approximation (17.10) follows immediately from (3.6) and (3.7) of Corollary 3.1 using the approximation $E(r_*^2) \approx (p + B^2/V)^{-1}$ derived in Section 4.1 of Rubin and Thomas (1992b). It is a refinement of (17.8) in Rubin and Thomas (1992b) based on the additional assumption of close variance matching.

---

[1] Reprinted as equation (16.28) in this volume.
[2] Reprinted as equation (16.29) in this volume.
[3] Reprinted as equation (16.30) in this volume.

18. Combining Propensity Score Matching with Additional Adjustments for Prognostic Covariates

Donald B. Rubin and Neal Thomas

Abstract: Propensity score matching refers to a class of multivariate methods used in comparative studies to construct treated and matched control samples that have similar distributions on many covariates. This matching is the observational study analog of randomization in ideal experiments, but is far less complete as it can only balance the distribution of observed covariates, whereas randomization balances the distribution of all covariates, both observed and unobserved. An important feature of propensity score matching is that it can be easily combined with model-based regression adjustments or with matching on a subset of special prognostic covariates or combinations of prognostic covariates that have been identified as being especially predictive of the outcome variables. We extend earlier results by developing approximations for the distributions of covariates in matched samples created with linear propensity score methods for the practically important situation where matching uses both the estimated linear propensity scores and a set of special prognostic covariates. Such matching on a subset of special prognostic covariates is an observational study analog of blocking in a randomized experiment. An example combining propensity score matching with Mahalanobis metric matching and regression adjustment is presented that demonstrates the flexibility of these methods for designing an observational study that effectively reduces both bias due to many observed covariates and bias and variability due to a more limited subset of covariates. Of particular importance, the general approach, which includes propensity score matching, was distinctly superior to methods that focus only on a subset of the prognostically most important covariates, even if those covariates account for most of the variation in the outcome variables. Also of importance, analyses based on matched samples were superior to those based on the full unmatched samples, even when regression adjustment was included.

1. INTRODUCTION

1.1. Combining Propensity Score Matching and Prognostic Covariate Adjustment

Matched sampling is a methodology for reducing bias due to observed covariates in comparative observational studies. In a typical matching setting, covariate data are available for a large sample of potential control subjects but only a relatively small treated sample, and the goal of matching is to select a subset of the control sample that has covariate values similar to those in the treated sample.

When there are several covariates, one approach is to select control subjects based on their Mahalanobis distance from the treated subjects (Carpenter 1977; Cochran and Rubin 1973; Rubin 1976b, 1980b), defined for covariate values $\mathbf{X}_1$ and $\mathbf{X}_2$ as $\{(\mathbf{X}_1 - \mathbf{X}_2)' \mathbf{S}_c^{-1}(\mathbf{X}_1 - \mathbf{X}_2)\}^{1/2}$, where $\mathbf{S}_c$ is the control sample covariance matrix. It is defined analogously for subsets of the covariates using corresponding submatrices of $\mathbf{S}_c$. Carpenter (1977) and Rubin (1980b) found that Mahalanobis matching with bivariate normal data is effective, especially when combined with regression adjustment on the matched pair differences (Rubin 1979b). As the number of covariates increases, however, the ability of this matching to find close matches on any specific covariate decreases, as does its ability to achieve similar treated and matched control means (Gu and Rosenbaum 1993; Rosenbaum and Rubin 1985a).

An alternative method, proposed by Rosenbaum and Rubin (1983a), which is closely related to discriminant matching (Rubin 1976b,c), is to match on the propensity score, $e = e(\mathbf{X})$, defined for each subject as the probability of receiving treatment given the covariate values, $\mathbf{X}$, and thus a scalar function of $\mathbf{X}$. Rosenbaum and Rubin showed that if exact matching on $e(\mathbf{X})$ can be achieved, then $\mathbf{X}$ will have the same distribution in the treated and matched control samples, thus eliminating bias in treatment effect estimates due to differences in $\mathbf{X}$. In typical practice, the propensity scores are not known, and even when they are, exact matches are not available. A common practical approach, which we study, is to estimate $e(\mathbf{X})$ using the sample linear discriminant or a linear logistic regression model of the treatment/control indicator variable on $\mathbf{X}$. The estimated propensity score, $\hat{e}$, is then used in place of the propensity score.

We consider both propensity score and Mahalanobis metric matching, and a combination of these methods using caliper matching (Rosenbaum and Rubin 1985a). For the combined method, all control subjects within intervals (calipers) surrounding each treated subject's $\hat{e}$ are identified as potential matches, and then Mahalanobis metric matching is applied to a subset of key covariates, $\mathbf{X}^{(s)}$, to make final selections from these potential matches, where $\mathbf{X}' \equiv (\mathbf{X}^{(s)\prime}, \mathbf{X}^{(r)\prime})$. Additional adjustments may also be performed using regression methods applied to the matched datasets.

Analytic and simulation results, which extend earlier results of Rubin and Thomas (1992a, 1992b, 1996), show that combining propensity score matching with matching on $\mathbf{X}^{(s)}$ can eliminate most of the bias with respect to a linear model relating outcome variables to $\mathbf{X}$ and can substantially reduce bias with respect to more general models with outcomes nonlinearly related to $\mathbf{X}^{(s)}$. Metric matching using $\mathbf{X}^{(s)}$, which may be functions of several original covariates, and model-based regression adjustments, should thus be viewed as complementary methods rather than competitive alternatives to propensity score matching and subclassification, as evaluated, for example, by Drake (1993).

The theory, example, and simulations that we present lead to two broader conclusions. First, the general approach to multivariate matching, which includes successful balancing of $\hat{e}$, is superior to methods that focus only on a subset of the prognostically most important covariates, even if those covariates account for most of the variation in the outcome variables. Second, analyses based on multivariate matched samples are superior to those based on the full unmatched samples, even if regression adjustment is used.

Table 18.1. *Matching variables in the hormone exposure data*

Label	Description
SEX	Male indicated by a 1, female by a 0
MOTAGE	mother's age
SES	Socioeconomic status (10 ordered categories)
ANTIH	Exposure to antihistamines
BARB	Exposure to barbiturates
CHEMO	Exposure to chemotherapy agents
CIGAR	Mother smoked cigarettes
MBIRTH	Multiple births
PSYDRUG	Exposure to psychotherapeutic drugs
RESPIR	Respiratory illness
SIB	First child indicated by a 1; other children, by a 0
CAGE	Calendar time of birth
LGEST	Length gestation (10 ordered categories)
PBC415	Pregnancy complication index
PBC420	Pregnancy complication index
MOTHT	Mother's height (4 ordered categories)
MOTWT	Mother's weight gain (6 ordered categories)
PSC	Estimated linear propensity score defined in Section 2.2

Note: Additional matching is performed on the variables in the top portion of the table. The variables in the middle portion of the table are dichotomous; those in the bottom portion are ordinal or continuous.

1.2. Example

We illustrate the effect of matching and regression adjustments on a dataset with 7,847 control children and 96 treated children whose mothers were exposed to a hormone during pregnancy (Reinisch, Sanders, Lykke-Mortensen, and Rubin 1995). The covariates, X, in Table 18.1 were collected from medical records during a previous study and are derived from the same database utilized by Rosenbaum and Rubin (1985b). The $X^{(s)}$ covariates chosen by behavioral researchers to receive special treatment in the matching are SEX of the child, age of the mother (MOTAGE), and an ordinal measure of socioeconomic status (SES).

Because collecting follow-up data is expensive and reducing the covariate differences between the treated and control samples is desirable, a subset of 96 control subjects was selected by matching for follow-up. The matching presented here illustrates the methods with real data but differs somewhat from that actually used. We also illustrate 1 (treated)-to-5 (control) matching, which can be practical in situations in which both the covariate and outcome data are available for the full treated and control samples for little or no additional cost. More complex methods that assign a variable number of control subjects to different treated subjects, or more than one treated subject to the same control subject, are alternative methods that can successfully extend matching techniques to difficult matching settings in which subsets of the treated subjects have very few potential matches (Rosenbaum 1989, 1991).

We present the results of four matching methods, each of which applies exact matching to the SEX covariate, in strict analogy with blocking in a randomized experiment, before additional matching criteria are considered:

1. Matching on the estimated linear propensity scores, $\hat{e}$.
2. Identifying all potential matches within relatively coarse $\hat{e}$ calipers, followed by Mahalanobis metric matching on two special matching covariates, MOTAGE and SES.
3. Mahalanobis metric matching on the two special covariates, $\mathbf{X}^{(s)}$, only.
4. Mahalanobis metric matching on all of the covariates, $\mathbf{X}$.

The matched datasets produced by these four methods are compared in Section 4.

The propensity scores were estimated using logistic regression with linear terms for each covariate listed in Table 18.1, and the matching was done on the logistic scale, so that matching on $\hat{e}$ means matching on the logit of $\hat{e}$.

All of the matching was performed using nearest-remaining-neighbor (i.e., nearest available) matching, beginning with the treated subject with the highest (and thus most difficult to match propensity score) and proceeding to the subject with the lowest propensity score, selecting m ($= 1$ or 5) matches for each treated subject. Rubin (1973a) found that this simple matching method produces near-optimal balancing of the sample means of a single matching covariate. The theoretical results in Section 3 predict that there is essentially no additional benefit for multivariate balancing from obtaining closer individual matching on $\hat{e}$ provided that close mean matching on the $\hat{e}$ is attained; this is consistent with results of Gu and Rosenbaum (1993) showing that optimal matching, although producing closer $\hat{e}$ than the nearest-neighbor algorithm, did not improve multivariate balance.

With the second method, calipers were formed around $\hat{e}$ for each treated subject by identifying all remaining unmatched control subjects within $\pm.2$ of the treated $\hat{e}$, with $\hat{e}$ standardized to have unit variance in the control sample; the choice of .2 is based on Cochran (1968a) and Cochran and Rubin (1973).

Figure 18.1 contains a histogram of the $\hat{e}$ in the full treated sample with a partial histogram of the $\hat{e}$ in the full control sample overlaid. The control sample histogram is truncated on the left because control subjects with $\hat{e}$ much lower than the lowest treated $\hat{e}$ are not relevant to the matching; the histogram is truncated at the top because there are so many more control subjects than treated subjects; the number of control subjects in each bin is indicated above it. The histogram bin width is .4, corresponding to the caliper matching width for matching method 2, although the bins used for caliper matching are centered around each treated value and are not fixed intervals. The large number of control subjects with $\hat{e}$ near most of the treated subjects demonstrates the considerable potential for matching on covariates other than $\hat{e}$.

Rosenbaum and Rubin (1985a) and Gu and Rosenbaum (1993) found good performance for coarse propensity score caliper matching followed by Mahalanobis matching on $\mathbf{X}$, but the latter found that the method did not perform as well as propensity score matching alone with high dimensional $\mathbf{X}$ and large initial biases, implying few potential matches for many of the treated subjects. Matching methods that treat the $\hat{e}$ as just one of many matching covariates may fail to eliminate bias in

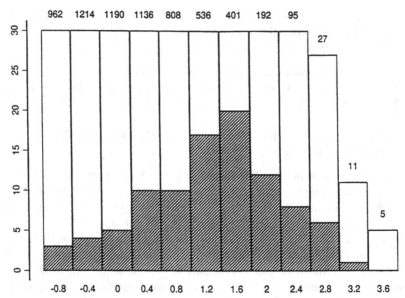

Figure 18.1. The logit of the estimated propensity scores in the treated sample are displayed in the shaded histogram. The logit of the estimated propensity scores in the control sample are displayed in the unshaded histogram. The control histogram is truncated on the left at -1.0, and above at 30. The number of control subjects in each bin is listed above the bin.

some covariates because close mean matching may not be attained on $\hat{e}$ (Rosenbaum and Rubin 1985a).

1.3. The Analogy Between Propensity Score Matching and Randomization, and Between Special Matching and Experimental Blocking

When close matching on $\hat{e}$ is attained, the treated and matched control samples are "balanced" in the sense that the differences between the treated and matched control means of the covariates included in $\hat{e}$ are even smaller than would be expected if the data had been generated by randomization (Rubin and Thomas, 1992b, 1996). Propensity score matching is also comparable to randomization in that it does not distinguish between covariates strongly related to the outcome variables and covariates poorly correlated with the outcome variables.

The results of Rubin and Thomas (1992b, 1996) and Gu and Rosenbaum (1993), combined with the results of Cochran (1968a) and Cochran and Rubin (1973), show that much of the balance resulting from matching on $\hat{e}$ alone can be achieved with relatively coarse matching. Consequently, when the control subjects outnumber the treated subjects, additional matching on a subset $\mathbf{X}^{(s)}$ will often be possible. Mahalanobis metric matching on $\mathbf{X}^{(s)}$ within $\hat{e}$ calipers can achieve the balancing properties of matching on $\hat{e}$, while improving the comparability of the distributions of $\mathbf{X}^{(s)}$. This method is roughly analogous to blocking in a randomized study to achieve improved balance on key covariates. The choice of covariates is a "design" issue determined by the knowledge of subject area researchers. A method directly analogous to blocking in a randomized study is to perform propensity score matching on $\mathbf{X}^{(r)}$ separately within blocks defined by $\mathbf{X}^{(s)}$; we did this with the covariate SEX. Within each block, standard results for propensity score matching apply, but the number of blocks can be excessive when some $\mathbf{X}^{(s)}$ have many distinct values.

Another analogy to randomized experiments is that once outcome variables are observed, model-based (regression) adjustments can be performed on the matched samples. Because of the reduced extrapolation involved with matched samples, such model-based adjustments also perform better than when applied to the unmatched samples, and the combination is effective at reducing bias due to $\mathbf{X}$, as demonstrated by our multivariate results in Section 5 and by the univariate and bivariate results of Rubin (1973b, 1979b).

The analogy between an observational study with matching on both $\hat{e}$ and a subset of prognostic covariates, and a randomized block experiment is limited. In particular, matching on $\hat{e}$ in an observational study adjusts only for linear trends due to *observed* $\mathbf{X}$, is not invariant to nonlinear transformations of the $\mathbf{X}$, and is successful only when there exists an adequately large control pool of subjects from which to select the matches. In contrast, randomization stochastically balances the distribution of *all* covariates.

1.4. Overview

Herein we extend the theoretical approximations in Rubin and Thomas (1992a,b) to produce distributional results describing the samples produced when matching on both $\hat{e}$ and a subset of covariates, $\mathbf{X}^{(s)}$. Section 2 presents terminology and notation. Section 3 contains analytic approximations describing the sampling properties of matched datasets formed by combining matching on $\hat{e}$ with close matching on $\mathbf{X}^{(s)}$, and concludes with a discussion of the practical implications. The datasets resulting from the different matching methods are compared in Section 4 and evaluated assuming various nonlinear models in Section 5. Settings in which 1–1 matched sampling is a financial necessity are considered, as well as settings in which costs are low enough that five control subjects can be selected for each treated subject.

2. NOTATION AND TERMINOLOGY

2.1. Covariate Distributions

Suppose that N_t treated and N_c control subjects are available and that there are p fully recorded covariates, $\mathbf{X}' = (X_1, \ldots, X_p)$, for each subject, where $R = N_c/N_t$. The covariates are arranged with the s special covariates subject to additional matching listed first and the r remaining covariates listed second, $\mathbf{X}' = (\mathbf{X}^{(s)\prime}, \mathbf{X}^{(r)\prime})$, where $s + r = p$. The $\mathbf{X}$ values among the treated and control subjects are assumed to be independent with finite moments and mean vectors and variance–covariance matrices denoted by

$$\mu_t = \begin{pmatrix} \mu_t^{(s)} \\ \mu_t^{(r)} \end{pmatrix}, \qquad \Psi_t = \begin{bmatrix} \Psi_{t(s,s)} & \Psi_{t(s,r)} \\ \Psi_{t(r,s)} & \Psi_{t(r,r)} \end{bmatrix}, \tag{18.1}$$

and

$$\mu_c = \begin{pmatrix} \mu_c^{(s)} \\ \mu_c^{(r)} \end{pmatrix}, \qquad \Psi_c = \begin{bmatrix} \Psi_{c(s,s)} & \Psi_{c(s,r)} \\ \Psi_{c(r,s)} & \Psi_{c(r,r)} \end{bmatrix}. \tag{18.2}$$

For the analytic approximations in Section 3, the distributions of $\mathbf{X}$ are modelled by a conditional normal distribution, similar to the general location model (Olkin and Tate 1961). Specifically, the treated and control distributions of $\mathbf{X}^{(r)}$ conditional on $\mathbf{X}^{(s)}$ are assumed to be multivariate normal with parallel linear regressions on $\mathbf{X}^{(s)}$, with common s by r matrix of coefficients denoted by β. The residuals from these regressions, $\mathbf{X}^{(r|s)} = \mathbf{X}^{(r)} - \beta'\mathbf{X}^{(s)}$, have means given by $\mu_t^{(r|s)} = \mu_t^{(r)} - \beta'\mu_t^{(s)}$ and $\mu_c^{(r|s)} = \mu_c^{(r)} - \beta'\mu_c^{(s)}$ conditional on $\mathbf{X}^{(s)}$ in the treated and control groups. The treated and control variance–covariance matrices of $\mathbf{X}^{(r|s)}$ (or $\mathbf{X}^{(r)}$) conditional on $\mathbf{X}^{(s)}$ are denoted by $\Psi_{t(r|s)}$ and $\Psi_{c(r|s)}$ and are assumed to be constant (i.e., not dependent on $\mathbf{X}^{(s)}$) and proportional, $\Psi_{t(r|s)} = \sigma^2 \Psi_{c(r|s)}$, for some positive scalar σ^2. The Mahalanobis distance in the control metric between the means of the $\mathbf{X}^{(r|s)}$ in the treated and control populations is given by

$$B_{r|s}^2 = \left(\mu_t^{(r|s)} - \mu_c^{(r|s)} \right)' \Psi_{c(r|s)}^{-1} \left(\mu_t^{(r|s)} - \mu_c^{(r|s)} \right).$$

This model has been used extensively in the discrimination literature (Daudin 1986; Krzanowski 1975, 1980). Rubin and Thomas (1996) showed that the analytic expressions obtained with stringent assumptions similar to these can still provide useful approximations for matching even when the normality and proportionality assumptions are clearly violated.

Although we estimate the propensity score using linear logistic regression, the distributional assumptions in this section need not give rise to a linear logit model for the population propensity score (e.g., if the variance matrices differ, then the logit model contains linear and quadratic terms). Models that include quadratic and interaction terms should also be explored. As all efforts to select a correct model may be only partially successful, it is important to understand performance when propensity scores are imperfectly modeled. Here we examine the common practice of matching on estimated linear propensity scores under non-ideal conditions.

2.2. Matching Methods Using Linear Propensity Score Estimators

Estimators of the linear coefficients motivated by normal-theory discriminant analysis are $\mathbf{S}_p^{-1}(\bar{\mathbf{X}}_t - \bar{\mathbf{X}}_c)$ and $\mathbf{S}_c^{-1}(\bar{\mathbf{X}}_t - \bar{\mathbf{X}}_c)$, where $\bar{\mathbf{X}}_t$, $\mathbf{S}_t$ and $\bar{\mathbf{X}}_c$, $\mathbf{S}_c$ are the means and variance–covariance matrices of $\mathbf{X}$ in the treated and control samples and $\mathbf{S}_p$ is the pooled sample variance–covariance matrix. The logistic regression estimator is also often used. Theoretical approximations are based on the "semi-estimated" discriminant coefficient, $\Psi_c^{-1}(\bar{\mathbf{X}}_t - \bar{\mathbf{X}}_c)$, which replaces population means with sample means. Simulation experience of Rubin and Thomas (1996), as well as asymptotic calculations, demonstrate that approximations for matching methods based on the semi-estimated coefficient are quite accurate for matching methods that use the sample linear discriminant or logistic regression estimates.

Denote the standardized population discriminant with respect to the control variance–covariance matrix by

$$Z = \{(\mu_t - \mu_c)' \Psi_c^{-1} (\mu_t - \mu_c)\}^{-1/2} \{\Psi_c^{-1}(\mu_t - \mu_c)\}' \mathbf{X}, \tag{18.3}$$

with $Z \equiv 0$ when $\mu_t = \mu_c$, and denote the corresponding standardized semi-estimated discriminant by

$$Z^* = \{(\bar{\mathbf{X}}_t - \bar{\mathbf{X}}_c)' \Psi_c^{-1} (\bar{\mathbf{X}}_t - \bar{\mathbf{X}}_c)\}^{-1/2} \{\Psi_c^{-1}(\bar{\mathbf{X}}_t - \bar{\mathbf{X}}_c)\}' \mathbf{X}. \tag{18.4}$$

Our analytic results apply to methods that match on $\mathbf{X}^{(s)}$ and $Z^*(\approx \hat{e})$, such as methods 1 and 2 of Section 1.2. Such matching methods are invariant to affine transformations of $\mathbf{X}$ that fix $\mathbf{X}^{(s)}$, which permits considerable analytic simplification when combined with conditionally ellipsoidal matching covariates (Rubin and Thomas 1992a).

2.3. Close Mean and Variance Matching on Special Covariates and the Estimated Propensity Score

For the analytic approximations that follow, we assume that the matching has produced approximate equality of the treated and matched control sample means and variances of $(\mathbf{X}^{(s)}, Z^*)$; although not all of our analytic approximations require close matching of these sample moments, we assume it throughout to simplify the presentation. After matching is completed, the sample moments of $(\mathbf{X}^{(s)}, Z^*)$ can be examined to determine if close matching has been achieved.

A useful guide for determining the feasibility of close matching of the sample moments of $(\mathbf{X}^{(s)}, Z^*)$, developed by Rubin (1976c) and Rubin and Thomas (1992b, 1996), can be obtained from the moments of the unmatched data assuming proportional multivariate normality of $\mathbf{X}$. Specifically, consider the maximum possible fractional bias reduction,

$$\theta_{\max}^* = \Omega(R/m)(pV + B^2)^{-1/2}, \tag{18.5}$$

where $V = (\sigma^2 + R^{-1})N_t^{-1}$, $B^2 = (\mu_t - \mu_c)' \Psi_c^{-1}(\mu_t - \mu_c)$, $\Omega(R/m)$ is the expectation of the R/m upper tail of a standard normal variate, approximated by

$$\Omega(R/m) \approx 2^{\pi/2}(2\pi)^{-1/2}(R/m)^{(1-\pi/4)}(1 - m/R)^{\pi/4},$$

and σ^2 in V is the ratio of the treated to control variance of the best linear discriminant (with respect to Ψ_c) and the constant in Section 2.1 when Ψ_t and Ψ_c are proportional (Rubin 1976c).

The condition $\theta_{\max}^* > 1$ ensures that matching methods that select exclusively from the upper tail of the distribution of Z^* in the control sample can eliminate the expected difference in the treated and matched control sample means of Z^*, which in turn implies that the expected differences in the sample means of $\mathbf{X}$ are also eliminated. Unless $\theta^*_{max} > 1$, close matching of the sample moments of $\mathbf{X}^{(s)}$ and Z^* is not possible. Consequently, if $\theta^*_{max} < 1$, it may not be worthwhile to expend resources collecting outcome data or continuing investigation in this setting, at least without limiting the goals of the study. Even when $\theta^*_{max} > 1$, however, the additional matching on $\mathbf{X}^{(s)}$ will not necessarily produce closely matched first or second sample moments of $\mathbf{X}^{(s)}$ or second moments of Z^*, and thus a value of θ^*_{max} larger than 1 is typically required to yield good matching with special covariates.

Figure 18.1 exemplifies the typical situation with most treated subjects having numerous potential $\hat{e}$ matches and the treated subjects with the largest $\hat{e}$ having little opportunity for additional matching. In the hormone matching example described in Section 4, where matching on $\hat{e}$ and additional covariates was successful, estimates of θ^*_{max} are 2.13 with 1–1 matching and 1.45 for 1–5 matching. Based on Figure 1 of Rubin and Thomas (1996), which displays bias reductions achieved for nearest-remaining-neighbor matching on $\hat{e}$ for a variety of different treated and control distributions, we anticipate that θ^*_{max} of 1.5 or larger indicates potential for successful matching on $\hat{e}$ along with additional matching on $\mathbf{X}^{(s)}$.

Note that the moments required to evaluate θ^*_{max} can sometimes be estimated using the unmatched sample moments even before any individual subject's outcome data are collected. As illustrated in Section 4, this is also true for the approximations developed in Section 3, which can provide useful information when designing an observational study, especially when deciding whether to go forward and collect outcome data.

3. ANALYTIC APPROXIMATIONS

3.1. Decomposition for a Generic Outcome Variable

Consider a linear combination of $\mathbf{X}$, $Y = \gamma' \mathbf{X}$, standardized to have unit variance in the control distribution; Y can be thought of as the conditional expectation of a generic outcome variable given X. Let $\mathcal{Z}$ be the standardized regression of Y on $(\mathbf{X}^{(s)}, Z)$ in the control distribution and let $\mathcal{W}$ be the standardized residual of this regression, where $\mathcal{W}$ is a linear combination of X orthogonal to $(\mathbf{X}^{(s)}, Z)$. The variable Y can be represented as

$$Y = \rho \mathcal{Z} + (1 - \rho^2)^{1/2} \mathcal{W}, \tag{18.6}$$

where ρ is the multiple correlation coefficient of Y on $(\mathbf{X}^{(s)}, Z)$; $\mathcal{Z} \equiv 0$ when $\rho = 0$, and $\mathcal{W} \equiv 0$ when $\rho = 1$. The means of $\mathcal{Z}$ and $\mathcal{W}$ in the treated and control samples are denoted by $\bar{\mathcal{Z}}_t$, $\bar{\mathcal{W}}_t$, and $\bar{\mathcal{Z}}_c$, $\bar{\mathcal{W}}_c$, and the corresponding means for the $m N_t$ matched controls are denoted by $\bar{\mathcal{Z}}_{mc}$, $\bar{\mathcal{W}}_{mc}$. The sample variances of a variable are denoted by $v_t(\cdot)$, $v_c(\cdot)$, and $v_{mc}(\cdot)$, in the treated, control, and matched control samples.

3.2. The Moments of X in Matched Samples with Close Matching

Assuming the distributions and matching methods described in Sections 2.1 and 2.2, and close mean and variance matching on $\mathbf{X}^{(s)}$ and $\hat{e}$ as described in

Section 2.3, we use the decomposition in (18.6) to provide analytic approximations for the first and second sample moments of Y in terms of the corresponding moments of $\mathcal{Z}$ and $\mathcal{W}$. Without close matching, the relationship between the moments of $\mathcal{Z}$ in the treated and control samples are not easily characterized because they depend on the distributions of $\mathbf{X}^{(s)}$, which can differ arbitrarily in the treated and control samples under the assumptions in Section 2.1. The approximations for the effect of the matching on the first two sample moments of each linear combination of $\mathbf{X}$ yield a complete characterization of all of its matched sample means, variances, and covariances; derivations are in the Appendix.

Results for the Expectation of the Sample Means.

$$\frac{E(\bar{Y}_t - \bar{Y}_{mc})}{E(\bar{Y}_t - \bar{Y}_c)} = \frac{E(\bar{\mathcal{Z}}_t - \bar{\mathcal{Z}}_{mc})}{E(\bar{\mathcal{Z}}_t - \bar{\mathcal{Z}}_c)} \approx 0; \tag{18.7}$$

also, $E(\bar{Y}_t - \bar{Y}_{mc}) \approx 0$ when $E(\bar{\mathcal{Z}}_t - \bar{\mathcal{Z}}_c) = 0$; that is, when there is no initial bias, the matching does not create bias.

Results for the Variance of the Difference in the Sample Means.

$$\text{var}(\bar{Y}_t - \bar{Y}_{mc}) = \rho^2 \, \text{var}(\bar{\mathcal{Z}}_t - \bar{\mathcal{Z}}_{mc}) + (1 - \rho^2)\text{var}(\bar{\mathcal{W}}_t - \bar{\mathcal{W}}_{mc}); \tag{18.8}$$

also, when $\rho \neq 0$,

$$\text{var}(\bar{\mathcal{Z}}_t - \bar{\mathcal{Z}}_{mc}) \approx \frac{r-1}{r + B_{r|s}^2/V}(m^{-1} - R^{-1})N_t^{-1}; \tag{18.9}$$

and when $\rho \neq 1$,

$$\text{var}(\bar{\mathcal{W}}_t - \bar{\mathcal{W}}_{mc}) \approx \left(1 - \frac{1}{r + B_{r|s}^2/V}\right)(m^{-1} - R^{-1})N_t^{-1}. \tag{18.10}$$

The quantity $\text{var}(\bar{\mathcal{W}}_t - \bar{\mathcal{W}}_{mc})$ is the same for each choice of the linear combination Y; that is, for all γ.

Results for the Expectation of Sample Variances.

$$E\{v_{mc}(Y)\} = \rho^2 E\{v_{mc}(\mathcal{Z})\} + (1 - \rho^2)E\{v_{mc}(\mathcal{W})\}; \tag{18.11}$$

also, when $\rho \neq 0$,

$$\left|\frac{E\{v_{mc}(\mathcal{Z})\}}{E\{v_t(\mathcal{Z})\}} - 1\right| \approx \left|\frac{(1 - \sigma^2)(r-1)}{r + B_{r|s}^2/V}\right| / E\{v_t(\mathcal{Z})\}; \tag{18.12}$$

and when $\rho \neq 1$,

$$E\{v_{mc}(\mathcal{W})\} \approx 1 + \frac{\sigma^2 - 1}{r + B_{r|s}^2/V}. \tag{18.13}$$

The quantity $E\{v_{mc}(\mathcal{W})\}$ is the same for each choice of the linear combination Y.

3.3. Comments on the Results

Remark A. Result (18.7) shows that (a) the bias in the matched sample means of any linear combination of $\mathbf{X}$ is determined by the bias in the matched sample means of $\mathcal{Z}$, a linear combination of $\mathbf{X}^{(s)}$ and Z, and (b) this bias is eliminated when close matching is achieved. Note that (18.7) implies that the expected difference in the sample means of any linear combination of $\mathbf{X}$ uncorrelated with $(\mathbf{X}^{(s)}, Z)$ is 0.

Remark B. Provided that close matching on $\hat{e}$ is possible, (18.7)–(18.10) show that methods based on propensity score matching are very effective in reducing the variability of differences in the sample means of many variables. Typical applications have numerous covariates, so ρ tends to be modest, implying that the variance of the difference in the sample means of most Y variables is heavily weighted by the contribution of the $\mathcal{W}$ component in (18.8). Moreover, the variance in (18.9) is less than or equal to the variance in (18.10). With numerous covariates and moderate or large sample sizes, $B_{r|s}^2 / V$ tends to be large, so (18.9) is near 0 and the first factor in (18.10) approaches 1. Thus $\text{var}(\bar{Y}_t - \bar{Y}_{mc})$ is dominated by (18.10) for most linear combinations of $\mathbf{X}$. The term $(m^{-1} - R^{-1})$ in (18.9) and (18.10) is a finite sampling correction arising from the selection of the matched controls from the full control sample. When $\sigma^2 \approx 1$ and the control sample is relatively large so that R^{-1} is small, (18.10) reduces to $1/(m N_t)$, which is no more than one-half of the variance that results from a corresponding randomized design (see also Hill, Rubin, and Thomas, 1999).

Remark C. As with the variance of the difference in the sample means, the expected value of the matched control sample variance of most linear combinations of $\mathbf{X}$ is determined largely by $E\{v_{mc}(\mathcal{W})\}$. The approximation in (18.12) shows that the matched control sample variance of the component of each linear combination of $\mathbf{X}$ determined by $\mathbf{X}^{(s)}$ and Z is similar to the corresponding treated sample variance, because the numerator in the righthand side of (18.12) is typically close to 0 and $E\{v_t(\mathcal{Z})\}$ is close to 1. This is not surprising in view of the close matching assumed for $\mathbf{X}^{(s)}$ and Z^* (or $\hat{e}$). But the approximation in (18.13) demonstrates that the matching does little to reduce differences in sample variances for the $\mathcal{W}$ component; the expected value of $v_c(\mathcal{W})$ is exactly 1 by construction. Any differences between the matched control and the treated sample variances are thus reduced by only a small amount, so the matching is not expected to reduce the bias (conditional or unconditional) due to nonlinear relationships in $\mathbf{X}^{(r)}$ for most covariates, other than due to $\mathbf{X}^{(s)}$.

Remark D. The multivariate normal assumption for the $\mathbf{X}^{(r)}$ distributions ensures the approximately constant expectation of each covariate uncorrelated with the Z^* and $\mathbf{X}^{(s)}$, conditional on the sample means. Multivariate normality also provides simplification by ensuring constant conditional variances for these covariates. The sensitivity of conclusions based on these approximations to violations of normality increases as the matching on $\hat{e}$ becomes increasingly different from random sampling of the full control sample, that is, as the treated and control distributions become more different. The robustness of the approximations based on the proportional normal model was examined by Rubin and Thomas (1996) in the case without special covariates, where the general conclusion is that the approximations work quite well.

Remark E. The approximation for the moments of $\mathcal{Z}$ in (18.9) and (18.12) are upper bounds provided that close matching is attained. The bounds are increasingly conservative as Y becomes a function of $\mathbf{X}^{(s)}$ alone; the conservative nature of the approximations is explained in the Appendix, where the results are derived.

Table 18.2. *Standardized bias in the treated and matched control sample means*

Variable	Method 1		Method 2		Method 3		Method 4		Unmatched
	1–1	1–5	1–1	1–5	1–1	1–5	1–1	1–5	
SEX	.00	.00	.00	.00	.00	.00	.00	.00	−.02
MOTAGE	.07	−.11	−.02	−.08	.00	.00	.13	.20	.92
SES	.02	−.03	−.01	−.02	.00	.00	.04	.19	.74
ANTIH	.03	.00	.03	.01	.28	.24	.02	.03	.18
BARB	.02	.02	.02	.00	.17	.12	.01	.01	.11
CHEMO	.00	.02	−.02	.00	.11	.09	.00	.01	.04
CIGAR	−.05	.02	−.06	−.02	−.28	−.19	.03	−.01	−.17
MBIRTH	.03	−.01	−.01	−.01	.01	.01	.00	.00	.00
PSYDRUG	.04	.02	.05	.00	.19	.15	.00	.00	.12
RESPIR	−.01	.01	−.03	.00	.04	.02	.00	.01	.02
SIB	.04	.01	.01	.04	.42	.25	.01	.03	.01
CAGE	.01	−.08	−.19	−.06	−.10	−.05	.06	.00	−.05
LGEST	.02	.01	.11	.06	−.02	−.03	−.10	−.04	−.03
PBC415	.02	−.03	.13	.10	.70	.54	.12	.10	.23
PBC420	−.09	.10	.06	−.01	2.71	1.62	.19	.25	.94
MOTHT	−.15	−.02	−.10	.00	.49	.46	−.03	.09	.34
MOTWT	−.01	.02	.00	−.04	−.07	.04	−.15	−.08	.01
PSC	.00	.01	.10	.06	2.31	1.49	.21	.39	1.44

Remark F. The $\mathbf{X}^{(s)}$ should be included when computing $\hat{e}$, as assumed by the theoretical results in Section 3. If they are not included, then there can be differences in the expected values of the treated and matched control sample means of some covariates even if close matching is attained for $\mathbf{X}^{(s)}$ and $\hat{e}$.

4. EXAMPLE

The mothers of children exposed to the hormone are substantially different than those of the unexposed children with respect to the covariates in Table 18.1. The Mahalanobis distance between the sample means measured with respect to S_c is 1.44 standard deviations, which is primarily due to three covariates, as summarized in the rightmost column of Table 18.2. The correlations between covariates are small to moderate (< 0.4). There are also differences in the variances of some of the covariates, as displayed in the rightmost column of Table 18.3. Because of the very large matching ratio, $R = 82$, the approximate value of $\theta^*_{\max}$ is 2.13 when $m = 1$, and 1.45 when $m = 5$, so the control pool is adequate to produce a matched control sample with means similar to the treated sample despite the large initial differences.

Table 18.2 summarizes the differences in the treated, control, and matched control sample means. The differences in the means of the continuous covariates are standardized by their treated sample standard deviations; the differences in the binary covariates are proportions. The differences in the $\hat{e}$ are reported at the bottom of the table. All of the matching methods except method 3, which uses only $\mathbf{X}^{(s)}$, produce large reductions in the differences in the sample means, which determines the

Table 18.3. *Difference in treated and matched control sample variances:* $(V_t - V_{mc})/V_t$

| Variable | Method 1 | | Method 2 | | Method 3 | | Method 4 | | |
	1–1	1–5	1–1	1–5	1–1	1–5	1–1	1–5	Unmatched
MOTAGE	−.80	−.77	−.02	−.09	.00	.01	−.25	−.35	−.83
SES	−.06	.07	.03	.06	.00	.01	−.01	.13	.14
CAGE	−.01	.12	.09	.15	.23	.17	.28	.29	.17
LGEST	.13	.12	−.04	−.05	−.20	−.05	.49	.17	−.20
PBC415	.23	.29	.28	.21	.15	−.01	.23	.20	.19
PBC420	.15	−.15	−.42	−.09	−.38	−.88	−.08	−.07	−.71
MOTHT	−.01	−.20	−.26	−.24	−.62	−.71	.10	.03	−.54
MOTWT	−.10	.14	.12	.18	.56	.34	.31	.39	.23
PSC	−.01	.03	−.03	.01	−.09	−.52	.02	.06	−.23

bias due to linear trends between outcome variables and the covariates. Method 3 produces almost exact matches for the $\mathbf{X}^{(s)}$ it uses for Mahalanobis metric matching (the lowest treated–matched control correlation for a special variable was .98), but results in large differences in several other covariates potentially related to the outcome variables.

From Table 18.2, matching methods 1 and 2, which use $\hat{e}$, successfully reduce the differences in $\hat{e}$, although the remaining standardized bias of .10 for the 1–1 matched control sample produced by method 2, which includes matching on $\mathbf{X}^{(s)}$, may be larger than desired. The result in (18.7) predicts that in expectation, the difference in the sample means of each covariate is essentially eliminated by these two propensity score methods. The differences in the means are generally consistent with our predictions for the performance of the propensity score methods based on the approximations in (18.8)–(18.10). If a much larger difference occurs than is predicted by its standard error, then that covariate should be examined, particularly for the possibility of adding interaction and quadratic terms in the linear propensity score model. The values of the independent-sample Z statistics vary considerably less than a standard normal deviate (not displayed), consistent with the theoretical prediction that their variances in the matched samples are less than one half that in a corresponding randomized design. (Hill et al. 1999 has an application in this setting.)

From Table 18.2, the Mahalanobis matching on all of the covariates, method 4, did a reasonable job eliminating differences in individual sample means. But the remaining difference in the means of $\hat{e}$ is large enough to be of concern and is reflected in the unacceptably large remaining mean differences in $\mathbf{X}^{(s)}$, especially for the 1–5 matching. The other notable feature of method 4 is its tendency to place very heavy weight on binary covariates representing rare events; this failing of Mahalanobis metric matching has also been noted by Rosenbaum and Rubin (1985a) and Gu and Rosenbaum (1993).

Table 18.3 summarizes the differences in the sample variances of the continuous and ordinal covariates, which are important because, along with the means, they determine the bias due to quadratic trends between the outcome variables and the covariates. The MOTAGE covariate has a much larger variance in the control sample, as does the pregnancy complication covariate, PBC420. The results in (18.11)–(18.13) predict that matching on $\hat{e}$ will produce only modest improvements in the sample variances. The results in Table 18.3 are typically consistent with this prediction; in particular, matching only on $\hat{e}$, method 1, did not reduce the differences in the variance of MOTAGE, although both methods 1 and 2 reduced those differences in PBC420, MOTHT, and MOTWT somewhat more than predicted. Method 2, matching on $\hat{e}$ and $\mathbf{X}^{(s)}$, successfully eliminated the differences in the variances of $\mathbf{X}^{(s)}$ as well as the differences in their covariances (not displayed).

Mahalanobis matching on all of the covariates, method 4, reduced the largest differences in the variances, but the differences for $\mathbf{X}^{(s)}$ are still unacceptably large. As with the sample means, by attempting to closely match all of the covariates, Mahalanobis matching produces some improvement in most covariates, but substantial differences remain, sometimes in the most prognostically important covariates.

5. BIAS REDUCTION DUE TO THE MATCHED SAMPLING – SIMULATED DATA

We evaluate the bias and the mean squared error (MSE) of some common estimators based on the different matched samples in Section 4 using data simulated from several linear and nonlinear models relating outcomes to the covariates $\mathbf{X}$. Sections 5.1 and 5.2 present results from evaluations that sample a large number of potential simulated models and use MSE and bias to summarize the bias reduction for the large class of models. Section 5.3 focuses on bias reduction in the most important prognostic covariates.

5.1. Simulated Outcomes Nonlinear Function of the Covariates

Following Rubin (1973b, 1979b), we use the exponential function to create models with varying amounts of curvature and nonadditivity. For a generic vector $\mathbf{X}$ of dimension J, we form a multiplicative model, $c_1(\lambda)e^{\lambda'\mathbf{X}}$, and an additive model, $c_2(\kappa)\sum_{j=1}^{J}\lambda_j e^{\pm\kappa X_j}$, where $c_1(\lambda)$, $c_2(\kappa)$, and $\|\lambda\|$ are chosen to produce additive and multiplicative functions of the covariates with variances close to 1 in the treated sample. These functions are applied to both $\mathbf{X}^{(s)}$ and $\mathbf{X}^{(r)}$ and are denoted by $f_s(\mathbf{X}^{(s)})$ and $f_r(\mathbf{X}^{(r)})$. The functions are combined additively to form expected values given $\mathbf{X}$ of an outcome variable, $Y = w_s f_s(\mathbf{X}^{(s)}) + (1 - w_s)f_r(\mathbf{X}^{(r)})$, where w_s is a weight that determines the relative contribution of $\mathbf{X}^{(s)}$ to the outcome. An additive treatment effect is used for each model, and there is no bias associated with unmeasured covariates.

We report results with both f_s and f_r having the additive form and with both having the multiplicative form. The amount of nonlinearity in the multiplicative function is determined by the magnitude of λ. We select a magnitude $\|\lambda\|$ that results in a model with Y nearly linear in $\mathbf{X}$, and a $\|\lambda\|$ that produces a model with linear $r^2 \approx .85$ when Y is linearly regressed on $\mathbf{X}$; this amount of nonlinearity is difficult to detect in realistic multivariate settings. The nonlinearity in the additive function is determined by κ. We select a positive κ to produce a nearly linear function and a function with linear r^2 of .85. The sign of the exponent for each covariate is randomly selected with probability one-half of a negative multiplier. The two levels of nonlinearity are applied to f_s and f_r separately, producing 2^2 combinations, and these functions are combined using two choices of weights, w_s, selected so that the r^2 between Y and $f_s(\mathbf{X}^{(s)})$ is high ($> .90$) or moderate ($.4 < r^2 < .6$). Thus eight types of additive and eight types of multiplicative models are evaluated.

The λ are selected uniformly over parameter spaces of appropriate dimension (s or r) by drawing λ from the standard multivariate normal distribution and then norming λ to have the specified $\|\lambda\|$ to achieve the appropriate nonlinearity and variance. The λ for f_s are generated independently of the λ for f_r. Five hundred λ were generated for each type of model to obtain accurate estimates of the expected performance of estimators within the uniform parameter space.

Each function is used to produce Y values for each subject in the treated and control samples, which are the mean responses of the subjects conditional on λ and the covariate values. Residual variation is not actually generated and added to the Y values; rather, the effect of residual variance is included in the evaluation of the estimators by specifying the proportion of the variance in the outcome variable due to the variation in the regression functions (model r^2) and then computing the residual variance required to produce the specified model r^2. The residual variances change for different λ because the variance in the generated mean responses vary with λ. After the residual variance is determined, the outcome variable and bias computed from it are rescaled so that the unconditional variance of the response variable equals 1.0 among the treated subjects; that is, the residual variance plus the variance of Y among the treated subjects is rescaled to 1.0. For the 1–1 matching, we consider integrated squared bias only (the model r^2 is one), following Rubin (1979b), because the variance of the estimators that we study are approximately the same for each matched dataset. For the 1–5 matching, we report results for a model r^2 of 1.0 and for a lower model r^2 of .25.

5.2. Results of Simulation

The two estimators that we evaluate are the difference between sample means and the difference between the sample means adjusted by linear least squares regression estimated by pooling the assumed parallel linear regressions in the treated and control, or matched control, samples (i.e., the common "covariance-adjusted" estimate). The bias of the difference in the sample means conditional on the covariate values is $\bar{Y}_t - \bar{Y}_c$ for the control sample and $\bar{Y}_t - \bar{Y}_{mc}$ in each matched control sample, where $\bar{Y}_t$, $\bar{Y}_c$, and $\bar{Y}_{mc}$ are the average of the Y values generated for the treated, control, and matched control samples. Letting Γ denote the coefficients of the least squares regression of Y on the covariates in the treated and control samples, $\bar{Y}_t - \bar{Y}_c - \Gamma'(\bar{\mathbf{X}}_t - \bar{\mathbf{X}}_c)$ is the bias of the linear regression estimator applied to the unmatched samples, and similarly for each matched sample using the Y values of the treated and matched control subjects, $\bar{Y}_t - \bar{Y}_{mc} - \Gamma'_{mc}(\bar{\mathbf{X}}_t - \bar{\mathbf{X}}_{mc})$.

We summarize the performance of the estimators in Table 18.4, which gives the square root of the average squared bias with respect to the uniform distribution of the λ parameters for each type of model estimated using the response function computed with each of the 500 λ values. Because the results have very little variation across the levels of nonlinearity in $\mathbf{X}^{(r)}$, the reported results are averages across these

Table 18.4. *Average conditional bias*

	High-weight special variables			Low-weight special variables		
	Approximate linear	Nonlinear additive	Nonlinear multiplicative	Approximate linear	Nonlinear additive	Nonlinear multiplicative
1–1 matching						
Method 1	.04	.11	.09	.07	.09	.08
Method 2	.04	.05	.05	.08	.09	.09
Method 3	.23	.27	.29	.55	.53	.63
Method 4	.07	.05	.08	.08	.07	.08
Unmatched	.60	.55	.57	.47	.44	.44
Method 1 + reg	.02	.12	.09	.03	.08	.06
Method 2 + reg	.01	.03	.03	.03	.04	.04
Method 3 + reg	.03	.18	.10	.06	.13	.09
Method 4 + reg	.01	.04	.04	.03	.04	.03
Unmatched + reg	.02	.13	.09	.03	.09	.06
1–5 matching						
Method 1	.05	.15	.10	.05	.11	.08
Method 2	.04	.05	.05	.05	.05	.05
Method 3	.15	.19	.18	.35	.35	.38
Method 4	.14	.15	.13	.11	.13	.10
Unmatched	.58	.52	.55	.45	.43	.39
Method 1 + reg	.01	.12	.08	.02	.09	.05
Method 2 + reg	.01	.04	.03	.02	.04	.03
Method 3 + reg	.02	.10	.06	.04	.08	.05
Method 4 + reg	.01	.09	.05	.03	.07	.04
Unmatched + reg	.02	.13	.09	.03	.09	.06

Note: Square root of the average conditional squared bias. The treated population standard deviation is 1, because the matching leaves the treated group fixed. Simulation error is beyond the reported decimal value.

settings of the simulation study. The results from the approximately linear model generated using the additive and multiplicative forms of the response function are similar, so these too are averaged with one column of results for these settings. The average squared bias for the linear regression–based estimators are very small with the data generated from the approximately linear models, as expected, but the squared biases are greater than 0 primarily because the results are averaged across nonlinear functions of $\mathbf{X}^{(r)}$.

The results in Table 18.4 demonstrate that the estimators based on the matched samples, with the exception of method 3, are more accurate than corresponding estimates based on the unmatched samples, and they are more robust to monotone nonlinearity in the regression functions. Linear regression adjustment consistently improved the performance of simple differences in sample means, although the performance of the simpler estimates is very good, particularly when based on the sample formed by the second matching method. The superior performance of regression adjustments applied to matched samples is consistent with previous theoretical and simulation studies (Rubin 1973b, 1979b). We expect the improvements resulting from linear regression adjustments to be much smaller for nonmonotone regression functions. As anticipated, matching and regression adjustment provide less improvement when the covariates are weakly predictive of the outcome variable. Results with $r^2 = .25$ based on the MSE (not displayed), which includes the variance of the estimators and the squared conditional bias, show that the differences in means based on the matched samples are still much more accurate than those based on the unmatched samples, and that the better matching methods are never substantially worse than linear regression based on the unmatched control sample.

There are some important and consistent differences across the various matched control samples, although the differences are small when contrasted with the unmatched control sample. Of particular importance, the matching using only $\mathbf{X}^{(s)}$ is inferior in this simulation, even when $\mathbf{X}^{(s)}$ accounts for 90–95% of the variation in the outcome. These results should discourage the approach of ignoring potentially relevant covariates. Matching on $\hat{e}$, method 1, is very effective when the response function is linear in $\mathbf{X}^{(s)}$, but it does not perform as well when the response function is nonlinear, consistent with the theoretical predictions of Section 3. Matching on $\mathbf{X}^{(s)}$ within $\hat{e}$ calipers, method 2, performed well in all settings, especially those with dominant $\mathbf{X}^{(s)}$ in a nonlinear regression. Mahalanobis matching, method 4, also performed well, although the 1–1 matched sample formed with it is noticeably better than the corresponding 1–5 matched sample. Mahalanobis matching can result in substantial bias when it fails to eliminate differences in the means of $\hat{e}$, as happened with 1–5 matching.

5.3. The Effect of Matching on Different Covariate Components

The simulation models described in Section 5.1 were also used to assess the reduction in bias due to specific components of the covariate space. The covariates evaluated are the ones most likely to be prognostic of outcome. Results are presented for matching method 2, Mahalanobis metric matching on special covariates within propensity score calipers.

Table 18.5. *Conditional bias for specific variables: method 2*
with 1–1 matching

	Approximately linear				Non-linear			
	Matched		Unmatched		Matched		Unmatched	
	Unadj	Adj	Unadj	Adj	Unadj	Adj	Unadj	Adj
PSC	.10	.00	1.38	.04	.10	.01	.89	.25
MOTAGE	.02	.00	.88	.01	.03	.01	.47	.16
SES	.01	.00	.73	.01	.00	.01	.61	.09
PBC415	.15	.02	.25	.02	.24	.14	.29	.15
PBC420	.04	.01	.89	.02	.00	.05	.53	.12

Note: Square root of the conditional squared bias. "Adj" results are for the estimator which includes covariate regression adjustment. "Unadj" results are without the adjustment.

Because the propensity score is the combination of covariates that the subject-matter experts (e.g., clinicians, educators, program administrators) and subjects judged important when selecting appropriate treatments, it will often be prognostically important. The first row of Table 18.5 contains the bias due to approximately linear and exponential trends applied to the propensity scores. (The additive and multiplicative model coincide with a univariate input variable.) The matching, especially when combined with linear regression, effectively eliminates the very large initial bias due to linear and nonlinear models. Similar results were obtained for the continuous and ordinal special matching covariates, mother's age and SES. The linear covariance adjustment substantially reduces the bias due to monotone nonlinear functions, but these linear adjustments perform much better when applied to the matched dataset. The unadjusted matched differences compare favorably to the covariance adjusted unmatched results.

Results for PBC420 were similar to those for $\hat{e}$; PBC420 had a large initial bias and was thus correlated (.65) with $\hat{e}$. In contrast, PBC415 had a moderate initial bias and is not highly correlated (.18) with the propensity score or $X^{(s)}$. The reduction in its initial difference between the treated and unmatched means is consistent with theory, but as predicted, the bias due to nonlinear trend was not eliminated by the matching and linear covariance adjustment. Although our results are specific to the models and the dataset in this section, we anticipate that similar results will attain more broadly.

6. SUMMARY

Our example demonstrates the importance of multivariate matching in observational studies, even when many of the covariates have relatively limited prognostic value. Estimators based on multivariate matched samples consistently perform better than estimators based on the unmatched samples. Multivariate matching methods should be designed to eliminate the difference in the sample means of the estimated propensity score, because there will be remaining potential bias due to some covariates if a method fails to eliminate such differences. Matching within propensity score calipers followed by additional matching on key prognostic covariates is an effective method for ensuring that differences in the propensity scores are eliminated while allowing researchers to use information about the relative prognostic value of different covariates. Combined with Mahalanobis matching on special covariates and regression adjustments, propensity score methods can effectively reduce bias due to a large number of measured covariates in an observational study. With prognostically important discrete covariates with a small number of distinct values, the methodology can be applied separately within blocks defined by these covariates.

APPENDIX: OVERVIEW OF THE DERIVATIONS

A.1. Reduction to a Canonical Form

Following Rubin and Thomas (1992a) we consider a matching method that is conditionally affinely invariant given $\mathbf{X}^{(s)}$ and Z^* in the sense that it is applied such that after any affine transformation that leaves $\mathbf{X}^{(s)}$ fixed, it produces the same matches as when matching on the untransformed $\mathbf{X}$. Matching methods that use only $\mathbf{X}^{(s)}$ and Z^* are conditionally affinely invariant. By regressing $\mathbf{X}^{(r)}$ on $\mathbf{X}^{(s)}$ and applying an appropriately rotated form of $\Psi_{c(r|s)}^{-1/2}$ to the residuals, we can assume without loss of generality that the distribution of $\mathbf{X}$ has moments of the form $\mu_c^{(r)} = 0$, $\mu_t^{(r)} = B_{r|s} r^{-1/2} 1_r$, $B_{r|s} \geq 0$, $\Psi_{c(r,r)} = \mathbf{I}_r$, $\Psi_{c(r,s)} = 0$, $\Psi_{t(r,s)} = 0$ and $\Psi_{t(r,r)} = \sigma^2 \mathbf{I}_r$, where 1_r and $\mathbf{I}_r$ are the r-dimensional equal angular vector and identity matrix. The $\mathbf{X}^{(s)}$ are unchanged with moments $\mu_t^{(s)}$, $\mu_c^{(s)}$, $\Psi_{t(s,s)}$, and $\Psi_{c(s,s)}$, which have no special conditions placed on them. In the canonical form, $\mathbf{X}^{(r)}$ are multivariate normal and independent of $\mathbf{X}^{(s)}$ in the treated and control distributions, and the standardized discriminant Z with respect to the control distribution is proportional to $(\mu_t^{(s)} - \mu_c^{(s)})' \Psi_{c(s,s)}^{-1} \mathbf{X}^{(s)} + B_{r|s} r^{-1/2} 1_r' \mathbf{X}^{(r)}$. Similar simplifying transformations have been used by Efron (1975), Rubin (1976b,c), Ashikaga and Chang (1981), and others.

We consider a linear combination, $U = \alpha' \mathbf{X} = \alpha^{(s)'} \mathbf{X}^{(s)} + \alpha^{(r)'} \mathbf{X}^{(r)}$, which is scaled to have variance equal to 1 in the control distribution. We specialize U to have one of two forms. The first corresponds to a standardized regression of some arbitrary linear combination of $\mathbf{X}$ onto $\mathbf{X}^{(s)}$ and Z, which we denoted by $\mathcal{Z}$ in Section 3.1, so $\alpha = (\alpha^{(s)'}, \alpha 1_r')'$ such that $\alpha^{(s)'} \Psi_{c(s,s)} \alpha^{(s)} + r\alpha^2 = 1$, with $\mathbf{X}$ in canonical form. The second form of U corresponds to an $\mathcal{W}$, also defined in Section 3.1, which is a standardized variable uncorrelated with $(\mathbf{X}^{(s)}, Z)$; thus $\alpha = (0', \alpha^{(r)'})'$ such that $\alpha^{(r)'} \alpha^{(r)} = 1$, $\alpha^{(r)'} 1_r = 0$.

A.2. Decomposition Based on the Semi-Estimated Discriminant

A decomposition analogous to (18.6) is now formed for U based on $\mathbf{X}^{(s)}$ and Z^*. With the covariates in canonical form,

$$Z^* \alpha (\bar{\mathbf{X}}_t^{(s)} - \bar{\mathbf{X}}_c^{(s)})' \Psi_{c(s,s)}^{-1} \mathbf{X}^{(s)} + (\bar{\mathbf{X}}_t^{(r)} - \bar{\mathbf{X}}_c^{(r)})' X^{(r)}. \tag{A.1}$$

The decompositions are computed with respect to the control distribution conditional on $(\bar{\mathbf{X}}_t^{(r)}, \bar{\mathbf{X}}_c^{(r)})$; the distribution of the $\mathbf{X}^{(r)}$ in the control sample conditional on $(\bar{\mathbf{X}}_t^{(r)}, \bar{\mathbf{X}}_c^{(r)})$ is multivariate normal with mean $\bar{\mathbf{X}}_c^{(r)}$ and variance–covariance matrix $(1 - N_c^{-1}) \mathbf{I}_r$. All of the correlations, moments, and regressions are computed with respect to the control distribution conditional on $(\bar{\mathbf{X}}_t^{(r)}, \bar{\mathbf{X}}_c^{(r)})$. We use the accurate approximation $\text{var}(U) = \boldsymbol{\alpha}^{(s)\prime}\boldsymbol{\Psi}_{c(s,s)}\boldsymbol{\alpha}^{(s)} + \boldsymbol{\alpha}^{(r)\prime}\boldsymbol{\alpha}^{(r)}(1 - N_c^{-1}) = 1$ throughout.

A conditional orthogonal decomposition for U is

$$U = \rho_* Z^* + (1 - \rho_*^2)^{1/2} W^*. \tag{A.2}$$

The component Z^* is the standardized linear regression of U on $\mathbf{X}^{(s)}$ and Z^* in the control distribution,

$$Z^* = c(\boldsymbol{\alpha}^{(s)}, \boldsymbol{\alpha}^{(r)}) \times \left\{ \boldsymbol{\alpha}^{(s)\prime}\mathbf{X}^{(s)} + \frac{\boldsymbol{\alpha}^{(r)\prime}(\bar{\mathbf{X}}_t^{(r)} - \bar{\mathbf{X}}_c^{(r)})}{\|(\bar{\mathbf{X}}_t^{(r)} - \bar{\mathbf{X}}_c^{(r)})\|^2} \left(\bar{\mathbf{X}}_t^{(r)} - \bar{\mathbf{X}}_c^{(r)}\right)' \bar{\mathbf{X}}^{(r)} \right\}, \tag{A.3}$$

where

$$c(\boldsymbol{\alpha}^{(s)}, \boldsymbol{\alpha}^{(r)}) = \left[\boldsymbol{\alpha}^{(s)\prime}\boldsymbol{\Psi}_{c(s,s)}\boldsymbol{\alpha}^{(s)} + \left(1 - N_c^{-1}\right) \left\{ \frac{\boldsymbol{\alpha}^{(r)\prime}(\bar{\mathbf{X}}_t^{(r)} - \bar{\mathbf{X}}_c^{(r)})}{\|(\bar{\mathbf{X}}_t^{(r)} - \bar{\mathbf{X}}_c^{(r)})\|} \right\}^2 \right]^{-1/2}.$$

The component W^* is the standardized residual of the regression; direct calculations show that the correlation of U and Z^* is $\rho^* = c^{-1}(\boldsymbol{\alpha}^{(s)}, \boldsymbol{\alpha}^{(r)})$.

When the general formula in (A.2) is applied to some choice of Z, the correlation, ρ_*, is denoted by a_*, and when (A.2) is applied to some choice of W, ρ_* is denoted by r_*. It is easy to check that $a_*^2 \to 1$ and $r_*^2 \to 0$ as $N_t, N_c \to \infty$ when $\mu_t^{(r)} \neq \mu_c^{(r)}$.

A.3. Distribution of W^*

The following theorem characterizes the distribution of the matched control sample moments of any variable W^* uncorrelated with Z^* and $\mathbf{X}^{(s)}$ in the control sample conditional on $(\bar{\mathbf{X}}_t^{(r)}, \bar{\mathbf{X}}_c^{(r)})$. The sample means and variances of Z^* and W^* are denoted analogous to those of Z and W in Section 3. The theorem is used extensively in the derivation of the results in Section 2. The proof of the theorem closely parallels the proof of theorem 3.1 of Rubin and Thomas (1992b), using the additional fact that $\mathbf{X}^{(s)}$ and $\mathbf{X}^{(r)}$ are independent in the canonical representation, and is not presented here.

Theorem A.1. For matching methods that depend only on $\mathbf{X}^{(s)}$ and Z^*, and covariate distributions satisfying the conditions of Section 2.1,

$$\mathrm{E}\left(\bar{W}_t^* - \bar{W}_{\mathrm{mc}}^* \mid \bar{\mathbf{X}}_t^{(r)}, \bar{\mathbf{X}}_c^{(r)}\right) = 0, \tag{A.4}$$

$$\text{var}\left(\bar{W}_t^* - \bar{W}_{\mathrm{mc}}^* \mid \bar{\mathbf{X}}_t^{(r)}, \bar{\mathbf{X}}_c^{(r)}\right) = n_{\mathrm{mc}}^{-1}\left(1 - \frac{n_{\mathrm{mc}}}{N_c}\right), \tag{A.5}$$

$$\mathrm{E}\left\{v_{\mathrm{mc}}(W^*) \mid \bar{\mathbf{X}}_t^{(r)}, \bar{\mathbf{X}}_c^{(r)}\right\} = 1, \tag{A.6}$$

and

$$\mathrm{E}\left\{v_t(W^*) \mid \bar{\mathbf{X}}_t^{(r)}, \bar{\mathbf{X}}_c^{(r)}\right\} = \sigma^2. \tag{A.7}$$

In addition, for any linear combination, $\mathcal{Z}^*$, of $\mathbf{X}^{(s)}$ and Z^*, and corresponding uncorrelated variable $\mathcal{W}^*$, conditional on $(\bar{\mathbf{X}}_t^{(r)}, \bar{\mathbf{X}}_c^{(r)})$, $(\bar{\mathcal{W}}_{mc}^*, v_{mc}(\mathcal{W}^*), \bar{\mathcal{W}}_t^*, v_t(\mathcal{W}^*))$ are independent of $(\bar{\mathcal{Z}}_{mc}^*, v_{mc}(\mathcal{Z}^*), \bar{\mathcal{Z}}_t^*, v_t(\mathcal{Z}^*))$.

The following corollary of Theorem A.1 is analogous to corollary 3.1 of Rubin and Thomas (1992b), except that the corollary applies to any standardized linear combination of $\mathbf{X}$.

Corollary A.1. *Consider a standardized linear combination, U, and its representation in (A.2). For matching methods that depend only on $\mathbf{X}^{(s)}$ and Z^*, and covariate distributions satisfying the conditions of Section 2.1,*

$$\mathrm{var}(\bar{U}_t - \bar{U}_{mc}) = \mathrm{var}\{\rho_*(\bar{\mathcal{Z}}_t^* - \bar{\mathcal{Z}}_{mc}^*)\} + \{1 - \mathrm{E}(\rho_*^2)\}n_{mc}^{-1}\left(1 - \frac{n_{mc}}{N_c}\right), \quad \text{(A.8)}$$

$$\mathrm{E}\{v_{mc}(U)\} = \mathrm{E}\{\rho_*^2 v_{mc}(\mathcal{Z}^*)\} + \{1 - \mathrm{E}(\rho_*^2)\}, \quad \text{(A.9)}$$

and

$$\mathrm{E}\{v_t(U)\} = \mathrm{E}\{\rho_*^2 v_t(\mathcal{Z}^*)\} + \sigma^2\{1 - \mathrm{E}(\rho_*^2)\}. \quad \text{(A.10)}$$

The equality in (A.8) is obtained by applying (A.2) to the sample means,

$$\bar{U}_t - \bar{U}_{mc} = \rho_*(\bar{\mathcal{Z}}_t^* - \bar{\mathcal{Z}}_{mc}^*) + (1 - \rho_*^2)^{1/2}(\bar{\mathcal{W}}_t^* - \bar{\mathcal{W}}_{mc}^*),$$

and computing $\mathrm{var}(\bar{U}_t - \bar{U}_{mc})$ by first conditioning on $(\bar{\mathbf{X}}_t^{(r)}, \bar{\mathbf{X}}_c^{(r)})$, then applying (A.4), (A.5), and the conditional independence of the moments of $\mathcal{Z}^*$ and $\mathcal{W}^*$,

$$\mathrm{E}\left(\bar{U}_t - \bar{U}_{mc} \mid \bar{\mathbf{X}}_t^{(r)}, \bar{\mathbf{X}}_c^{(r)}\right) = \mathrm{E}\left\{\rho_*(\bar{\mathcal{Z}}_t^* - \bar{\mathcal{Z}}_{mc}^*) \mid \bar{\mathbf{X}}_t^{(r)}, \bar{\mathbf{X}}_c^{(r)}\right\}$$

and

$$\mathrm{var}\left(\bar{U}_t - \bar{U}_{mc} \mid \bar{\mathbf{X}}_t^{(r)}, \bar{\mathbf{X}}_c^{(r)}\right) = \mathrm{var}\left\{\rho_*(\bar{\mathcal{Z}}_t^* - \bar{\mathcal{Z}}_{mc}^*) \mid \bar{\mathbf{X}}_t^{(r)}, \bar{\mathbf{X}}_c^{(r)}\right\}$$
$$+ (1 - \rho_*^2)n_{mc}^{-1}\left(1 - \frac{n_{mc}}{N_c}\right),$$

establishing (A.8). Similarly, (A.9) and (A.10) follow from (A.6), (A.7), and the independence conditions in Theorem A.1 applied to the representations

$$v_{mc}(U) = \rho_*^2 v_{mc}(\mathcal{Z}^*) + (1 - \rho_*^2)v_{mc}(\mathcal{W}^*) + 2\rho_*(1 - \rho_*)^{1/2}\mathrm{cov}_{mc}(\mathcal{Z}^*, \mathcal{W}^*)$$

and

$$v_t(U) = \rho_*^2 v_t(\mathcal{Z}^*) + (1 - \rho_*^2)v_t(\mathcal{W}^*) + 2\rho_*(1 - \rho_*)^{1/2}\mathrm{cov}_t(\mathcal{Z}^*, \mathcal{W}^*),$$

where cov_t and cov_{mc} are the treated and matched control sample covariances.

A.4. Derivation of the Results in Section A.3

Using the representation in (A.1) with $U = \mathcal{Z}$ and $\rho_* = a_*$,

$$\bar{\mathcal{Z}}_t - \bar{\mathcal{Z}}_{mc} = a_*(\bar{\mathcal{Z}}_t^* - \bar{\mathcal{Z}}_{mc}^*) + (1 - a_*^2)^{1/2}(\bar{\mathcal{W}}_t^* - \bar{\mathcal{W}}_{mc}^*). \quad \text{(A.11)}$$

So from (A.4),

$$E(\bar{Z}_t - \bar{Z}_{mc}) = E\{a_*(\bar{Z}_t^* - \bar{Z}_{mc}^*)\} \, ,$$

and the latter term is approximately 0 when close matching of the sample moments of $\mathbf{X}^{(s)}$ and Z^* is achieved, yielding (18.7).

Approximations for $E(a_*^2)$ and $E(r_*^2)$

The results in Section 3.2 involving variances require approximations for the expectations of the conditional correlations defined at the end of Section A.2. Specializing the formula for ρ_* to a_* and noting that $\alpha^2 = r^{-1}(1 - \alpha^{(s)'}\Psi_{c(s,s)}\alpha^{(s)})$, it follows that

$$E(a_*^2) \geq E\left[r^{-1}\left\{ \frac{1_r'(\bar{X}_t^{(r)} - \bar{X}_t^{(r)})}{\|\bar{X}_t^{(r)} - \bar{X}_c^{(r)}\|} \right\}^2 \right] \approx \frac{1 + B_{r|s}^2/V}{r + B_{r|s}^2/V}. \tag{A.12}$$

The latter approximation is obtained from (A.2) of Rubin and Thomas (1996), with r and $B_{r|s}^2$ substituted for p and B^2. The approximation for $E(a_*^2)$ is very conservative if $\mathcal{Z}$ is a function of the the the special matching variables, $\mathbf{X}^{(s)}$. An approximation for $E(r_*^2)$ is obtained directly from Section 4.1 of Rubin and Thomas (1992b) with the same substitutions,

$$E(r_*^2) \approx \frac{1}{r + B_{r|s}^2 V}. \tag{A.13}$$

Results Involving Variances of Matched Sample Means

Applying (A.8) of Corollary A.1 to $\mathcal{Z}$, approximation (A.12), and the approximate close matching of the means, $\bar{Z}_t^* \approx \bar{Z}_{mc}^*$,

$$\text{var}(\bar{Z}_t - \bar{Z}_{mc}) = \text{var}\left\{ a_*\left(\bar{Z}_t^* - \bar{Z}_{mc}^*\right)\right\} + \left\{1 - E\left(a_*^2\right)\right\} n_{mc}^{-1}\left(1 - \frac{n_{mc}}{N_c}\right)$$

$$\approx \{1 - E(a_*^2)\} n_{mc}^{-1}\left(1 - \frac{n_{mc}}{N_c}\right)$$

$$\approx \left(\frac{r-1}{r + B_{r|s}^2/V}\right) N_t^{-1}(m^{-1} - R^{-1}).$$

A similar derivation is used for $\text{var}(\bar{W}_t - \bar{W}_{mc})$, along with (A.13).

Results Involving Expectations of Matched Sample Standard Deviations

Applying (A.9) with $U = \mathcal{Z}$, and using the close matching assumption $v_t(\mathcal{Z}^*) \approx v_{mc}(\mathcal{Z}^*)$,

$$E\{v_{mc}(\mathcal{Z})\} = E\{a_*^2 v_{mc}(\mathcal{Z}^*)\} + \{1 - E(a_*^2)\} \approx E\{a_*^2 v_t(\mathcal{Z}^*)\} + \{1 - E(a_*^2)\}.$$

Substituting (A.10) into $E\{a_*^2 v_t(\mathcal{Z}^*)\}$ yields

$$E\{v_{mc}(\mathcal{Z})\} - E\{v_t(\mathcal{Z})\} = (1 - \sigma^2)\{1 - E(a_*^2)\},$$

and using (A.12),

$$\left| \frac{E\{v_{mc}(\mathcal{Z})\}}{E\{v_t(\mathcal{Z})\}} - 1 \right| \approx |(1 - \sigma^2)\{1 - E(a_*^2)\}|/E\{v_t(\mathcal{Z})\}$$

$$\approx \left| (1 - \sigma^2)\frac{r - 1}{r + B_{r|s}^2/V} \right| \bigg/ E\{v_t(\mathcal{Z})\}.$$

A similar derivation using (A.13) and the fact that $E\{v_t(\mathcal{W})\} = \sigma^2$ establishes the approximation for $E\{v_{mc}(\mathcal{W})\}$.

Each of the chapters in Part VI has a real application as its focus, but each one also includes some new methodological contribution. This is in some contrast to the chapters in Part VII, where the focus is almost entirely on the application and/or providing some general advice based on work done in other articles.

Chapter 19, Holland and Rubin (1988), consider case-control or retrospective studies (e.g., see Breslow and Day, 1980). Such studies, which are often used to investigate rare diseases, do not match treated and control subjects, as with all the methods we have studied here that generate the "prospective" study. Rather, the pair-matched retrospective study matches cases (e.g., infected with a particular disease) to controls (free of that disease) on the basis of background variables. Although this is a very helpful design when searching for possible causes of a disease, such matching is inherently "backward" and is fully correct only under the null hypothesis of no effect of treatment on the disease. Retrospective matching of cases and controls, therefore, does not generally control for the background variables used to create the matches – additional stratification or model-based adjustment is still needed. This fact is pointed out in the article, which also illustrates the point using data from a case-control study of the effect of coffee drinking on myocardial infarction.

This conclusion was previously known but not stated within the context of the "Rubin Causal Model" using potential outcomes and an assignment mechanism. This model was mentioned in the initial introduction to this book and is described in more detail in the final, concluding section. The fact that retrospective matching still requires further adjustment for the covariates used to create matches appears earlier, for example, in Seigel and Greenhouse (1973).

Chapter 20, Hill, Rubin, and Thomas (1999), was an article that I was invited to write for a memorial book for Donald T. Campbell, one of the most creative social science researchers working on causal inference and methodology. I asked Neal Thomas and Jennifer Hill, a very able PhD student with strong social science interests and great intuition, to join me. This seemed particularly appropriate because we were all involved in aspects of designing a "school choice" randomized experiment. In particular, the design was a "propensity score matched pair design" based on the theoretical results of the Rubin and Thomas articles (in Part V) on the superiority of the estimated propensity score over the population propensity score. Jennifer cleverly applied software provided by Neal to the real data. In the relevant part of this study, about 400 treated subjects were chosen by a lottery from a pool of about 4,000 applicants. The study could afford to follow only about 800 students for the entire period of the experiment. In order to increase precision of estimation, instead of choosing 400 controls at random, the 400 controls were chosen by propensity score matching, using the estimated propensity score based on the 400 treated versus the 3,600 controls.

The final matching confirmed the theoretically derived "root 2" standard-error reducing benefits – see Figure 20.1. This result from Chapter 16 (Rubin and Thomas, 1992b) effectively says that when matching can remove all bias along the estimated propensity score, creating matched samples using the estimated propensity scores leads to half the variance in mean differences (i.e., between the treatment and control groups in each covariate) that would be found when matching on the true propensity scores. The school choice issue eventually was decided by the U.S. Supreme Court in 2002 (*Zelman et al. v. Simmons-Harris et al.*). An analysis of the first-year data from the study is reported in Barnard et al. (2003), with discussion.

Chapter 21, D'Agostino and Rubin (2000), dealt with the practical problem of missing values in the covariates, and its method was in fact used in Chapter 20 (Hill, Rubin, and Thomas, 1999). The material in Chapter 21 followed from Ralph D'Agostino's PhD thesis (D'Agostino, 1994) on using propensity score methods when there are missing data in the X's and not enough subjects to find exact matches on the pattern of missing data, as would be ideal according to Chapter 12 (Rosenbaum and Rubin, 1984b). The key idea was to build a multivariate "general location model," and match in one of two ways: with nonignorable missing data, use the propensity score to match on the pattern of missing data and the values of the observed data, and with ignorable missing data, use the propensity score to match just on the values of the data. This is a tricky topic, especially because of the needed diagnostics for balance, and requires more study. Both methods of estimation were illustrated with a real data set concerning post-term pregnancy. Also, the use of the ECM algorithm (Meng and Rubin, 1993) to implement the estimation is a useful contribution. Incidentally, Ralph has gone on to make his own important contributions to propensity score methodology, primarily in bio-medical applications.

The final contribution in Part VI is Chapter 22, Rubin (2001b). This applied article deals with the creation of matched data sets for the U.S. tobacco litigation. The resultant data sets, one for males and one for females, matched current smokers with never smokers and matched former smokers with never smokers, using about 100 covariates. The basic data set used was the 1987 National Medical Examination Survey, which has been used in much of the tobacco litigation. Almost all analyses of these data relied on simple linear regression adjustment to control for the covariates, even though such methods had been shown to be notoriously unreliable decades earlier, as stated here in the introduction to Part II. Smokers and nonsmokers have quite different distributions of the covariates (e.g., typically nearly a standard deviation apart with variance ratios differing appreciably from one), and so matching really must be used before any reliable regression adjustment can be applied. Here Mahalanobis-metric matching within propensity score calipers was applied to create one–one matches, and after that, subclassification on the matched samples was applied to reduce the remaining distributional differences.

The most novel methodological feature of this article is the use of simple but revealing diagnostic tables to display the resulting balance in the matched samples when there are many covariates. These tables were based on the geometrical insights developed in Part V. Of major practical importance, especially in such a high profile and litigious problem, was the fact that the matching was completed without any access to any outcome data: the observational study was designed analogously to the design of a randomized experiment.

An aside here is that better diagnostics than reported in this article would have been based on the eigen decomposition orthogonal to the discriminant. The reason is that all of these components are mutually orthogonal, and thus affinely invariant, whereas the diagnostics presented in the article are for the original covariates orthogonal to the discriminant, and thus generally not affinely invariant. The choice made, however, was deliberate: the work was done for litigation, and I wanted to produce something that would be understandable to the statisticians and economists who were acting as experts for both sides, as well as possibly understandable to the more sophisticated attorneys working on the case. An article discussing the ethics of my involvement in the tobacco litigation is Rubin (2002).

19. Causal Inference in Retrospective Studies

Paul W. Holland and Donald B. Rubin

Abstract: The problem of drawing causal inferences from retrospective case-control studies is considered. A model for causal inference in prospective studies is reviewed and then applied to retrospective studies. The limitations of case-control studies are formulated in terms of the level of causally relevant parameters that can be estimated in such studies. An example using data from a large retrospective study of coffee-drinking and myocardial infarctions is used to illustrate the ideas of the article.

Philosophical discussions of causality often emphasize the *meaning* of causation. Scientists are usually concerned with *understanding* causal mechanisms. Purely statistical discussions of causality are substantially more limited in scope, because the unique contribution of statistics is to the *measuring* of causal effects and not to the understanding of causal mechanisms or to the meaning of causation. This distinction is sometimes expressed as "statistics can establish correlation, but not causation." We feel our emphasis on *measurement* is more appropriate, because it focuses on what statistical theory *can* contribute to discussions of causality. Measuring causal effects accurately without any understanding whatsoever of the causal mechanisms involved or of the meaning of causation is not only possible but is, of course, a commonplace experience of everyday life; that is, people are quite capable of using automobiles, ovens, calculators, and typewriters safely and effectively without any knowledge of how these devices work. Of course, careful measurements of causal effects often lead to a better understanding of the causal mechanisms involved.

In this article we first review a mathematical model for causal inferences in prospective studies that is based on the work of Rubin (1974, 1977a, 1978a, 1980a), and developed further in Holland and Rubin (1983), Rosenbaum (1984a, 1984b, 1984c), Rosenbaum and Rubin (1983a, 1983b, 1984a, 1984b, 1985a, 1985b), Holland (1986a, 1986b) and Rubin (1986). We then apply this model to study causal inference in retrospective case-control studies.

Reprinted from *Evaluation Review*, 1988, **12**, 3, 203–231, by permission of Sage Publications, Inc.

1. CAUSAL INFERENCE IN PROSPECTIVE STUDIES

The logic of measuring causal effects is clearest in prospective studies, so we begin with that case. The essential elements of a prospective study are the following:

(1) a *population* of units, U
(2) a set, K, of well-defined causal agents (also called treatments or causes) to which each unit u can be exposed. (For notational simplicity, we consider only two causal agents, K = {t,c}.)
(3) a *response* Y that can be recorded for each unit after exposure to a causal agent in K.

In a prospective study, a sample of units from U is obtained and each unit is assigned to a treatment in K. The causal agents are then applied, and later the response of each unit in the study is recorded. The intuitive notion of causal effect that we wish to describe with our model is the difference between the response measured on a unit that is exposed to cause or treatment t and the response that would have been measured *on the same unit* had it been exposed to treatment c. Thus our notion of the causal effect of a causal agent will always be relative to another causal agent, and is defined for each unit in U.

This meaning of causal effect is not foreign to statistical thinking, and is evident in the writings of R. A. Fisher (1935), Kempthorne (1952), Cochran (1965), and Cox (1958), for example (see Holland, 1986b). Although this notion of a causal effect can be *defined* for each unit in U, in general we are not able to directly *measure* a causal effect for a single unit because having exposed a unit to t, we cannot return in time to expose the same unit to c, instead. This is the Fundamental Problem of Causal Inference, to which we shall return in the next section.

Before turning to the formal model we need to clarify the nature of the *response* Y. For our discussion we will assume that Y is dichotomous, taking on only the values 0 or 1. The extension to a general Y is straightforward. We have chosen to restrict Y to be dichotomous because it is the situation of common interest in retrospective studies.

1.1. The Formal Model and the Definition of Unit-Level Causal Effects

In the model, instead of a single dependent variable Y, we have a dependent variable, Y_k, for each of the treatments to which the unit could have been exposed. Thus if the unit is exposed to causal agent t we will record the value of Y_t for that unit. If that same unit had been exposed to causal agent c instead of t, then we would record the value of Y_c for that unit and not the value of Y_t. More formally, for two treatments, we associate the following vector with each unit in U,

$$(Y_t, Y_c), \tag{19.1}$$

where $Y_k(u)$ = the response made by unit u if it is exposed to cause $k \epsilon K$.

The novel feature of this model is the introduction of several versions of the response variable Y. There is a version of Y for each of the causal agents in K, because our definition of causal effect compares Y_t (the response made if exposed to t) to Y_c (the response made if exposed to c). Rubin (1980a, 1986) refers to the assumption that the vector (19.1) fully represents the possible values of Y under all pairings of $k \epsilon K$ with $u \epsilon U$ as the "stable unit-treatment value assumption," or the SUTVA.

The fact that each unit has a value for both Y_t and Y_c is very important because it allows us to *define* causal effects at the level of individual units. On unit u in U, the causal effect of t relative to c is a comparison of $Y_t(u)$ and $Y_c(u)$, for example the difference $Y_t(u) - Y_c(u)$.

A question that immediately arises is whether or not it is *ever possible* to expose a unit to more than one treatment and thereby directly observe more than one component of the vector in (19.1). One can argue that this is never possible in principle, because once a unit has been exposed to a treatment, the unit is different from what it was before. As mentioned earlier, this is the Fundamental Problem of Causal Inference (Holland, 1986b). However, the propriety of this extreme position depends on the nature of the treatments and the units under study. We will not pursue this issue further here, but will simply make the "worst-case" assumption that a unit can be exposed to at most one treatment condition. For our application to retrospective studies this assumption is adequate, since in these studies units are exposed to only one of the causal agents.

In order to relate the vector in (19.1) to the data that are actually observed, we introduce the variable S, where S = k if the unit is exposed to cause k; S is the "causal indicator" variable that indicates to which $k \epsilon K$ each unit is exposed.

The observed data from a unit u is the vector

$$(Y_S(u), S(u)). \tag{19.2}$$

The notation Y_S is used because it indicates that we can observe only the response of a unit to the treatment to which it is exposed, that is,

$$Y_S = Y_k \quad \text{if } S = k, \quad \text{for } k \epsilon K. \tag{19.3}$$

The quantity Y_S is the observed value of the response and is therefore what is usually called the "dependent variable" in statistical discussions (e.g., in an ANOVA of an experiment). We never can observe Y_k if $S \neq k$. Since we can observe only the value of Y_t or Y_c but not both, it is a consequence of the model that causal effects for individual units *are not directly measurable*.

In summary, our idealized model for a prospective causal study can be viewed as based on the following sequence of steps:

(1) determination of the population U under study
(2) determination of the set K composed of causal agents, treatments, or causes under study
(3) determination of the response variable Y to be observed
(4) consequent definition of the vector (Y_t, Y_c) for every unit in U
(5) determination of the causal indicator S for every unit in the study
(6) consequent definition of the vector of observable data (Y_S, S) for every unit in the study
(7) observation of (Y_S, S) for each unit in the study

1.2. The Three Levels of Causal Inference

There are three levels or "strengths" of causal inferences that arise in practice. These are: unit-level, subpopulation-level, and population-level causal inferences. These levels are ordered by decreasing strength in the sense that knowledge of all unit-level causal inferences implies knowledge of all sub- and population-level causal inferences, and knowledge of all subpopulation causal inferences for a partition of U implies knowledge of population-level causal inferences, but not vice versa. We briefly describe each of these levels. Due to the Fundamental Problem of Causal Inference, all of these involve *indirect* estimation of causal effects.

Unit-level causal inference: The *definition* of causal effects is at the unit level and is the difference

$$Y_t(u) - Y_c(u).$$

Units are called *homogeneous* when $Y_k(u_1) = Y_k(u_2)$ for every pair of units, u_1, u_2. Homogeneous units may be encountered in laboratory research in the physical sciences. When unit homogeneity can be assumed, unit-level causal inferences are easy, since $Y_t(u) - Y_c(u) = Y_t(u_1) - Y_c(u_2)$ for every u, u_1, and $u_2 \in U$. Hence unit-level causal inference only requires the observation of Y_t on u_1 and Y_c on u_2 for one pair of units, u_1, u_2. Such situations do not require statistical methods for the estimation of causal effects.

Population-level causal inference: The population distribution of Y_t and Y_c over U are, in the dichotomous case, specified by $P(Y_t = 1)$ and $P(Y_c = 1)$. A population causal inference is a comparison of these two probabilities – or, more generally, of the distributions of Y_t and Y_c over U. A population causal inference is weaker than a unit-level causal inference because it only describes how t or c affects the distribution of Y over all of U rather than how it affects the value of Y on a given unit u in U. There are many ways to compare distributions, and two important ones for the dichotomous case are the *difference*

$$P(Y_t = 1) - P(Y_c = 1), \tag{19.4}$$

and the *odds-ratio*,

$$\alpha = \frac{P(Y_t = 1)}{P(Y_t = 0)} \bigg/ \frac{P(Y_c = 1)}{P(Y_c = 0)}. \tag{19.5}$$

Although (19.4) and (19.5) are equally good ways of comparing dichotomous distributions in general, they are quite different from a causal point of view. The difference (19.4) may be interpreted in two ways, first as a difference in probability, and second as an *average causal effect*, or ACE,

$$P(Y_t = 1) - P(Y_c = 1) = E(Y_t - Y_c) = ACE. \tag{19.6}$$

An ACE is the average of all the unit-level causal effects over U, and sometimes, for example when Y is continuous, an ACE can be shown to approximate *all* of the unit-level causal effects in U. This fortunate state of affairs occurs when we have a case of constant effects, i.e.,

$$Y_t(u) - Y_c(u) = T \text{ for all } u \in U.$$

This form of additivity does not often occur for dichotomous Y's except under very special circumstances.

The odds ratio in (19.5), while a useful comparison of probabilities, does not have an interpretation as an ACE, but we shall see that it arises naturally in the study of retrospective research designs. The odds ratio (19.5) is a comparison of the distribution of Y_t over U with that of Y_c, with no concern with how Y_t and Y_c might be related for particular units in U.

Subpopulation-level causal inference: Subpopulations of U may be defined in many ways, but we shall use only one method. It is to define subpopulations by the values of *covariates.*

A covariate is a value of a variable that is defined on each unit u in U but that is not affected by the exposure of units to causes in K. In our model, the introduction of a *variable*, X, defined on the units of U, requires the notation $X_k(u)$ to indicate that, in general, X can depend both on the unit u and on which cause, k, to which u is exposed. A *covariate* is a special type of variable for which $X_k(u) = X(u)$ for all $k \epsilon K$. Variables that are *measured prior to* the exposure of units to causal agents are always covariates, but sometimes variables measured after exposure are also covariates – for example, a drug treatment is not likely to change the value of variables such as an adult's height.

Subpopulations defined by a covariate, X, allow us to consider probabilities of the form

$$P(Y_t = 1 \mid X = x) \text{ and } P(Y_c = 1 \mid X = x).$$

Subpopulation-level causal inferences may be based on such probabilities. For example, we may consider the *conditional difference*

$$P(Y_t = 1 \mid X = x) - P(Y_c = 1 \mid X = x), \tag{19.7}$$

or the *conditional odds ratio*

$$\alpha(x) = \frac{P(Y_t = 1 \mid X = x)}{P(Y_t = 0 \mid X = x)} \bigg/ \frac{P(Y_c = 1 \mid X = x)}{P(Y_c = 0 \mid X = x)}. \tag{19.8}$$

Just as population-level causal inferences distinguish (19.4) from (19.5), so, too, do subpopulation-level causal inferences distinguish (19.7) from (19.8) for the same reasons. Similarly, the difference (19.7) also can be interpreted as a conditional ACE, that is,

$$P(Y_t = 1 \mid X = x) - P(Y_c = 1 \mid X = x) = E(Y_t - Y_c \mid X = x), \tag{19.9}$$

whereas the odds ratio (19.8) has no such interpretation. Yet (19.8) is still useful as an intermediate type of causal inference between the unit level and population level because it can describe the way that the causal agent t changes the distribution of Y-values relative to that of the causal agent c for the subpopulation of U for which $X(u) = x$. If this is a relatively homogeneous population, such information can be tantamount to a unit-level causal inference, and thus a conditional ACE.

1.3. The Role of Randomization in Prospective Studies

It is well known that randomization aids one substantially in drawing causal inferences, yet why is this so? The model developed above gives an easy answer to this
question.

When randomization is used to assign units to exposure to the causal agents in
K, the variable S is made statistically independent of *all* other variables defined on
U. Hence in particular,

$$E(Y_k \mid S = k) = E(Y_k) \tag{19.10}$$

for $k = t$ or c (or for all choices of $k \in K$ when K has more than two members).

Now let us consider the data that can be observed in a prospective study, that is,
$(Y_S(u), S(u))$ for u in the study.

From a set of observations of (Y_S, S) we can calculate the distribution of Y_S
given the observed value of S. In particular we can estimate the value of

$$E(Y_S \mid S = k). \tag{19.11}$$

The mean value in (19.11) is also equal to

$$E(Y_k \mid S = k) \tag{19.12}$$

by the standard rules of conditional probability. Now suppose randomization is
employed. Using (19.10), (19.11), and (19.12) we have this basic identity (that
holds only for randomized studies in general):

$$E(Y_S \mid S = k) = E(Y_k). \tag{19.13}$$

The difference

$$E(Y_S \mid S = t) - E(Y_S \mid S = c) \tag{19.14}$$

is called the *prima facie* ACE (or FACE) in general. In a randomized study the FACE
equals the ACE, that is,

$$E(Y_S \mid S = t) - E(Y_S \mid S = c) = E(Y_t - Y_c). \tag{19.15}$$

The FACE in (19.14) is a quantity that can be estimated from the data in *any*
prospective study. In a *randomized* prospective study the FACE has *causal relevance*,
since it equals the ACE.

1.4. The Role of Covariates in Prospective Studies

When a covariate is available in a study, the observed data for each unit in the study
is expanded from (Y_S, S) to

$$(Y_S, S, X) \tag{19.16}$$

where X is the covariate (possibly vector-valued). From observed values of these
data for the units in the study we can, in principle, estimate the regression of Y_S on
S and X, that is,

$$E(Y_S \mid S = k, X = x). \tag{19.17}$$

The quantity in (19.17), however, equals

$$E(Y_k \mid S = k, X = x) \tag{19.18}$$

from the standard rules of conditional probability. But (19.18), as it stands, is much like (19.12) in not having any causal relevance. Rosenbaum and Rubin (1983a) define a special condition that generalizes randomization and that gives (19.18) causal relevance when it holds. It is the condition of *strong ignorability*. Treatment assignment (i.e., the distribution of S given X and Y_k) is *strongly ignorable* if (a) given X, S is independent of Y_k for $k \epsilon K$, and (b) $P(S = k|X) > 0$ for all $k \epsilon K$. This is a stronger condition than *ignorability* defined by Rubin (1978a) for Bayesian inference.

If strong ignorability holds, then (19.18) becomes

$$E(Y_k \mid X = x), \tag{19.19}$$

and the difference

$$E(Y_t \mid S = t, X = x) - E(Y_c \mid S = c, X = x), \tag{19.20}$$

which is the conditional FACE, equals the conditional ACE; that is, (19.20) equals

$$E(Y_t \mid X = x) - E(Y_c \mid X = x). \tag{19.21}$$

Note that by averaging over the *conditional* ACE in (19.21) we obtain the ACE in (19.15), that is,

$$E[E(Y_t - Y_c \mid X)] = E(Y_t - Y_c). \tag{19.22}$$

Hence the condition of strong ignorability is less restrictive than that of randomization, but it still allows us to measure the average causal effect in a prospective study using the data that are available. Of course, the plausibility of the assumption of strong ignorability needs to be considered carefully in any real application. For a more detailed discussion of these issues see Rubin (1977a), Holland and Rubin (1983), and Rosenbaum (1984b).

2. Causal Inference in Retrospective Case-Control Studies

The structure of a retrospective case-control study is considerably different from the general prospective study discussed in the preceding section. In a case-control study, a population of units is divided into those who have a particular symptom or disease of interest (i.e., the "cases") and those who do not have the symptom or disease (i.e., the "controls"). Samples (random samples, in principle) of cases and controls are selected from this population, and information about each selected person is obtained to ascertain (a) the level of exposure to the particular causal agents of interest and (b) other medically relevant information that may be used to define subpopulations of units.

The response variable for a case-control study is the dichotomous variable that indicates whether or not the unit is a "case" or a "control," that is,

$$Y_S = \begin{array}{l} 1 \text{ if unit is a case} \\ 0 \text{ if unit is a control.} \end{array}$$

Case-control studies are *retrospective* because they begin at the endpoint of a prospective study (i.e., observations of the response variable for each unit in the study) and then look back in time to discover the causal agent to which each unit has been exposed (i.e., the value of the causal indicator S). In retrospective studies, the basic groups are cases ($Y_S = 1$) and controls ($Y_S = 0$) with S measured on each sampled unit, whereas in prospective studies the basic groups are exposed ($S = t$) and not exposed ($S = c$), with Y_S measured on each sampled unit. In addition to this fundamental difference between case-control and prospective studies, two other differences should be mentioned. First, since the investigator can only *collect* data on prior exposure to the causal agents of interest, it is impossible to use randomization to *assign* units to the causal agents. Thus case-control studies are never randomized. Prospective studies, on the other hand, may or may not employ randomization depending on the amount of control that is possible. Second, the populations studied in case-control studies usually consist of survivors only, because it is often impossible to obtain comparable data on individuals who are deceased. This limitation may have consequences for the interpretability of the results of a case-control study. An excellent reference for case-control studies is Breslow and Day (1980).

Although, in principle, it is almost always possible to design a prospective version of a case-control study, it is often much more expensive than the case-control study. There are several reasons for this: (a) prospective studies require large sample sizes in which the "cases" are rare (e.g., when $Y_S =$ represents a rare disease) and (b) prospective studies may involve long time spans before relevant data become available. Hence it is likely that case-control studies will always be an attractive possibility for many types of scientific investigations, especially in the early stages of the research. It is therefore important to know their limitations, to design them as well as possible and to analyze the data collected in such studies correctly. Our goal in the present article is to illuminate all of these points by applying the model for causal inference outlined in Section 1 to case-control studies.

2.1. The Standard Two-Way Table

When analyzing data from a case-control study, it is customary to form and draw conclusions from the two-way table of counts illustrated in Table 19.1. We assume that this table is formed by randomly sampling m_{1+} "cases" (units with $Y_s = 1$) from the subpopulation of cases, and randomly sampling m_{0+} "controls" (units with $Y_s = 0$) from the subpopulation of controls.

In Table 19.1, m_{yk} is the number of units in the study for which $Y_S = y$ and $S = k$. For example, m_{1c} is the number of "cases" in the study that were observed exposed to causal agent c. Before examining this table of sample data, let us consider the population table that underlies it. Table 19.2 gives the population proportion of people with exposure to t or c among all those who are cases or controls. These population values are denoted by

$$r_{yk} = P(S = k \mid Y_s = y). \tag{19.23}$$

The corresponding sample ratio

$$\hat{r}_{yk} = m_{yk}/m_{y+} \tag{19.24}$$

Table 19.1. *The standard two-way table in retrospective studies showing the sample distribution of cases and controls observed for each causal agent*

		Causal agent		Total
		$S = t$	$S = c$	
Cases	$Y_S = 1$	m_{1t}	m_{1c}	m_{1+}
Controls	$Y_S = 0$	m_{0t}	m_{0c}	m_{0+}
	Total	m_{+t}	m_{+c}	m_{++}

estimates r_{yk}. We shall call the $\{r_{yk}\}$ the *retrospective probabilities* of the study. They are "retrospective" because the conditioning is on an event that occurs *after* the event whose probability is being assessed.

In this development we must emphasize the importance of representing the observed value of the response as Y_S. For example, in (19.23) it would be incorrect to condition on $Y_k = y$ since Y_k is the response made if exposed to cause k, whereas Y_S is the observed response. Because Y_S is being conditioned on in Table 19.2, it is sometimes said that in a case-control study *exposure* is the dependent variable and *diagnosis* (i.e., case or control) is the independent variable. This description is neither helpful nor of scientific interest, and we will not describe the situation in these terms.

If we consider the weakest level of causal inference, that is, a population-level causal inference, then the causal parameters are the marginal probabilities $P(Y_t = 1)$ and $P(Y_c = 1)$. Thus the retrospective probabilities in (19.23) are not in themselves of any causal interest, because, at the very least, they do not address the correct events. However, by applying the usual rules of probability, we may reverse the roles of S and Y_S in (19.23) and obtain more interesting probabilities. The result of this reversal is the accepted justification for looking at Table 19.1 (see Cornfield, 1956).

2.2. Relating Retrospective and Prospective Probabilities

To reverse the roles of S and Y_S we make use of Bayes's theorem to obtain

$$P(Y_S = y \mid S = k) = P(S = k \mid Y_S = y)\frac{P(Y_S = y)}{P(S = k)}. \tag{19.25}$$

Table 19.2. *The population table of retrospective probabilities r_{yk} related to the sample table in Table 19.1*

		Causal agent		Total
		$S = t$	$S = c$	
Cases	$Y_S = 1$	$r_{1t} = P(S = t \mid Y_S = 1)$	$r_{1c} = P(S = c \mid Y_S = 1)$	1
Controls	$Y_S = 0$	$r_{0t} = P(S = t \mid Y_S = 0)$	$r_{0c} = P(S = c \mid Y_S = 0)$	1

Table 19.3. *The population table of prospective probabilities*

		Causal agents	
		$S = t$	$S = c$
Cases	$Y = 1$	$p_{1t} = p(Y_t = 1 \mid S = t)$	$p_{1c} = p(Y_c = 1 \mid S = c)$
Controls	$Y = 0$	$P_{0t} = P(Y_t = 0 \mid S = t)$	$P_{0c} = P(Y_c = 0 \mid S = c)$
	Total	1	1

However,

$$P(Y_S = y \mid S = k) = P(Y_k = y \mid S = k), \tag{19.26}$$

so it follows that

$$P(Y_k = y \mid S = k) = P(S = k \mid Y_S = y)\frac{P(Y_S = y)}{P(S = k)}.$$

The probability $P(Y_k = y \mid S = k)$ is "prospective" because the conditioning event occurs prior in time to the event whose probability is being assessed. We denote these prospective probabilities by

$$p_{yk} = P(Y_k = y \mid S = k). \tag{19.27}$$

Hence the retrospective and prospective probabilities are related by

$$p_{yk} = r_{yk}\frac{a_y}{b_k}, \tag{19.28}$$

where

$$a_y = P(Y_S = y),$$

and

$$b_k = P(S = k)$$
$$= \sum_y P(S = k \mid Y_S = y)P(Y_S = y),$$

or

$$b_k = \sum_y r_{yk}\, a_y. \tag{19.29}$$

Hence the prospective probabilities p_{yk} can be determined from the retrospective probabilities r_{yk} and the overall proportion of cases and controls in the population, a_y, via

$$p_{yk} = \frac{r_{yk}\, a_y}{\sum_z r_{zk}\, a_z}. \tag{19.30}$$

We illustrate the array of "prospective" probabilities of (19.26) and (19.27) in Table 19.3.

The cross-product ratio for Table 19.2 may be expressed as:

$$\alpha_{\text{ret}} = \frac{r_{1t}}{r_{1c}} \Big/ \frac{r_{0t}}{r_{0c}} = \frac{P(S = t \mid Y_S = 1)}{P(S = c \mid Y_S = 1)} \Big/ \frac{P(S = t \mid Y_S = 0)}{P(S = c \mid Y_S = 0)}. \qquad (19.31)$$

The cross-product ratio for Table 19.3 may be expressed as:

$$\alpha_{\text{pro}} = \frac{p_{1t}}{p_{0t}} \Big/ \frac{p_{1c}}{p_{0c}} = \frac{P(Y_t = 1 \mid S = t)}{P(Y_t = 0 \mid S = t)} \Big/ \frac{P(Y_c = 1 \mid S = c)}{P(Y_c = 0 \mid S = c)}. \qquad (19.32)$$

Because Tables 19.2 and 19.3 are related via row and column multiplication, see equation (19.28), it is well known (e.g., Bishop et al., 1975) and easily shown that the two cross-product ratios α_{ret} and α_{pro} are equal.

2.3. Why the Standard Two-Way Table is Misleading

Even though the equality of α_{ret} and α_{pro} is the usual reason one ever looks at the data in Table 19.1, the crucial question of how it relates to the causal parameters of interest remains, that is, $P(Y_t = 1)$, $P(Y_c = 1)$ or the average causal effects in (19.6) or (19.9). The answer is that, without an additional assumption, the cross-product ratio, α_{pro}, has *no causal relevance*.

The crucial assumption is randomization, because then S is independent of Y_k and

$$P(Y_k = 1 \mid S = k) = P(Y_k = 1),$$

and hence we have

$$\alpha_{\text{pro}} = \frac{P(Y_t = 1)}{P(Y_t = 0)} \Big/ \frac{P(Y_c = 1)}{P(Y_c = 0)} \quad (\text{i.e.}, = \alpha). \qquad (19.33)$$

Hence randomization implies that α_{pro} equals α defined in (19.5), which is a population-level causal parameter. However, as we stated earlier, retrospective studies are never randomized, so that the assumption that S is independent of Y_k is dubious in most cases. Thus there is generally no value to examining the data in Table 19.1 from the point of view of using it to estimate causal parameters.

2.4. The Role of Covariates in Retrospective Studies

If there is a covariate X (possibly a vector) that is measured on each unit in the study, then we may form a table like Table 19.1 for each value of X. Let m_{ykx} be the number of units in the study for which $Y_S = y$, $S = k$, and $X = x$. These are arrayed in Table 19.4 for $X = x$. We suppose that at each value of X, the data arise from a random sample of cases and a random sample of controls, not necessarily with the same sampling rates. This sampling scheme includes matched case-control pairs, where the cases are randomly sampled from the population of cases, and for each sampled case with $X = x$, a matching control with $X = x$ is found.

Table 19.4. *The distribution of cases and controls in the sample observed for each causal agent, at $X = x$*

		Value of $X = x$ Causal Agent		
		$S = t$	$S = c$	Total
Cases	$Y_S = 1$	m_{1tx}	m_{1cx}	m_{1+x}
Controls	$Y_S = 0$	m_{0tx}	m_{0cx}	m_{0+x}
	Total	m_{+tx}	m_{+cx}	m_{++x}

The sample ratios

$$\hat{r}_{ykx} = m_{ykx}/m_{y+x} \tag{19.34}$$

estimate the population *retrospective probabilities*

$$r_{ykx} = P(S = k \mid Y_S = y, X = x). \tag{19.35}$$

We may again apply Bayes's theorem to reverse the roles of S and Y_S in (19.35) as we did in (19.25). This yields

$$P(Y_S = y|S = k, X = x) = P(S = k \mid Y_S = y, X = x)\frac{P(Y_S = y \mid X = x)}{P(S = k \mid X = x)}. \tag{19.36}$$

However,

$$P(Y_S = y \mid S = k, X = x) = P(Y_k = y \mid S = k, X = x), \tag{19.37}$$

so that

$$P(Y_k = y \mid S = k, X = x) = P(S = k \mid Y_S = y, X = x)\frac{P(Y_S = y \mid X = x)}{P(S = k \mid X = x)}. \tag{19.38}$$

Again the probability $P(Y_k = y|S = k, X = x)$ is "prospective," and we denote it by

$$p_{ykx} = P(Y_k = y \mid S = k, X = x), \tag{19.39}$$

and, as before, the retrospective and prospective probabilities are related by

$$p_{ykx} = r_{ykx}\frac{a_{yx}}{b_{kx}}, \tag{19.40}$$

where

$$a_{yx} = P(Y_S = y \mid X = x),$$

and

$$b_{kx} = P(S = k \mid X = x) = \sum_y P(S = k \mid Y_S = y, X = x)P(Y_S = y \mid X = x),$$

or

$$b_y = \sum_y r_{ykx}\, a_{yx}.$$

Again, p_{ykx} can be determined from r_{ykx} and a_{yx}, the proportion of cases and controls among those units with $X = x$, via

$$p_{ykx} = \frac{r_{ykx}\, a_{yx}}{\sum_z r_{zkx}\, a_{zx}}. \qquad (19.41)$$

The odds ratio for the tables of retrospective probabilities underlying the sample in Table 19.4 is

$$\alpha_{ret}(x) = \frac{r_{1tx}}{r_{1cx}}\Bigg/\frac{r_{0tx}}{r_{0cx}}. \qquad (19.42)$$

The corresponding odds ratio for the prospective probabilities is

$$\alpha_{pro}(x) = \frac{p_{1tx}}{p_{0tx}}\Bigg/\frac{p_{1cx}}{p_{0cx}}. \qquad (19.43)$$

As before, these two odds ratios are equal, that is,

$$\alpha_{pro}(x) = \alpha_{ret}(x). \qquad (19.44)$$

Under strong ignorability we have

$$P(Y_k = y \mid S = k, X = x) = P(Y_k = y \mid X = x), \qquad (19.45)$$

so that

$$\alpha_{pro}(x) = \frac{P(Y_t = 1 \mid X = x)}{P(Y_t = 0 \mid X = x)}\Bigg/\frac{P(Y_c = 1 \mid X = x)}{P(Y_c = 0 \mid X = x)} \quad (\text{i.e., } = \alpha(x)). \qquad (19.46)$$

Hence, when strong ignorability holds, $\alpha_{pro}(x)$ equals $\alpha(x)$ as defined in (19.8), which is a subpopulation causal parameter.

It may happen that $\alpha(x)$ as defined in (19.8) and (19.46) does not depend on x – this is the case of no interaction of X with the effect of the causal agents on the distribution of Y-values. In such a situation we will denote this common value by α_0. Note that $\alpha_{pro}(x)$ can be constant in x regardless of the plausibility of strong ignorability – these are two, quite different, assumptions. The constancy of $\alpha_{pro}(x)$ is testable with the data (Y_S, S, X) whereas strong ignorability is not.

Unfortunately, there is no simple relation between the population causal parameter

$$\alpha = \frac{P(Y_t = 1)}{P(Y_t = 0)}\Bigg/\frac{P(Y_c = 1)}{P(Y_c = 0)} \qquad (19.47)$$

and α_0 even when strong ignorability and the assumption of no interaction between X and the effect of t holds. Nevertheless, α_0 is a causally interesting parameter itself: α_o is the amount by which the odds for $Y_t = 1$ are increased over the odds that $Y_c = 1$ in each X-stratum of U, and, thus, α_o is a causal parameter. Since α_o is specific to each X-stratum of U, it provides causal information about the effects of t relative to c in U that is at a more detailed level than the overall population level. However, in general it is not as strong a causal inference as a unit-level causal inference.

Our conclusion is that in a case-control study, the simple two-way table (Table 19.1) generally holds no causal interest even for a matched case-control study (i.e., $m_{1+x} = m_{0+x}$). It is crucial to stratify on covariates and to estimate $\alpha_{ret}(x) = \alpha_{pro}(x)$ – which is a causally relevant parameter, $\alpha(x)$, under strong ignorability. If the stratified table exhibits constant odds ratios, then, assuming strong ignorability, this parameter equals α_o and gives the amount that t increases the proportion of units in each X-stratum that are "cases" relative to c. This "amount of increase" is in terms of the odds corresponding to the proportions. Thus, for example, for a given value of the proportion $P(Y_c = 1 \mid X = x)$, we may calculate $P(Y_t = 1 \mid X = x)$ via the formula

$$P(Y_t = 1 \mid X = x) = \frac{\alpha_o P(Y_c = 1 \mid X = x)}{P(Y_c = 0 \mid X = x) + \alpha_o P(Y_c = 1 \mid X = x)}. \qquad (19.48)$$

Comparing (19.48) to the given value of $P(Y_c = 1 \mid X = x)$ leads to a causal inference about the effect of the causal agent when $X = x$. In general, however, the stratified table will not exhibit a constant odds ratio and then the values of $\alpha(x)$ are the causally relevant parameters (under strong ignorability), and can be used in place of α_o in (19.48) to calculate $P(Y_t = 1 \mid X = x)$ from $P(Y_c = 1 \mid X = x)$.

3. AN EXAMPLE

3.1. The Data

The following data are taken from a case-control study of the relationship between coffee drinking and the occurrence of myocardial infarctions (MI) by Jick et al. (1973). We use these data for illustrative purposes only. A total of 24,741 patients were classified as "cases" (had an MI) or "controls" (did not have an MI). Table 19.5 shows the standard two-way table that presents the cases and controls cross-classified by the potential causal agents under study – self-reported daily coffee consumption. Although our previous notation has considered only two causal agents, Table 19.5 presents four, a control (0 cups per day) and three levels of the amount of coffee drinking; the extensions needed to handle this extra complexity are simple. The odds ratios estimated in Table 19.5 are defined by

$$\alpha_{pro}^k = \frac{P(Y_k = 1 \mid S = k)}{P(Y_k = 0 \mid S = k)} \bigg/ \frac{P(Y_1 = 1 \mid S = 1)}{P(Y_1 = 0 \mid S = 1)}, \qquad (19.49)$$

for $k = 2, 3, 4$, that is, α_{pro}^k is the odds ratio for level k of coffee drinking relative to the control of no coffee drinking.

Table 19.5. *Cross-tabulation of self-reported daily coffee intake (S) by cases and controls (Y) for 24,741 patients*

		S = 1 0 cup/day	S = 2 1–2 cups	S = 3 3–5 cups	S = 4 6+ cups	Total
$Y_S = 1$	MI cases	128	269	147	86	630
$Y_S = 0$	non-MI controls	6918	9371	5290	2532	24111
	Total	7046	9640	5437	2618	24741

Note: Estimated raw odds-ratios, α_{pro}^k, relative to k = 1: $\alpha_{pro}^2 = 1.551$, $\alpha_{pro}^3 = 1.502$, $\alpha_{pro}^4 = 1.836$.

Table 19.5 suggests a modest increase in the risk of MI among persons who drink coffee. The odds ratios range from 1.5 to 1.8. The odds ratios exhibited in Table 19.5 are not monotone in the amount of self-reported coffee drinking, and the effect seems to be almost as strong for persons who drink 1–2 cups per day as for those who drink 6 or more cups per day.

Table 19.5, however, does not take various background variables into account and, as we have discussed earlier, is therefore likely to be misleading because it is not reasonable to believe that the drinking of coffee is randomly assigned and therefore independent of $Y_k, k = 1, 2, 3, 4$.

In addition to the variables

$$S = \text{level of self-reported coffee intake}$$

and

$$Y = \text{case or control,}$$

the following set of variables were also available on all patients in the study.

A = Age: 6 levels: 20–29, 30–39, ... , 70–79
G = Gender: 2 levels: male, female
C = Smoking: 3 levels: other, ex-smoker, current smoker
O = Other heart disease: 2 levels: yes, no

In addition, because the data were collected from 24 suburban Boston hospitals, a fifth variable, H = hospital, was included in the analysis (with 24 levels). This results in a covariate X, which takes on $6 \times 2 \times 3 \times 2 \times 24 = 1728$ values, so that when Table 19.5 is stratified on X we obtain a seven-way contingency table with $2 \times 4 \times 1728 = 13,824$ cells. With a total of 24,741 observations, this gives us about 1.9 observations per cell – a very sparse table indeed! Many approaches to simplifying this sort of situation are possible, for example, see Breslow and Day (1980). We shall use log-linear contingency table models: (a) because of their direct relationship to odds ratios, (b) because they allow us to adjust for the effect of all of the covariates simultaneously, and (c) because they allow us to smooth the sparse, seven-dimensional table.

3.2. Log-Linear Models for This Problem

Let $X = (A, G, C, O, H)$ denote our complete vector of covariates. The logs of the retrospective probabilities r_{ykx} from (19.35) may be expressed as

$$\log(r_{ykx}) = u + u_{1(y)} + u_{2(k)} + u_{3(x)} + u_{12(y,k)} + u_{13(y,x)} + u_{23(k,x)} + u_{123(y,k,x)},$$
$$(19.50)$$

where the u-terms in (19.50) are assumed to satisfy the usual ANOVA-like identifying constraints, $u_{1(+)} = u_{2(+)} = 0$, etc. (Bishop et al., 1975). We need to express the odds ratios

$$\alpha_{ret}^k(x) = \frac{r_{1kx}}{r_{11x}} \Big/ \frac{r_{0kx}}{r_{01x}}, \qquad (19.51)$$

in terms of the u-terms in (19.50). It is easy to show that the following equation holds:

$$\alpha_{ret}^k(x) = \alpha_{ret}^k \exp\{u_{123(1,k,x)} - u_{123(1,1,x)} - u_{123(0,k,x)} + u_{123(0,1,x)}\}, \qquad (19.52)$$

where

$$\alpha_{ret}^k = \exp\{u_{12(1,k)} - u_{12(1,1)} - u_{12(0,k)} + u_{12(0,1)}\}. \qquad (19.53)$$

From (19.52) it follows that the hierarchical log-linear model specified by setting all $u_{123} = 0$ corresponds to the assumption that

$$\alpha_{ret}^k(x) = \alpha_{ret}^k, \qquad (19.54)$$

for all x. Thus we may investigate the question of whether or not the odds ratio, $\alpha_{ret}^k(x)$, depends on x by testing for three-way interaction of the various covariates in X and with Y_S and S. Furthermore, if a model where all $u_{123} = 0$ is acceptable, the estimated u_{12}-terms may be used to obtain estimates of α_{ret}^k. If we are willing to make the assumptions necessary to ensure that α_{ret}^k is the causally relevant parameter discussed in Section 3.4, that is, α_0^k, then we may test $\alpha_{ret}^k = 1$ (i.e., no effect of different levels of the causal agent) by testing that $u_{12} = 0$. This test will adjust for the distribution of the covariates in the several exposure groups. In the remainder we assume strong ignorability and refer to α_0^k rather than α_{ret}^k.

3.3. Simplifying the Analysis

As described above it may seem as though we are considering the whole $2 \times 4 \times 1728$ table, but one important feature of the use of log-linear models is that they do not force this when there are insufficient data to do so. Instead we break up the joint distribution of $X = (A, G, C, O, H)$ into various marginal distributions and expand the model in (19.50) to make use of them. In the present example we expand the table to the full seven dimensions, fit all u-terms involving Y_S and/or S but not X, and only fit effects for the following pairs and triples of variables involving X:

$$(u_{23}) \quad HS/AS/GCS/GOS/COS/$$
$$(u_{13}) \quad HY/AGY/ACY/AOY/GCY/GOY/COY/$$
$$(u_3) \quad HA/HG/HC/HO/AGC/AGO/ACO/GCO \qquad (19.55)$$

Table 19.6. *Summary of study of dependence of* $\alpha_{ret}^k(x)$ *on* x

Interaction of YS with	df	LR statistic	Level attained
H	69	79.39	.20
A	15	10.31	.80
G	3	2.5	.47
C	6	8.83	.18
O	3	3.97	.25

The u-terms in parentheses indicate which terms in (19.50) have been expanded in the seven-way table.

3.4. Results

If we fit the log-linear model indicated by the pairs and triples of variables in (19.55) and then delete the YS terms and refit the model, we obtain a likelihood-ratio test of $\alpha_0^k = 1$ for $k = 2, 3, 4$. The value of the likelihood-ratio statistic is 12.3, which, under the null hypothesis, has three degrees of freedom. Thus this analysis supports the conclusion that at least one α_0^k is not 1, and thus that there is a relationship between coffee consumption and myocardial infarctions. The estimated values of α_0^k are

$$\hat{\alpha}_0^2 \quad \hat{\alpha}_0^3 \quad \hat{\alpha}_0^4$$
$$1.188 \quad 1.235 \quad 1.719 \tag{19.56}$$

as opposed to the raw odds ratios given in Table 19.5. These adjusted odds ratios are monotonic in the amount of coffee consumed, with the major effect for higher levels of coffee consumption. If we are willing to assume strong ignorability of the distribution of coffee consumption and diagnostic status (case or control) given the set of covariates in X, then these estimated odds ratios are the subpopulation-level causal effects described in Section 2.4.

To study the question of whether $\alpha_{ret}(x) = \alpha_{pro}(x)$ varies with x, we fit five additional models, each of which supplements (19.55) with one of these triples of variables: HSY, ASY, GSY, CSY, or OSY. The likelihood-ratio statistics for these models, the degree of freedom, and attained significance levels are given in Table 19.6.

None of these interactions are strong enough to be statistically significant at conventional levels. This result contradicts previous analysis of these data that found an interaction of the effect of coffee drinking on diagnostic status with these variables, (Miettinen, 1976).

4. DISCUSSION AND SUMMARY

4.1. Risk Factors: Causal Agents Versus Attributes

In medical studies the term "risk factor" is used to lump "true" causal agents such as smoking or coffee drinking, which can be altered, with individual attributes such as age and sex, which cannot be altered. We think it is wise to distinguish carefully between these two types of risk factors and to reserve discussions of causation to include only the former. The model used in this article presupposes a response value Y_t if the unit is exposed to t and a value Y_c if it is exposed to c. When t or c is construed to be an attribute of a unit (e.g., a person's sex) it is entirely unclear how to define both Y_t *and* Y_c on each unit. In this sense it is meaningless to speak of estimating the causal effects of attributes of units. This does not mean that attributes have no predictive value, since prediction is simply a consequence of association between variables, which does not necessarily involve notions of cause. Our definition of a causal agent is much stricter than some definitions used by economists, for example, Granger Causality (Granger, 1969). Granger labels as a *cause* any predictor of Y that adds independent information to the prediction. We believe this is too generous a definition of causality, not only misusing the language but possibly leading researchers away from the study of the effects of manipulations that are possible – see Holland (1986b) and Rubin (1986).

4.2. Randomization and Strong Ignorability

One useful feature of the model developed for prospective studies in Section 1 is that it clarifies the importance of randomization in causal studies. The statistical independence of Y_k and S that randomization ensures is very important but not always appreciated by writers on the subject. For example, it is often asserted that there is some sort of difficulty in resolving randomization of treatments to units with the Bayesian/likelihood/ modeling framework (Basu, 1980, Kempthorne, 1976, and Kruskal, 1980, but argued otherwise by Rubin, 1978a, 1980a). One possible source of confusion is that the independence of S and (Y_t, Y_c) does not imply that S is independent of the observed response Y_S except in very special circumstances, for example, when $Y_t = Y_c$ for all units. However, the equation

$$E(Y_S \mid S = t) - E(Y_S \mid S = c) = E(Y_t - Y_c) \tag{19.57}$$

which is a consequence of randomization, has an impact on both Bayesians and frequentists alike. This is simply because it states that a population parameter that can be estimated with observed data, that is, the FACE, $E(Y_S|S = t) - E(Y_S|S = c)$, equals a population parameter that has causal relevance, that is, the ACE, $E(Y_t - Y_c)$.

The assumption of strong ignorability is a crucial one for causal inferences in retrospective studies. Because such studies are never randomized, strong ignorability appears to be one of the few constructs available to us for using data from retrospective studies to draw the type of conclusions we might try to make in a prospective study. There are two reasons why we might be willing to assume strong ignorability even when the stronger assumption of randomization is absurd. First of all, if each X-stratum contains a very homogeneous set of units who tend to respond very similarly to t or c then it can be shown that strong ignorability will hold approximately. Second, we may be willing to make the assumption of ignorability because there is nothing in the observed data to contradict it. This is a subtle point and needs elaboration. Suppose that we assumed that S is randomized and, therefore, independent of Y_k and all other variables, including X. Then we could check this assumption by examining the observed distribution of X given S. Under randomization X and S are independent, so

$$P(X = x \mid S = k) = P(X = x). \tag{19.58}$$

Equation (19.58) can be checked with a simple chi-square test for homogeneous proportions, and rejection indicates that S is *not* independent of X and therefore not randomized. However, if we assume that S is *conditionally* independent of Y_k given $X = x$ then we cannot use the observed distribution of X given $S = k$ to disprove this assumption. Hence, strong ignorability is the *strongest* independence assumption that we can make that is not contradictable by the data if we restrict ourselves to (Y_S, S, X).

One point that should be emphasized is that the causal parameters that can be estimated in retrospective studies are more limited than those that can be estimated in prospective studies, even when we are willing to make the strong ignorability assumption in both cases. For retrospective studies we can estimate

$$\alpha(x) = \frac{P(Y_t = 1 \mid X = x)}{P(Y_t = 0 \mid X = x)} \Big/ \frac{P(Y_c = 1 \mid X = x)}{P(Y_c = 0 \mid X = x)}.$$

However in prospective studies we may estimate $\alpha(x)$ but also

$$E(Y_t - Y_c),$$

or

$$E(Y_t - Y_c \mid X = x),$$

and

$$\alpha = \frac{P(Y_t = 1)}{P(Y_t = 0)} \Big/ \frac{P(Y_c = 1)}{P(Y_c = 0)}.$$

Thus when comparing the results of prospective and retrospective studies of the same causal agents it is important to be sure that estimates of comparable parameters are being considered.

4.3. The Role of Matching in Prospective and Retrospective Studies

We close with a comment on an alternative way to describe the difference between prospective and retrospective studies from the point of view of matching.

In prospective matching, units exposed to t and to c are matched on X, whereas in retrospective matching a unit that is a case is matched on X with a unit that is a control. Suppose for simplicity that S is independent of Y_k given X so that at each level of X we have a randomized experiment, that is, the experiment is a randomized block design with the blocks defined by X. Prospective matching reconstructs the randomized block experiment by creating matched pairs of units exposed to t and to c. The average matched-pair difference is an unbiased estimate of the treatment effect for the population defined by the values of X in the matched pairs. Thus prospective matching on X perfectly controls for X in this population whenever both members of each matched pair have the same values of X.

In contrast, retrospective matching on X in general cannot perfectly control for X because it does not reconstruct the randomized block experiment. In each matched pair, one member is a case and one member is a control; to reconstruct the randomized block experiment, one member must be exposed and one unexposed, which generally does not occur when one member is a case and the other a control. Thus summaries from the case-control matched sample, such as the odds ratio, do not represent an estimate of a causal effect for which X has been controlled, even when all matched pairs are exactly matched with respect to X. With retrospective matches, we really need to estimate the odds ratio in each matched pair, and this requires building a model relating Y_1, Y_2 to X and S such as we have illustrated in Section 3.

HISTORICAL NOTE

A version of this article titled "Causal Interference in Prospective and Retrospective Studies" was delivered at the Jerome Cornfield Memorial Session of the American Statistical Association, August 1980, in Houston. The topic of the article was especially appropriate for that session since many important contributions to the study of health effects from prospective and retrospective studies were made by Jerome Cornfield.

20. The Design of the New York School Choice Scholarships Program Evaluation

Jennifer L. Hill, Donald B. Rubin, and Neal Thomas

1. INTRODUCTION

One of Don Campbell's many influential contributions was to the design of studies to estimate causal effects (e.g., Campbell & Stanley, 1966). He had particular interest in the trade-offs between matching and covariance adjustments (e.g., Campbell & Erlebacher, 1970; Cook & Campbell, 1979). One of the authors (Rubin), in fact, had his first conversation with Don on the topic, more than a quarter of a century ago, having recently completed his Ph.D. thesis under the direction of W. G. Cochran on the potential benefits of matching in observational studies. That author believes that the topic of this chapter, using matching in randomized experiments, would have been of great interest to Don and that this chapter would have benefited from his insightful comments. Moreover, we hope that he would have been pleased to see our example of an educational evaluation that did not have to rely on quasi-experimental techniques.

Randomized designs have been recognized since the ground-breaking work of R. A. Fisher in the early part of the 20th century as the most principled way to identify empirically causal relationships between treatments and outcomes. The strength of the randomized design lies in its ability to create treatment groups that have similar background characteristics on average. Randomization balances not only the observed characteristics but also the unobserved characteristics of the experimental units. If relevant characteristics are not balanced across treatment groups, differences in outcomes can be attributed either to effects of the treatments or to those characteristics that systematically vary across treatment groups.

The design literature has expanded over the years to comprise a wide range of different techniques to balance explicitly (through stratification or blocking) or implicitly (through randomization) characteristics of treatment groups. Extensions also have been made to these standard designs to accommodate practical constraints such as timing issues (see, e.g., Efron, 1971) and large numbers of covariates (e.g., Morris, 1979; Pocock & Simon, 1975).

From *Research Designs: Donald Campbell's Legacy*, 155–180. Edited by Leonard Bickman. Copyright © 2000. Reprinted with permission of Sage Publications, Inc.

Experimenters are often prevented, for ethical or logistical reasons, from using one of these standard randomized designs. For the New York School Choice Scholarships Program (NYSCSP), which awarded private school scholarships to disadvantaged New York City public school students, only the simplest type of randomization, a lottery, could be performed because of equity considerations. A fair and nondiscriminatory method of deciding who would receive the limited number of scholarships was considered necessary for public acceptance of the program. Each eligible applicant was given an equal chance of receiving a scholarship, with the exception that students from the worst public schools were assigned a higher probability in the lottery.

Enthusiasm for the scholarships generated more than 20,000 applicants, more than 5,000 of whom eventually were deemed eligible and thus permitted to participate in the lottery. NYSCSP funding, however, was adequate for only approximately 1,300 families to receive scholarships. Applicants who were randomly denied a scholarship are to be used to evaluate the effect of offering a scholarship to the lottery winners.

Evaluation funding allowed for collection of extensive baseline data on those applicants who participated in the lottery. In addition, the outcomes of most scholarship recipients, as measured by test scores and school satisfaction levels, will be tracked. Limited funding for evaluation, however, restricted the number of control members whose outcomes are being tracked to be approximately equal to the number of scholarship awardees. The existing experimental design literature does not adequately address the best choice of a control group in a situation such as this where the control reservoir (all those not awarded scholarships) is several times larger than the eventual control group (the subset of the control reservoir that will be followed).

A simple random sample of eligible applicants who were not awarded scholarships is certainly an option for the choice of control group. This method, however, ignores all the important background information already gathered about the lottery participants through surveys and pretesting, which could be used to improve the balance between the comparison groups. Any balance achieved in addition to that expected through randomization has efficiency implications. For instance, the more balance in covariate means across treatment groups, the lower the standard errors for coefficients in a linear model with those covariates (for discussions, see Campbell & Stanley 1966; Haggstrom, 1976; Rubin, 1984c; Student, 1937). For this reason, multivariate stratified random sampling of the controls to ensure balance on key covariates should be preferable to simple random sampling. This approach, however, is generally limited to only a few covariates, and they must be discrete or be made discrete to create a reasonably small number of strata.

A new experimental design, which couples simple treatment/control randomization with a technique called propensity score matching (Rosenbaum & Rubin, 1983a), will be explored as a solution to this problem. We will show that this design produced, in the NYSCSP, better balance than that achieved by either a simple random sample or a stratified random sample approach. The new design is referred to as a propensity matched pairs design (PMPD).

The NYSCSP is described in Section 2. Section 3 explains the methodology behind the PMPD. Section 4 presents the results of the design implementation, including comparisons to standard designs. Several nuances of both the use of and implications of propensity score methodology in the randomized experiment context are discussed in Section 5. Some concluding comments are given in Section 6.

2. EXPERIMENTAL DESIGN FOR THE NYSCSP

The plan for the lottery included three stages.

1. Interested families would submit applications to the program. For administrative purposes, applications were batched by the date received into five time periods.
2. All potentially eligible families would be invited to a half day of screening, which would include confirmation of eligibility, pretesting of children, and completion of a survey regarding the family's relevant background characteristics. This initial plan was followed for the first batch of applicants; however, because of a variety of logistical constraints, coupled with an overwhelming response to the program, not all potentially eligible families were screened in the next four waves. Sampling of applicants had to be performed to reduce the number invited to participate in the screening stage. To keep the aggregate probability of receiving a scholarship equal across the time periods, the probability of receiving a scholarship among those screened had to be increased to offset the reduced probabilities of being invited to a screening session.
3. Families who completed the screening and whose eligibility was confirmed would be allowed into the final lottery.

In addition, in accordance with the goals of the School Choice Scholarships Foundation, applicants from "bad" schools (85% of the winners) were given a higher chance of winning a scholarship than those from "good" schools (15% of the winners). "Bad" schools are defined as those for which the average test scores were below the median test scores for the city. This oversampling took place during the lottery for those who applied in the first wave (because there was no sampling performed at the *screening* stage). In the second through fifth waves, however, the differential selection of those from good or bad schools was largely accomplished in the sampling at the screening stage. The implication of this difference is that the treatment and control groups in the last four waves are balanced on the bad/good variable, whereas the treatment and initial control groups (i.e., those who didn't win a scholarship) from the first wave are unbalanced on the bad/good variable as well as variables correlated with this variable.

2.1. Multichild Families

The NYSCSP was set up so that all eligible siblings of scholarship winners were also offered scholarships. Because of this, families are the unit of randomization, and all matching and subsampling took place at the family level. Because covariate data were collected not only at the family level but also at the student level, the set of these variables is somewhat different for the families in which more than one child applied to the program ("multichild" families). That is, because our units of observation are families, yet some data are collected at the student level, multichild families have more information than single-child families, so the variable "reading test score," for instance, cannot mean the same thing for all families.

For families with more than one child applying, new family variables were created. These variables were computed across all family members applying. For each family, the average and standard deviation of continuous variables were calculated for initial test scores, age, education expectations, and grade level. The mean and standard deviation are based on available values; if only one value is available for a multichild family, then the standard deviation is missing. For the majority of multichild families, which are two-child families, the original values can be derived from the mean and standard deviation. Binary variables (e.g., bad/good, gifted, special education, and sex) were recoded as 1 if all responding children in the family responded negatively, 3 if all responding children responded positively, and 2 if responses were mixed. Indicators for the presence of any missing data among all family members for each variable were also created.

2.2. PMPD Versus Randomized Block

The study design provides an opportunity to test empirically the performance of the PMPD. In the first application lottery, in which all apparently eligible applicants were invited to be screened, the ratio of eligible nonwinners (control families) to winners (treatment group families) is approximately five to one, an ideal situation for the PMPD. In the second through fifth waves, however, which had smaller control groups because of the limits placed on how many families were invited to be screened, the groups are more nearly equal in size. This latter scenario is more appropriate (given the study design) for a randomized block experiment, with time periods (waves) serving as blocks. Implementing both designs concurrently allows for an empirical comparison of efficiency. The PMPD, however, has a more difficult setting in which to achieve balance because of the initial imbalance on the bad/good variable and other baseline covariates correlated with it.

2.3. Implementation

The implementation of the two designs proceeded as follows. The data can be conceptualized as being divided into four subgroups based on family size (single vs. multiple children) and design (PMPD vs. randomized block). The initial sample sizes,[1] further broken down by time period, are displayed in Table 20.1.

The goal was to equalize sample sizes across treatment groups and then, if possible, across blocks, including across single- versus multichild families. It was apparent that we would only be able to approximate this goal in the stratified study. The limiting factor is the number of multichild control families (137).

---

[1] These are the sample sizes after removal of 100 families randomly chosen from the control group to receive scholarships for the following academic year, and 100 for the year after that. The additional scholarship offers were used as incentives to increase participation in the follow-up data-collection process. New winners were announced following the second and third follow-up testing visits. The Appendix describes how these winners were chosen.

Table 20.1. *Initial sample sizes (unit is a family)*

| Family size | Treatment | PMPD | Randomized block | | | | | |
			2	3	4	5	Subtotal	Total
Single	Scholarship	404	115	67	82	192	456	860
	Control	2,626	72	65	87	135	359	2,985
Multichild	Scholarship	147	44	27	31	75	177	324
	Control	969	27	23	33	54	137	1,106

Because of financial constraints, we could follow up only 2,000 study participants (a "participant" is a family), and thus some random subsampling of lottery winners was performed. Because we had very similar numbers of lottery winners in each design, we targeted a similar number of control families in each design, given in Table 20.2.

2.3.1. PROPENSITY MATCHED PAIRS DESIGN. The strategy for the PMPD was to match 500 subsampled scholarship winners from the first time period to 500 controls from the same time period, with separate matching for single- and multiple-child families. As a consequence of the data set being split into two parts (single- versus multichild families), all matching takes place within family size categories. This exact matching on family size produces perfect balance for this variable; thus, it was implicitly treated as the most important variable.

Determinations had been made by the evaluators as to the relative "importance" of the remaining covariates. As described further in Section 3.1.3, importance is judged by a combination of the initial imbalance of a covariate across treatment groups and the perceived strength of the predictive relationship of it to postrandomization outcome measures, which include the primary outcomes themselves, noncompliance behavior (referring to whether or not a family uses an offered scholarship), attrition from the study, and other types of missing data.

After family size, the most important variable by this definition was judged to be the binary variable for bad versus good school, because we think it will be correlated with the outcomes and because of the imbalance that occurred in the first time period as a result of its use in determining lottery winners. It is closely followed in importance by grade level and initial test scores. The remaining covariates are ranked in order as ethnicity, mother's education, participation in special education,

Table 20.2. *Target sizes for both scholarship and control samples*

Family size	PMPD	Randomized block	Total
Single	353	323	646
Multi	147	177[a]	354
Overall	500	500	1,000

[a] Only 137 available in control group.

Table 20.3. *Final sample sizes*

Family size	Treatment	PMPD	Randomized block				Subtotal	Total
			2	3	4	5		
Single	Scholarship	353	72	65	82	104	323	676
	Control	353	72	65	82	104	323	676
Multichild	Scholarship	147	44	27	31	75	177	324
	Control	147	27	23	33	54	137	284
	Total	1,000					960	1,960

participation in a gifted and talented program, language spoken at home, welfare receipt, food stamp receipt, mother's employment status, educational expectations, number of siblings (includes children not eligible because of age), and an indicator for whether the mother was foreign born. The final propensity score models, presented in Sections 4.1 and 4.2, were chosen based on the balance created in these variables' distributions across treatment groups. Identification of special variables and the overall ranking of the covariates informed decisions regarding which variables might be appropriate for exact matching, which should receive special treatment in the propensity score method, and what trade-offs to make in terms of the resulting balance.

The ranking of the variables can be helpful in implementing the propensity score methodology; however, correlations among the variables diminish the importance of the ordering chosen. The specific ordering chosen therefore may not have a major impact on the creation of matched pairs and should not be viewed as an assumption required for successful implementation.

2.3.2. SUBSAMPLING FOR THE RANDOMIZED BLOCK DESIGN. We randomly subsampled from the cells of the randomized block design to arrive at the final sample sizes, which met the limitation of 1,000 families per design. The numbers subsampled were selected to equalize the number of scholarship and control families within blocks and the number of families across blocks.

1. One hundred thirty-three original single-child lottery winners were randomly withheld for the randomized block design: 43 in time period 2, 2 in time period 3, and 88 in time period 5.
2. Thirty-six single-child eligible controls were randomly withheld from randomized block design: 5 in time period 4 and 31 in time period 5.

The final sample sizes are displayed in Table 20.3.

3. GENERAL PROPENSITY SCORE METHODOLOGY

Propensity score matching was introduced by Rosenbaum and Rubin (1983a) as a means of creating better balance in observational studies, thereby allowing for valid causal inference under the assumption of strongly ignorable treatment assignment, that is, treatment assignment on the basis of the covariates being used to estimate the propensity score. Matching is used as a way of alleviating the biases that can be created by self-selection as well as other "threats to validity" described in Cook and Campbell (1979). As documented in a variety of places (e.g., Rubin 1973b, 1979b), the combination of matching and regression adjustment is typically far superior to either technique alone for controlling bias in observational studies. Not only does matching reduce bias created by the self-selection into treatment groups that occurs in observational studies, but it also increases efficiency in randomized experiments, such as the one in this study. The extra payoff from matching mostly arises when the linear model underlying regression adjustment is not entirely correct.

The propensity score methods that we use are well documented and, in the case of no missing data, quite straightforward (Rosenbaum & Rubin 1984b). When missing data exist, as they do in this study, extensions of the general methodology (D'Agostino & Rubin, 2000) can be implemented. The goal is to balance closely all covariates and patterns of missing data across the treated and matched control groups.

3.1. Complete Data

In the case of complete data, the general strategy is to calculate a "propensity score" for each study participant. This score represents a participant's chance or "propensity" of receiving the treatment (e.g., a scholarship offer),

$$P(Z = 1|X), \tag{20.1}$$

where Z denotes treatment assignment and X denotes all the measured covariates (recall, here, fully observed). This probability is straightforward to estimate using logistic regression or linear discriminant techniques.

3.1.1. MATCHING ON THE PROPENSITY SCORE. The propensity scores can be regarded as defining a new covariate value for each individual that is a function of all the covariates potentially correlated with the outcomes. In practice, the logits of these estimated probabilities are often used because they are linear in the covariates. Balancing this new covariate generally has the effect of improving the balance of all the other covariates that went into its estimation. A good way to balance propensity scores when the treatment group is much smaller than the control reservoir is to match on propensity scores. Procedurally, this can be accomplished by sorting the treatment group members by their propensity scores and then, one by one, finding for each treated subject the control group member who has the closest score. Once a match has been made, the chosen control group member is removed from the control reservoir so it cannot be chosen again (Cochran & Rubin, 1973). This is called nearest remaining neighbor, or nearest available, matching.

3.1.2. NEAREST AVAILABLE MAHALANOBIS MATCHING WITHIN PROPENSITY SCORE CALIPERS. The Mahalanobis metric (or distance) between a treatment group member with vector covariate values X_t and a control group member with covariate values X_c (the same set of variables for both) is

$$(X_t - X_c)^T \Sigma^{-1} (X_t - X_c), \tag{20.2}$$

where Σ is the variance-covariance matrix for these variables; in practice, we substitute the pooled sample variance-covariance matrix. A combination of propensity score matching and matching based on the Mahalanobis metric has many of the advantages of each method (Rubin & Thomas, 1996). The combination has been shown to be superior to either technique used on its own (Rosenbaum & Rubin, 1985a). With this refinement, as before, propensity scores are calculated for all study participants and then treatment participants are ordered by their propensity scores. Each treatment group member in turn will be initially "matched" to a subset of the control reservoir members whose scores are no more than c propensity score units (e.g., $c = 0.10$ propensity score standard deviations) away from the treatment member's propensity score. Thus, the initial matches must fall within a $2c$ length propensity score caliper, symmetric about that treatment group member's score.[2] Mahalanobis matching is used to choose a "nearest neighbor" within this subset of study participants with respect to several special covariates. The control group member whose values, X_c, of the special covariates minimize the distance from the values, X_t, of the special covariates for the treatment member is chosen from the subset of controls who fall within the caliper. We include only the continuous covariates most predictive of the outcome variables in the Mahalanobis metric, as discussed in Section 3.1.3.

3.1.3. SPECIAL VARIABLES. The more predictive a covariate is of the outcomes of interest, the more crucial is the balance of this covariate across treatment groups. For example, controlling for a covariate (e.g., by balancing) that is uncorrelated with the outcomes plays no useful role, whereas controlling for one that is highly correlated with the outcome will play a crucial role for precise estimation.

Covariates that evaluators are most concerned about balancing receive special treatment in one of two ways. When feasible, exact matches can be required for the most critical of these variables. For instance, if sex were deemed to be the most important variable to balance, when looking at matches for a female treatment group member, no males would be considered. It is only possible to exact match on discrete variables and only desirable to match on one or two of these. (For an example of exact matching in a propensity score context, see Rosenbaum and Rubin [1984b].) Recall that in this study, we exact match on family size.

As an alternative to, or in addition to, this exact matching, the Mahalanobis matching within propensity score calipers can be constrained to only a chosen few variables considered more important to balance than the others. Mahalanobis matching is most effective when applied to a small number of essentially continuous covariates (Gu & Rosenbaum, 1993; Rosenbaum & Rubin, 1985a). Matching within propensity score calipers attempts to improve balance for all the covariates, whereas

---

[2] This technique is described and illustrated in the context of a real-life example in Rosenbaum and Rubin (1985a).

Mahalanobis matching within calipers attempts to achieve close pair matches on the few special covariates.

3.2. Advantages Over ANCOVA (Analysis of Covariance) Adjustments

We have already mentioned the benefits of using matching in addition to ANCOVA (regression adjustments) for both bias reduction and precision of estimation. There is another benefit of matching relative to regression adjustment. Adjusting for covariate differences after the experiment has the disadvantage that researchers could settle on the "best" model solely by choosing the one that best supports their a priori biases regarding the issue in question. Matching, on the other hand, uses only covariate balance as a diagnostic; outcomes are not even included in the model, nor are they often even available at the time of matching, as in our application. No such researcher bias, therefore, can occur in the selection of the propensity score model.

3.3. Diagnostics

A variety of combinations of the above techniques each will yield "matched" treatment and control groups. The estimation of the propensity score alone could be accomplished by numerous models, depending on what variables are included and what interactions or nonlinear terms are added. Diagnostics, which compare the treatment and control groups with respect to the covariates, help the researcher determine which matched control group is superior. Because the goal of the matching is balanced groups, the adequacy of a model or procedure can be judged by treatment versus control group comparisons of sample moments of the joint distribution of the covariates, primarily means and variances but also correlations. It is often helpful at this stage to have a ranking of covariates in order of perceived importance, beyond just the few selected to be "special" variables. Such a ranking, as described for this study in Section 2.3.1, can help the researcher choose between models with good overall balance that have slight trade-offs in terms of more or less exceptional balance on specific variables.

3.4. True Versus Estimated Propensity Scores

A surprising fact about the use of propensity scores is that, in general practice, the use of the estimated propensity score typically results in more precise estimates than the use of the "true" population propensity score. This is especially true when the treatment and control groups are relatively similar initially. The logic is as follows. Two types of errors can result from estimates of treatment effect. The first involves systematic biases, which occur when, in expectation, the two groups differ on important characteristics. The second involves conditional biases, which refer to the random differences between groups that average to zero over repeated samples but are nonetheless present in any given sample. Both population and estimated propensity scores effectively reduce the systematic bias in samples, but estimated propensity scores more effectively reduce sample-specific randomly generated bias (Rubin & Thomas, 1992b). Because a randomized lottery was held to determine scholarship receipt, there is no systematic bias, so the estimated propensity scores would more effectively reduce conditional bias.

3.5. Incomplete Data

Rarely is a study performed that generates no missing data. Techniques to estimate propensity scores in the presence of missing data have been proposed by D'Agostino and Rubin (1996). The type of strategy that is optimal depends on how the missing data were generated and the relationship of this missingness to the outcomes of interest.

The NYSCSP study starts from the advantageous position of a randomized design, within which incomplete baseline data are less problematic than in the case of an observational study. The goal is simply to get the best possible balance on all covariates that we expect to be predictive of outcomes. To the extent that the "missingness" of our covariates is predictive of outcomes, we want propensity score models that include information about the missing data mechanisms (e.g., indicators for the missingness of a particular variable) so as to balance the missingness across treatment groups better than it would be balanced by chance alone. If we believe that this missingness is predictive of the outcomes, then this balance has efficiency implications for our inferences about treatment effects, just as better balance on any other covariate improves efficiency of estimation. In addition, to the extent that we will be forced to model other outcome phenomena, such as compliance behavior, it will be important to consider which covariate-specific missing data mechanisms are correlated with these outcomes.

As an example, in the NYSCSP there were single mothers in the study who refused to fill out the part of the application survey pertaining to the father of the child. The missingness of these variables could be viewed as a proxy measure for the strength of the relationships in the family and so was hypothesized a priori to be predictive of the outcomes. This missingness "variable" therefore was included in our propensity model so that we could try to improve its balance across treatment groups.

The other missingness indicator chosen by evaluators as important in this study was that corresponding to mother's education. Investigators think that a missing response to this question reflects a mother's attitude toward education, which could be predictive of educational outcomes, compliance behavior, or subsequent missing data.

The techniques appropriate for including missing data mechanisms in a model are more complicated than those we discussed in Section 3.1. We used a computer program written by one of the authors (Thomas) to implement the technique developed by D'Agostino and Rubin (1996), which relies on the ECM algorithm (Meng & Rubin, 1993) to calculate propensity scores for each subject, including those with missing covariate values. The ECM algorithm is a variant of the standard EM algorithm, which is used in situations where the maximization step is computationally awkward. It replaces the M-step with two or more conditional maximization (CM) steps, each of which has a straightforward solution.[3]

---

[3] For the general location model (often used with missing data; e.g., Little and Rubin [1987] and Schafer [1997]), one CM-step gets maximum likelihood estimates for the parameters in the normal distributions conditional on the parameters for the log-linear model (cell probabilities for the contingency table), and a second CM-step obtains estimates for the log-linear model conditional on the parameters of all the multivariate normal distributions.

The Mahalanobis matching within propensity score calipers in the NYSCSP project was modified for missing covariate values as follows. If possible, for the matched control, the same missing pattern was required. If no such matched control was found, we exact matched on the design variable bad/good school, which was fully observed. If a matched control still was not found, we would have matched on the propensity score alone; however, this situation never occurred.

4. RELATIVE STRENGTHS OF DESIGNS – DIAGNOSTICS

We can judge the relative strengths of our designs through diagnostics that measure balance in various ways. Results from the PMPD will be contrasted with results from both the randomized block design (second through fifth time periods) and a simple random sample chosen from the control reservoir in the first time period.

4.1. Single-Child Families

Following the criteria discussed in Section 2.3.1, a model for the propensity score was chosen. The contingency table for the categorical variables – ethnicity (Hispanic/Black/other), religion (Catholic/other), participation in gifted program, participation in special education, and winning a scholarship – is constrained by a log-linear model that allows for two-way interactions. The continuous portion of the general location model places an additive model across contingency table cells on the means of the following variables: language (Spanish/English), whether or not father's work status is missing, participation in food stamp program, participation in Aid to Families with Dependent Children (AFDC), bad/good school, mother's birth location (U.S./Puerto Rico/other), sex, number of eligible children in household, income, mother's education, math scores, and grade level. Mahalanobis matching was done in 0.10 calipers of propensity score standard deviations on the two test score variables and the grade level variable; the bad/good variable also played a special role in the Mahalanobis matching as described in Section 3.5. For algorithmic efficiency, discrete variables that are fully observed (such as bad/good), and any of their interactions, can be treated as continuous with no loss of generality. This is preferable because it reduces the dimensionality of the model.

The resulting balance is displayed in Table 20.4. In the table, the column heading Z is for the Z statistic corresponding to the difference in means between the two groups for a covariate;[4] "pmist" and "pmisc" stand for "percent missing" in the treatment and control groups, respectively. The results for the PMPD are compared to the results for the randomized block design and to the results of a simple random sample of the same size from the pool of all potential matching subjects.

---

[4] This is calculated for each covariate, x, as

$$\frac{\bar{x}_t - \bar{x}_c}{\sqrt{\hat{\sigma}_t^2/n_t + \hat{\sigma}_c^2/n_c}},$$

where t and c subscripts denote sample quantities from the treatment and control groups, respectively.

Table 20.4. *Balance: single-child families*

Variable	Simple random sample			PMPD			Randomized block		
	Z	pmist[a]	pmisc[b]	Z	pmist	pmisc	Z	pmist	pmisc
Bad/good	−5.41	0.00	0.00	0.11	0.00	0.00	0.21	0.00	0.00
Grade level	0.45	0.02	0.02	−0.03	0.02	0.01	−0.39	0.02	0.01
Reading score	−2.03	0.38	0.35	0.48	0.38	0.37	−1.05	0.35	0.32
Math score	−0.14	0.51	0.43	0.20	0.51	0.51	−1.37	0.36	0.34
Ethnicity	0.05	0.05	0.05	1.59	0.05	0.05	1.74	0.06	0.05
Mom's education	−0.78	0.06	0.05	0.09	0.06	0.07	1.67	0.09	0.03
Special education	2.27	0.01	0.01	−0.17	0.01	0.02	0.22	0.03	0.01
Gifted program	−0.31	0.01	0.03	−0.13	0.01	0.04	0.75	0.03	0.02
Language	2.15	0.06	0.08	−1.03	0.06	0.08	−0.44	0.07	0.04
AFDC	−0.51	0.10	0.10	0.83	0.10	0.11	−1.57	0.11	0.11
Food stamps	−1.39	0.06	0.07	0.94	0.06	0.06	−1.31	0.06	0.07
Mother works	−1.10	0.05	0.05	−1.18	0.05	0.08	0.40	0.07	0.05
Education expectations	−0.36	0.05	0.05	0.57	0.05	0.03	0.19	0.04	0.04
Eligible in household	−2.07	0.03	0.05	0.41	0.03	0.06	−1.02	0.07	0.06
Birth location	−0.15	0.03	0.03	−1.40	0.03	0.03	−0.69	0.07	0.05
Length of residence	1.10	0.03	0.03	0.66	0.03	0.02	−0.78	0.02	0.02
Dad's work missing	0.00	0.00	0.00	0.00	0.00	0.00	0.16	0.00	0.00
Religion	0.04	0.05	0.05	−0.74	0.05	0.06	−0.80	0.06	0.08
Sex	0.23	0.02	0.02	0.76	0.02	0.02	0.53	0.02	0.01
Income	−1.86	0.60	0.06	0.74	0.06	0.07	−1.21	0.08	0.08
Age as of 4/97	0.87	0.02	0.04	−0.47	−0.02	0.02	−0.87	0.02	0.02

[a] pmist = percentage missing in treatment group.
[b] pmisc = percentage missing in control group.

Overall, the resulting balance from the PMPD is quite good. Notice the large improvement in the balance of the bad/good variable as compared to that achieved using a simple random sample of the control reservoir from the first application lottery. Compared to the randomized block design, the PMPD has lower absolute Z scores for 16 variables and higher Z scores for 5. It is beaten by the simple random sample for 8 variables; however, the gains when PMPD wins over the simple random sample are generally larger than the gains when the simple random sample wins over PMPD.

Propensity score theory predicts a gain in efficiency for differences in covariate means over simple random sampling by a factor of approximately two (Rubin & Thomas 1992b, 1996). A general method for checking whether a data set appears to follow a given distribution is to plot the ordered observations against the corresponding quantiles from the baseline distribution; this is referred to as a quantile-quantile,

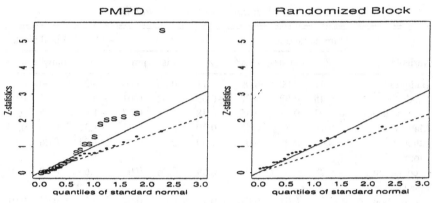

Figure 20.1. Half-normal plots of Z statistics for single-child families.

or q-q, plot (for a description, see Chambers, Cleveland, Kleiner, and Tukey [1983]). The closer the observations fall to a line with a slope of 1 running through the plot, the more evidence we have that they truly follow the distribution in question. We have constructed half-normal plots of the Z scores, which display the same information as a standard normal quantile plot without regard to signs, by taking the absolute value of each difference and comparing these to the χ_1 (the positive half of a standard normal) distribution.

We formed the Z statistics displayed in Table 20.4 by standardizing the differences in means for each covariate by a variance estimate that assumes random allocation to treatment groups. Therefore, we expect these Z statistics to follow the standard normal distribution when the assumptions of random allocation is true (thus, the Z statistics are expected to fall on the solid line with slope 1 in each diagram). If the observations fall above the line with slope 1, they originate from a distribution with larger variance than we are using to standardize the differences, because they are systematically more dispersed than the corresponding quantiles of the standard normal. If they fall below that line, they originate from a distribution with smaller variance than we are using to standardize the differences because they are systematically less dispersed than the standard normal.

In Figure 20.1, the dotted line has slope $1/\sqrt{2}$, corresponding to the normal distribution with variance 1/2. This figure thus reveals that in practice the gains predicted by Rubin and Thomas (1992b) are fairly closely achieved for the single-child families. In fact, compared to a simple random sample (shown by the "S" points), we see even greater gains because of the initial imbalance in the bad/good variable. These results can be contrasted with those from the randomized block experiment, which are consistent with the theoretically predicted slope of 1 given by the solid reference line.

Because the variance in the difference in means is reduced by a factor of two, this is equivalent to increasing the sample size by a factor of two for these variables. Although this holds for any linear combination of the measured covariates, an outcome variable may not be perfectly predicted by these variables, resulting in a less dramatic improvement in efficiency (Rubin & Thomas, 1996).

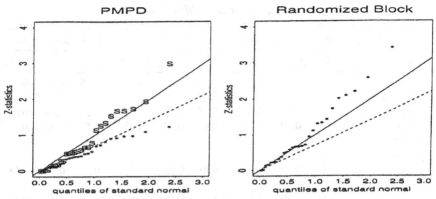

Figure 20.2. Half-normal plots of Z statistics for multichild families.

4.2. Multichild Families

Following the criteria discussed in Section 2.3.1, a propensity model was chosen. The contingency table for the categorical variables (ethnicity, religion, sex, birth location, and winning a scholarship) is constrained by a log-linear model that allows for two-way interactions. The continuous portion of the general location model places an additive model across contingency table cells on the means of the following variables: participation in gifted program, participation in special education, language, whether father's work status is missing, participation in food stamp program, participation in AFDC, bad/good, number of eligible children in household, income, mother's education, mother's length of residence, mother's work status, average and standard deviation of children's ages, average and standard deviation of educational expectations, average and standard deviation of math and reading scores, and average and standard deviation of grade. Mahalanobis matching was done in 0.10 calipers of propensity score standard deviations on the four test-score variables and the two grade-level variables; the bad/good variable also played a special role in the Mahalanobis matching as described in Section 3.5.

The resulting balance of the design as compared with the corresponding randomized block design and with a simple random sample of the potential matches is displayed in Table 20.5. The initial imbalance in the bad/good variable is also present with the multichild families, but the PMPD still achieves very good overall balance. Compared to the randomized block design, the PMPD has lower absolute Z scores for 17 variables and higher Z scores for 8. The win/loss ratio is 16 to 9 for the comparison with the simple random sample, although again the gains when PMPD wins over the simple random sample are generally larger than when simple random sample wins over PMPD.

Half-normal quantile-quantile plots for the multichild families in both experiments, displayed in Figure 20.2, are similar to those for single-child families. Gains in efficiency by a factor of two appear to be achieved by the PMPD over the randomized block design, with gains by a factor slightly greater than two over the simple random sample because of the initial imbalance in the bad/good variable.

Although the special test score variables are not quite as well balanced in the PMPD as in the randomized block design for the multichild families (probably

Table 20.5. *Balance: multichild families*

Variable	Simple random sample			PMPD			Randomized block		
	Z	pmist[a]	pmisc[b]	Z	pmist	pmisc	Z	pmist	pmisc
Bad/good	−1.68	0.00	0.00	−0.98	0.00	0.00	0.15	0.00	0.00
Average grade level	−0.03	0.00	0.06	0.38	0.01	0.01	0.23	0.03	0.03
sd grade level	−0.50	0.10	0.11	−0.40	0.07	0.05	0.58	0.06	0.06
Average reading score	1.14	0.22	0.25	0.91	0.21	0.20	−0.23	0.14	0.18
sd reading score	−0.50	0.49	0.63	1.23	0.59	0.57	−2.20	0.53	0.58
Average math score	0.14	0.27	0.25	0.82	0.25	0.22	0.32	0.11	0.18
sd math score	−0.15	0.59	0.70	0.33	0.65	0.65	−1.11	0.53	0.56
Ethnicity	−0.79	0.05	0.10	0.20	0.07	0.05	2.09	0.07	0.06
Mom's education	0.66	0.02	0.14	−0.21	0.06	0.03	−0.22	0.10	0.10
Special education	−0.55	0.00	0.06	−0.11	0.01	0.01	0.68	0.03	0.02
Gifted program	1.73	0.00	0.07	−0.07	0.03	0.01	−0.52	0.05	0.02
Language	0.52	0.15	0.11	0.92	0.11	0.07	−0.64	0.14	0.09
AFDC	0.27	0.05	0.15	0.13	0.10	0.10	3.42	0.14	0.10
Average age	−0.18	0.02	0.07	0.48	0.03	0.01	0.66	0.03	0.03
sd age	−0.15	0.10	0.11	0.00	0.09	0.05	0.38	0.08	0.07
Average educational experience	0.02	0.02	0.07	0.49	0.02	0.01	−0.71	0.06	0.06
sd educational experience	−1.54	0.15	0.15	−0.10	0.13	0.07	0.94	0.12	0.12
Eligible in household	−1.34	0.05	0.10	−0.40	0.08	0.04	−0.13	0.10	0.08
Religion	0.28	0.07	0.11	−0.97	0.07	0.04	−0.66	0.10	0.04
Length of residence	−1.94	0.02	0.07	0.54	0.03	0.03	1.31	0.06	0.04
Dad's work missing	−1.68	0.00	0.00	0.70	0.00	0.00	1.73	0.00	0.00
Food stamps	0.52	0.05	0.15	−0.35	0.07	0.06	2.58	0.09	0.10
Mom works	0.00	0.12	0.14	0.73	0.10	0.06	−0.49	0.10	0.06
Birth	1.26	0.02	0.06	−0.42	0.05	0.01	1.34	0.05	0.06
Sex	0.59	0.00	0.06	−0.17	0.02	0.01	−1.43	0.04	0.04

[a] pmist = percentage missing in treatment group.
[b] pmisc = percentage missing in control group.

because of correlations between these and the bad/good variable), they are still well balanced. Furthermore, the high correlation commonly seen between pre- and posttest scores makes this variable a prime candidate for covariance adjustments within a linear model to take care of the remaining differences between groups. For the single-child families, the PMPD is clearly superior in terms of test-score variable balance.

5. DISCUSSION

This section presents further discussion of the PMPD, including suggestions for future investigation and additional potential gains over standard experimental designs.

5.1. Comparison of Propensity Scores in Observational Studies and Randomized Experiments

Part of what we have learned from these results is that we need to investigate the differences in optimal implementation of the propensity score estimation methodology in observational studies (its traditional use) and in randomized experiments (its use here). In observational studies, those variables that are initially most highly imbalanced are those that help determine whether the individual sought "treatment." These often tend to be highly correlated with outcomes, so treating them as special variables is often good practice. In randomized studies, differences in magnitude of initial imbalance are purely random, so although we are still trying to choose those covariates most predictive of outcomes, more thought needs to go into the decision of how to choose special variables: In a randomized experiment, the covariates most important in the estimated propensity score may not be highly related to outcomes.

5.2. Other Potential Advantages of the PMPD

The gains expected for the PMPD relative to the simple random sample and randomized block design were realized. The primary benefit should be more precise estimation of treatment effects resulting from more closely matched covariates' means. This increased precision is especially important given that effect sizes in educational initiatives are notoriously small. In addition, treatment effect estimates for this project may be particularly noisy given that the effect of interest, "private school attendance," is actually a mixture of many different effects because each participating private school can be conceptualized as a different treatment.

In addition to the more precise estimation in a setting with fully observed values, we expect modest gains in settings with real-world complications including attrition, missing data, and noncompliance. These additional advantages are enumerated below. Experience will provide guidance regarding the strength of these effects in practice.

1. We expect a more equal dropout rate across the treatment groups (where "dropout" refers to someone who leaves our analysis sample, such as by moving). Because the variance of the estimated treatment effect is usually proportional to

$(1/n_1 + 1/n_2)$, where n_1 and n_2 represent the final sample sizes in each of the treatment groups, for fixed $n_1 + n_2$, $n_1 = n_2$ implies greater precision.

2. We expect more similar patterns of missing covariates and missing outcomes other than those due to dropout from the sample. This, again, has efficiency implications. In addition, naive procedures for the missing data problem, such as complete-case or available-case analyses, will be less likely to create the same amount of bias and efficiency loss as they would with a simple random sample of the control group. We have found more similar patterns of missing covariates across treatment groups in the PMPD relative to the other designs.

3. Subgroup analyses should have more power. The smaller the subgroup size, the more likely that simple random sampling will create imbalance. Consider, for example, a subgroup of 20 Hispanic, male second graders from single-parent homes. If 15 of these are from treatment and only 5 from control, the estimates will be very imprecise. Matching should help to alleviate this problem by creating more nearly equal sample sizes for these groups.

4. We would expect similar compliance patterns across treatment groups. Correct treatment of noncompliance issues requires disentangling mixtures components within treatment arms (e.g., see Angrist, Imbens, & Rubin, 1996). These mixture components are subgroups, which, as previously noted, are more easily unbalanced when using simple random sampling rather than matching.

There should also be advantages in terms of perceived equity. For instance, enhanced balance in subgroups will make the lottery appear more "fair" to outsiders doing post hoc analyses. As an example, it is more likely with matching that there will be a more similar number of African American children in treatment and control groups. This result did in fact occur.

6. CONCLUSION

This study provides a unique opportunity to assess empirically the effectiveness of a new design, the PMPD (propensity matched pairs design). It is advantageous that the design was introduced so early in the school choice debate that future evaluations, likely with similar logistical constraints, can benefit from what we have learned and from what we will continue to discover about it in this evaluation. In addition, we are in the position to assess the PMPD before receiving outcome data, thereby separating critics' views about the effectiveness of the methodology from their views about the subsequent findings regarding the effectiveness of school choice itself. We find that PMPD methodology, though requiring more complex theoretical development, is flexible, is easy to apply in practice, and can produce more efficient experimental designs. The PMPD produced a design with balance between measured covariates like that of a corresponding randomized design with twice as many subjects. The PMPD design is demonstrably better than the randomized block design of a similar size, which was also created for the evaluation. It achieved superior results despite being applied to a subset of the families with substantially larger initial differences as a result of the operational requirements of the NYSCSP lottery. More development, both theoretical and applied, is needed to determine what other benefits might result from the PMPD.

Table 20.A1. *Initial sample sizes before selection of new winners*

| Family size | Treatment | PMPD | Randomized block | | | | Subtotal | Total |
			1	2	3	4		
Single	Scholarship	404	115	67	82	192	456	860
	Control	2,697	86	76	106	156	424	3,121
Multichild	Scholarship	147	44	27	31	75	177	324
	Control	998	34	33	41	64	172	1,170

APPENDIX

New Winners

Concerns regarding possible noncompliance and missing outcome data led to an added complication in the implementation of the design. To provide an incentive for the controls to participate in the testing in the spring, controls desired for follow-up received a letter offering them admittance to a special lottery for next year's scholarships if they agreed to participate in the study (e.g., take a posttest this spring). They were informed that this special lottery would give them a greater probability of winning than they would have by applying to the regular lottery for those scholarships. A total of 100 families was slated to receive these special scholarships in each of the next 2 years. The second- and third-year special scholarship winners were randomly chosen from the control reservoir (all those in our files who applied, were eligible, and were invited to participate in the lottery but didn't win a scholarship). This occurred before implementation of either design.

The choice of new scholarship winners proceeded as follows. The data can be conceptualized as being divided into four subgroups based on family size (single vs. multiple children) and design (PMPD vs. randomized block). The sample sizes before new winners were chosen, further broken down by strata in the second design, are displayed in Table 20.A1.

Second-Year Lottery Winners

Using the target sample sizes from Table 20.2, described in Section 2.3, the relative proportion of each cell in our analysis should be approximately as shown in Table 20.A2. To give each student in the follow-up callback an equal chance to win the new lottery conditional on his or her study cell membership, we selected the 200 new winning families (100 for the 1998–1999 school year and 100 for the following year) randomly from these four cells with size proportional to their final representation (e.g., all multichild families were invited back, so a higher proportion

Table 20.A2. *Proportion of sample in each subgroup*

Family size	PMPD	Randomized block
Single	0.353	0.323
Multichild	0.147	0.177

Table 20.A3. *Number of winners in each subgroup*

Family size	PMPD	Randomized block
Single	71	65
Multichild	29	35

of them are winners than in the general population). The numbers of winners are displayed in Table 20.A3.

Calculations for Probability of Selection for "New Winners"

It is important to calculate the probability of winning a scholarship for new winners to ensure fulfillment of the promise given to those who agreed to participate in the study. This calculation is, not surprisingly, highly dependent on the interpretation of this phrase.

First, focus on those families who will win scholarships for next year (as opposed to the following year). Simplistically, one might think of the probability of winning a scholarship for next year in the special lottery as just 100 divided by the number of families in the control reservoir (approximately 5,000). The fact that the phrase "probability of winning a scholarship" will be used only in a letter received by a subset of this control reservoir, however, argues for a *conditional* probability, the probability of winning *given* that a family has been asked to be part of the control group.

Specifically, *if* they agree to be a member of this group, *then* they will have a certain probability of winning a scholarship. If they agree to participate, then their probability of winning becomes conditional on being a member of that group, and the number of people in that group will be the denominator in associated probability calculations. In that case, the conditional probability of a family winning given that it is part of the control group is 100/1,200, or 1/12; the denominator of 1,200 consists of the 1,000 core control group members and the 200 supplemental control group members chosen in the special lottery.

Other forms of conditional probabilities may be of interest. For instance, we could exclude from the denominator those people who have won scholarships for the third year. This would change the conditional probability of winning in a scholarship in the second year to 1/11. We could look at the probability of winning over the next 2 years (again conditional on being invited to be in the control group). This conditional probability would be 1/6.

ACKNOWLEDGMENTS

Thanks are due Paul Peterson and David Myers for involving us in this project and for their insights regarding the New York School Choice Scholarship Program. Thanks are also due William Howell for helping to address data questions.

21. Estimating and Using Propensity Scores with Partially Missing Data

Ralph B. D'Agostino, Jr., and Donald B. Rubin

Abstract: Investigators in observational studies have no control over treatment assignment. As a result, large differences can exist between the treatment and control groups on observed covariates, which can lead to badly biased estimates of treatment effects. Propensity score methods are an increasingly popular method for balancing the distribution of the covariates in the two groups to reduce this bias; for example, using matching or subclassification, sometimes in combination with model-based adjustment. To estimate propensity scores, which are the conditional probabilities of being treated given a vector of observed covariates, we must model the distribution of the treatment indicator given these observed covariates. Much work has been done in the case where covariates are fully observed. We address the problem of calculating propensity scores when covariates can have missing values. In such cases, which commonly arise in practice, the pattern of missing covariates can be prognostically important, and then propensity scores should condition both on observed values of covariates and on the observed missing-data indicators. Using the resulting generalized propensity scores to adjust for the observed background differences between treatment and control groups leads, in expectation, to balanced distributions of observed covariates in the treatment and control groups, as well as balanced distributions of patterns of missing data. The methods are illustrated using the generalized propensity scores to create matched samples in a study of the effects of postterm pregnancy.

1. INTRODUCTION

1.1. Background on Propensity Scores

Since they were introduced by Rosenbaum and Rubin (1983a), propensity scores have been used in observational studies in many fields to adjust for imbalances on pretreatment covariates, $\mathbf{X}$, between a treated group, indicated by $Z = 1$, and a control group, indicated by $Z = 0$ (e.g., D'Agostino 1998; Dehejia and Wahba 1999; Rubin 1997). Propensity scores are a one-dimensional summary of multidimensional covariates, $\mathbf{X}$, such that when the propensity scores are balanced across the treatment and control groups, the distribution of all the covariates, $\mathbf{X}$, are balanced in expectation across the two groups. Typically, matched sampling (e.g., Heckman, Ichimura, Smith, and Todd 1996; Lytle et al. 1999; Rosenbaum and Rubin 1985a; Takizawa et al. 1999; Willoughby et al. 1990) or subclassification (e.g., Barker et al.

1998; Conners et al 1996; Nakamura et al. 1999; Rosenbaum and Rubin 1984b; U.S. General Accounting Office 1994) on estimated propensity scores is used, often in combination with model-based adjustments (Curley, McEachern, and Speroff 1998; Lieberman et al. 1996; Rich 1998; Rubin and Thomas 2000; Smith et al. 1998).

The propensity score for an individual is the probability of being treated conditional on the individual's covariate values. To estimate propensity scores for all individuals, one must model the distribution of Z given the observed covariates, $\mathbf{X}$. There is a recent and large technical literature on propensity score methods with complete data (e.g., Gu and Rosenbaum 1993; Rubin and Thomas 1992a,b, 1996). In practice, however, typically some covariate values will be missing, and so it is not clear how the propensity score should be estimated. In addition, the missingness itself may be predictive about which treatment is received in the sense that the treatment assignment mechanism is ignorable (Rubin 1978c) given the observed values of $\mathbf{X}$ and the observed pattern of missing covariates but not ignorable given only the former.

Rosenbaum and Rubin (1984b) considered using a "pattern mixture" model (Little 1993; Rubin 1978c) for propensity score estimation with missing covariate data. Appendix B of Rosenbaum and Rubin (1984b) defined a "generalized" propensity score as the probability of treatment assignment given $\mathbf{X}^*$, the vector covariate with an asterisk indicating a missing component of the vector covariate $\mathbf{X}$ (as in Rubin 1976a). This is equivalent to conditioning on the observed values of $\mathbf{X}$, $\mathbf{X}_{obs}$, and a missing covariate indicator R ($R = 1$ for observed, $R = 0$ for missing); with discrete covariates, this is equivalent to adding an additional "missing" category to each covariate. Rosenbaum and Rubin (1984b) proved that adjustment for the "generalized" propensity score in expectation balances the observed covariate information and the pattern of missing covariates. They suggested that in large enough samples, one can estimate this generalized propensity score by estimating a separate logit model using the subset of covariates fully observed for each pattern of missing data. The practical problem is that typically there are many patterns of missing data with only a few individuals from each of the two treatment groups, thereby making the straightforward pattern-mixture approach infeasible. Thus estimating the generalized propensity score requires smoothing the parameters of Z given $\mathbf{X}_{obs}$ across the patterns of missing data.

Our approach is to model the joint distribution of $(Z, \mathbf{X}, R)$. The particular approach that we implement in our application is based on a general location model (Olkin and Tate 1961) accounting for the missing data (Schafer 1997). This modeling implies a conditional distribution for Z given $(\mathbf{X}_{obs}, R)$; that is, the generalized propensity score: probabilities of $Z = 1$ versus $Z = 0$ for each unit as a function of its observed covariate values $\mathbf{X}_{obs}$ and its missing-data pattern R. Because $\mathbf{X}$ is missing when $R = 0$, a saturated model for $(\mathbf{X}, R)$ cannot be fit, even with the general location model. We impose log-linear constraints on the categorical variables, which include the missing value indicators for covariates whose missingness is related to treatment assignment. In the special case of no missing data and only continuous covariates, the approach reduces to estimating propensity scores by discriminant analysis, which practically is very close to logistic regression (Rubin and Thomas 1992a). Our methods use as basic computational tools the EM (Dempster, Laird, and

Rubin 1977) and ECM (Meng and Rubin 1993) algorithms applied to the general location model. We estimate three different propensity score models and use these estimated propensity scores to select matched samples that have similar distributions of observed covariates and missing-value indicators. We also provide suggestions for diagnostic procedures to assess the success of the matching in creating balanced distributions of these observed covariates and missing-value indicators, and use the results from these diagnostics to compare the three propensity score models. We illustrate these procedures in the context of a matched-sampling study of the effects of postterm pregnancy.

It is important to note that our problem is different from most missing-data problems in which the goal is parameter estimation. We are not interested in obtaining one set of estimated parameters for a logistic regression or discriminant analysis, or a posterior distribution for these parameters, or even in drawing inferences about these parameters. Rather, parameters particular to each pattern of missing data serve only in intermediate calculations to obtain estimated propensity scores for each subject. Moreover, the propensity scores themselves serve only as devices to balance the observed distribution of covariates and patterns of missing covariates across the treated and control groups. Consequently, the success of the propensity score estimation is assessed by this resultant balance rather than by the fit of the models used to create the estimated propensity scores. This goal is not special to the case with missing values in covariates, but rather has been the goal with propensity score estimation from the start.

Because the major goal of this article is to present a method for handling missing covariates when estimating propensity scores and to illustrate its use in a real application that involves finding matches for participants in an observational study, we do not consider other matching methods that do not use propensity scores. Such other methods have been addressed by others (e.g., Rosenbaum and Rubin 1985a; Rubin 1976b). Here we focus on nearest available matching on the estimated propensity score.

The remainder of Section 1 presents details of the motivating example, which led to the development of our methods. Section 2 provides notation and describes our method. Section 3 applies our methodology to the motivating example and evaluates the resulting balance of three different models.

1.2. The Database and Problem

The original motivation for our work was a particular March of Dimes observational study examining the effects of postterm birth versus term birth on neuropsychiatric, social, and academic achievements of children age 5–10 years. This study is still ongoing, and to date there has not been a published report of any results. Because the focus of this study has been only on the effects of postterm birth versus normal-term birth, children born preterm are not of interest. At the onset of the study, the investigators had available a collection of 9,275 birth records of children born term or postterm at Beth Israel Hospital, Boston, with prenatal and birth history information, including gestational age. About 92% of these children were born term (37–41 weeks), whereas about 8% were born postterm (43+ weeks). Children with gestational ages of 42 weeks are not included, because the medical investigators felt that these children may potentially be a mixture of term and postterm children due to the variability in assessing gestational age, whereas children with gestational ages between 37 and 41 weeks were clearly term and those with gestational ages over 43 weeks were clearly postterm. The investigators were interested in selecting a sample of term and postterm children from this population to be part of their study because it was financially infeasible to follow up all children in the database. Therefore, the initial issue that they faced was how to select the sample to facilitate inference for the effect of being postterm. Because postterm children were relatively rare, essentially all postterm subjects could be followed up, and a matched sample of term children was desired; that is, matched with respect to covariates. A complication was that for some children, some covariates had missing values.

We illustrate our methodology for estimating propensity scores with missing values, and its application to obtain matched samples, using a random sample of 4,500 of the 9,275 subjects in this dataset. The remaining 4,775 subjects were used in validation studies that extend beyond the scope of this article. Of the 4,500 subjects chosen for analysis, 4,155 (92.3%) were term babies and 345 (7.7%) were postterm babies. We estimate several propensity score models using 25 of the covariates that are thought to be scientifically significant for predicting postterm birth and prognostically important for predicting outcomes, and thus if left uncontrolled could confound estimated treatment effects. At this point, it is scientifically important to point out that many of these covariates are not truly proper covariates in the sense of taking their values before "treatment assignment" to term or postterm conditions. That is, in the hypothetical experiment underlying the observational study, before week 42 a decision could have been made to induce labor for the postterm babies, and the effect of not doing so is the effect we seek. Formally, any covariate measured after 41 weeks thus is an improper covariate, because it could be affected by treatment.

For example, infant's weight had the largest initial imbalance, but can be considered an outcome of being postterm and not a proper covariate. Despite this, as with the other improper covariates, the investigating physicians felt strongly that this variable needed to be controlled. Another example, induction, measures whether there was some form of medical induction performed on the woman during labor (0 = no induction, 1 = elective induction, and 5 = induction due to ruptured membranes, a medical disorder, or a fetal disorder). It too is formally an improper covariate, because it is possibly affected by the hypothesized treatment assignment. Because these improper covariates can be thought of as proxies for unmeasured proper pretreatment covariates that predict fetal disorders, the physicians and investigators felt that they needed to explicitly control these variables as if they were proper covariates, if useful inferences were to be drawn about policy-relevant advice concerning postterm pregnancies. Regardless, we acknowledge that the inclusion of such improper covariates may actually adjust away part of the true treatment effect. However, this limitation occurs regardless of which method for control is used (i.e., matching or covariate modeling). In any case, the example illustrates the estimation and use of propensity scores with missing covariate data, which could have been applied using only proper covariates.

Tables 21.1 and 21.2 present descriptive statistics for the covariates and fitted propensity scores, separately for the term and postterm groups. It is important

Table 21.1. *Means (standard deviations), standardized differences (based on available cases) in percent, and variance ratios for continuous covariates in both groups before matching*

Covariate	Term mean (SD)	Postterm mean (SD)	Initial standardized difference (%)[b]	Variance ratio[c]
Antepartum complications (yes/no)	.72 (.45)	.72 (.45)	1	1.00
Previous obstetrical history (yes/no)	.47 (.50)	.40 (.49)	−14	.96
Vaginal bleeding (yes/no)	.12 (.33)	.11 (.31)	−4	.88
Second-stage indicator[a]	.81 (.39)	.77 (.42)	−10	1.16
Delivery mode	1.26 (.51)	1.30 (.51)	8	1.00
Labor complications	.58 (.63)	.66 (.59)	14	.88
Class[a]	2.37 (.77)	2.31 (.77)	−8	1.00
Diabetes[d]	.15 (1.05)	.11 (.82)	−4	.61
Fetal distress	.04 (.64)	.15 (1.2)	11	3.51
Induction	.17 (.88)	.41 (1.2)	23	1.85
Pelvic adequacy (clinic)[a]	.19 (.68)	.19 (.67)	0	.97
Pelvic adequacy (X-Ray)[a]	1.71 (.94)	1.69 (.79)	−3	.70
Placental problems	.11 (1.04)	.09 (.93)	−2	.80
Previous perinatal mortality[a]	.22 (1.49)	.15 (1.13)	−6	.57
Urinary tract disorders	.11 (.51)	.13 (.53)	4	1.08
Child's age (months from 1980, range 0–48)	23.4 (13.0)	23.9 (11.4)	4	.77
Infant's weight (grams)[a]	3,338 (461)	3,626 (533)	58	1.33
Length of first stage (min)[a]	784 (571)	910 (665)	20	1.35
Length of second stage (min)[a]	53.8 (65)	59.5 (66)	9	1.03
Time since membranes ruptured (min)[a]	454 (791)	414 (651)	−6	.68
Mother's age (years)	28.8 (5)	28.2 (5)	−12	1.00
Parity	.77 (1.0)	.66 (1.1)	−10	1.20
Total length of labor (min)[a]	841 (589)	968 (688)	20	1.37

[a] Covariate suffers from some missing data.

[b] The standardized difference is the mean difference as a percentage of the average standard deviation: $\{[100(\bar{x}_p - \bar{x}_t)]/\sqrt{[(s_p^2 + s_t^2)/2]}\}$, where for each covariate $\bar{x}_p$ and $\bar{x}_t$ are the sample means in the postterm and term groups and s_p^2 and s_t^2 are the corresponding sample variances.

[c] The variance ratio is s_p^2/s_t^2.

[d] Diabetes: 0 = none, 1 = diabetes insipidus or glucosuria, 5 = abnormal glucose tolerance test, and 10 = diabetes mellitus.

[e] Previous perinatal mortality: 0 = no previous child deaths, 5 = previous late death (in the first year of life), 10 = previous stillbirth or neonatal death, and 20 = previous stillborn and previous neonatal death (or any combination of 2 or more perinatal mortalities).

to emphasize that these statistics are descriptive and not inferential, in the sense that they do not purport to estimate relevant population parameters, but rather simply describe the two samples and their differences. Table 21.1 presents, for each continuous covariate, the mean and standard deviation using available cases;

Table 21.2. *Table of observed proportions and percent differences for categorical covariates and missing-value indicators for initial data*

		Term	Postterm	Difference (in %)
Covariate				
Race	White	.70	.72	2
	Nonwhite	.30	.28	−2
Gender	Male	.49	.51	2
	Female	.51	.49	−2
	Vertex	.77	.72	−5
Delivery mode	Cesarean	.21	.27	6
	Other	.02	.01	−1
	No labor (Cesarean)	.08	.06	−2
Labor	No complications	.26	.21	−5
complications	Some complications	.66	.73	−7
Missing-value indicators (proportion observed)				
Pelvic adequacy (x-ray)		.05	.10	5
Length of second stage of labor		.78	.74	−4
Race		.95	.95	0
Second stage of labor indicator		.99	1.00	1
Class		.99	.99	0
Pelvic adequacy (clinic)		.85	.90	5
Infant's weight		.99	1.00	1
Length of first stage of labor		.89	.91	2
Time that membranes ruptured		.97	.97	0
Length of labor		.89	.91	2

also presented are standardized percentage differences, defined as the mean difference between postterm and term groups as a percentage of the standard deviation $\{[100(\bar{x}_p - \bar{x}_t)] / \sqrt{[(s_p^2 + s_t^2)/2]}\}$, where $\bar{x}_p$ and $\bar{x}_t$ are the sample means in the postterm and term groups and s_p^2 and s_t^2 are the corresponding sample variances, again based on available cases. Also presented are the variance ratios, s_p^2/s_t^2.

The first two columns of Table 21.2 present, using available case data for the categorical covariates (first four rows), the proportion of women in each category in the term and postterm groups. Also presented are the corresponding results for the missing-data indicators (last 10 rows) for the 10 covariates with any missing values (either continuous or categorical). The third column displays the differences in percent between the term and postterm groups for each of the categorical covariates and missing-data indicators. Two covariates, delivery mode and labor complications, were considered to be either continuous or categorical, depending on the specific propensity score model and thus appear in both Tables 21.1 and 21.2.

Another diagnostic assessment compares the pairwise available-case correlations between the 23 continuous covariates in Table 21.1. Suppose that we plotted the $253 = 23 \times 22/2$ pairwise correlations in the initial term group against the pairwise correlations for the postterm group. If the two groups had similar distributions of their pairwise correlations, then they would have approximately the same means and variances, and we would see a roughly linear relationship. The mean correlation is .0143 in the term group and .0161 in the postterm group, and the corresponding variances are .016 and .022, indicating slightly larger and more variable correlations in the postterm group; for example, most of the pairwise correlations are between −.5 and .5 for the postterm group, but between −.4 and .4 for the term group. Moreover, the R^2 value of .61 is not particularly high. Note that we are not recommending the use of available-case correlations to estimate population correlations – rather, we are using them only to summarize aspects of the observed data for comparison across treatment groups.

The initial term versus postterm group differences summarized in Tables 21.1 and 21.2 and by the R^2 value indicate the possible extent of biased comparisons of outcomes due to different distributions of observed covariates and patterns of missing data in the initial term and postterm groups. That is, ideally all such descriptive statistics should suggest the same distribution in the term and postterm groups, as they would be in expectation if the treatment indicator (term vs. postterm) had been randomly assigned. As can be seen from these tables, there exists considerable initial bias between the term and postterm groups. For instance, nine of the continuous covariates have initial standardized differences larger than 10%. In addition, there is substantial differences between the groups based on the estimated propensity scores. Among categorical covariates, we see that labor complications and delivery mode are different between the postterm group and the term group. The missingness rates appear similar, except that there seems to be a trend for some indicators of potential complications to be observed more often in the postterm group (e.g., pelvic adequacy, both x-ray and clinical), suggesting a greater need for such medical tests among the postterm subjects. In addition, the missing-data indicator for length of second stage of labor shows that more individuals had this variable observed in the term group than in the postterm group (78% versus 74%).

2. NOTATION

2.1. Estimation of Propensity Scores

With complete data, Rosenbaum and Rubin (1983a) introduced the propensity score for subject $i(i = 1, \ldots, N)$ as the conditional probability of receiving a particular treatment ($Z_i = 1$) versus control ($Z_i = 0$) given a vector of observed covariates, $\mathbf{x}_i$,

$$e(\mathbf{x}_i) = \text{pr}(Z_i = 1 \mid \mathbf{X}_i = \mathbf{x}_i), \tag{21.1}$$

where it assumed that, given the $\mathbf{X}$'s, the Z_i are independent,

$$\text{pr}(Z_1 = z_1, \ldots, Z_N = z_N \mid \mathbf{X}_1 = \mathbf{x}_1, \ldots, \mathbf{X}_N = \mathbf{x}_N)$$
$$= \prod_{i=1}^{N} e(\mathbf{x}_i)^{z_i} \{1 - e(\mathbf{x}_i)\}^{1-z_i}. \tag{21.2}$$

Rosenbaum and Rubin (1983a) showed that for a specific value of the propensity score, the difference between the treatment and control means for all units with that value of the propensity score is an unbiased estimate of the average treatment effect at that propensity score, if the treatment assignment is strongly ignorable given the covariates. Thus matching, subclassification, or regression (covariance) adjustment on the propensity score tends to produce unbiased estimates of the treatment effects when treatment assignment is strongly ignorable, which occurs when the treatment assignment, Z, and the potential outcomes, Y, are conditionally independent given the covariates $\mathbf{X}$: $\Pr(Z \mid \mathbf{X}, Y) = \Pr(Z \mid \mathbf{X})$.

2.2. Propensity Scores With Incomplete Data

Let the response indicator be R_{ij}, ($j = 1, \ldots, T$), which is 1 when the value of the jth covariate for the ith subject is observed and 0 when it is missing; R_{ij} is fully observed by definition. Also, let $\mathbf{X} = (\mathbf{X}_{\text{obs}}, \mathbf{X}_{\text{mis}})$, where $\mathbf{X}_{\text{obs}} = \{\mathbf{X}_{ij} \mid R_{ij} = 1\}$ denotes the observed parts and $\mathbf{X}_{\text{mis}} = \{\mathbf{X}_{ij} \mid R_{ij} = 0\}$ denotes the missing components of $\mathbf{X}$.

The generalized propensity score for subject i, which conditions on all of the observed covariate information, is

$$e_i^* = e_i^*(\mathbf{X}_{\text{obs},i}, R_i) = \text{pr}(Z_i = 1 \mid \mathbf{X}_{\text{obs},i}, R_i). \qquad (21.3)$$

Rosenbaum and Rubin (1985a) showed that with missing covariate data and strongly ignorable treatment assignment given $\mathbf{X}_{\text{obs}}$ and R, the generalized propensity score e_i^* in (21.3) plays the same role as the usual propensity score e_i in (21.1) with no missing covariate data. Treatment assignment is strongly ignorable given $(\mathbf{X}_{\text{obs}}, R)$ if $\Pr(Z \mid \mathbf{X}, Y, R) = \Pr(Z \mid \mathbf{X}_{\text{obs}}, R)$. If in addition, the missing-data mechanism is such that $\Pr(R \mid \mathbf{X}, Z) = \Pr(R \mid \mathbf{X}_{\text{obs}})$, then $\Pr(Z \mid \mathbf{X}, Y, R) = \Pr(Z \mid \mathbf{X}_{\text{obs}})$, and R itself can be ignored in the modeling. It is important to emphasize that, just as with propensity score matching with no missing data, the success of a propensity score estimation method is to be assessed by the quality of the balance in the $(\mathbf{X}_{\text{obs}}, R)$ distributions between term and postterm groups that has been achieved by matching on it. Consequently, the usual concerns with the fit of a particular model (i.e., the general location model) are not relevant if such a balance is achieved.

2.3. General Location Model With Complete Data

The distribution of $(\mathbf{X}, Z)$ is defined by the marginal distribution of the dichotomous variable, Z, and the categorical covariates, U, and the conditional distribution of continuous covariates, say V, given (U, Z). U_{ij}, Z_i locates the ith subject in one of the m cells of the table formed by (U, Z).

We assume that (U, Z) are iid multinomial random variables, and conditional on U_i, Z_i, we assume that V_i is K-variate normal with mean that depends on the cell but with a common covariance. This is the general location model (Olkin and Tate 1961) with parameters Π = cell probabilities from the multinomial distribution, $\mathbf{\Gamma}$ = the matrix of cell means, and $\mathbf{\Omega}$ = the positive-definite covariance matrix common to all cells; $\theta = (\Pi, \mathbf{\Gamma}, \mathbf{\Omega})$.

Krzanowski (1980, 1982), Little and Rubin (1987), and Little and Schlucter (1985), have described restricted general location models with fewer parameters.

One way to reduce the number of parameters to be estimated is to constrain Π by a log-linear model (Bishop, Fienberg, and Holland, 1975; Goodman 1968); for example, three-way and higher-order interactions are set to 0. Maximum likelihood (ML) estimates of the parameters for these models have closed-form solutions for many configurations, but if they do not have a closed form, they can be found by using an iterative procedure such as iterative proportional fitting (IPF; Bishop et al. 1975).

A second way to reduce the number of parameters to be estimated in the general location model is to impose analysis of variance (ANOVA)-like restrictions on the means, Γ, using a known design matrix A to define Γ in terms of a lower-dimensional matrix of unknown regression coefficients β; if A is the identity matrix, then there are no restrictions. With standard models and complete data, the parameter estimates for β and Ω can be found using standard regression techniques (Anderson 1958, Chap. 8).

Although the restrictions described previously reduce the number of parameters to be estimated in the model, we could also generalize the model to increase the number of parameters to be estimated. For instance, the assumption of a common covariance Ω across all cells of the contingency table can be relaxed to allow for possibly different covariance matrices to be estimated in different cells. But this can require substantial sample sizes in each cell. A more useful extension may be to estimate separate covariance matrices only for the treated and the control groups. Other extensions involve proportional covariance matrices and more general ellipsoidal distributions (e.g., t distributions as in Liu and Rubin 1995, 1998).

2.4. Fitting the General Location Model With Missing Data

The basic method for finding estimates for the parameters of the general location model when there are ignorably missing data has been outlined by Little and Rubin (1987, Chap. 10) and is based on the EM algorithm (Dempster et al. 1977) and the ECM algorithm (Meng and Rubin 1993), which is used when log-linear restrictions have been placed on the general location model such that IPF is needed with complete data. Of particular importance for our situation, where we want to include explicitly the response indicator R in the modeling, is that R is a fully observed collection of categorical (in fact, binary) covariates. For notational convenience, let $U^* = (U, R)$, with corresponding changes to the other notation. Because X_{ij} is missing when $R_{ij} = 0$, some restrictions are needed to obtain unique ML estimates of parameters for the joint distribution of (Z, X, R), which we need to obtain unique ML estimates of the conditional distribution of Z given X_{obs} and R.

At each iteration of the EM algorithm, the E step computes the expected values of the complete-data sufficient statistics given the observed data and the current estimates of the parameters, $\theta^{*(t)} = (\Gamma^{*(t)}\Omega^{*(t)}, \Pi^{*(t)})$, where t indexes iterations. The M step computes the ML estimates for the parameters using the estimated values of the sufficient statistics. These become the current estimates of the parameters to be used in the next E-step calculations. The E and M steps are repeated until convergence. When there are log-linear constraints on the categorical covariates, ECM is used instead of EM; the M step of EM is replaced by CM steps, which perform one cycle of IPF. The complete-data sufficient statistics for this model are the raw sums of squares and cross-products of the V's ($\Sigma V_i^T V_i$), the sums of the V's in each cell (cell totals), and the cell frequencies from the table defined by (Z, U^*).

Once EM or ECM has converged, we have ML estimates, $\hat{\theta}^*$, for the parameters $\theta^* = (\Pi^*, \Gamma^*, \Omega^*)$ for the joint distribution of (V, Z, U^*), which we use to calculate an estimated propensity score for each subject, $\hat{e}_i^*$, as in (21.3) with $\mathbf{X}_{\text{obs},i} = (V_{\text{obs},i}, U_{\text{obs},i})$:

$$\hat{e}_i^* = \text{pr}(Z_i = 1 \mid V_{\text{obs},i}, U_{\text{obs},i}^*, R_i, \hat{\theta}^*). \tag{21.4}$$

To find the estimated propensity score (21.4) from $\hat{\theta}^*$ and the observed data, we simply run one E step using the converged MLE $\hat{\theta}^*$, but now treating Z_i as missing.

2.5. Generalized Propensity Score Estimates: An Illustration

To illustrate the calculation outlined in Section 2.4, consider a March of Dimes example in which there is one categorical covariate, gender (two levels), U_1, with some missing values indicated by R_1; one continuous covariate, mother's age, V_1, with some missing values indicated by R_2; and the two-level treatment indicator Z. Consider the case where the only missing-value indicator that needs to be balanced between the groups is R_1. To estimate the propensity score, we need to estimate the parameters for the joint distribution of (U_1, R_1, V_1, Z). To estimate the cell proportions defined in the $2 \times 2 \times 2$ contingency table defined by (Z, U_1, R_1), we could fit a log-linear model with no three-way interaction. To estimate parameters for V_1 conditional on (Z, U_1, R_1), the mean and variance of V_1 in each of the eight cells, we could place an additive structure on the means so that the linear regression of V_1 on Z, U_1, and R_1 has a constant, a term for Z, a term for U_1, and a term for R_1, and a common variance.

To fit such models with complete data, we would use IPF to estimate the cell probabilities for the three-way table using the sufficient statistics that are the observed cell counts for the $2 \times 2 \times 2$ table of (Z, U_1, R_1), and ordinary least squares for the regression of V_1 on (Z, U_1, R_1), where the sufficient statistics are the within-cell totals and raw sums of squares of V_1. With missing data, we use the ECM algorithm to estimate the parameters for these models.

The E step of ECM gets the expectations of the sufficient statistics given the observed data and the current estimates of the parameters for each pattern of missing data (U_1 and V_1 observed, U_1 observed and V_1 missing, U_1 missing and V_1 observed, and U_1 and V_1 both missing). When both covariates are observed, the estimates of the sufficient statistics are the observed cell counts, cell totals, and sum of squares of V_1. When V_1 is missing, the expected cell totals and raw sum of squares are based on the additive regression model described earlier and involve the expected value both of V_1 and of its square within each cell. When U_1 is missing, the expected cell counts are found by performing a discriminant analysis between male and female to determine the probabilities of being in each cell given the observed value of V_1, the observed value of R_1, and the observed treatment indicator Z. When both covariates are missing, the expected sufficient statistics are found conditionally given $R_1 = 0$ and the observed value of Z. The expected cell counts are found using the conditional probability for being in each cell defined by U_1, based on the current estimates of cell probabilities. The expected within-cell totals and raw sum of squares are based on the regression equations for V_1 conditional on being in one and then the other cell defined by U_1.

Once we have the updated expectations of the sufficient statistics, we compute the complete-data estimates from these sufficient statistics using the CM step of the ECM algorithm. First, we perform one iteration of IPF to get new estimates of the cell proportions based on the log-linear model specified. Then we get estimates for the cell means and within-cell variance parameters from fitting the regression model specified.

We continue iterating through ECM to find the ML estimates for the joint distribution of (Z, U_1, V_1, R_1). Finally, we run the E step of the algorithm one more time with Z considered missing to estimate the propensity scores. This E step, conditional on the final parameters estimates, essentially performs a discriminant analysis to determine the probabilities of being treated or control given $(\mathbf{X}_{obs}, R_1)$. [Recall that $\mathbf{X}_{obs}$ corresponds to the observed parts of (U_1, V_1).] The propensity score is then found by summing the probabilities across the four cells of the three-way table (defined by U_1, R_1, and Z) that correspond to $Z = 1$, the treated group.

3. PROPENSITY SCORE MODELS FOR MARCH OF DIMES DATA

3.1. Specific Models

All three propensity score models we fit used all 23 continuous covariates in Table 21.1 and all 4 categorical covariates in Table 21.2. However, the models differed in their inclusion of missing-data indicators.

Among the 25 covariates used in our analyses, the missingness on 2 were differentially distributed in the treatment groups and believed to be prognostically important: results of a pelvic x-ray and length of the second stage of labor. These were determined to be prognostically important for the following reasons. During the years 1981–1984, when the birth data for this study were recorded, pelvic x-rays were performed on women who were having difficulties during labor. The x-ray would usually be performed if the obstetrician suspected that the baby was too large to fit through the mother's pelvis. These x-rays were rarely ordered, and the existence of an x-ray order was an indication that the labor would be different than an ordinary one. In our data, out of the 4,500 women, only 234 (or 5.2%) had pelvic x-rays recorded, but their frequency was twice as great among postterm subjects (10.4% vs. 4.8%); the missing-value indicator for a pelvic x-ray may be prognostically important, even as important as the outcome of the x-ray itself. Of course, this is not a proper covariate inasmuch as it may be an outcome of allowing the fetus to continue to grow, but the physicians wanted it to be controlled.

The other missingness indicator considered prognostically important was the length of the second stage of labor; that is, the length of time from full dilation of the cervix to delivery. If a woman does not have a cesarean section, then she must have a second stage of labor. In our data, 1,002 out of 4,500 women did not have a second stage of labor recorded. Of these, 872 had a cesarean section, whereas 130 did not.

We fit three generalized propensity score models with missing-value indicators; the first included the missingness indicator for the pelvic x-ray, the second included the missingness indicator for length of the second stage of labor, and the third included both indicators. The first and second models without restrictions each have 659 parameters. In each model, log-linear constraints were placed on the cell probabilities so that the three-way and higher interactions were set to 0, and thus we estimated four main effects and six two-way interactions. The design matrix relating the means of the continuous variables to the categorical variables includes an intercept, main effects for each of the categorical variables, and terms for each of the two-way interactions of the categorical variables. These constrained models have 539 total parameters: 10 for the contingency table, 253 regression coefficients, and 276 variances and covariances. The third model without any constraints has 1,043 parameters. Here we fit a log-linear model with only main effects and two-way interactions. The design matrix relating the continuous variables to the categorical variables includes an intercept, main effects for each of the categorical variables, and terms for each of the two-way interactions of the categorical variables. The resulting 659 parameters are 15 for the contingency table, 368 regression coefficients, and 276 variances and covariances.

3.2. Matching Using Estimated Propensity Scores

We estimate propensity scores for each of the models using the ECM algorithm as illustrated in Section 2.5. We then use nearest-available matching on these estimated propensity scores to choose matches for the postterm subjects. We randomly order the term and postterm subjects, and then select the term subject with the propensity score closest to the first postterm subject. Both subjects are then removed from the pools of subjects. We repeat this procedure for each postterm subject, which results in selecting a total of 345 term subjects from the 4,150 available ones. We could have used many other approaches to select the matches, but chose this straightforward approach using propensity scores to focus on how well matching based on the different propensity score models succeeds in balancing the distribution of observed covariates and missing-data indicators between the term and postterm groups.

3.3. Resultant Distributions of Propensity Scores and Their Logits

Figure 21.1 gives, for each model, associated boxplots of estimated propensity scores in the term group ($n = 4,150$), the postterm group ($n = 345$), a group of randomly selected term subjects ($n = 345$), and the matched term group ($n = 345$) for that model. In this figure, for each boxplot the interquartile range is displayed by a rectangular box, with a solid horizontal line within the box representing the median propensity score for each group. Above and below the interquartile ranges are small notches representing the upper and lower values for the 5th and 95th percentiles. Above the upper percentile notch, individual circles represent propensity scores for single subjects beyond the 95th percentile. For example, in model 1 there is a postterm subject with a propensity score nearly equal to 1. With all models, there are two striking features: (1) The median propensity score for the postterm group is

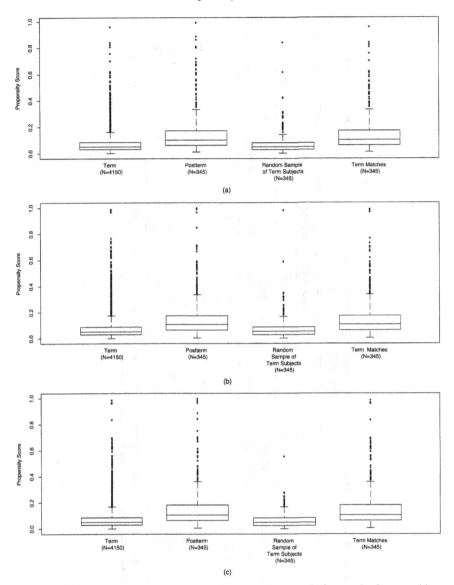

Figure 21.1. Boxplots comparing propensity score balance before and after matching. (a) Model 1; (b) Model 2; (c) Model 3.

larger than the 75th percentile propensity score in the unmatched term group and the randomly selected term group, but nearly equal to the median propensity score in the matched term group, and (2) the spread of propensity scores for the unmatched term group is wide enough to cover most of the propensity scores for the postterm group in each model. This second feature allows us to find matches from the term group with propensity scores close to the propensity scores of the subjects in the postterm group. In Figure 21.1 the distribution of propensity scores for the random

Table 21.3. *Standardized differences (in %) and variance ratios for propensity scores and their logits, before and after matching using each model*

Propensity score	Initial standard difference (%)[a]	Initial variance ratio[b]	Results after matching					
			Matching model 1		Matching model 2		Matching model 3	
			Standard difference (%)	Variance ratio	Standard difference (%)	Variance ratio	Standard difference (%)	Variance ratio
Model 1	67	5.24	2	1.13	15	1.58	8	1.22
Model 2	67	3.85	5	1.07	1	1.01	1	.99
Model 3	69	5.00	7	1.28	11	1.46	2	1.19
Logit model 1	85	1.61	2	1.1	13	1.27	8	1.14
Logit model 2	83	1.39	11	1.10	3	1.12	5	1.03
Logit model 3	86	1.59	7	1.26	11	1.28	2	1.18

[a] The standardized difference in % is the mean difference as a percentage of the average standard deviation: $\{[100(\bar{x}_p - \bar{x}_t)]/\sqrt{[(s_p^2 + s_t^2)/2]}\}$, where for each covariate, $\bar{x}_p$ and $\bar{x}_t$ are the sample means in the postterm and term groups, and s_p^2 and s_t^2 are the corresponding sample variances.

[b] The variance ratio is s_p^2/s_t^2.

sample of term subjects has few values near those of postterm subjects, whereas the propensity scores for the matched term subjects for each model covers most propensity scores in the postterm group.

Table 21.3 compares the distributions of the estimated propensity scores and their logits that result from each of the three propensity score models. Columns 1 and 2 present the standardized differences in percent and variance ratios prior to matching. Columns 3–8 present the same statistics after matching for model 1 (columns 3 and 4), model 2 (columns 5 and 6) and model 3 (columns 7 and 8). The initial standardized differences in percentage were quite large, with model 3 having the largest initial difference, 69%, which suggests that this model may be the most revealing by maximally separating the term and postterm groups. The initial variance ratios were also quite large with the propensity scores from models 1 and 3 each having ratios of 5.00 or greater. When we compare the standardized differences after matching (columns 3, 5, and 7), we see that matching using any of the three propensity score models performed well in reducing the standardized differences, with model 3 performing best; matching based on propensity scores from this model reduced the standardized differences to below 10% for all models. More explicitly, when we selected matches based on propensity scores estimated from model 3 and then compared the distribution of propensity scores estimated from model 1, we find that the standardized difference was 8% (from column 7). When we compared variance ratios (columns 4, 6, and 8) across the three models, we again found that model 3 performed best; this model produced variance ratios that were on average closer to 1 than the other two models.

3.4. Resultant Covariate Balance After Matching

To further assess the relative success of the propensity score models for creating balanced matched samples, we compare balance on observed covariates and missing-data indicators in the matched samples created by each model. It is important for practice to realize that, as done by Rosenbaum and Rubin (1984b, 1985a), these assessments can be made before any resources have been committed to collecting outcome data on the matched controls. Also, it is important to realize that because these comparisons involve only observed covariates and their missing-data indicators and not outcome variables, there is no chance of biasing results in favor of one treatment condition versus the other through the selection of matched controls.

Figure 21.2 compares the standardized differences in percent, after matching, for the continuous covariates. The matching using any of the models performs well in reducing the bias of the background covariates with moderate-to-large initial standardized differences. For instance, the initial standardized difference for the length of first stage of labor variable is 20%, and all models were able to reduce this significantly, ranging from 3% (models 1 and 3) to 5% (model 2). Even the initial standardized difference for infant's weight is substantially reduced by the matching using any of the models, although the difference is 11% using model 3.

From Figure 21.3, which compares the available-case cell proportions for the categorical covariates and missing-value indicators between the term and postterm groups for the three models, we find that the initial imbalance in delivery mode and labor complications was moderate, with 26% of postterm babies being cesarean

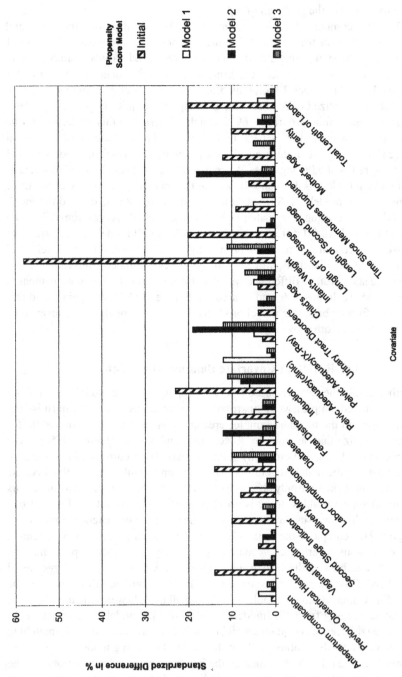

Figure 21.2. Comparison of standardized difference (in %) for covariates between term and postterm women, based on available case means for each propensity score model.

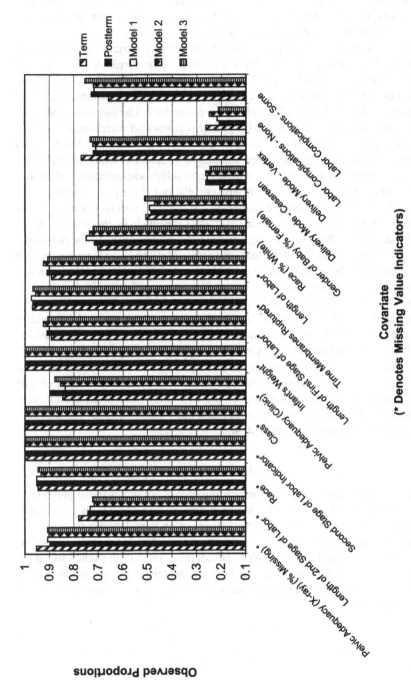

Figure 21.3. Comparison of observed proportions for categorical covariates, including missing-data indicators, between term and postterm women, for each propensity score model.

births versus 20% of term pregnancies and 73% of postterm pregnancies having some complications versus 66% of term pregnancies. These differences were reduced by the matching for all models. For missing-value indicators, we found that all models improved the balance between the term and postterm groups.

When we compared the pairwise correlations among covariates for the postterm and matched term groups, we found that the means of the correlations for matched samples obtained by models 2 and 3 were closer to the postterm values than those obtained by model 1. The mean in the postterm group was .0161, compared to .0108 with model 1, .0145 with model 2, and .0155 with model 3. The variances in all three models were closer to the variance in the postterm group after matching. In addition, we found that the R^2 values for the samples matched by models 2 and 3 were higher (.85 and .82) than those for the samples matched by model 1, which indicates that the pairwise correlations among the matched term subjects selected using these two models more closely resemble the pairwise correlations in the postterm group.

Model 3 (with race, gender, missingness for length of second stage, and missingness for pelvic x-ray as categorical covariates) appears to be the best propensity score model that we fit for producing balanced matched samples on the propensity scores, their logits, and the individual covariates. This model reduces the bias on all covariates with large or moderate initial bias and on the missing-value indicator for the pelvic x-ray exam. Moreover, it revealed the largest initial bias based on the propensity score (69% initial standardized difference).

We acknowledge that many other plausible propensity score models could be constructed using the 25 covariates and their missing-data indicators, and that among these there may exist models that produce better balance than our model 3. Still, this model did give the investigators what they wanted – propensity scores that were used to select matches for the postterm babies from the available pool of term babies, where the bias observed between the term and postterm groups on many covariates and their missingness prior to matching was substantially reduced (and often essentially removed) by the matching.

4. CONCLUSIONS

We have presented an approach for estimating propensity scores in the presence of missing data using the EM and ECM algorithms as computing tools. The framework allows the investigator to impose a structure on the relationships among the covariates in the model, including missing-value indicators for specific effects. We illustrated our approach using March of Dimes data. Simulation studies are underway to examine the effects of specifying different missing-data mechanisms on the data. In addition, we are in the process of developing user-friendly software to perform these analyses.

22. Using Propensity Scores to Help Design Observational Studies: Application to the Tobacco Litigation

Donald B. Rubin

Abstract: Propensity score methodology can be used to help design observational studies in a way analogous to the way randomized experiments are designed: without seeing any answers involving outcome variables. The typical models used to analyze observational data (e.g., least squares regressions, difference of difference methods) involve outcomes, and so cannot be used for design in this sense. Because the propensity score is a function only of covariates, not outcomes, repeated analyses attempting to balance covariate distributions across treatment groups do not bias estimates of the treatment effect on outcome variables. This theme will be the primary focus of this article: how to use the techniques of matching, subclassification and/or weighting to help design observational studies. The article also proposes a new diagnostic table to aid in this endeavor, which is especially useful when there are many covariates under consideration. The conclusion of the initial design phase may be that the treatment and control groups are too far apart to produce reliable effect estimates without heroic modeling assumptions. In such cases, it may be wisest to abandon the intended observational study, and search for a more acceptable data set where such heroic modeling assumptions are not necessary. The ideas and techniques will be illustrated using the initial design of an observational study for use in the tobacco litigation based on the NMES data set.

1. INTRODUCTION – THE IMPORTANCE OF DESIGNING AN OBSERVATIONAL STUDY

In many contexts, the objectivity of a statistical design is of critical importance. For example, consider Phase III randomized trials, which are required before marketing pharmaceuticals. These studies are carefully designed and their structure is subject to the prior approval of the FDA (Food and Drug Administration); the initial design, which includes data collection and organization, and primary analyses are established before any outcome data from the trials are available. There still can be issues of analysis after the outcome data are available, such as conflicting results in subgroups or surprising patterns of side effects, but the design of these randomized trials attempts to minimize the impact of such complications. Arguably, the most important feature of experiments is that we must decide on the way data will be collected before observing the outcome data. If we could try hundreds of designs and for each see the resultant answer, we could capitalize on random variation in answers

Reprinted from *Health Services & Outcomes Research Methodology*, **2**, 2001, 169–188, ©2002 Kluwer Academic Publishers, with kind permission of Springer Science and Business Media.

and choose the design that generated the answer we wanted! The lack of availability of outcome data when designing experiments is a tremendous stimulus for "honesty" in experiments and can be in well-designed observational studies as well.

Thus analogous care should be exercised in the design of important observational studies, such as ones attempting to assess the causal effects of cigarette smoking on health outcomes or of the causal effects of the conduct of the tobacco industry on medical expenditure outcomes. Of course, it is easier to control statistical bias in a randomized trial than in an observational study, where extraneous factors are not balanced by the randomization, but it is still possible to duplicate one crucial feature of a randomized experiment: one can design an observational study without access to the outcome data. In fact, one of the common maladies of observational studies for causal effects is that there is no real design, just repeated model-based analyses that produce answers. It is essentially impossible to be objective when a variety of analyses are being done, each producing an answer, either favorable, neutral, or unfavorable to the investigator's interests. But this situation is the norm in observational studies where typically the data base includes the outcome data as well as covariates and treatment indicators.

Study design, as conceptualized in this article, includes the organization of data, such as by matching and/or subclassification, that can be done without any examination of outcome data whatsoever. Thus, study design explicitly excludes doing any analyses that require access to any outcome data, e.g., least squares analyses of health-care-expenditure outcomes in the tobacco litigation example (Rubin, 2000a), or difference of differences estimates of schemes in a study of the effect of the minimum wage (Card and Kreuger, 1994), or the methods described in Heckman and Hotz (1989), or covariance adjustment for the "propensity score" (Rosenbaum and Rubin, 1983a). *Initial* study design, as used here, also excludes the specification of any analyses that use outcome data. *Entire* study design, in contrast, should include the specification of the analyses to be performed once outcomes are available. These specifications are important for objectivity but are different from the efforts discussed here, which can be completed and evaluated before outcomes are available.

A key tool for implementing initial observational-study design is the use of propensity score methods, whether for the purpose of constructing matched samples of treated-control pairs, subclasses of similar groups of treated and control units, or weighting adjustments. Propensity score methods use solely the covariates, not the outcomes, and propensity scores will be the topic of this article. The focus will be on bias reduction in design because in observational studies that typically is more important than variance reduction. Similar ideas based on the propensity score can be used to increase the precision of the design, as shown in the context of the design of a randomized experiment in education in Hill, Rubin, Thomas (1999).

After motivating this perspective on using propensity scores to help design observational studies, it will be illustrated using the primary data set appearing in much of the tobacco litigation, NMES (the National Medical Expenditure Survey, AHCPR, 1992), which is a large nationally representative data base of nearly 30,000 adults. NMES has many features that make it a natural candidate for estimating quantities that play a role in estimating the causal effects of smoking and the effect of the tobacco companies' alleged misconduct (see Rubin, 2000a,b, 2001a). Central to the design of an objective observational study on these effects of smoking is the assembling of groups of never smokers and smokers with similar distributions of covariates, without allowing any examination of health outcomes or health-care expenditure outcomes. In particular, because I have been involved in the tobacco litigation (Rubin, 2000a,b, 2001a), all of the analyses and results of the design phase have been conducted without access to any outcome data.

2. PROPENSITY SCORES – NO OUTCOME VARIABLES IN SIGHT

Propensity score methodology was introduced by Rosenbaum and Rubin (1983a). The propensity score is the probability of being treated ($W_i = 1$ vs. $W_i = 0$) given the observed value of a vector of observed covariates X_i, where i indexes the units in the study ($i = 1, \ldots, N$), and W_i is the indicator for received treatment. An extension also allows conditioning on the observed pattern of missingness in covariates (Rosenbaum and Rubin, 1984b; D'Agostino and Rubin, 2000), but this extension was not used here. In a randomized experiment, the propensity scores are known, whereas in an observational study, they must be estimated from the data on W_i and X_i. No outcome data are required or desired; even if available in the data set, they should be set aside when designing the study, as was done here.

With no missing data in the covariates $\{X_i\}$, the propensity score e_i is defined as the probability that the ith unit is treated given that its vector of covariates is X_i,

$$e_i \equiv e(X_i) \equiv Pr(W_i = 1 \mid X_i), \tag{22.1}$$

a scalar summary of vector X_i. Thus, the mapping from X_i to e_i is generally a many-one function. The central result in Rosenbaum and Rubin (1983a) is that if a group of treated units and control units have the same value of the propensity score, e_i, then they have the same distribution of multivariate X_i, no matter what the dimension of X_i. Thus, having these groups of treated and control units with matching propensity scores automatically controls for all the observed covariates, at least in big samples: if there are differences in outcomes between the treated and control units, these differences cannot be due to these observed covariates. At the risk of being overly repetitive, if treatment and control groups have the same distribution of propensity scores, they have the same distribution of all observed covariates, just like in a randomized experiment.

Of course, propensity score technology can only attempt to achieve balance in observed covariates whereas randomization in experiments can stochastically balance all covariates, both observed and unobserved. Also, a randomized experiment offers the advantage that it provides an unambiguous definition of what constitutes a proper covariate – a variable that can be measured before the actual assignment of treatments. Variables measured after treatment assignment may "proxy for" proper covariates, but sometimes their status is rather ambiguous, and decisions whether to include them or not as covariates challenging.

With respect to NMES and the tobacco litigation, there is reason to include more than the usual list of proper covariates because the role of improper covariates in analyses involving outcome data will be different from the role of proper covariates; see Rubin (2000a, Section 4) for the logic and Rubin (2001a, Section 5) for a simple illustration. These two general topics, adjusting for unmeasured proper covariates and correctly adjusting for measured but improper covariates, are both beyond the scope of this paper because they require analyses involving outcome data.

3. HISTORY OF PROPENSITY SCORES – DISCRIMINANT MATCHING IN 1973 AND BEYOND; OBSERVED AND UNOBSERVED COVARIATES

Given the simplicity of reducing a large space of covariates, X_i, to a one-dimensional summary, the probability of treatment assignment, e_i, it is interesting to review the development of the idea. The most direct path comes from matched sampling, which can also be accomplished without having outcomes available. Although there exists an extensive literature of its early use in applications (e.g., in sociology or education, Peters, 1941), statisticians seem to have mostly eschewed the topic for many years. An early formal statistical investigation in the context of treatment and control groups with different distributions was Rubin (1970), followed by Cochran and Rubin (1973), Rubin (1973a, 1976b,c), and Carpenter (1977).

A key bridge between matching and propensity scores is the use of a one-dimensional summary for matching, specifically "discriminant matching" (Cochran and Rubin, 1973; Rubin, 1976b,c, 1979b, 1980b). Under normality with a common covariance matrix in the treatment and control groups, the best linear discriminant is the linear version of the propensity score ($\text{logit}(e_i) = \log[e_i/(1 - e_i)]$). Discriminant matching summarizes the covariates by the discriminant and uses this scalar variable for matching.

Of historical importance, the propensity score is not in the same class as any of the "confounder scores" of Miettinen (1976) nor the "selection models" of Heckman (1976). These attempts at dealing with the probability of treatment assignment either directly involve the observed dependent variable or indirectly through instrumental variables assumptions, and use this probability in a different way. In particular, they cannot be used to create balance in the observed covariates.

To be clear, I am most certainly not condemning parametric modeling based on relating the outcome variable to observed covariates or even hidden ones. Typically, analyses based on models for the outcome variables are important for obtaining the best final inferences. After the initial design stage, and after obtaining measured outcomes, models relating outcomes to observed covariates are very likely to reach improved (reduced bias, more precise) inferences, in analogy with covariance adjustments in randomized block experiments (Rubin, 1970, 1973b, 1979b; Roseman, 1998; and Rubin and Thomas, 2000).

Differences due to unobserved covariates should be addressed after the balancing of observed covariates in the initial design stage, using models for sensitivity analyses (e.g., Rosenbaum and Rubin, 1983b) or models based on specific structural assumptions. When the list of observed covariates is rich, adjustment for hidden covariates may not be necessary. When the list of observed covariates is less rich, the conclusion after the initial design stage may be that the samples still require substantial adjustment for unobservables to reach scientifically plausible conclusions (e.g., Smith and Todd, 2001). This topic too is beyond the scope of this article.

4. MAJOR TECHNIQUES FOR INITIAL DESIGN – MATCHING, SUBCLASSIFICATION, AND WEIGHTING

The initial design of an observational study attempts to assemble groups of treated and control units such that within each group the distribution of covariates is balanced. This balance allows the initial attribution of any observed difference in outcomes to the effect of the treatment vs. control rather than differences in covariates. The techniques can be usefully, although imperfectly, classified into one of three types: matching, subclassification, and weighting. All rely on the propensity score as a fundamental component.

Propensity score matching refers to the pairing of treatment and control units with similar values of the propensity score, and possibly other covariates, and the discarding of all unmatched units (see Rosenbaum and Rubin (1983a, 1985a); Rosenbaum (1989, 1991); Gu and Rosenbaum (1993); Rubin and Thomas (1992a,b, 1996, 2000); also see Cochran and Rubin (1973); and Rubin (1976b,c, 1979b, 1980b)) on discriminant matching. Typically, the matching finds for each treated unit one control unit, but sometimes more than one match is found (Rosenbaum, 1989, 1991; Rubin and Thomas, 2000). One-one Mahalanobis metric matching within propensity score calipers (Rosenbaum and Rubin, 1985a) is a popular method illustrated here in Section 7.

Subclassification on the propensity score ranks all units by their propensity score and then uses boundaries to create subclasses with treated and control units with similar values of the propensity score – typically five or six subclasses are used, with approximately the same total number of units within each subclass (Rosenbaum and Rubin, 1983a, 1984b; U.S. GAO, 1994). We illustrate the use of this technique on matched samples in Section 8.

Weighting methods use the inverse of the propensity score as a weight to apply to each treated unit and the inverse of one minus the propensity score as the weight to apply to each control unit (Czajka et al., 1992; Imbens, 2000). Such weighting methods are sometimes viewed as a final method of analysis based on ideas of Horvitz-Thompson (1952) estimation. However, this weighting can also be viewed as the limit of subclassification as the number of observations and subclasses tend to infinity. This technique is illustrated in Section 9.

The reasons for working this hard in the initial design stage of an observational study are twofold. First, as already mentioned, since no outcome data are available, none of these design efforts can inappropriately slant estimation of treatment effects on outcomes. Second, these design efforts, which result in more balanced distributions of covariates across treatment groups, make subsequent model-based adjustments (e.g., covariance adjustments, logistic regression odds ratio adjustments, instrumental variables models) more reliable. This second point arises because such model adjustments can be extremely unreliable when the treatment groups are far apart on covariates. This unreliability of model-based adjustments in such cases, although critical, seems to have been frequently ignored in many applied fields, even though documented for at least a half-century as we shall see in the next section. Recent work confirming this unreliability in practice appears in Lalonde (1986); also see Dehija and Wahba (1999), where this unreliability of these models with Lalonde's data is rectified using propensity score methods.

5. TRADITIONAL BENCHMARKS FOR THE USE OF REGRESSION ADJUSTMENT – WHEN IS IT RELIABLE?

The statistical literature has, for many years, warned that regression analysis cannot reliably adjust for differences in observed covariates when there are substantial differences in the distribution of these covariates in the two groups.

For example, William G. Cochran wrote extensively on methods for the analysis of observational studies, as summarized in Rubin (1984c). In Cochran (1957), he wrote:

> ... when the x-variables [i.e., covariates] show real differences among groups – the case in which adjustment is needed most – covariance adjustments [i.e., regression adjustments] involve a greater or less degree of extrapolation. To illustrate by an extreme case, suppose that we were adjusting for differences in parents' income in a comparison of private and public school children, and that the private-school incomes ranged from $10,000–$12,000, while the public-school incomes ranged from $4,000–$6,000. The covariance would adjust results so that they allegedly applied to a mean income of $8,000 in each group, although neither group has any observations in which incomes are at or near this level.

And later, in Cochran (1965), he wrote:

> If the original x-distributions diverge widely, none of the methods [e.g., regression adjustment] can be trusted to remove all, or nearly all, the bias. This discussion brings out the importance of finding comparison groups in which the initial differences among the distributions of the disturbing variables are small.

And in the same article:

> With several x-variables, the common practice is to compare the marginal distribu-tions in the two groups for each x-variable separately. The above argument makes it clear, however, that if the form of the regression of y on the x's is unknown, identity of the whole multivariate distribution is required for freedom from bias.

In particular, there are three basic distributional conditions that in general prac-tice must simultaneously obtain for regression adjustment (whether by ordinary lin-ear regression, linear logistic regression, or linear-log regression) to be trustworthy. If any of these conditions is not satisfied, the differences between the distributions of covariates in the two groups must be regarded as substantial, and regression adjustment will be unreliable and cannot be trusted. These conditions are:

1. The difference in the means of the propensity scores in the two groups being compared must be small (e.g., the means must be less than half a standard devi-ation apart), unless the situation is benign in the sense that: (a) the distributions of the covariates in both groups are nearly symmetric, (b) the distributions of the covariates in both groups have nearly the same variances, and (c) the sample sizes are approximately the same.

2. The ratio of the variances of the propensity score in the two groups must be close to one (e.g., 1/2 or 2 are far too extreme).
3. The ratio of the variances of the residuals of the covariates after adjusting for the propensity score must be close to one (e.g., 1/2 or 2 are far too extreme); "residuals" precisely defined shortly.

These three guidelines also address regression adjustments on the logit or linear log scale because they too rely on linear additive effects in the covariates (for discussion of this point, see e.g., Anderson et al., 1980).

Specific tabulations and calculations relevant to these guidelines can be found, for example, in Cochran and Rubin (1973); Rubin (1973b); Rubin (1979b); and Rubin and Thomas (2000). In particular, Cochran and Rubin (1973) state at page 426 that "linear regression on random samples gives wildly erratic results . . . , sometimes markedly overcorrecting [percentage bias reduction $\gg 100\%$], or even . . . greatly increasing the original bias [percentage bias reduction $\ll 0\%$]." Tables in that article, summarized here in Table 22.1, imply that, when the ratio of the variances of any covariate is two or one-half, linear regression can grossly overcorrect for bias or grossly undercorrect for bias; B is the number of standard deviations between the means of the groups and R is the ratio of treatment variance to control variance. When there is a large initial bias, the remaining bias, even if most of it has been removed, can still be substantial (e.g., the $B = 1$, $R = 1$, marked nonlinearity condition of Table 22.1).

The reasons why conditions, 1,2,3, are relevant in the general situation with many covariates are the following. All mean bias is, by definition, along the propensity score. Thus, both the bias along the propensity score and its variance ratio are relevant to assessing the degree of extrapolation involved in regression adjustment, especially when the propensity score is transformed to approximate normality. Orthogonal to the propensity score, the variables have the same mean, and thus the variance condition is relevant to assessing the degree of extrapolation involved with regression adjustment. Operationally, regress each of the original variables on the estimated linear propensity score (i.e., project each original covariate on the linear combination of the covariates that defines the estimated propensity score), and then take the residual of this regression (i.e., the part of the original variable orthogonal to – uncorrelated with – the propensity score). The variance ratios of these residuals are what are referenced in the third condition.

These conditions implicitly assume normally distributed covariates, or at least variables whose distributions can be adequately summarized by means and variances. With markedly nonnormal covariates, analogous conditions for reliability of regression adjustment can be more complex. An obvious condition with nonnormally distributed propensity scores is the overlap of distributions of the propensity scores in the two groups. This point was made and illustrated a quarter-century ago in Rubin (1977a) and is critical in applications such as the Lalonde (1986) data. When there are some treated subjects with propensity scores outside the range of the control subjects, no inference can be drawn concerning the effect of treatment exposure for these treated subjects from the data set without invoking heroic modeling assumptions based on extrapolation.

Table 22.1. *Percent reduction in bias using regression adjustment*

| | B = 1/4 | | | | B = 1/2 | | | | B = 3/4 | | | | B = 1 | | | |
| | Moderate | | Marked | | Moderate | | Marked | | Moderate | | Marked | | Moderate | | Marked | |
R	y = exp (x/2)	exp −(x/2)	exp (x)	exp (−x)	exp (x/2)	exp (−x/2)	exp (x)	exp (−x)	exp (x/2)	exp (−x/2)	exp (x)	exp (−x)	exp (x/2)	exp (−x/2)	exp (x)	exp (−x)
2	62	298	48	−304	80	146	72	292	90	129	88	170	96	113	102	139
1	100	100	101	101	101	101	102	102	101	101	104	104	102	102	108	108
1/2	298	62	−304	48	146	80	292	72	123	90	170	88	113	96	139	102

x is normally distributed and outcome y is related to x by one of the moderately nonlinear or markedly nonlinear relationships; B is the number of standard deviations between the means of the x distributions in the two groups, and R is the ratio of the variances of x in the two groups.

Note: If all bias were removed by regression adjustment, then all tabled values would be 100%. A negative number means that the adjustment, instead of removing bias, creates more bias in the same direction as the original bias; 0% means that the adjustment does not accomplish any bias reduction; a value larger than 200% indicates that the adjustment increases bias beyond the original amount but in the opposite direction.

Sources of values: Cochran and Rubin (1973). "Controlling Bias in Observational Studies: A Review." *Sankhyā,* Series A, Vol. 35, 4, Tables 3.2.1, 3.2.2, and 3.2.3, pp. 427–429.

6. COMPARING SMOKERS AND NEVER SMOKERS IN NMES

We illustrate these ideas for observational study design with NMES, where the outcome variables, which we never see in initial design, are health-care expenditures of various types and the occurrence of various smoking-related diseases. The covariates in NMES are numerous and include age, sex, race, marital status, education, etc. Also available are detailed smoking information that allows us to classify people first, as "never smokers," "former smokers," and "current smokers," and then to classify all smokers further by length and density of smoking behaviors, and former smokers also by years since quitting.

The objects of inference are "smoking attributable fractions," "conduct attributable fractions," "relative expenditure risks," etc., all of which are based on a comparison of specific health-care expenditures (or disease rates) for a particular type of smoker with never smokers with the same values of covariates, as a function of dosage and covariates (see Rubin (2000a) for the definition of these quantities). The comparisons are typically based on linear regressions or part linear regression/logistic regression models (e.g., see Harrison, 1998; Zeger et al., 2000) with no initial design effort. Here we focus solely on initial design and try to create samples of smokers and never smokers in NMES with the same multivariate distribution of covariates. We evaluate the success of these efforts using the benchmarks of Section 5.

Males and females were considered separately. Two treatment groups were defined as current smokers – 3,510 males and 3,434 females, and former smokers – 3,384 males and 2,657 females. The control group for both current and former smokers consisted of never smokers – 4,297 males and 7,691 females. All four propensity scores (male/female × current/former) were estimated by logistic regression using the 146 covariates defined by main effects, quadratic effects, and interaction effects; see Display 22.1 for a listing of these covariates. These analyses used the NMES sampling weights.

All propensity scores (which are estimated probabilities) were then transformed to the logit scale so that they were linear in the original covariates and their squares and products. This transformation was done for three reasons. First, relative to the raw propensity (probability) scale, the linear propensity (logit probability) is more relevant for assessing the efficacy of linear modeling adjustments (including those based on linear regression, logistic regression, and linear-log models). Second, the linear propensity scores tend to produce more benign distributions with more similar variances and more symmetry, because they are weighted averages of the original covariate values. And third, the linear propensity scores are more directly related to the benchmarks in the literature on adjustments for covariates based on linearity assumptions.

To assess the degree of overlap in distributions using the guidelines of Section 5, the following quantities were then calculated for all estimated linear propensity scores: (1) the standardized difference in the means of the propensity scores between smokers and never smokers, B; (2) the ratio of the variances of the propensity scores for smokers and never smokers, R; and (3) for each of the covariates, the ratio of the variance of the residuals orthogonal to the propensity scores for smokers to the variance of these residuals for never smokers (i.e., the residuals after adjusting for the

Display 22.1

Variables used in propensity model	Description
Seatbelt	5 levels of reported seatbelt use
Arthritis	Whether reported suffering from arthritis
Census Division	9 census regions
Champ Insurance	Whether have military insurance
Diabetes	Doctor ever told having diabetes
Down time	6 levels of reported emotional down time
Dumps time	6 levels of reported in the dumps time
Employment	Indicating employment status each quarter
English	English is a primary language
Retirement	Indicator for retirement status
Number of Friends	7 levels measuring the number of friends
Membership in Clubs	6 levels measuring memberships in clubs
Education	Completed years of education
HMO coverage	Indicating HMO coverage each quarter
High blood pressure	Doctor ever told having high blood pressure
Industry Code	14 Industry codes
Age	Age of the respondent
Labor Union	Indicator for a member of labor union
Log Height	Natural Logarithm of height
Log Weight	Natural Logarithm of weight
Marital Status	Marital status in each quarter
Medicaid	On medicaid (each quarter)
Medicare	On medicare (each quarter)
Occupation	Occupation code (13 levels)
Public Assistance	Other public assistance program (each quarter)
Friends over	Frequency of having friends over (7 levels)
Physical Activity	Indicator variable for physically active
Population density	3 levels
Poverty Status	6 levels
Pregnant 1987	Pregnancy status in 1987 (women)
Private Insurance	Other private insurance (each quarter)
Race	4 levels
Rated Health	5-point self rating of health status
Home ownership	Indicator for owning home
Rheumatism	Indicator for suffering from rheumatism
Share Life	Indicator variable for having somebody to share their life
Region	4 levels of region of the country
MSA	4 levels indicating types of metropolitan statistical area
Risk	General risk taking attitude (5 levels)
Uninsured	Indicator for lack insurance (each quarter)
Veteran	Indicator for veteran status
Incapler	Survey weight in NMES database
Agesq	Age*Age
Educat.sq	Education*Education
Age_wt	Age*Logwt
Age_educt	Age*Education
Age_ht	Age*Loght
Educat_wt	Education*Logwt
Educat_ht	Education*Loght
Loght_logwt	Loght*Logwt
Loghtsq	Loght*Loght
Logwtsq	Logwt*Logwt

Table 22.2. *Estimated propensity scores on the logit scale for "smokers" versus never smokers in full NMES*

| | | | Percent of covariates with specified variance ratio orthogonal to the propensity score | | | |
| | | | >1/2 and | >4/5 and | >5/4 and | |
Treated group	B	R	≤1/2	≤4/5	≤5/4	≤2	>2
Male current N = 3,510	1.09	1.00	3	9	57	26	5
Male former N = 3,384	1.06	0.82	2	15	61	15	7
Female current N = 3,434	1.03	0.85	1	15	59	23	2
Female former N = 2,657	0.65	1.02	5	7	85	7	5

B = Bias, R = Ratio of "smoker" to never-smoker variances; also displayed is the distribution of the ratio of variances in the covariates orthogonal to the propensity score.

propensity scores). The results of those calculations are found in Table 22.2 for the four comparisons (smokers vs. never smokers for male/female × current/former). The left two-thirds of Figure 22.1 displays the corresponding histograms of the propensity scores in the initial (unmatched) samples.

Comparing the results in Table 22.2 to the benchmarks in Section 5, it is clear that any linear (or part linear) regression model cannot be said to adjust reliably for these covariates, even if they were perfectly normally distributed. All values of B are greater than 1/2, and many of the values of R for the residuals of the covariates are outside the range (4/5, 5/4).

7. MAHALANOBIS METRIC MATCHING OF SMOKERS AND NEVER SMOKERS WITHIN PROPENSITY SCORE CALIPERS

For males and females, and current and former smokers, a one-one matched sample of never smokers was then selected. That is, for example, for the 3,510 male current smokers, 3,510 "matching" male never smokers were chosen from the pool of 4,297 male never smokers. The technique used was Mahalanobis metric matching (Rubin, 1976b) within propensity score calipers as defined in Rosenbaum and Rubin (1985a). The caliper width was 0.2 of a linear propensity score standard deviation, and the variables included in the metric were: age, education, body mass index, and sampling weight. The use of the metric matching after propensity score matching is the observational study equivalent of blocking in a randomized experiment; see Rosenbaum and Rubin (1985a) for an example and Rubin and Thomas (2000) for further explanation.

More specifically, for each "smoker," a "donor" pool of available matches was defined to include all never smokers who were within ± 0.2 standard deviations on the estimated propensity score; in our case, all such donor pools had never smokers. Starting with the hardest to match smoker (i.e., the one with the largest

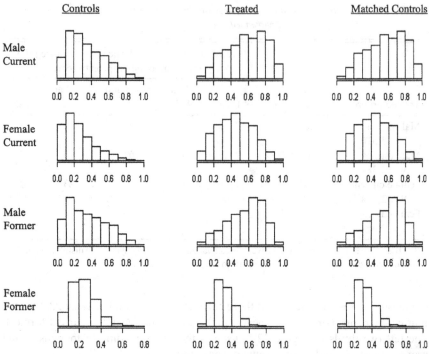

Figure 22.1. Histograms of propensity scores in: Full NMES controls, NMES treated, Matched NMES controls.

propensity score), individual matches were chosen. Specifically, the match chosen for the smoker was the never smoker with the smallest Mahalanobis distance with respect to the few key continuous covariates mentioned in the previous paragraph. The smoker and his match were then set aside. This process continued until each smoker had been assigned a match. The unchosen never smokers were discarded (e.g., 4,297 − 3,510 = 787 never smokers were discarded in the male current-smoker matching). Other versions of matching can certainly be defined, such as with-replacement of donors, or several donors to each treatment.

Table 22.3 displays the resultant matched samples with respect to the original estimated propensity score and the orthogonal components. Notice the dramatic reduction in bias along this estimated propensity score and the improved variance ratios in the orthogonal components. The right third of Figure 22.1 displays histograms in the matched control samples. Clearly, discarding the unmatched controls has created similar distributions of the propensity score in the matched samples. More refinement is still desirable, but the initial matching has done well at bringing the smokers and their control groups into balance.

The issue of what we would have done if some donor pools had been empty now arises. That is, what if there is a treated unit who has a propensity score not close to any control units' propensity scores? The correct answer is that inferences for the causal effects of treatment on such a unit cannot be drawn without making relatively heroic modeling assumptions involving extrapolations. Usually, such a unit should

Table 22.3. *Estimated propensity scores from full NMES on the logit scale for "smokers" versus never smokers in matched NMES*

Treated group	B	R	≤1/2	>1/2 and ≤4/5	>4/5 and ≤5/4	>5/4 and ≤2	>2
			Percent of covariates with specified variance ratio orthogonal to the propensity score				
Male current N = 3,510	0.08	1.16	1	3	90	6	0
Male former N = 3,384	0.04	0.99	1	1	94	3	1
Female current N = 3,434	0.04	0.94	1	1	93	5	0
Female former N = 2,657	0.06	1.02	0	2	91	7	0

B = Bias, R = Ratio of "smoker" to never-smoker variances; also displayed is the distribution of the ratio of variances in the covariates orthogonal to the propensity score.

be explicitly excluded from the analysis. In the tobacco litigation, however, such a unit legally cannot be excluded, and so here would have been included with an attendant acknowledgment of the inferential difficulty of drawing causal inferences for such a unit.

8. FURTHER SUBCLASSIFICATION OF MATCHED SAMPLES OF SMOKERS AND NEVER SMOKERS

Although Table 22.3 demonstrates dramatic improvement in balance relative to Table 22.2, more improvements are still desirable for the following reason. The estimated propensity score being summarized in Tables 22.2 and 22.3 (and Fig. 22.1) is the estimated propensity score in the full samples of Table 22.2, which does not equal the estimated propensity score in the matched samples themselves. That is, with respect to the linear combination of covariates that is the full sample estimated propensity score, there is now excellent balance in the matched samples. But we could now look at the linear combination of the covariates that is the estimated propensity score in the matched samples, and we might see some imbalance unless all individual covariates have extremely similar distributions in the matched samples; Table 22.3 reveals that there still are some differences. The difference between these two estimated propensity scores is due theoretically to small sample variation, just like random imbalance in covariates in a completely randomized experiment.

Table 22.4 gives the same information as Table 22.3 but now for the re-estimated propensity score, that is, estimated in the matched samples (again, using the NMES sampling weights). Although the differences between the treatment and control groups in Table 22.4 are minor relative to the differences displayed in Table 22.2 (e.g., all values of B are now less than 0.4), and arguably minor enough to satisfy the criteria set forth in Section 5, we seek even better balance within the matched samples on all covariates through the use of subclassification on the estimated propensity score within the matched samples.

Table 22.4. *Estimated propensity scores on the logit scale for "smokers" versus never smokers in matched NMES*

| Treated group | B | R | | Percent of covariates with specified variance ratio orthogonal to the propensity score | | | |
			≤1/2	>1/2 and ≤4/5	>4/5 and ≤5/4	>5/4 and ≤2	>2
Male current N = 3,510	0.39	1.33	0	4	88	8	0
Male former N = 3,384	0.32	1.33	0	1	95	3	1
Female current N = 3,434	0.35	1.18	1	1	92	6	0
Female former N = 2,657	0.31	1.09	0	2	91	7	0

B = Bias, R = Ratio of "smoker" to never-smoker variances; also displayed is the distribution of the ratio of variances in the covariates orthogonal to the propensity score.

Specifically, we rank the smokers and their matched sample of never smokers on the estimated propensity score in the matched samples. First, we create two equal-size (weighted) subclasses, low and high on the propensity score. The treated and control units with low propensity scores are to be compared against each other, and those with high propensity scores are to be compared against each other, and the two estimates of treatment effect averaged. When doing this comparison for any covariate (including treating the propensity score as a covariate), the answer, ideally, should be zero since there is no effect of treatment on the covariates. Effectively, this subclassification creates new weights within each subclass: for the treated, the new weights are equal to the total number (weighted) of treated and controls in that subclass divided by the number (weighted) of treated in that subclass, and for the controls, the total number (weighted) of units in that subclass divided by the number (weighted) of controls in that subclass. Because the eventual goal is to draw inferences about the smokers if they had been never smokers, it can be argued that these high and low subclasses should be equal size within smokers. For the points in this article, however, this refinement makes little difference.

The same idea for weighting works no matter how many subclasses we have. Now the subclass specific weights can be attached to each unit, treated or control, and we can then do a new weighted propensity score analysis that reflects the extra balance created by the subclassification. The results of these weighted subclass propensity score analyses (for k = 2, 4, 6, 8, 10 subclasses) are displayed in Table 22.5A–22.5D. Table 22.4 serves as a comparison for these results because it reflects only one subclass – no subclassification adjustment.

The results in Table 22.5 show the reduction in initial bias that occurs with further subclassification. It is rather dramatic, both along the propensity score (re-estimated in matched samples) and orthogonal to it. The choice of the number of subclasses can be made by intense examination of specific covariates without fear of "biasing" any result involving the outcomes because there is no outcome variable being used. Once the outcome variables are available, modeling adjustments

Table 22.5A. *Propensity subclassification analyses on the logit scale for current vs. never-smoker males in matched NMES*

Number of subclasses	B	R		Percent of covariates with specified variance ratio orthogonal to the propensity score			
			≤1/2	>1/2 and ≤4/5	>4/5 and ≤5/4	>5/4 and ≤2	>2
1*	0.39	1.33	0	4	88	8	0
2	0.18	1.36	0	2	98	0	0
4	0.10	1.25	0	1	99	0	0
6	0.09	1.30	0	0	100	0	0
8	0.08	1.16	0	0	100	0	0
10	0.07	1.12	0	0	100	0	0

B = Bias, R = Ratio of "smoker" to never-smoker variances; also displayed is the distribution of the ratio of variances in the covariates orthogonal to the propensity score.
* Results from Table 22.4, row 1.

Table 22.5B. *Propensity subclassification analyses on the logit scale for former vs. never-smoker males in matched NMES*

Numbers of subclasses	B	R		Percent of covariates with specified variance ratio orthogonal to the propensity score			
			≤1/2	>1/2 and ≤4/5	>4/5 and ≤5/4	>5/4 and ≤2	>2
1*	0.32	1.33	0	1	95	3	1
2	0.16	1.38	0	1	98	1	0
4	0.09	1.32	0	1	98	1	0
6	0.07	1.30	0	0	99	1	0
8	0.07	1.37	0	0	99	1	0
10	0.07	1.31	0	0	99	1	0

B = Bias, R = Ratio of "smoker" to never-smoker variances; also displayed is the distribution of the ratio of variances in the covariates orthogonal to the propensity score.
* Results from Table 22.4, row 2.

Table 22.5C. *Propensity subclassification analyses on the logit scale for current vs. never-smoker females in matched NMES*

Number of subclasses	B	R		Percent of covariates with specified variance ratio orthogonal to the propensity score			
			≤1/2	>1/2 and ≤4/5	>4/5 and ≤5/4	>5/4 and ≤2	>2
1*	0.35	1.18	1	1	92	6	0
2	0.14	1.26	0	1	98	1	0
4	0.08	1.44	0	1	99	0	0
6	0.06	1.69	0	1	99	0	0
8	0.05	1.69	0	0	100	0	0
10	0.05	1.70	0	0	100	0	0

B = Bias, R = Ratio of "smoker" to never-smoker variances; also displayed is the distribution of the ratio of variances in the covariates orthogonal to the propensity score.
* Results from Table 22.4, row 3.

Table 22.5D. *Propensity subclassification analyses on the logit scale for former vs. never-smoker females in matched NMES*

Number of subclasses	B	R	Percent of covariates with specified variance ratio orthogonal to the propensity score				
			$\leq 1/2$	$>1/2$ and $\leq 4/5$	$>4/5$ and $\leq 5/4$	$>5/4$ and ≤ 2	>2
1*	0.31	1.09	0	2	91	7	0
2	0.13	1.09	0	0	97	3	0
4	0.08	0.85	0	0	99	1	0
6	0.07	0.85	0	0	100	0	0
8	0.06	0.77	0	0	100	0	0
10	0.06	0.92	0	0	100	0	0

B = Bias, R = Ratio of "smoker" to never-smoker variances; also displayed is the distribution of the ratio of variances in the covariates orthogonal to the propensity score.
* Results from Table 22.4, row 4.

within the subclasses should be performed and then combined across subclasses. This approach allows for nonlinear relationships between outcomes and covariates through a separate adjustment model within each subclass; see Benjamin (1999) for a specific example.

9. OTHER APPROACHES TO DESIGNING AN OBSERVATIONAL STUDY IN NMES

We could let the number of subclasses continue to grow until each subclass had at most one treated or control unit – maximal subclassification (Rosenbaum, 1989). This, however, would preclude further standard modeling adjustments, although they could be performed on "matched" pair differences formed within each maximal subclass – like a classical matched pair analysis. This is an unstudied approach to the best of my knowledge.

Another approach is to form weights directly from the estimated propensity score without subclassification. Thus, a treated unit's weight is the inverse of its propensity score (times its NMES weight) and a control person's weight is the inverse of one minus its propensity score (times its NMES weight). This process can generate unrealistically extreme weights when an estimated propensity score is near zero or one, something that is avoided in the subclassification approach.

Table 22.6 displays results from a weighted propensity score analysis using these inverse probabilities as multipliers of the NMES weights. The odd result for the variance ratio of the propensity score for the male former-smoker group is due to an extreme weight – a propensity score near zero. Another approach, unstudied to the best of my knowledge, would be to use subclasses for the highest and lowest estimated propensity score ranges (e.g., the upper and lower 2.5%) and the weights for the remaining interior 95% of the propensity scores.

A summary of all analyses can be displayed in one table for each treatment group comparison, as illustrated for male-current smokers in Table 22.7. This table documents the nice progression of increasing multivariate balance that can

Table 22.6. *Weighted propensity score analyses based on inverse probabilities for weights in matched NMES*

| Treated group | B | R | | Percent of covariates with specified variance ratio orthogonal to the propensity score | | | |
			≤1/2	>1/2 and ≤4/5	>4/5 and ≤5/4	>5/4 and ≤2	>2
Male current N = 3,510	0.03	1.19	0	0	100	0	0
Male former N = 3,384	0.08	0.22	0	0	100	0	0
Female current N = 3,434	0.03	1.70	0	0	100	0	0
Female former N = 2,657	0.03	0.66	0	0	100	0	0

B = Bias, R = Ratio of "smoker" to never-smoker variances; also displayed is the distribution of the ratio of variances in the covariates orthogonal to the propensity score.

be obtained using the combined techniques of matching and finer and finer subclassification.

10. DISCUSSION

The analyses we have done here are really just an indication of the types of analyses that can be done to help design an observational study using propensity scores, all without any fear of opportunistically biasing estimates of treatment effects. Care

Table 22.7. *Estimated propensity scores on the logit scale for male current smokers (N = 3,510) versus male never smokers in NMES*

| Analysis | B | R | | Percent of covariates with specified variance ratio orthogonal to the propensity score | | | |
			≤1/2	>1/2 and ≤4/5	>4/5 and ≤5/4	>5/4 and ≤2	>2
Full N = 4,297	1.09	1.00	3	9	57	26	5
Matched N = 3,510	0.08	1.16	1	3	90	6	0
Matched							
$K = 1$	0.39	1.33	0	4	88	8	0
$K = 2$	0.18	1.36	0	2	98	0	0
$K = 4$	0.10	1.25	0	1	99	0	0
$K = 6$	0.09	1.30	0	0	100	0	0
$K = 8$	0.08	1.16	0	0	100	0	0
$K = 10$	0.07	1.12	0	0	100	0	0
$K = \infty$	0.03	1.19	0	0	100	0	0

B = Bias, R = Ratio of "smoker" to never-smoker variances; also displayed is the distribution of the ratio of variances in the covariates orthogonal to the propensity score.

must be taken when estimating standard errors, as for the purpose of obtaining confidence intervals. Typically the standard errors and confidence intervals calculated assuming estimated propensity score are conservative (see Rubin and Thomas, 1992b). Refinements in our example could include defining finer categories of smoking (e.g., heavy or light) and selecting matches from the never smokers for each such treatment group.

The most principled strategy requires, in addition to the type of initial study design described here, the full specification of all analyses to be performed and assessment of their calibration. That is, ideally we may also wish to specify particular model-based analyses we intend to apply, again before observing any outcomes, to encourage complete objectivity in the analyses. For example, the sequence of "primary" and "secondary" analyses could be specified a priori, as in an FDA submitted protocol for a pharmaceutical company's randomized experiment. As mentioned earlier, if substantial balance in covariates has been obtained at the initial design stage, the exact form of the modeling adjustment is not critical because the similar treated and control covariate distributions implies only limited model-based sensitivity. Of course, repeated subgroup analyses, looking for "action" somewhere, are subject to more complicated interpretations, just as they are in a randomized experiment. Also, analyses involving adjustments for unobserved covariates are nearly always quite subjective, although recent progress on understanding adjustments for intermediate outcomes using principle strata (Frangakis and Rubin, 2002), shows promise for clarifying underlying assumptions.

It is hoped that this article will encourage the development of designed observational studies.

ACKNOWLEDGMENTS

I thank T. E. Raghunathan for carrying out the computations described in this paper, and Guido Imbens and three reviewers for extremely helpful comments on an earlier draft.

The chapters in Part VII are more pure applications than the others in this volume. Chapter 23, Witkin et al. (1976), was a study of criminality in XYY and XXY men in Copenhagen, stimulated by proposals by John Mitchell, the U.S. Attorney General under President Richard Nixon, to identify such babies at birth and warn their parents of the potential antisocial tendencies; see the introductory section of this article for the underlying reasoning. This study matched XXY and XYY males to normal XY males on height, IQ, education, and parental socioeconomic status. The method used was simple stratified matching. Further model-based adjustments were also made. This article was a true committee effort, with the different authors of the different parts not always agreeing on what was important to say or how to say it. A careful reading is therefore more confusing and interesting than usual.

Chapter 24, Rubin (1991b), was the written version of a President's Invited Address to the Biometric Society, and reported some matched sample analyses that I was doing for Sandoz Pharmaceuticals. The issue concerned the effects of switching from a name-brand drug to a generic version after the original drug's patent expired. In particular, Sandoz had an antipsychotic drug for which it was investigating evidence that switching from the name-brand to a generic version was potentially harmful. Much of the article concerns how to formulate this causal inference problem correctly, especially how to deal with the post-intervention information. I now believe that the "principal stratification" framework of Frangakis and Rubin (2002) would have been clarifying.

Chapter 25, Reinisch et al. (1995), is a medical publication resulting from the matched samples created using methods very similar to those studied by Paul Rosenbaum, Neal Thomas, and me. The study, on which I started consulting in the mid-1970s, used Mahalanobis-metric matching within propensity score calipers, with intuitive refinements by the first author, and furthermore, used regression adjustment on the matched samples. Although the paper is very applied, there are some nice tables showing the bias reduction due to the matching (Table 25.1) and the resultant estimated causal effects (Tables 25.2 and 25.3). There is also some relatively confusing discussion of matching on posttreatment "covariates" in footnote * to Table 25.1 about which variables were, and were not, included in the matching, and related confusing discussion at the bottom of page 430; the proper conceptualization of these issues uses the framework of principal stratification (Frangakis and Rubin, 2002), as with similar issues in Chapter 24.

Chapter 26, Rubin (1997), on propensity scores, appeared in a special edition of the *Annals of Internal Medicine* based on presentations at the Sixth Annual Regenstrief Conference. I tried to write this article to communicate the key idea behind propensity scores to physicians. I apparently succeeded because it seems to be highly cited. It also is the only generally accessible reference where the results of a U.S. General Accounting Office (GAO) study of treatments for breast cancer is reported and compared to the observational study results. GAO wanted to go beyond the randomized clinical trials reported in Table 26.2 of this article, because of their concern that the medical care in those carefully controlled clinical trials may not fairly represent the medical care patients would receive in common practice. Moreover, they were concerned that the women and the doctors who would agree to participate in randomized trails of mastectomy versus breast conservation were not typical of U.S. women and their doctors. Hence, they wanted to conduct a careful observational study of common medical practice in the United States for breast cancer to "replicate" the international randomized experiments. I was a consultant on the project for GAO, which used propensity scores, and the results are summarized in Table 26.3 of that article. The observational study results are consistent with the conclusions of the randomized experiments, but there are some interesting differences. For example, both the experiments and the observational studies provide little evidence of treatment differences on mortality for the types of women studied, but the overall survival rates in the experiments were better, presumably due to more experienced care in the hospitals participating in the experiments, as anticipated by GAO.

The final chapter in this part, Chapter 27, McIntosh and Rubin (1999), is an editorial from *Medical Care*, concerning estimating the effect of DNR (do not resuscitate) orders. Martin McIntosh was an excellent PhD student at Harvard University before moving to the University of Washington at Seattle and the Fred Hutchinson Cancer Center, and he was quite sophisticated about such medical problems. This short editorial note provides some sensible advice and perspective on the study of the effects of DNR orders.

23. Criminality in XYY and XXY Men

Herman A. Witkin, Sarnoff A. Mednick, Fini Schulsinger, Eskild
Bakkestrøm, Karl O. Christiansen, Donald R. Goodenough, Kurt
Hirschhorn, Claes Lundsteen, David R. Owen, John Philip,
Donald B. Rubin, and Martha Stocking

Few issues in behavior genetics have received more public and scientific attention than that given to the possible role of an extra Y chromosome in human aggression. Soon after the literature began to suggest an elevated frequency of the XYY genotype among inmates of institutions for criminals and delinquents, interest in this issue had a meteoric rise; and it has been sustained ever since. This happened for several reasons. Stories about a few men who had or were presumed to have an extra Y chromosome and who had committed serious crimes were given prominent attention in the press, suggesting the intriguing idea that the single Y chromosome normally found in males contributes to "aggressive tendencies" in that sex and that an extra Y carries these tendencies beyond their usual bounds. Reports of antisocial behavior in XYY men, often based on a single case, soon began to appear in the scientific literature[1] and were taken as evidence of an XYY-aggression linkage. The serious moral and legal implications of such a linkage attracted the interest of social scientists and legal groups to the XYY phenomenon (Shah, 1970), and students of genetics and psychology saw in it, as Lederberg (1973) has said, "one of the most tangible leads for connecting genetic constitution with behavior in man."

A number of studies have supported the earlier finding of an elevated frequency of cases with an XYY complement among men in institutions, particularly in penal-mental institutions.[2,3] At the same time, these studies have not provided clear evidence of whether or not there exists an "XYY syndrome" of which antisocial behavior is a prominent component. Neither have they provided a definitive answer to

[1] See, for example, Forssman, 1967; Goodman, Smith, and Migeon, 1967; Kelly, Almy, and Bernard, 1967; Persson, 1967; Richards and Stewart, 1966.

[2] For recent comprehensive reviews of this evidence see Borgaonkar and Shah, 1974; Hook, 1973; Owen, 1972. See also Jarvik, Klodin, and Matsuyama, 1973; Shah and Borgaonkar, 1974.

[3] In view of institutional variations in admissions policies, not always clearly specified in the published reports, it is difficult to judge whether the tendency of XYY men to appear more frequently in penal-mental than in straight penal institutions reflects the involvement of a mental aspect as well as a penal aspect. No clear tendency toward higher representation of XYY's in mental institutions has been found; however, studies of such institutions have been few.

The elevated crime rate of XYY males is not related to aggression. It may be related to low intelligence.

Reprinted from *Science*, 1976, **193**, 547–555, with permission of the American Association for the Advancement of Science.

the question of why men with an extra Y chromosome are at higher risk for institutionalization than XY men. For the sake of identifying the kind of research that is needed to clarify these issues, it is worth reviewing the main limitations of the studies on which our present information about XYY men is based, and also the lacunae in our knowledge.

First, the search for XYY men has often been conducted in selected groups presumed to be likely to contain them, such as institutionalized men and tall men. Second, a number of reports now in the literature are based on observations of a single case or just a few cases.[1] Third, many studies of XYY's have not included control XY's; and in those that did, comparisons were often made without knowledge of the genotype of the individuals being evaluated (Borgaonkar and Shah, 1974). The control groups used have varied in nature, and comparison of results from different studies has therefore been difficult. There has been a dearth of psychological, somatic, and social data obtained for the same individual XYY men. Finally, there do not yet exist adequate prevalence data for the XYY genotype in the general adult population with which the XYY yield of any particular study may be compared. Though incidence data on neonates are available for a fairly large number of subjects, incidence studies of neonates are still few, and there are potential problems in the practice of pooling highly variable incidence findings from studies with populations that are quite different from each other or whose characteristics have not been adequately specified.

With the evidence in its present state, it is not surprising to find divergent views about the support it provides for a link between XYY and aggression, as for example in the contradictory conclusions reached in recent reviews of the XYY literature.[2] Whatever the interpretation of the available evidence, however, most investigators concerned with the XYY problem are agreed on the research that is now required to determine whether the XYY complement has any behavioral or social consequences. What is needed, ideally, is an ascertainment study of a large population unselected with regard to institutionalization or height, and a comparison of the XYY cases identified with control XY's in psychological, somatic, social, and developmental characteristics, the evaluation of these characteristics in the two groups being made according to a double-blind procedure.

The study we undertook, done in Denmark because of the excellent social records kept there, was designed to meet as many of these specifications as our financial resources allowed. It was already evident at the time we undertook the study that XYY's tend to be very tall.[4] As a way of maximizing the chances of obtaining a sufficient sample of XYY's for intensive individual study, we decided to do chromosomal determinations of all men in the top 15 percent of the height distribution of our Danish male population. A sampling of men in the bottom 85 percent would also have been desirable, but the probability of finding XYY's among shorter men is so small as to make an effective study of shorter XYY's too expensive to conduct.

---

[4] This has been shown both by height data from XYY ascertainment studies of unselected male populations (for example, Bartlett et al., 1968; Crandel, Carrell, and Sparkes, 1972; DeBault, Johnston, and Loeffelholz, 1972; Hook and Kim, 1971; Jacobs et al., 1965; Jacobs et al., 1971; Nielsen, 1971; Vianna et al., 1972); and by studies which compared height of XYY's to that of their siblings and parents (Owen, 1973, p. 222).

A first aim of this research project was to determine whether XYY's from the total general population have an elevated crime rate. A second aim, if an elevated crime rate appeared, was to identify intervening variables that may mediate the relation between an extra Y chromosome and increased antisocial behavior. Three variables of particular interest are aggressiveness, intelligence, and height.

A common interpretation of the finding that XYY's tend to be over-represented in criminal institutions is that aggressiveness is an intervening variable. In this view, an extra Y chromosome increases aggressive tendencies and these, in turn, lead to increased criminal behavior. If this interpretation is correct, we may expect crimes committed by XYY men to be aggressive in nature, involving assaultive actions against other persons. We designate this the aggression hypothesis.

Concerning intelligence the reasoning is as follows: In common with most genetic aberrations, an extra Y in the male chromosomal complement is likely to have an adverse effect on development. Among possible dysfunctions, of particular interest for the XYY question is dysfunction in the intellectual domain. There is some evidence, although it is hardly consistent, of an intellectual impairment in XYY men (Owen, 1972; Baughman and Mann, 1972; Finley et al., 1973; McKerracher, 1971; Nielsen and Christiansen, 1974; Noël et al., 1974). Intellectual impairment may contribute to antisocial behavior. It seems plausible also that when individuals with impaired intellectual functioning commit crimes, they are more likely to be apprehended than are criminals of normal intelligence. This conception of intelligence as an intervening variable, mediating the relation between the presence of an extra Y chromosome and antisocial behavior, may be designated the intellectual-dysfunction hypothesis.

The extreme height of XYY's may facilitate aggressive acts on their part. In addition, it may cause them to be perceived by others as dangerous, with the possible consequence that they are more likely than shorter men to be suspected of crimes, to be pursued and apprehended, and, when tried, to be convicted. The view that tallness may serve as an intervening variable we designate the height hypothesis.

Because XXY's (Klinefelter males) also tend to be tall, we could expect our case-finding effort in the top 15 percent height group to identify a number of XXY's as well as XYY's. Most studies of XXY's have suffered from essentially the same limitations as those mentioned earlier for studies of XYY's; a study of XXY's is therefore of value in its own right. In addition, there is some evidence that XXY's also appear in institutions with disproportionate frequency. This raises the possibility that any sex chromosome aberration in males, and not particularly an extra Y, may be associated with increased risk of institutionalization (Borgaonkar and Shah, 1974). Comparison of XYY's and XXY's may help to assess this possibility.

This article deals with the results of the case-finding study among tall men and with the evidence obtained to this point from the social records available for the XYY and XXY men who were found. An intensive individual study is now being conducted with these men and their controls.[5]

---

[5] Subjects in the individual case study are the XYY's and XXY's identified in the case-finding study described below, and two XY controls for each. Both controls are matched to the proband in height, age, and parental social class, and one is matched as well in intellectual level, as judged from scores on an army selection test. The tests given these subjects include a battery, approximately 16 hours in duration, of cognitive tests, personality tests, an interview, and physical, neurological, and neuropsychological examinations.

CASE-FINDING PROCEDURE

The population from which we sampled consisted of all male Danish citizens born to women who were residents of the municipality of Copenhagen, Denmark, between 1 January 1944 and 31 December 1947, inclusive. Not only did Copenhagen afford a very large source population, but available demographic data indicated that most men born in the city at that time would still be living there. The parish records, in which all births are registered, were used to identify the males born there in the chosen period. They numbered 31,436. The *folkeregister* (the Danish national register) provided current addresses and other information.

Information about the height of these men was obtained from draft board records where possible. The use of the 4-year period between 1944 and 1947 to define the target population provided us with a group of men who were at least 26 years old, the age by which Danish men are required to report to their draft boards for a physical examination, at which time their heights are recorded. For the small group of men who for some reason had never visited their draft boards, height data were obtained from other sources, such as institutions and the civil defense. In the very few cases where such sources were not available, cards were addressed directly to the men themselves requesting that they send us their heights.

By these methods, a group was composed consisting of 28,884 men who were still alive when the study began and for whom height information could be obtained. This group numbered 2552 fewer cases than the target population of 31,436 Copenhagen-born men. Of these 2552 cases 1791 were dead; 37 could not be traced for height determination; 21 could not be located at all in the *folkeregister* (probably for such reasons as name changes, death at birth, address change at birth); and 703 had emigrated and no record of their height was available. Of the 664 emigrants for whom we were able to determine age at emigration, 85.2 percent had emigrated before age 18, so that it is probable that in many of those cases the decisions about emigration were made by parents.

A cutoff point of 184 centimeters was used in composing the tall group in which the search for sex chromosome anomalies was to be conducted. The resulting group consisted of 4591 men, the top 15.9 percent of the height distribution in the total group of 28,884. Deaths reduced the tall group by 33 during the case-finding period, leaving a total of 4558 men to be searched for sex chromosome anomalies. An attempt was made to visit the homes of all 4558 for the purpose of obtaining blood samples and buccal smears to be used in determining chromosomal constitution.

Before home visiting began, members of the study staff were interviewed by the news media. These interviews provided an opportunity to publicize the purpose and nature of the study, with the result that most of the men who were asked to participate in the study had already heard of it when first approached. The initial individual contact was made by a letter, which mentioned the nature of the study and indicated that someone from the study staff would visit at a specified time. If the subject expected not to be at home at that time, he was asked to return a card on which he could suggest an alternative time. Men not found at home, and who had not asked for an alternative date, were subsequently revisited, up to a total of 14 times in the most extreme instance.

When the subject was seen, he was shown a newspaper clipping reporting the interview with our staff members and any questions he had were answered. The subject was also assured that his anonymity would be maintained. He was then asked whether he would be willing to give a buccal smear and, if he agreed, whether he would be willing to give a few drops of blood from an earlobe as well. The home visitor also asked the subject to fill out a questionnaire, and told him that at a later time he might be asked to participate further in the study. As the case-finding effort progressed, various methods were adopted to facilitate and encourage participation, such as setting up a station at one of the centrally located hospitals in Copenhagen to which the men were invited to come during a wide range of daytime and evening hours, and offering small financial inducements.

The 4139 men for whom sex chromosome determinations were made constituted 90.8 percent of the starting group of 4558 living tall men.[6] Of the 419 unexamined cases, 174 men declined to participate; 138 men emigrated in the course of the study or were sailors and away from Denmark; 25 were destitute men without identifiable homes; and 82 men, on repeated visits, were not found at the official addresses listed for them in the *folkeregister*.[7] Some characteristics of these cases, and of the 174 men who declined to take part, are given below.

The buccal smears and blood samples were taken to the Chromosomal Laboratory of Rigshospitalet, in Copenhagen, for analysis. The buccal smears were stained with hematoxylin for the detection of X chromatin (Barr and Bertram, 1949) and with quinacrine dihydrochloride for the detection of Y chromatin (Pearson and Bobrow, 1970). The peripheral blood was treated by a micromethod modification (Hirschhorn, 1965) of the method of Moorhead et al. (1960). Chromosome preparations were stained conventionally with orcein and by the method of Seabright (1971) for G-banding and the method of Caspersson et al. (1970) for Q-banding and identification of the Y chromosome.[8]

DOCUMENTARY DATA

A variety of records was available for almost all the men in the study. The present report is limited to data from these records for five variables: height, convictions for criminal offenses, level of intellectual functioning as indicated by scores on an army selection test and by educational attainment, and parental social class at the time of the subject's birth.

The sources of information about height have already been described.

The source of data on convictions for criminal offenses was penal certificates (*straffeattest*) obtained from penal registers (*strafferegistrene*) maintained in the offices of local police chiefs. These certificates are extracts of court records of trials and cover all violations of the penal code that resulted in convictions. Offenses in the Danish penal code include such acts as these, among others: forgery, intentional arson, sexual offenses, premeditated homicide, attempted homicide, manslaughter, assault and battery, housebreaking, larceny, receiving stolen goods, and damage to property belonging to others. The penal certificates contain highly reliable

---

[6] The identity of two of the sampled cases found to be XY's was lost to us in our records and could not be considered in the data analysis.

[7] It is likely that many of these 82 men had moved in with friends and did not register the move as an official change of residence.

[8] A more detailed account of the chromosomal studies is given in Philip et al. (1976).

Table 23.1. *Crime rates and mean values for background variables of XY's, XYY's, and XXY's. Significance level pertains to comparison with the control group (XY) using a two-sided test. For criminality rate an exact binomial test was used; for all other variables a t-test was used.*

Group	Criminality		Army selection test (BPP)			Educational index			Parental SES			Height		
	Rate (%)	N	Mean	S.D.	N	Mean	S.D.	N	Mean	S.D.	N	Mean	S.D.	N
XY	9.3	4096	43.7	11.4	3759	1.55	1.18	4084	3.7	1.7	4058	187.1	3.0	4096
XYY	41.7*	12	29.7†	8.2	12	0.58*	0.86	12	3.2	1.5	12	190.8†	4.6	12
XXY	18.8	16	28.4†	14.1	16	0.81‡	0.88	16	4.2	1.8	16	189.8†	3.6	16

* *P* < .01.   † *P* < .001.   ‡ *P* < .05.

390

information concerning the section of the penal law violated and the penalty imposed. A subject was considered to have a criminal record if he was convicted of one or more criminal offenses.

For evaluation of level of intellectual functioning, two kinds of measures were used. One was scores from the test employed in screening army recruits for intelligence, the *Børge Priens Prøver* (BPP), available from the draft-board records. Because the BPP was constructed as a screening device, it covers only a limited number of cognitive dimensions. The BPP scores are accordingly only rough indicators of intellectual level. The scores could not be obtained for some men; most frequent in this group were men who had never taken the test and men whose records were not available because they were in the army.

The second measure of intellectual functioning was educational level achieved. In Denmark examinations are given at the end of the 9th, 10th, and 13th years of schooling. From the available social records it was possible to determine which, if any, of these examinations was passed. For our "educational index" subjects who passed no examination were given a score of 0, and those who passed the first-, second-, and third-level examinations were given scores of 1, 2, and 3, respectively. It should be noted that the maximum rating assigned was 3 regardless of how many additional years of education the individual may have had.[9] In a very small number of cases information needed for determining the educational index could not be obtained.

Parental socioeconomic status (SES) was classified primarily according to father's occupation at the time of the subject's birth. In a small number of cases the father or his occupation was not known; in some of these instances mother's occupation was known and was used instead. A seven-point SES classification was used, modified from a nine-point classification devised by Svalastoga (1959).

FREQUENCY OF XYY'S AND XXY'S AND OF CRIMINALS AMONG THEM

Among the 4139 men for whom sex chromosome determinations were made, 12 XYY's and 16 XXY's were identified. These frequencies represent prevalence rates of 2.9/1000 and 3.9/1000, respectively. Thirteen men were identified as XY's with other chromosomal anomalies. The remainder, all identified as having the normal XY complement, constituted the control group.[6]

A search in the penal registers showed that 41.7 percent of the XYY's (5 of 12 cases), 18.8 percent of the XXY's (3 of 16 cases), and 9.3 percent of the XY controls had been convicted of one or more criminal offenses (Table 23.1). The difference between the percentages for the XYY's and the XY controls is statistically significant ($P < .01$, exact binomial test). The rate for the XXY's is somewhat higher than the rate for the XY controls, but that difference is not significant; neither is the difference in rates between the XYY's and XXY's.

A first approach to evaluation of the aggression hypothesis was to examine the nature of the crimes of which the five XYY's had been convicted (Table 23.2). Their offenses were not particularly acts of aggression against people. Only in case 2 do

---

[9] Not a single one of the XYY's or XXY's reached the 13-year examination, but a number of the XY controls did and may have gone on for additional education.

Table 23.2. *Nature of offenses of XYY's convicted on one or more criminal charges.*

Case No. 2

This man is a chronic criminal who, since early adolescence, has spent 9 of 15 years in youth prisons and regular prisons. By far his most frequent criminal offense, especially in his youth, has been theft or attempted theft of a motor vehicle. Other charges included burglary, embezzlement, and procuring for prostitution. On a single occasion he committed a mild form of violence against an unoffending person; for this together with one case of burglary he received a sentence of around three-quarters of a year. This aggressive act was an isolated incident in a long period of chronic criminality. Except for this act, and the charge of procuring, all his nearly 50 offenses were against property, predominantly larceny and burglary. His single most severe penalty was somewhat less than a year in prison. Most of his crimes were committed in the company of other persons (BPP, 27).

Case No. 3

This man committed two thefts, one in late adolescence, the second in his early 20's. The penalties for both were mild – a small fine for the first, and less than 3 months in prison for the second. His last offense was 7 years ago (BPP, 37).

Case No. 5

This man committed two petty offenses as a young adult, within a short time of each other (one the theft of a motor-assisted cycle, the other a petty civil offense), for which the penalties were detentions of approximately 2 weeks and less than 2 weeks, respectively. His last offense was committed 10 years ago (BPP, 28).

Case No. 7

This man committed his only criminal offenses in his 20's, within a short period of time: falsely reporting a traffic accident to the police and causing a small fire. On both occasions he was intoxicated. The penalty was probation. His last offense was committed 5 years ago (BPP, 25).

Case No. 12

This man was under welfare care as a child and has spent only three to four of the last 20 years outside of institutions for the retarded. He is an episodic criminal. When very young he committed arson. Later his crimes included theft of motor vehicles, burglary, larceny, and embezzlement. His more than 90 registered offenses were all against property, mostly theft and burglary. For crimes committed while he was out of an institution, the penalty imposed was placement in an institution for the mentally retarded. For crimes committed while he was in such an institution – once theft of a bicycle, another time theft of a quantity of beverage – he was continued in the institution (BPP, 18).*

* Since this man was mentally retarded and spent many years in an institution for the retarded, he was not given a BPP at the draft board. The BPP of 18 was estimated by a stepwise linear regression, using a double cross-validity design, from the correlation between BPP scores and scores for the Wechsler Adult Intelligence Scale for the men in the individual case study.

we find an instance of such an act. The difference between the XYY's and XY's in percentage of cases with one or more convictions for crimes of violence against another person – 8.4 percent (one man out of the 12) versus 1.8 percent (71 out of 4096) – is not statistically significant (one-tailed exact binomial test). If we compare only those who had criminal convictions, the XYY's and XY's are again very similar in percentage of convictions that involved crimes of violence against a person (20.0

Table 23.3. *Nature of offenses of XXY's convicted on one or more criminal charges.*

Case No. 17

This man's only criminal offense, committed when he was well into his 20's, was that he attacked his wife in an exceptionally brutal way, without any provocation from her. This happened twice, within a very short interval, while he was under the influence of liquor. For this he was imprisoned for somewhat more than a year (BPP, 26).

Case No. 25

The criminal career of this man consisted of two offenses: the first, in late adolescence, a theft of edibles from a food store, for which he was placed on probation, the second the theft of a motor vehicle, for which he was given less than 3 weeks of simple detention. Both crimes were committed in company with others. The last occurred 7 years ago (BPP, 11).

Case No. 27

This man had a short period of juvenile delinquency. His offenses included attempted theft and theft of a motor vehicle and a bicycle, burglary, and theft from a vending machine. On his first offense, in early adolescence, the charge was withdrawn and he was put under the care of child welfare authorities. His two other penalties consisted of withdrawal of charge on payment of a fine. Several of his offenses were committed in company with another person. The last occurred 10 years ago (BPP, 16).

and 19.4 percent, respectively). These data provide no evidence that XYY's are more likely to commit crimes of violence than XY's.[10]

The generally mild penalties imposed on the convicted XYY's (Table 23.2) indicate that their crimes were not extremely serious. By far the most severe sentence was imprisonment for somewhat less than a year,[11] imposed on case 2. Also suggesting that the XYY's with records of criminal convictions are not serious criminals is the fact that for one (case 5) the last conviction was 10 years ago, for another (case 3) 5 years ago, and for a third (case 7) 5 years ago. Of the remaining two XYY's, both of whom had extensive criminal records, one (case 12) is mentally retarded. In fact, all five of these XYY's have BPP's below the average of 43.7 (Table 23.1) for the XY controls, all but one of them well below that average.

There is a suggestion in our data that several of the crimes of XYY's were committed under circumstances which made detection of the crime and apprehension of the perpetrator particularly likely. Thus, one man sent in a false alarm directly to the police about a presumably serious traffic accident. Another man committed many burglaries of homes while the owners were on the premises.

Turning to the three XXY's with criminal records (Table 23.3), we find that one, in the single crime he committed, assaulted his wife in an extremely brutal way, while under the influence of alcohol. The other two had short periods of juvenile

---

[10] Of the 149 offenses of which the five XYY's were convicted (most of them attributable to two of these men), 145 were against property, many of them thefts of a motor vehicle, a motor-assisted cycle, or a bicycle. The four exceptional offenses include the one instance of aggression already cited, one case of procuring for prostitution, one case of a false alarm to the police, and one case of calling the fire department after setting a small fire.

[11] As an aid to maintaining anonymity of the cases and yet conveying the essence of their criminal records, approximate rather than actual values of penalties (as in this instance) and ages are given in the text and in Table 23.2.

delinquency, and the penalties imposed on them were slight. The last conviction of one was 7 years ago, of the other 10 years ago. All three have BPP scores well below the XY average. Finally, in percentage of cases with one or more convictions for crimes of violence against another person, the difference between the XXY's and the XY controls (6.2 percent versus 1.8 percent) was not statistically significant (one-tailed exact binomial test).[12]

As a first step toward evaluating the intellectual-dysfunction hypothesis, the intellectual level of each of the two proband groups was compared to that of the XY controls, scores on the army selection test (BPP) and the educational index being used for this purpose. For both kinds of measures the mean for the control group (Table 23.1) is significantly higher than the means for either the XYY or the XXY group ($P < .05$ in each instance, t-test). The means for the two proband groups are not significantly different from each other. (It should be noted that the two indicators of intellectual level we used are highly related; for the control group the correlation between BPP scores and educational index is .59 [$P < .00005$, t-test].)

Having established that the proband groups are significantly lower than the XY controls on both measures of intellectual level, we next examine the relation of these measures to criminality. Our data show, first, that BPP scores are significantly related to frequency of occurrence of registered crimes leading to convictions. In the control group, men with no record of such crimes had a mean BPP of 44.5, whereas those with one or more such crimes had a mean BPP of 35.5 ($P < .00001$, t-test). The educational index showed a similar relation to criminality (means of 1.62 for noncriminals and 0.74 for criminals [$P < .00001$, t-test]).

Overall, then, the pattern of results on intellectual functioning provides support for the intellectual-dysfunction hypothesis.

Both proband groups are significantly taller than the XY control groups (Table 23.1). Within the restricted height range of the XY's, however, noncriminals (mean height 187.1 centimeters) were slightly taller ($P = .0013$, t-test) than criminals (mean height, 186.7 centimeters), a finding contrary to the hypothesis that tallness may mediate the relation between the extra Y chromosome and the likelihood of criminal convictions.

In neither proband group is parental SES significantly different from that of the XY control group (Table 23.1). As expected, parental SES was significantly higher ($P < .0001$, t-test) for noncriminals (mean $= 3.71$) than for criminals (mean $= 3.02$).

---

[12] Our case-finding procedures provided an opportunity to obtain some additional evidence on real-life social behavior of XYY's and XXY's. As noted, repeated visits were paid, when necessary, to the homes of men from whom we were seeking buccal smears and blood samples. It is reasonable to speculate that men who required many visits before contact was made with them, under circumstances where the visit was expected and the opportunity given to propose an alternative time, were thereby showing a lack of cooperativeness. The mean number of home visits required to obtain a specimen was not significantly different between the XY controls and the XYY's or XXY's. Being at home or not seems to be related to marital status. A comparison of randomly selected samples of 150 men found at home on the first visit and 150 men not at home showed that in the first group 35 percent were single and 65 percent were married, whereas in the second group 60 percent were single and 40 percent married.

In the individual case study, the XYY's and XXY's were again not significantly different from their controls in frequency of refusals to participate on the first invitation.

CRIMINAL RATES AFTER ADJUSTMENT FOR
BACKGROUND VARIABLES

We next compare the criminal rates of XYY's and XXY's with that of XY's equivalent to them in level of intellectual functioning, height, and parental SES. Only subjects for whom complete data were available are included (all 12 XYY's, all 16 XXY's, and 3738 XY's). The analysis consists of three stages: the first step establishes the probability that an XY with a particular set of values for the background variables is a criminal; the second step establishes for each XYY or XXY the probability that he would be a criminal if he were an XY with his background-variable values; the third step compares the observed frequency of criminals in the proband group with the frequency predicted in the second step.

To permit use of existing programs for the log-linear analysis we employed, it was necessary to make the two continuous variables, BPP and height, categorical. Past work (Cochran, 1968a) with simpler problems suggests that five or six categories often provide a very good representation of a continuous variable. Because height showed a very low relation to criminality in our sample of tall men, we used only five categories for height. The finding that BPP is strongly related to criminality, particularly within the restricted range of low BPP scores, made it advisable to use more categories for BPP. Accordingly, seven categories were employed for this variable.

In the 980-cell contingency table representing all combinations of the $5 \times 7 \times 4 \times 7$ categories for height, BPP, educational index, and parental SES, respectively, let us consider a particular cell. If $n_1 + n_2$ of the 3738 XY's (n_1 being criminals and n_2 noncriminals) fall in that cell, the proportion $n_1/(n_1 + n_2)$ gives a good estimate of the probability that a new XY in that cell will be a criminal, provided the value of $n_1 + n_2$ is large. In many of the 980 cells, however, both n_1 and n_2 are small; in fact in 376 cells both n_1 and n_2 are zero. But it is not necessary to consider each cell independently. Instead, a model may be built for these probabilities which lets us "borrow strength" from similar cells. Critical in building such a model is the definition of similarity. The method we used in constructing a model for our contingency table has been described by Bishop, Fienberg, and Holland (1975). This method is based on log-linear models that are analogous to the usual linear models (that is, regression and analysis of variance). The difference between them is that log-linear models are appropriate for categorical dependent variables (such as criminal/noncriminal), whereas linear models are appropriate for continuous dependent variables. For those familiar with linear models, it is appropriate to think of the analyses performed here as if they were regressions, with criminality taken as the dependent variable and BPP, height, parental SES, and educational index as independent variables, even though in fact the computations required for the log-linear analyses are quite different. We present here only the final results of our analyses.

For the 3738 tall XY's used in the analyses, the proportion of criminals in each of the 980 cells of the contingency table can be very accurately predicted from a model with only six parameters: a grand mean parameter, reflecting the overall level of criminality in the population; one regression parameter reflecting the tendency of criminality to increase as parental SES decreases; two regression parameters (for linear and quadratic components) reflecting the tendency of criminality to increase more and more rapidly as educational index decreases; and two regression parameters (for linear and quadratic components) reflecting the tendency of

Table 23.4. *Individual XYY's. Values of background variables, observed criminality*
(1 = record of one or more convictions, 0 = no record), and probability of criminality
predicted from the XY model.

Case no.	Educational index	Parental SES	BPP	Height (cm)	Criminality Observed	Predicted Probability	Standard error
1	0	2	41	201	0	.14	.013
2	0	1	27	188	1	.21	.018
3	2	6	37	194	1	.05	.008
4	1	3	23	188	0	.22	.026
5	0	3	28	191	1	.18	.012
6	0	3	19	191	0	.27	.022
7	0	4	25	193	1	.16	.013
8	2	4	44	192	0	.05	.005
9	0	4	24	187	0	.25	.024
10	2	2	37	184	0	.07	.009
11	0	5	33	196	0	.15	.016
12	0	1	18*	185	1	.31	.027
					$N = 5$	$N = 2.06$	

* Estimated (see Table 23.2, footnote).

criminality to increase more and more rapidly as BPP decreases. The absence of two kinds of parameters should be noted. There are no parameters relating criminality to height, indicating that in the presence of the components for parental SES, BPP, and educational index, there is no additional effect attributable to height. Also, there are no parameters reflecting interactions among these variables, because their effects on criminality are independent.

The fit of this six-parameter model to the full contingency table is extremely good. Globally, the adequacy of the fit is indicated by the log-likelihood ratio criterion divided by its degrees of freedom. (In fact, this corresponds to an F test in the analysis of variance or multiple regression. Under the null hypothesis that the model reflects the true state of nature the ratio should be about unity, and it should be larger than unity if the model does not fit.) This criterion for the six-parameter model and our four-way table is 491.65/(980-376-6) or 0.82. The significance test of the adequacy of the model follows from the fact that under the six-parameter model the log-likelihood criterion is distributed as χ^2 with 598 degrees of freedom. Hence, the significance level for the test of the adequacy of the six-parameter model, as opposed to the alternative one-parameter-per-cell model, is $P > .99$. The six-parameter model thus cannot be rejected.

Locally, the fit of the six-parameter model is also very good. In individual cells with reasonably large $n_1 + n_2$ values, the estimated probability is close to the observed proportion; and in collections of cells in which the total $n_1 + n_2$ is large, the estimated probability is also close to the observed proportion.

We now apply this model to the XYY and XXY probands. Tables 23.4 and 23.5 show the educational index, parental SES, BPP, and height of each proband and

Table 23.5. *Individual XXY's. Values of background variables, observed criminality (1 = record of one or more convictions, 0 = no record), and probability of criminality predicted from the XY model.*

Case no.	Educational index	Parental SES	BPP	Height (cm)	Observed	Predicted Probability	Standard error
						Criminality	
13	1	4	27	188	0	.13	.014
14	2	5	47	185	0	.04	.005
15	0	4	35	191	0	.16	.013
16	0	3	23	184	0	.27	.022
17	1	3	26	187	1	.14	.014
18	2	4	50	195	0	.05	.005
19	0	5	14	192	0	.37	.058
20	2	7	49	186	0	.04	.006
21	0	3	14	188	0	.41	.054
22	0	5	9	197	0	.37	.057
23	1	1	39	191	0	.12	.015
24	2	7	33	188	0	.07	.013
25	0	4	11	190	1	.39	.055
26	0	1	15	195	0	.31	.027
27	0	4	16	190	1	.25	.024
28	2	7	46	189	0	.04	.006
					$N = 3$	$N = 3.16$	

whether or not he had a criminal record. Also shown is the probability, predicted from the six-parameter model, that he would be a criminal if he were an XY. The last column gives the estimated standard errors of the estimated proportions. Since the sample size for the model (3738) is very large compared to the number of parameters used, these estimated standard errors are probably very accurate. It should be noted that for both XYY's and XXY's, the standard error is less than 16 percent of the estimated proportion except in one case (an XXY), in which it is 19 percent. Hence, the predicted probabilities are very accurate, particularly for the XYY's and may be considered exact for our purposes.

The number of criminals to be expected in each of the proband groups if their crime rate was the same as that of XY's equivalent in parental SES, educational index, BPP, and height is given by the sum of the predicted probabilities (column 7 of each table); the observed number of criminals is given by the sum of the 0 and 1 values (column 6).

The result of adjustments for BPP, parental SES, and educational index indicates that these variables account for some of the raw difference in criminality between the XYY and XY groups. However, an elevation in crime rate among the XYY's remains ($P = .037$; one-sided exact binomial test) even after these adjustments are made, the observed and the predicted number of criminals in the XYY group being 5 and 2.06, respectively. The XXY's are not significantly different in criminality from the XY group ($P = .41$) after the adjustment for background variables is made, as is shown by the agreement between observed and predicted number of criminals (3 versus 3.16).

DISCUSSION AND CONCLUSIONS

A first question to consider is the validity of the ascertainment data of 2.9/1000 XYY's and 3.9/1000 XXY's among the tall men sampled in the study.

For XYY's other prevalence data are available from two recent large-scale studies of tall men, but comparison with ours is made difficult by differences in source populations and in the heights taken as cutoff points. In one of these studies (Zeuthen and Nielsen, 1973) the prevalence rate for the tall normal men sampled appears to be a good deal higher than our own, whereas in the other (Noël et al., 1974) it seems not very different.

Studies of the incidence of XYY in neonates are still few in number and their results variable. Two recent reviews have tentatively suggested an incidence rate of about 1/1000 live-born males (Hook, 1973; Owen, 1972) and another (Borgaonkar and Shah, 1974) a rate in the range of 1/1500 to 1/3000. On the basis of these values the expected number of XYY's in our total population at birth would be about 9 to about 29. For several reasons it is difficult to determine the proportion of these who would be likely to be present in the tall segment we sampled in adulthood. There are uncertainties about the rate of attrition through death among XYY's.[13] There are also uncertainties about how many XYY's there might be among the unexamined tall men in our study (those who refused to participate, who emigrated, or who could not be located)[14] and among the shorter men we did not examine. Though our obtained frequency falls well within the range estimated from neonatal studies, given all the limitations and uncertainties involved in generating those estimates the correspondence must be interpreted with caution. At the same time, there seem to be no grounds in these figures for suspecting a bias in our prevalence findings.

---

[13] There is some suggestion in our data on reasons for draft-board rejection that XYY's may be vulnerable to somatic disorders, so that a correction factor derived from attrition data for the general population may not be appropriate for XYY's. The draft-board rejection data are not considered in detail here because of some ambiguities in the rejection classification used. Even if somatic difficulties are more common among XYY's, however, this in itself is probably not a contributing factor to their elevated frequency of criminal offenses. In our XY's a χ^2 test showed that the difference in frequency of criminal offenses between men rejected by their draft boards for health reasons (9.8 percent) and men not rejected on such grounds (12.4 percent) was not significant.

[14] This question deserves particular attention because some of the subgroups in the unexamined sample have unusually high crime rates, as might be expected. The rates were 17.2 percent for men who declined to participate in the study, 17.4 percent for those who emigrated or were away at sea; 44.0 percent for the destitute men; and 31.7 percent for those who were not living at their address of record. The differences between these values and the 9.3 percent rate of the tall XY men examined are all statistically significant (χ^2, $P < .003$). If XYY's do tend to have higher crime rates, then according to these data XYY prevalence rates may be higher in the unexamined group than in the examined group. The number of unexamined tall men is relatively small, however. Further, if we apply to the unexamined group the finding for the examined group of a 1.3 percent XYY rate among all tall men with criminal convictions, then the probability is that among the 97 men with one or more criminal convictions in the unexamined tall group there is only one XYY. Because the crime rate of the unexamined group may understate the case, and because of uncertainties about some of the extrapolations made, this pattern of results cannot of course be taken as definitive.

An evaluation of our XXY prevalence figure of 3.9/1000 presents the same problems as our XYY data: there are no satisfactory prevalence data for comparison, and incidence data are not yet adequate.

As to the rate of criminality, our finding is consistent with past findings from studies of institutionalized populations, in that the XYY's we identified had a higher mean rate of criminal convictions than the XY controls.

With regard to the possible correlates of the elevated XYY crime rate, the hypothesis we considered that height may be an intervening variable was not confirmed. In fact, within our tall XY group height showed a small but statistically significant negative relation to criminality.

On the other hand, the evidence from this study is consistent with a second hypothesis we considered, the intellectual-dysfunction hypothesis. The XYY's had an appreciably lower mean on the BPP, the army selection intelligence test, than did the XY's, and they also had a substantially lower mean on the related index of educational level attained, although some of the XYY's were within or not far out of the normal range on these variables. Moreover, in our XY sample criminality showed a substantial relation to both measures of level of intellectual functioning.

While intellectual functioning is thus clearly implicated as an important mediating variable, we cannot at this time say whether it is the only factor involved. When the two intelligence indicators were controlled, along with parental SES and height, in order to determine how these variables account for the observed XYY-XY difference in crime rate, an elevation in the XYY crime rate remained, though the difference was reduced. However, the BPP is not a comprehensive test of intelligence. It is possible that there are areas of cognitive dysfunction in XYY's that it does not tap and that a more comprehensive battery of cognitive tests would increase the explanatory power of the intellectual-dysfunction hypothesis.[15]

In evaluating that hypothesis it is important to recall that the crime data we used were derived from records of individuals who were actually apprehended. People of lower intelligence may be less adept at escaping detection and so be likely to have a higher representation in a classificatory system based on registered crimes. The elevated crime rate found in our XYY group may therefore reflect a higher detection rate rather than simply a higher rate of commission of crimes.

It should be stressed that finding a relation between the presence of an extra Y chromosome and impaired intellectual functioning does not mean that the Y

---

[15] In fact, in our individual case study we do have data from a wider assortment of cognitive tests for the probands and their matched XY controls which indicate that this may be true. These data suggest that more of the difference in rate of criminality between the XYY's and the XY's might have been accounted for had they been tested for an additional cognitive factor, figure fluency (FF) [Ekstrom, French, and Harman, 1975]. A test of the FF factor was included in the battery used in the individual case study, but this factor is not well represented in the BPP, as is shown by a low, nonsignificant correlation of .28 between scores from the test of this factor and BPP scores. After adjusting for BPP and other background variables by the analysis of covariance, we find FF means of 32.5, 24.0, and 26.0 for XY's, XYY's, and XXY's, respectively. The F test for difference of means is significant at the .05 level. Especially significant ($P = .02$, two-tailed) is the t-test for the difference between the XYY adjusted mean and the XY adjusted mean, indicating that the average scores of XYY's on FF are lower even with BPP and the other background variables controlled. Furthermore, among the 49 XY controls the 8 men with records of convictions for one or more criminal offenses had a mean FF score of 29.0, while the mean for the 41 noncriminal XY's was 36.0 ($P < .05$, one-tailed t-test).

chromosome is ordinarily implicated in intellectual functioning and that a specific genetic basis for intelligence has thereby been established. That is no more true than would be the conclusion that, because trisomy 21 is associated with the markedly impaired intellectual functioning found in Down's syndrome, autosome 21 must make a direct genetic contribution to ordinary intellectual development. Chromosomes and genes exert their influence on development in concert; altering any one of them may accordingly affect the overall organization of the individual's genetic material, with consequences beyond the specific contribution each component may make individually. The potentially serious consequences of altering the organization of genetic material is reflected in the finding that chromosomal abnormalities are evident in about half of all spontaneous abortions in the first trimester of pregnancy, the period when such abortions are most frequent (Boue, Boue, and Lazar, 1975). In the case of the XYY complement, as in the case of Down's syndrome, it seems more plausible that the intellectual deficit found is one manifestation of altered ontogenetic development, resulting from a change in overall organization of genetic material, than that the particular chromosome involved (the Y chromosome or autosome 21) is directly implicated in ordinary intellectual functioning. A finding that has already emerged from the individual case studies we are conducting of the XYY's and the XY controls seems consistent with the view that the aberrant XYY complement may have broad adverse developmental consequences. In waking electroencephalograms (EEG's) the XYY's showed a significantly lower average frequency of alpha rhythm than matched XY controls (Volavka et al., 1976). Slower EEG frequencies are normally predominant at an earlier age (Lindsley, 1939; Matousek and Petersen, 1973); our finding can therefore be viewed as suggesting a developmental lag.

The third hypothesis we examined, the aggression hypothesis, received little support in an examination of the criminal records of the XYY's. Among all offenses committed by XYY's there was only a single instance of an aggressive act against another person; and in that case the aggression was not severe. Thus the frequency of crimes of violence against another person was not statistically significantly higher in the XYY's than in the XY's. The elevated crime rate in our XYY sample reflects an elevated rate of property offenses. This picture is in keeping with results of previous studies, most of which have also found that XYY's are not more likely to commit crimes against people than are XY's (Borgaonkar and Shah, 1974). The infrequency of violent criminal acts among our XYY's is in line as well with the observation that XYY's show less aggressive behavior while in prison than do XY prisoners (Price and Whatmore, 1967; Street and Watson, 1969). Also consistent is our finding that XYY's were no more likely to decline to participate in this study than XY's. The aggression hypothesis cannot be ruled out by the analyses done thus far, but the evidence from the personality evaluations and the social-developmental histories in the individual case studies now being analyzed will allow a further and more direct assessment of that hypothesis.

We did not examine shorter XYY's in this study, but such men appear to be uncommon in institutions, even when ascertainment has been done with men unselected for height. Further, a recent study by Owen (1972), based on the entire group of approximately 28,000 men who served as the source population for this study, has shown a slight inverse relation between height and criminal offenses. Thus height differences would not explain why short XYY's should appear less frequently in institutions than tall ones. Whether they do not appear because they are uncommon or because they do not commit detectable crimes, with regard to aggression shorter XYY's need be of no greater concern to society than the general run of men.

In addition to the variables of height, intellectual functioning, and aggression thus far examined as possible mediators in the relation between the XYY complement and an elevated crime rate, other variables are being considered in the individual case studies. These include characteristics of endocrine, neurological, and neuropsychological functioning.

The picture of the XXY's that has emerged to this point is in most ways similar to that of the XYY's. The XXY's showed a somewhat elevated crime rate compared to the XY's, but below that of the XYY's. The difference in crime rate between the XXY's and XYY's was not statistically significant. Though the XXY crime rate was slightly higher than that of the XY's, the difference was not statistically significant and the elevation disappeared when background variables were controlled. As to aggression, only one of the XXY's was convicted for an act of aggression, which was severe in nature. The XXY's were not significantly different from the XY's in frequency of crimes of violence against other persons. The XXY evidence thus does not provide any more impressive support for the aggression hypothesis than the XYY evidence does. With regard to intelligence, the XXY's, like the XYY's, had a substantially lower mean BPP and mean intellectual index than XY's did. The similarities between the XYY's and the XXY's suggest that, with regard to the characteristics considered thus far, the consequences of an extra Y chromosome may not be specific to that chromosomal aberration but may result from an extra X chromosome as well.

The data from the documentary records we have examined speak on society's legitimate concern about aggression among XYY and XXY men. No evidence has been found that men with either of these sex chromosome complements are especially aggressive. Because such men do not appear to contribute particularly to society's problem with aggressive crimes, their identification would not serve to ameliorate this problem.

ACKNOWLEDGMENTS

For advice at different points in the conception and conduct of the study, in analyses of the data, and in the preparation of this article we are grateful o L. L Cavalli-Sforza, P. Holland, J. Lederberg, R. Schiavi, S. A. Shah, and A. Theilgaard. We are indebted to the Danish Ministry of Justice for making available the criminal records that were used in this article.

24. Practical Implications of Modes of Statistical Inference for Causal Effects and the Critical Role of the Assignment Mechanism

Donald B. Rubin

Abstract: Causal inference in an important topic and one that is now attracting serious attention of statisticians. Although there exist recent discussions concerning the general definition of causal effects and a substantial literature on specific techniques for the analysis of data in randomized and nonrandomized studies, there has been relatively little discussion of modes of statistical inference for causal effects. This presentation briefly describes and contrasts four basic modes of statistical inference for causal effects, emphasizes the common underlying causal framework with a posited assignment mechanism, and describes practical implications in the context of an example involving the effects of switching from a name-brand to a generic drug. A fundamental conclusion is that in such nonrandomized studies, sensitivity of inference to the assignment mechanism is the dominant issue, and it cannot be avoided by changing modes of inference, for instance, by changing from randomization-based to Bayesian methods.

1. INTRODUCTION

1.1. Causal Inference

Causal inference is a topic that statisticians are addressing more vigorously and rigorously in recent years. This is a desirable development for statistics, as supported by Cox's (1986) comment on Holland (1986b) that "...the issues explicitly and implicitly raised by the article seem to me more important for the foundations of our subject than the discussion of the nature of probability...". Many recent articles now exist on foundational issues regarding how to define causal effects in situations beyond randomized experiments (e.g., see Rubin, 1978a; Holland, 1986b; Pratt and Schlaifer, 1988; and their references), and on specific practical issues regarding the design and analysis of nonrandomized studies for causal effects such as matching and subclassification (e.g., see Cochran and Rubin, 1973; Rubin, 1984c; Rosenbaum and Rubin, 1984b, 1985a). There appear to be few references, however, describing

This article is the written version of the President's Invited Address, Biometric Society Conference, Boston, Massachusetts, March 29, 1988. The invited topic was "...important methodological developments that have potential or demonstrated application to biometrics...causality, matching/ propensity, missing data/EM, and Bayesian ideas in relation to randomization or other issues is certainly fair game"; this presentation was an attempt to cover all these bases, albeit briefly, in the context of a biomedical example.

and contrasting modes of statistical inference for causal effects and relating the distinctions to practical issues.

This presentation (i) briefly describes and contrasts four basic modes of statistical inference for causal effects, (ii) emphasizes the common underlying causal framework and the common need to specify an assignment mechanism, and (iii) describes practical implications in the context of a real-world example involving the effects of switching from a name-brand version of a prescription drug to a generic version. This example is of interest in its own right because simple conceptualizations of the assignment mechanism lead to the conclusion that either (a) the treatment received is sometimes itself missing data or (b) there is a fully missing covariate, or (c) both problems are present.

1.2. Four Modes of Statistical Inference for Causal Effects and the Existence of a Common Framework

The four formal modes of causal inference that are considered conceptually distinct in Rubin (1990a) are:

(1) Randomization-based tests of sharp null hypotheses;
(2) Bayesian predictive inference for causal effects – or, more descriptively, perhaps – full probability modelling for causal effects;
(3) Repeated-sampling randomization-based inference;
(4) Repeated-sampling model-based inference.

The general thrust concerning these modes can be summarized quite briefly. Randomization-based tests of sharp null hypotheses (essentially due to Fisher, 1925, 1935) comprise an elegant and very appealing method of drawing causal inferences, but is too limited to address the full complexity of practical problems. Full probability modelling for causal effects (Bayes' predictive modelling – apparently first presented in generality in Rubin, 1978a) is on target in principle but far more easily abused in practice than randomization-based tests of sharp null hypotheses. Repeated sampling modes of inference, both randomization-based and model-based, are really evaluations of operating characteristics of procedures rather than modes of statistical inference per se; the model-based mode is the most commonly used method in current practice. The four modes of inference will be implicitly defined by being applied to our example. More formal discussion is given in Rubin (1990a).

A more important message than their differences, however, is that all modes share a common conceptual framework in which causal inferences can be drawn, and that a clear formulation of this framework is an essential ingredient of a valid statistical inference in a practical problem. In particular, this common framework requires the specification of a posited *assignment mechanism* describing the process by which treatments were assigned to units; it is required for each mode of inference in the sense that causal answers generally change if the posited assignment mechanism is changed. The major source of uncertainty in the analysis of an observational study is generally not the mode of inference but rather the specification of this assignment mechanism, as will be emphasized in our biomedical example.

2. THE EXAMPLE

2.1. Thioridazine

Our example uses a data set from a study concerning the propriety of generic substitution of presumptively chemically equivalent (bioequivalent) drugs. Current Food and Drug Administration standards for declaring bioequivalence are generally regarded as loose: Using a single-dose study with 18–24 normal subjects, drugs can be declared bioequivalent if their mean rate and extent of absorption are within ±20%. Perhaps as a result of these relatively loose standards, anecdotal stories of real differences between supposedly bioequivalent drugs are not uncommon, especially with respect to certain prescribed psychoactive drugs.

One such psychoactive drug is thioridazine, whose generic form became available March 18, 1983, and our example concerns the possible effects of switching from name-brand to generic thioridazine. The data base available to estimate the effects of switching is proprietary and not fully available to the data analysts, and thus the description of the data base provided here is somewhat simplified and not absolutely precise. Nevertheless, the data analyses presented aptly emphasize both the critical role of the assignment mechanism when drawing causal inferences and the potential difficulties in positing an acceptable assignment mechanism in an observational study. This sort of discussion should be an important component of the analysis of any observational study for causal effects but is often entirely absent.

2.2. Basic Data

The data used for this study were gleaned from a large multipurpose data base consisting of detailed information on Medicare patients during the period from January 1, 1982 to April 15, 1985. Many types of information were recorded. Demographic variables included age, gender, race, urban/rural, state, and so on. Drug-use information was very detailed, including dates, types, and dosage levels for all types of prescription drugs. Furthermore, detailed records of medical encounters were kept, classified by type such as mental health versus other, and inpatient versus outpatient.

In one state, which served as the focus of this study, there were approximately $N = 10,000$ name-brand thioridazine users *pre-intervention date* – that is, during the period from January 1, 1982 to March 18, 1983. Of the 10,000 pre-intervention-date thioridazine users, about 7,000 were nonswitchers in the sense that they never used generic thioridazine according to the information in the data base, and about 3,000 were switchers in the sense that they substituted generic for name-brand thioridazine sometime during the post-intervention-date period, March 18, 1983 to April 15, 1985. Almost all 3,000 switchers switched for the first time within 12 months after intervention date, thus leaving at least one year to observe post-switch outcomes. Reasons for switching are only vaguely understood, but are generally thought to be related to whether the prescribing doctor's habit was to use brand or generic names and how the pharmacist handled generic prescriptions; the cost to the patient was not an issue due to their Medicare status.

2.3. The Initial Matching

The initial task, as defined by the biomedical researchers, was to create matched pairs of switchers–nonswitchers from which inferences about the effect of switching could be drawn free of the effects of the variables used for matching. The reasons for creating a matched sample were that (i) it was exceedingly expensive to assemble the data base, prohibitively so for all 10,000 thioridazine users, and (ii) it was thought that by discarding the 4,000 nonswitchers least like the 3,000 switchers, biases could be reduced with little loss of precision. Of course the target population would be thereby reduced, effectively to the 3,000 switchers and their matches. For the purposes of creating matched pairs, let X = all pre-intervention-date information. Table 24.1 lists some of the potential matching variables. The method used for matching is a variant of the propensity/Mahalanobis metric matching method described in Rosenbaum and Rubin (1985a), and can be a useful procedure in many studies. Parenthetically, if all 10,000 subjects could have been retained for data analysis, it might have been preferable to begin with a variant of propensity-score subclassification described in Rosenbaum and Rubin (1984b).

First, it was decided to exact-match on "heavy-dose," defined as "six or more pre-intervention date thioridazine prescriptions," because any effects of switching might be quite difficult to estimate for occasional users of thioridazine; there were 2,000 heavy-dose switchers and 4,500 heavy-dose nonswitchers. Second, a propensity score was estimated as a function of X, including some nonlinear functions of some variables (e.g., logs, squares, some interactions), by performing a logistic regression predicting switcher/nonswitcher status from X. The propensity score, say $e(X)$, is the probability of being a switcher rather than a nonswitcher given X. For each switcher, an acceptable reservoir of potential nonswitcher matches was defined by requiring (i) an exact match on "heavy dose" and (ii) being within a caliper width of $\pm\frac{1}{4}$ standard deviation on the propensity score (on the logit scale, i.e., logit[$\hat{e}(X)$]). If this reservoir had ever been empty, the closest propensity-score nonswitcher would have been declared the only acceptable match. When the reservoir had more than one nonswitcher, the best match was chosen as the nearest available with respect to the Mahalanobis metric based on key functions of X: the propensity score, log(# days), log(# prescriptions), log(dose per day), log(total medical encounters), etc.

The use of the propensity-score calipers is an observational-study analogue of randomization in a controlled experiment: Within these calipers, switchers and nonswitchers have approximately the same predicted probability (given X) of being a switcher. Randomization in a paired comparison experiment does better, of course, in that the truth of this distributional claim does not depend on the conditioning on X but also applies given any set of observed or unobserved covariates. The use of Mahalanobis metric matching within calipers is analogous to blocking in a randomized experiment: Given units with equal probability of exposure to treatment and control, blocking on important background factors increases precision.

2.4. Results of the Matching

The effectiveness of such a scheme can appear to be rather dramatic, even in cases such as this one, where the ratio of treated units (switchers) to nontreated units

Table 24.1. *Pre-intervention variables (X) for thioridazine study – first 14 variables explicitly used for matching*

Description of variable	Transform	Standardized bias Pre-matching	Post-matching
Sex	0/1	.020	.015
Age as of 1985	$\log(X)$	.030	.059
Rural/Urban	0/1	−.104	−.009
Total days of potential exposure	$\log(X)$	−.179	−.032
Number of thioridazine prescriptions	$\log(X)$	−.250	.000
Total days at risk	$\log(X)$	−.256	.005
Total dose/Total days	$\log(X)$	.000	−.008
Total mental health psychotic conditions	$\log(X+1)$	.053	−.004
Non-thioridazine neuroleptics	0/1	.080	−.025
Total medical encounters	$\log(X+1)$	−.047	−.026
Total mental health encounters during days at risk	$\log(X+1)$	.081	−.007
Inpatient mental health encounters during days at risk	$\log(X+1)$	.047	−.031
More than 5 prescriptions of thioridazine	0/1	−.179	−.001
Propensity score	$\text{logit}^{-1}(\hat{e})$	.384	−.007
Race	3 categories	.120	.087
Presence of non-thioridazine neuroleptics	0/1	.095	.008
Total dose	$\log(X)$	−.162	−.002
Total mental health encounters	$\log(X+1)$	−.012	−.016
Inpatient mental health encounters	$\log(X+1)$	.030	−.023
Outpatient mental health encounters	$\log(X+1)$	−.013	−.015
Inpatient psychotic mental health encounters	$\log(X+1)$	.040	−.003
Outpatient psychotic mental health encounters	$\log(X+1)$	.054	−.003
Total mental health non-psychotic conditions	$\log(X+1)$	−.042	−.011
Inpatient mental health encounters	$\log(X+1)$	.009	−.033
Outpatient mental health encounters	$\log(X+1)$	−.042	−.010
Nursing home transaction	0/1	.000	.000
Cerebral stimulants	0/1	.063	.022
Antidepressants	0/1	.028	.020
Anxiolytics	0/1	.043	.011
Sedatives/Hypnotics	0/1	.014	−.014
Anti-Parkinson agents	0/1	.056	−.002
Other CNS drugs	0/1	.108	.077
Organic psychosis	0/1	.038	.024
Schizophrenia	0/1	.027	.021
Affective disorders	0/1	.014	−.040
Paranoid states	0/1	−.035	−.057
Neurotic disorders	0/1	.027	.022
Personality disorders	0/1	−.025	.026
Adjustment reactions	0/1	.000	−.056

Table 24.1. *(Continued)*

Description of variable	Transform	Standardized bias	
		Pre-matching	Post-matching
Other special psychological symptoms	0/1	−.003	−.011
Other medical conditions	0/1	−.011	−.017
Outpatient mental health encounters per days at risk	$\log(X + 1)$	.078	−.004
Total mental health psychotic conditions per days at risk	$\log(X + 1)$	.025	−.010
Inpatient psychotic mental health encounters per days at risk	$\log(X + 1)$	.021	−.042
Outpatient psychotic mental health encounters per days at risk	$\log(X + 1)$	.075	−.004
Total non-psychotic mental health conditions per days at risk	$\log(X + 1)$	.044	−.003
Inpatient non-psychotic mental health encounters per days at risk	$\log(X + 1)$	.044	−.008
Outpatient non-psychotic mental health encounters per days at risk	$\log(X + 1)$	.039	−.002
Inpatient medical encounters	$\log(X + 1)$	.043	−.006
Outpatient medical encounters	$\log(X + 1)$	−.050	−.027
Inpatient medical encounters per days at risk	$\log(X + 1)$	−.021	.012
Outpatient medical encounters per days at risk	$\log(X + 1)$	−.175	−.028
Total medical encounters per days at risk	$\log(X + 1)$	−.172	−.026

(nonswitchers) is modest (e.g., 7 to 3). Table 24.2 summarizes the standardized biases for 53 covariates (including some transformations and interaction variables) before and after matching, where standardized bias here is the number of standard deviations between the means of the covariate in the switcher and nonswitcher groups: $\bar{x}_d / \sqrt{(s_1^2 + s_2^2)/2}$, where $\bar{x}_d$ is the mean difference (before matching or after matching, depending on context) and s_1^2 and s_2^2 are the variances within group before matching. Notice that before matching, two covariates had standardized biases greater than .25, and 10 of the 50 had biases greater than one-tenth of a standard deviation, whereas after matching no covariates had biases this big. Thus the matching, which selected 3,000 nonswitchers from the pool of 7,000, was viewed as being very effective, especially considering the relatively modest ratio for the reservoir. Of course, the matching variables are often highly correlated, some being monotone functions of each other, but the resulting balance is still impressive. Nevertheless, despite this success, it was thought cost-effective at this time to focus attention on the heavy-dose thioridazine users – that is, the 2,000 matched pairs with at least six prescriptions during the pre-intervention period. All 6,500 (=2,000 + 4,500) subjects in the data base with at least six pre-intervention thioridazine prescriptions used some form of thioridazine during the post-intervention period.

Table 24.2. *Summary of standardized bias for 53*
covariates of Table 24.1

Size of bias	# Pre-matching	# Post-matching
>.25	2	0
.25> >.20	0	0
.20> >.15	5	0
.15> >.10	3	0
.10> >.05	13	4
.05>	30	49

3. A CONCEPTUAL FRAMEWORK FOR DRAWING FIRST-PASS CAUSAL INFERENCES

3.1. Structure of Observables

Before proceeding with any initial data analysis on the observed outcome data from the resultant matched samples, a conceptual framework is described in which causal effects can be defined and causal inferences can be drawn from the observed values. A crucial feature of this formulation is the description of a process by which some units were exposed to the "switch" treatment and the others were exposed to the "nonswitch" treatment. In order to have precise definitions of the assignment mechanism and the causal effects to be estimated, notation is needed for the observables (the values that might be observed) in the study.

Table 24.3 displays the observables in the study. Each row corresponds to one of the $N = 10,000$ units in the study. As previously, X represents the pre-intervention variables. The variable W is an indicator for treatment, s (switch) and ns (not switch). Outcomes, which are basically the nondemographic X variables observed during the post-intervention period, are labelled Y, where Y_s are the outcomes if everyone switched and Y_{ns} are the outcomes if no one switched. The question marks (?) in the Y column, indicating missing data, arise because for each unit, only Y_s or Y_{ns} is actually observed, as indicated by W, the other Y being missing. Causal effects, the estimands, are comparisons of Y_s and Y_{ns}; for example, the average causal effect is $\overline{Y}_s - \overline{Y}_{ns}$, where these involve the missing values because the averages are over all units. Because of our restriction to heavy-dose subjects, we focus attention on the 2,000 heavy-dose switchers and the 4,500 heavy-dose nonswitchers. For most of the remaining discussion, we narrow the focus to the 2,000 heavy-dose nonswitchers most similar to the 2,000 heavy-dose switchers as determined by the matched sampling. Notice that the selection of these 4,000 matched units from the 10,000 is solely on the basis of X and W.

3.2. SUTVA

The stable-unit-treatment-value assumption (SUTVA) asserts that the representation in Table 24.3 is complete. That is, the values of $Y = (Y_s, Y_{ns})$ for the units are not affected by the values of W – the set of components of (Y_s, Y_{ns}) we observe is determined by the value of W, but the values of (Y_s, Y_{ns}) are the same no matter what

Table 24.3. *Observables in the study:* $\sqrt{}$ = *data observed;*
? = data missing; s = switch; ns = not switch

			Outcomes Y	
		Switching	Under	Under not
	Covariates	indicator	switch	switch
Units	X	W	Y_s	Y_{ns}
1	$\sqrt{}$	s	$\sqrt{}$	?
2	$\sqrt{}$	ns	?	$\sqrt{}$
3	$\sqrt{}$	s	$\sqrt{}$	?
4	$\sqrt{}$	s	$\sqrt{}$	?
5	$\sqrt{}$	ns	?	$\sqrt{}$
$\vdots$	$\vdots$	$\vdots$	$\vdots$	$\vdots$
10,000	$\sqrt{}$	ns	?	$\sqrt{}$

the value of W is. Thus, causal estimands such as the average causal effect, $\overline{Y}_s - \overline{Y}_{ns}$, take the same value no matter which units were assigned to which treatments. In particular, SUTVA implies that there can be no interference between units and no versions of treatments; more discussion of these points is given in Cox (1958, Chap. 2) and Rubin (1978a, 1980a, 1986, 1990a, 1990b). A commonly assumed special case of SUTVA is additivity, under which the causal effect is to add a constant value for all units: $Y_{s,i} - Y_{ns,i} = $ constant.

SUTVA is not *needed* for defining causal effects and drawing causal inferences, and can be replaced by other more general or complicated assumptions. But some such assumption is needed, and whatever it is, critical aspects of it will be untestable from the data at hand, even in the best controlled randomized experiment – it will be an assumption about the smoothness of our world, the sort that allows us to use today's data to say something about tomorrow's events.

Because switch dates vary across the switchers, in order to satisfy SUTVA we assume that associated with each unit is a fixed switch date, so that if assigned to switch, switching will take place on that date. The fact that switch dates have to be observed indirectly and are unknown for nonswitchers will haunt a later analysis.

3.3. The Posited Assignment Mechanism

The structure in Table 24.3 with its SUTVA (or a similar assumption) is needed before any causal effects can even be defined. For causal inferences to be drawn, no matter which mode of inference is to be used, we also need to posit an assignment mechanism, a probability model for the process by which we got to see the Y values we did see; that is, we need to specify the process that leads to the observed values of W. Each mode of inference uses the posited assignment mechanism to draw causal inferences, and for fixed data and fixed other assumptions, the causal inferences generally change as the posited assignment mechanism changes.

Suppose each of the 6,500 heavy-dose subjects received the switching treatment with probability $1 > e(X) > 0$ and the nonswitching treatment with probability $1 - e(X)$, where $e(X)$ is an unknown function of X. Although $e(X)$ is unknown, it can be estimated by, for example, a logistic regression; logit$[\hat{g}(X)]$ is in fact the propensity score used for the matching described in Section 2.3. This perspective leads to the assignment mechanism

$$\Pr(\mathbf{W} \mid \mathbf{X}, \mathbf{Y}_s, \mathbf{Y}_{ns}) = \prod_{i=1}^{6,500} e(X_i)^{W_i}[1 - e(X_i)]^{1-W_i}, \tag{24.1}$$

where the boldface for the random variables is used to emphasize that the symbols represent the entire collection of values (e.g., $\mathbf{W}$ is a column vector with 6,500 components, $W_i = 1$ indicating switching and $W_i = 0$ indicating not switching). Within the context of this assignment mechanism, the matching simply discarded 2,500 heavy-dose nonswitchers to create 2,000 switcher–nonswitcher matched pairs with similar values of X. For notational simplicity, we assume the pairings are unambiguously known from X.

A crucial feature of this assignment mechanism that makes inference relatively straightforward is that it is an *unconfounded* assignment mechanism in the sense that it is free of dependence on $\mathbf{Y} = (\mathbf{Y}_s, \mathbf{Y}_{ns})$. Dependence of the assignment mechanism on $\mathbf{X}$ can be addressed in analysis, usually in straightforward ways, whereas dependence on $\mathbf{Y}$, especially on unobserved components of $\mathbf{Y}$, is generally difficult to handle in a truly satisfactory way.

Another important feature of this specification for the assignment mechanism is that it is a *probability* assignment mechanism, in the sense that each unit has a positive probability of being assigned s or ns:

$$1 > \Pr(W_i = s \mid \mathbf{X}, \mathbf{Y}_s, \mathbf{Y}_{ns}) > 0 \quad \text{for each unit } i = 1, \dots, 6,500.$$

If some unit has no chance of being assigned some treatment, it is often wise to consider why, and whether such a unit then really belongs in the conceptualized study.

3.4. A Simplified Assignment Mechanism for the 4,000 Units

The focus on the 4,000 units in the 2,000 matched pairs can lead to a simple random-ized paired-comparison treatment assignment mechanism instead of (24.1), which involves unknown probabilities. Specifically, suppose the matching on X is very good in the sense that the matched-pair differences on X are small enough to make $e(X)$ nearly the same for both members of each matched pair; in fact, this is plau-sible, since the estimated probabilities, $\hat{e}(X)$, within a matched pair are very close. Then, both members of a matched pair are essentially equally likely to be switchers. Furthermore, condition on the outcome that each matched pair has one switcher and one nonswitcher, which is presumably ancillary in the sense that it conveys no infor-mation about causal effects. Then for the 4,000 units in the 2,000 matched pairs, the assignment mechanism (24.1) reduces to the randomized paired-comparison mechanism:

$$\Pr(\mathbf{W} \mid \mathbf{X}, \mathbf{Y}_s, \mathbf{Y}_{ns}) = \begin{cases} (.5)^{2,000} & \text{if paired units get different treatments,} \\ 0 & \text{otherwise.} \end{cases}$$

$$\tag{24.2}$$

Thus each of the $2^{2,000}$ values of **W** that assigns s to one member of a pair and ns to the other member of the pair is equally likely, and any **W** that assigns both members of a pair to the same treatment has zero probability.

When the assignment mechanism is considered to be given by (24.1), the ways the modes of inference use the observed data to draw inferences generally differ more substantially than under the paired-comparison assignment mechanism (24.2), and moreover, can be substantially more complicated for the non-Bayesian modes than under (24.2). Because the matching on X is very good, we provisionally accept the paired-comparison conceptualization for the 4,000 paired units.

3.5. The Objective of Statistical Inference

The objective when analyzing the data is to learn about the question marks in Table 24.1 – that is, the switchers' Y values if they had not switched and the non-switchers' Y values if they had switched – and thereby to learn about the causal effects, such as $\overline{Y}_s - \overline{Y}_{ns}$ for the 4,000 units in the matched pairs. Another estimand is $\overline{Y}_s - \overline{Y}_{ns}$ for the 2,000 switchers – that is, the average effect of switching for the 2,000 heavy-dose subjects who chose to switch. Extensions involve learning about (Y_s, Y_{ns}) for all 10,000 units or even for units not in the study for whom neither Y_s nor Y_{ns} is observed, such as future units who might be exposed to the treatments. Under common models that assume additivity, the estimands are the same. The various modes of statistical inference introduced in Section 1 differ in how they use the observed data to learn about the unobserved values that define the causal estimands.

In this example with the structure of Table 24.3 and the randomized paired-comparison assignment mechanism of (24.2), two particular statistics – that is, two functions of the observed values – play important roles in common analyses where the average causal effect, $\overline{Y}_s - \overline{Y}_{ns}$, is the estimand. These statistics are the mean, $\bar{y}_d$, and variance, s_d^2, of the matched-pair differences: Y_s for the switcher in the pair minus Y_{ns} for the nonswitcher. Intuitively, these are important statistics because if the matching were perfect in the miraculous sense that (X, Y_s, Y_{ns}) were the same for both members of a matched pair, the matched-pair $s - ns$ difference in observed Y values would in fact equal the causal effect $(Y_s - Y_{ns})$ for both members of the matched pair, and then the statistics $\bar{y}_d$ and s_d^2 would equal the mean causal effect, $\overline{Y}_s - \overline{Y}_{ns}$, and variance of the causal effects, respectively, across the 4,000 units.

4. APPLICATION OF MODES OF INFERENCE TO THE PAIRED-COMPARISON FORMULATION

4.1. Randomization-Based Tests of Sharp Null Hypothesis

Suppose that there is absolutely no effect of switching so that $Y_s \equiv Y_{ns}$ for each unit, that is, $\mathbf{Y}_s \equiv \mathbf{Y}_{ns}$ in Table 24.3, which is a special case of additivity. This is called a sharp null hypothesis because a known value is thereby specified for each question mark in Table 24.3. That is, the sharp null hypothesis coupled with the observed data implies a specific value for each question mark, and thus under the sharp null hypothesis, all values and all causal effects are known. Furthermore, the values of statistics such as $\bar{y}_d$ and s_d^2 are known for each possible assignment W; in particular, under the randomized paired-comparison assignment mechanism, the values of $\bar{y}_d$ and s_d^2 are known for the $2^{2,000}$ a priori equally likely allocations of treatments to units, not only for the one allocation that actually took place. The general requirement that sharp null hypotheses must be specified, which allow the values of the statistics to be known for all possible randomizations, is unique to this mode of inference.

A randomization-based test of a sharp null hypothesis is based on the idea that if the observed value of a statistic is unusual in the collection of possible values, then either (i) an unusual (rare) event took place; or (ii) $\mathbf{Y}_s \not\equiv \mathbf{Y}_{ns}$ (the sharp null hypothesis is incorrect); or (iii) the posited assignment mechanism is wrong.

Consider the randomization distribution of the statistic $\bar{y}_d$ – that is, its $2^{2,000}$ a priori equally likely values under the null hypothesis. By construction, $\bar{y}_d = 0$ (the value of $\bar{y}_d$ under the null hypothesis), and by common convention, values of $\bar{y}_d$ more extreme than the observed $\bar{y}_d$ are considered more unusual; the proportion of values as usual as the observed $\bar{y}_d$ or more unusual is the P-value (one-sided) or significance level of the null hypothesis. Computational problems can be overcome by sampling the randomization distribution or approximating it analytically using normal-theory results. In particular, it has been well known since Fisher (1935) that the P-value can often be well approximated by treating $\bar{y}_d/[s_d/\sqrt{2,000}]$ as a standard normal deviate – that is, by treating $\bar{y}_d$ as an estimate with standard error $s_d/\sqrt{2,000}$.

Such P-values can be calculated for: any sharp null hypothesis (e.g., each Y_s value equals $1.2 \times$ the corresponding Y_{ns} value), any test statistic (e.g., the median difference in matched-pair differences), and any a priori definition of unusualness (e.g., squared value greater than observed squared value). Consequently, there is a great deal of arbitrariness (flexibility in the hands of a knowledgeable and wise analyst) except by convention. The situation becomes especially complicated in the presence of nuisance parameters such as arise in factorial experiments, split-plot designs, and multiple comparison problems, because sharp hypotheses generally need to be formulated for the nuisance parameters in order to yield specific values for the unobserved Y values [e.g., compare Snedecor and Cochran (1974, p. 373) with Kempthorne (1952, §19.1) or Cochran and Cox (1957, §7.15) on the proper error terms for split-plot designs].

Nevertheless, Fisher's idea for obtaining P-values (and thereby, confidence intervals for additive effects by inverting tests) is elegant. It provides real direct inference for causal effects based on the assignment mechanism, and for fully specified sharp null hypotheses provides P-values for any assignment mechanism, even confounded nonprobability ones (although such P-values can be inferentially useless as when they equal 1 for all null hypotheses).

4.2. Bayesian Predictive Inference

Bayesian predictive inference differs from tests of sharp null hypotheses in that (i) instead of postulating a fixed known value for $\mathbf{Y} = (\mathbf{Y}_s, \mathbf{Y}_{ns})$, it posits a "prior" probability distribution for $\mathbf{Y}$ given $\mathbf{X}$, thereby enriching the model specification, and (ii) instead of seeing whether the observed data are consistent with the postulated $\mathbf{Y}$, it derives the distribution of the causal estimands given the observed values under a joint model specification. An extended view of Bayesian inference (Rubin, 1981, 1984a) allows for model monitoring by comparing observed values of statistics with their distribution under hypothetical replications such as those of randomization theory. This extended view can thereby encompass Fisher's test, but the main object of Bayesian inference is the posterior distribution of causal estimands.

The phrase "Bayesian predictive inference" is used to emphasize the fact that the posterior distribution of causal effects follows from the posterior predictive distribution of the missing (or unobserved) values – that is, the distribution of the question marks in Table 24.3 given all observed values. This distribution can be written as

$$\Pr(\mathbf{Y}_{\mathrm{mis}} \mid \mathbf{X}, \mathbf{W}, \mathbf{Y}_{\mathrm{obs}}) = \frac{\Pr(\mathbf{W} \mid \mathbf{X}, \mathbf{Y}) \Pr(\mathbf{Y} \mid \mathbf{X})}{\int \Pr(\mathbf{W} \mid \mathbf{X}, \mathbf{Y}) \Pr(\mathbf{Y} \mid \mathbf{X}) \, d\mathbf{Y}_{\mathrm{mis}}},$$

where $\mathbf{Y} = (\mathbf{Y}_{\mathrm{mis}}, \mathbf{Y}_{\mathrm{obs}})$ partitions $\mathbf{Y}$ into missing and observed values.

The observed values of $(\mathbf{X}, \mathbf{W}, \mathbf{Y}_{\mathrm{obs}})$ coupled with this posterior distribution for $\mathbf{Y}_{\mathrm{mis}}$ imply a distribution for any function of $(\mathbf{X}, \mathbf{W}, \mathbf{Y})$ such as the individual causal effects $(Y_{s,i} - Y_{ns,i})$, the average causal effect $\overline{Y}_s - \overline{Y}_{ns}$, the average causal effects for the set of subjects who are male, etc. It is important to realize that the Bayesian framework has allowed the calculation of a distribution for a causal estimand, such as $\overline{Y}_s - \overline{Y}_{ns}$,

$$\Pr(\overline{Y}_s - \overline{Y}_{ns} \mid \mathbf{X}, \mathbf{W}, \mathbf{Y}_{\mathrm{obs}}),$$

and thus can be thought of as a procedure for generating typical values for causal estimands, as summarized in a data set, for example, by a stem-and-leaf display.

THE ROLE OF THE ASSIGNMENT MECHANISM IN BAYESIAN INFERENCE. The assignment mechanism plays a major role in Bayesian inference, as it did in randomization inference. In particular, if the assignment mechanism is unconfounded, then

$$\Pr(\mathbf{Y}_{\mathrm{mis}} \mid \mathbf{X}, \mathbf{W}, \mathbf{Y}_{\mathrm{obs}}) = \Pr(\mathbf{Y} \mid \mathbf{X}) \bigg/ \int \Pr(\mathbf{Y} \mid \mathbf{X}) \, d\mathbf{Y}_{\mathrm{mis}}, \tag{24.3}$$

so that for fixed observed values, the same posterior distribution of causal effects is obtained for all unconfounded assignment mechanisms. An assignment mechanism for which (24.3) holds is termed ignorable (Rubin, 1978a) because it does not explicitly appear in the posterior distribution for causal effects; all unconfounded assignment mechanisms are ignorable and some confounded ones are ignorable (e.g., with sequential designs). Nonignorable assignment mechanisms can be handled within Bayesian inference but typically require far more complex modelling tools with attendant heightened sensitivity to details of model specifications. Consequently, the form of the assignment mechanism plays a crucial role in Bayesian inference for causal inference, albeit a different role than in randomization-based inference. In particular, Bayesian inferences with a fixed choice for $\Pr(\mathbf{Y} \mid \mathbf{X})$ are identical for all ignorable assignment mechanisms, whereas randomization-based tests are, in principle, far more complex for some ignorable assignment mechanisms than others [see Rosenbaum (1984a, 1988) for examples of the randomization approach].

MODELS FOR THE DATA, Y GIVEN X. The need in randomization inference to deal directly with $\Pr(\mathbf{W} \mid \mathbf{X})$ and to specify null hypotheses, statistics, and directions for unusualness is replaced in Bayesian inference with the need to specify $\Pr(\mathbf{Y} \mid \mathbf{X})$. Usually this is done by appealing to deFinetti's theorem and writing

$$\Pr(\mathbf{Y} \mid \mathbf{X}) = \int \prod_{i=1}^{n} f(Y_i \mid X_i, \theta) P(\theta) d\theta,$$

where Y_i is the ith row of $\mathbf{Y}$, and $f(Y_i \mid X_i, \theta)$ is the common model for the conditional distribution of Y given X for each row of $(\mathbf{Y}, \mathbf{X})$ given the unobservable parameter θ with prior distribution $P(\theta)$. If $f(\cdot \mid \cdot)$ is assumed to be a normal linear regression model for Y_i given X_i (where θ includes the multivariate regression coefficients for $Y = (Y_s, Y_{ns})$ on X and the residual covariance matrix of Y given X), very neat normal-theory answers are obtained. Even when answers are not analytically tractable, it is often simple to simulate posterior distributions of causal effects, and thus causal estimands. Recent work on simulating posterior distributions includes that by Tanner and Wong (1987), Rubin (1981, 1987), and Gelfand and Smith (1990), and some aspects are reviewed in Rubin (1991).

In our paired-comparison problem with relatively close matching on X, under a variety of simple models with approximately additive treatment effects, the posterior distribution of the average causal effect can be approximated as normal to yield the familiar result

$$(\overline{Y}_s - \overline{Y}_{ns} \mid \mathbf{X}, \mathbf{W}, \mathbf{Y}_{\text{obs}}) \sim N\left(\bar{y}_d, s_d^2/2{,}000\right). \tag{24.4}$$

The result in (24.4) follows from straightforward extensions of results in Rubin (1987, Chap. 2). Despite this simple answer, which states that the estimated average causal effect is $\bar{y}_d$ with standard error $s_d/\sqrt{2{,}000}$, in basic agreement with the randomization result under additivity, the Bayesian predictive framework has access to a richness of models that can be used to create estimators and standard errors in all sorts of complicated situations in practice.

4.3. Randomization-Based Repeated-Sampling Inference

The randomization-based repeated-sampling mode of inference is due to Neyman in 1923, recently translated as Neyman (1990) with discussion (Rubin, 1990b). This mode of inference treats $\mathbf{X}$ and $\mathbf{Y}$ as fixed but $\mathbf{Y}$ as unknown, and obtains the operating characteristics of procedures over repeated draws from the assignment mechanism, $\Pr(\mathbf{W} \mid \mathbf{X}, \mathbf{Y})$. For example, when $\Pr(\mathbf{W} \mid \mathbf{X}, \mathbf{Y})$ is given by (24.2), standard conclusions are that (i) over the $2^{2,000}$ randomizations, the statistic $\bar{y}_d$ is unbiased for $\overline{Y}_s - \overline{Y}_{ns}$, that is,

$$E(\bar{y}_d \mid X, Y) = \overline{Y}_s - \overline{Y}_{ns},$$

(ii) the statistic $s_d^2/2{,}000$ is unbiased for $\text{var}(\bar{y}_d \mid X, Y)$ under additivity, that is,

$$E\left(s_d^2/2{,}000 \mid \mathbf{X}, \mathbf{Y}; Y_{s,i} - Y_{ns,i} = \Delta \; \forall i\right) = \text{var}(\bar{y}_d \mid \mathbf{X}, \mathbf{Y}; Y_{s,i} - Y_{ns,i} = \Delta \; \forall i)$$

(and conservative otherwise); and (iii) the higher moments of the distribution of $\bar{y}_d$ converge to those of a normal distribution as sample size increases (e.g., see Pitman, 1937; Welch, 1937). In contrast to the situation with randomization-based tests of sharp null hypotheses, no null hypotheses need to be formulated with randomization-based repeated-sampling evaluations; however, the method may not work, because in general, the expectation of any statistic will depend on $(\mathbf{X}, \mathbf{Y})$ (e.g., for confounded assignment mechanisms, there is no guarantee that there exists an unbiased estimator). Armed with these simple distributional results about $\bar{y}_d$ and s_d^2 in the paired-comparison situation, an appeal is made to the central limit theorem effect for survey inferences on means to conclude that the Bayes normal answer $([(\bar{Y}_s - \bar{Y}_{ns}) - \bar{y}_d]/[s_d/\sqrt{2{,}000}] \sim N(0, 1))$ quite generally gives interval estimates with essentially correct (or conservative) confidence coverage. Analogous conclusions hold in standard experimental design settings for standard procedures.

This mode of inference can take any procedure, however, and evaluate its operating characteristics as a function of observed X values, $\mathbf{X}$, observed Y values, $\mathbf{Y}_{\text{obs}}$, and missing Y values $\mathbf{Y}_{\text{mis}}$. For example, interesting comments concerning estimates on transformed outcomes and their properties from the randomization-based repeated sampling perspective are offered in Gail, Wieand, and Piantadosi (1984). Although there do not exist strong guiding principles for deriving procedures, this perspective provides a collection of useful ways of evaluating the quality of inferential procedures (e.g., mean squared error, power, etc.) under repeated draws of $\mathbf{W}$ from $\Pr(\mathbf{W} \mid \mathbf{X}, \mathbf{Y})$ with $\mathbf{X}, \mathbf{Y}$ fixed. Consequently, this mode is more a way to evaluate causal inference procedures, than as a direct mode of causal inference per se.

4.4. Model-Based Repeated-Sampling Inference

This mode of inference is similar in spirit to the randomization-based repeated-sampling mode in that operating characteristics of procedures are calculated, but now not only over repeated $\mathbf{W}$ from the assignment mechanism $\Pr(\mathbf{W} \mid \mathbf{X}, \mathbf{Y})$ for fixed $(\mathbf{X}, \mathbf{Y})$, but also over repeated data $\mathbf{Y}$ from the sampling distribution $\Pr(\mathbf{Y} \mid \mathbf{X}, \theta)$,

$$\Pr(\mathbf{Y} \mid \mathbf{X}, \theta) = \prod_{i=1}^{N} f(Y_i \mid X_i, \theta),$$

with θ fixed at some true but unknown value. At times, $\mathbf{W}$ or even various functions of $(\mathbf{W}, \mathbf{Y})$ are regarded as fixed according to "conditioning-on-ancillary" arguments. This mode of inference accepts the Bayesian's exchangeable specification for $f(Y_i \mid X_i, \theta)$, but stops short of using the prior distribution on θ, $P(\theta)$, that makes life inferentially direct by allowing posterior distributions of estimands to be calculated.

Nevertheless, this approach leads to many well-known mathematical statistical results about θ, especially when $f(\cdot \mid \cdot)$ is normal (e.g., $\bar{y}_d$ is the best estimator, confidence intervals using $\pm s_d/\sqrt{2{,}000}$ are the shortest, etc.); one view of these results is simply to say that they establish the large-sample repeated-sampling optimality of Bayes procedures when $f(\cdot \mid \cdot)$ is correctly chosen.

This mode of inference is the most commonly used one in practice and in mathematical statistical work. Although it is highly useful for evaluating Bayesian procedures when it is assumed that the Bayesian specification for $\Pr(Y|X)$ is incorrect, it is unappealing conceptually as a mode of inference. Even William Cochran may not have been enthusiastic about it as a mode of inference with a specific real data set. He was willing to interpret confidence intervals as Bayes probability intervals, and even seemed to think that if they could not be so interpreted, they were probably of little interest to the investigator of a data set (oral communications in the context of a graduate course on observational studies at Harvard University, 1970).

My major specific reason for a lack of affinity for model-based repeated-sampling inference is that it makes nearly all the assumptions of a Bayesian predictive inference, which is much more satisfactory, and the assumptions it avoids are of little practical importance. Extending a model-based repeated-sampling inference to include a prior distribution on θ, $P(\theta)$, gives a direct and conceptually neat Bayesian inference without the vagaries of conditioning arguments; practical sensitivity to the specification of $P(\theta)$ is usually very slight, whereas practical sensitivity to the model $f(Y_i|X_i, \theta)$ is often nontrivial, and in nearly all observational studies, and even in many real-world controlled experiments, sensitivity to the assignment mechanism $\Pr(\mathbf{W} \mid \mathbf{X}, \mathbf{Y})$ is paramount. Why balk at using the final part of a model specification that makes life easy, when there is essentially no deleterious effect on answers by doing so? The reluctance to accept any model at all for $\Pr(\mathbf{Y} \mid \mathbf{X})$ is understandable, and so the effort to try to make randomization inference work is also understandable, but the primary justification for model-based repeated-sampling inference appears to be in its richness of mathematical results rather than its practical relevance. This discussion supports doing a Bayesian analysis with particular attention paid to model sensitivity.

5. BACK TO THE THIORIDAZINE DATA

5.1. The Matched Sample Analyzed as a Paired-Comparison Experiment

We now return to the matched switcher–nonswitcher data armed with $\bar{y}_d$ as the estimated average causal effect and $s_d/\sqrt{2{,}000}$ as its standard error. Analyses of most outcome variables were done in the log scale so that additive effects correspond to fractional increases; Table 24.4 summarizes the largest results in terms of (i) the estimated percentage effects for nonswitchers relative to switchers, and (ii) the associated t statistics, $\bar{y}_d/[s_d/\sqrt{2{,}000}]$ on the log scale, which can be viewed either as the number of posterior standard errors the posterior means are from the null value of zero or as input to t tests to obtain P-values. Outcomes not reported in Table 24.4 differed by less than 1% between switchers and nonswitchers. The conclusion from Table 24.4 is clear: Nonswitchers fare better than switchers; nonswitchers have 6% fewer prescriptions than switchers during the post-intervention period, and, of real importance, less than nearly one-sixth the total dose! Also, they use fewer other drugs and experience fewer medical encounters, and the estimates are several standard errors from zero. (In fact, these results reflect covariance adjustment for X on matched-pair differences, which leads to essentially the same conclusions as the simple matched-pair analysis because of the closeness of the matches.) Should we trust these answers, in particular, the very large effect for total dose per day?

Table 24.4. *Summary of analyses of 2,000 matched pairs with at least 6 pre-intervention thioridazine prescriptions. Results based on covariance adjustment of matched-pair differences on log scale*

Conclusions for nonswitchers on key outcomes	% reduction for nonswitchers	t statistic
Fewer total thioridazine prescriptions	6	15
Smaller total dose	84	35
Fewer other prescriptions	2	7
Fewer medical encounters	1	6

Certainly, given the previous discussion of modes of inference, the mode seems fairly irrelevant as do any issues of assumed distributions for $\Pr(\mathbf{Y} \mid \mathbf{X})$ or lack thereof. The relevant issue is the conceptualization of the study and in particular, the posited assignment mechanism, which is common to the inferences summarized in Table 24.4 for all inferential modes.

5.2. Some Second Looks at the Data and Posited Assignment Mechanism – What Happens Post-Intervention, Pre-Switch?

In any situation, it is generally unwise to use a model contradicted by observed data. Consequently, in an observational study in which the assignment mechanism is questionable, it is wise to try to find ways to check its plausibility. Several authors have proposed specific examinations, for instance, Cochran (1965) more informal ones (also discussed in Rubin, 1984c), and Rosenbaum (1984b) more formal ones. Here, the period between intervention date and the switch date for the switcher of each matched pair is relevant to addressing the veracity of the conceptualization of a randomized paired comparison with assignment to treatment at intervention date (24.2), or for that matter, (24.1) with close matches within pairs on $g(X)$. (In fact, switch date is known only to be between some point prior to first generic prescription and after the last name-brand prescription for the switcher of the pair; several different plausible definitions were used in the actual data analyses with little sensitivity of conclusions.) Before switch date, both members of a pair used only name-brand thioridazine and, consequently, this period really should be viewed as a control period during which there should be no systematic differences between switchers and nonswitchers. Unfortunately for this conceptualization, however, there do appear to exist some real differences; for example, the nonswitchers have 12% lower total dose than the switchers ($t = 2.5$), and there are too many other outcomes with large t statistics favoring the nonswitchers. This result leads us either to believe that a rare event has occurred or to disbelieve both simple models for the assignment mechanism, (24.1) and (24.2). (Another possibility is that the assignment mechanism is appropriate but that the data model is incorrect, but this position is not well supported by model-free randomization theory arguments.) A more satisfactory conceptualization should posit an assignment mechanism that can lead, without resort to a rare event story, to the post-intervention pre-switch data that have been observed.

Two more simple conceptualizations for the assignment mechanism will now be considered. Both modify the mechanism (24.1), which by reasonable conditioning arguments leads to (24.2) for the 2,000 matched pairs. Both, however, lead to new problems of missing data, either missing treatment assignments or a fully missing covariate that should have been included in the matching, neither of which reduce to (24.2).

6. A MODIFIED FRAMEWORK WITH AN IGNORABLE TREATMENT ASSIGNMENT MECHANISM

6.1. Treatment Assignment Dependent on Post-Intervention Pre-Switch Outcomes

As before, X represents pre-intervention data. For simplicity in this section, accept the tenuous assumption that both members of a matched pair have the same switch date; that is, whichever subject had been assigned to switch would do so at the same time. The critical new assumption is that treatment assignment was determined at switch date with probabilities determined, not only by X as in (24.1), but also by post-intervention pre-switch outcomes, Z. That is, the primary modification arises from splitting the old outcome Y into Z and a new outcome Y^*, where Z represents post-intervention pre-switch values (scored as intensities, i.e., per day of the period), and Y^* represents post-switch values (also defined as intensities). The new assignment mechanism is still unconfounded, but is written

$$\Pr(\mathbf{W} \mid \mathbf{X}, \mathbf{Z}, \mathbf{Y}^*) = \Pr(\mathbf{W} \mid \mathbf{X}, \mathbf{Z}) = \prod_{i=1}^{6,500} h(X_i, Z_i)^{W_i}[1 - h(X_i, Z_i)]^{1-W_i},$$

(24.5)

to allow dependence of treatment assignment on $\mathbf{Z}$ as well as $\mathbf{X}$, where $h(X_i, Z_i)$ is the probability of being assigned to switch as a function of X and Z. In particular, it is posited that at switch date, switch/not switch is assigned in such a way that the worse-off subjects (with respect to Z) are more likely to switch, as seen with the observed data. The assignment mechanism is still an unconfounded probability one, but now can generate a systematic switcher–nonswitcher difference in Z and thus presumably a systematic difference in Y^*, which needs to be addressed in the analysis.

6.2. Results Adjusted for Post-Intervention Pre-Switch Outcomes

Without thinking too hard, accept that conditioning arguments analogous to those in Section 3.4 applied to the assignment mechanism (24.5) lead to a paired-comparison assignment mechanism for the 4,000 matched pairs, where the worse-off member of each matched pair (with respect to Z) is more likely to switch, that is, a biased-coin randomized paired comparison, where the bias is a function of the relative values of Z in a matched pair. Under this conceptualization, all modes of inference can be used to provide answers that adjust post-switch outcomes, Y^*, for imbalance in the post-intervention pre-switch outcomes, Z.

Table 24.5. *Summary of analyses for outcomes Y* during post-switch period.*
Estimated average percentage reduction for nonswitchers (and t statistic below
in parentheses) obtained from an analysis of covariance of log(Y*) on (X, Z)
using matched-pair differences

Number of matched pairs	2,000	1,000	600
Number of pre-intervention prescriptions	$\geqslant 6$	$\geqslant 6$	$\geqslant 10$
Number of post-intervention pre-switch prescriptions		$\geqslant 3$	$\geqslant 5$
Conclusions for nonswitchers on key outcomes			
Fewer thioridazine prescriptions per day	6%	2%	2%
	(14)	(6)	(4)
Smaller total thioridazine dose per day	83%	55%	37%
	(39)	(12)	(7)
Fewer other prescriptions	2%	1%	1%
	(6)	(4)	(4)
Fewer medical encounters	1%	<1%	<1%
	(6)	(<1)	(<1)

Predictive Bayesian inference is direct and under standard normal linear models leads to analysis of covariance (ANCOVA) results. For unrealistically simple choices of specific assignment mechanisms of the form (24.5), repeated sampling evaluations of ANCOVA are standard textbook fare (e.g., Snedecor and Cochran, 1974), and randomization tests of the ANCOVA statistics can be performed. The work to formulate a specific version of (24.5), however, does not seem to be worth the effort for several reasons. First, some experience and theory support the claim that the practical conclusions will be close to the Bayesian conclusions. Second, the ANCOVA-adjusted results are still about the same as in Table 24.4; specifically see column 1 in Table 24.5. And third, there exist problems with the posited assignment mechanism (24.5) and the reduction to the biased-coin, paired-comparison, conditional assignment mechanism.

6.3. A Problem with This Treatment Assignment Mechanism – Missing Treatment Assignments

A serious problem with the current formulation is that the observation of treatment assignment, W, is too indirect even assuming switch date is known and the same for both members of a matched pair. That is, we do not *know* $W = 1$ (*s*) or 0 (*ns*) for each unit, but rather *s* or *ns* was inferred as a result of observing outcomes involving thioridazine prescriptions. In particular, 400 of the 2,000 nonswitchers in the switcher–nonswitcher matched pairs had no post-treatment thioridazine prescriptions. Consequently, under the current formulation with treatment assignment at switch date, at least 400 of the 2,000 "nonswitchers" in the data set should not be viewed unambiguously as nonswitchers but as missing treatment assignment – under (24.5), they might have been assigned to switch but quit early.

A parallel issue can arise in a proper randomized experiment when the treatment assignment labels are lost and have to be reconstructed from the observation of covariates and outcomes. In particular, accept for now the unconfounded treatment assignment mechanism (24.5) where switch date is known for each unit, and at switch date, treatment is assigned to each subject independently as a stochastic function of X and Z. Also define the variable Q that indicates a decision to quit thioridazine before switch date: $Q = 1$ means "did not quit before switch date," $Q = 0$ means "quit before switch date."

Neither Q nor W is directly observed; both are partially missing because all we always observe is $Q \cdot W$, which is 1 if the subject is assigned to switch *and* did not quit before switch date, and is 0 if the subject is assigned to not switch *or* if the subject quit before switch date. Thus, among the 6,500 subjects with six or more thioridazine prescriptions pre-intervention date, the 2,000 switchers have $Q = W = 1$, but the 4,500 others that were observed to not switch ($Q \cdot W = 0$) are a mixture of early quitters assigned to not switch ($Q = 0$, $W = 0$), early quitters assigned to switch ($Q = 0$, $W = 1$), and nonquitters assigned to not switch ($Q = 1$, $W = 0$). Assuming switch dates, and thus treatment assignment dates, are known for everyone, members of this last group ($Q = 1$, $W = 0$) can be unambiguously identified because of post-switch-date name-brand prescriptions. But even assuming switch dates are known, switchers with $Q = 0$, $W = 1$, cannot be distinguished from nonswitchers with $Q = 0$, $W = 0$.

The crucial issue in the current formulation is to eliminate the early quitting switchers ($Q = 0$, $W = 1$ subjects) from the $Q \cdot W = 0$ group and put them into the switcher group before doing the matching or any of the analyses that have been discussed. In particular, the implication is that some of the matched pairs in the current analysis really might be switcher ($Q = 1$, $W = 1$)–switcher ($Q = 0$, $W = 1$) matched pairs. Even if exactly matched on (X, Z), the member of such a pair with no post-switch dose is labelled the nonswitcher and thus automatically has no post-switch prescriptions, whereas the switching member of the pair has at least the switching prescription, and this automatically generates a bias indicating more pre-scriptions for the switcher. This bias exists in randomized experiments with lost treatment assignments handled this way for any outcome that is not conditionally independent of Q given (X, Z). Even the apparently ubiquitous EM algorithm (Dempster, Laird, and Rubin, 1977) for handling missing data does not seem to be of help here with missing treatment assignments.

6.4. Diagnostic Analyses Focusing on Those Less Likely to Quit

Since those subjects who look least healthy with respect to (X, Z) are presumably less likely to quit before switch date, a simple procedure designed to address the effect of missing treatment assignments is to perform an analysis on a heavier-use stratum of matched pairs. Specifically, we examined the 1,000 matched pairs of the 2,000 that had at least three prescriptions during the post-intervention pre-switch period (the Z period). Table 24.5, column 2, displays ANCOVA results for the variables in Table 24.4, which still tend to favor nonswitchers; other outcomes not presented also tend to slightly favor nonswitchers, but all such effects round to 100%.

Quite small effects are obtained except for total dose intensity. Two possible explanations are: (i) doctors adjust dosage levels to "stabilize" results (some discussion to this effect has been presented regarding the more uniform bioavailability of name-brand thioridazine due to its superior coating), and (ii) 150 of the pairs still have no post-switch dose for the nonswitcher, thus suggesting that the problem of potentially missing treatment assignments still plagues this analysis.

In an attempt to extricate ourselves from the early quitter problem, a stratum of 600 pairs of extremely heavy thioridazine users was examined: 10 or more prescriptions pre-intervention-date and 5 or more Z prescriptions (post-intervention/pre-switch). The results appear in the last column of Table 24.5 and are similar to those in the second column of Table 24.5 but less favorable toward not switching; again, outcomes not reported tend to slightly favor not switching, although all effects, except for total dose, are very small. Could the results for total dose intensity favoring not switching still be due to an early quitter problem, which is still an issue because 70 of the matched pairs are ones where the nonswitcher had no post-switch thioridazine prescriptions?

One suggestion that was made was to discard the matched pairs with no post-switch thioridazine prescriptions for the nonswitcher; that is, essentially assume that $Q = 1$ and $W = 0$ for those observed to have $Q \cdot W = 0$ with no post-switch thioridazine prescriptions. This idea is not conceptually sound because it selects units on the basis of a post-treatment outcome variable rather than pre-treatment variables as in Tables 24.4 and 24.5. The resulting erroneous analysis still shows not switching to be slightly favorable to, although virtually indistinguishable from, switching.

Perhaps the "standard" conceptualization with an ignorable treatment assignment at switch date has not really helped, especially when we recall that we have yet to face the fact that switch date is not known precisely for anyone, and unknown (really, a fabrication) for all subjects not observed to switch.

7. A CONCEPTUALIZATION INVOLVING A NONIGNORABLE TREATMENT ASSIGNMENT MECHANISM

7.1. Return to Treatment Assignment at Intervention Date

Recall that with the original conditional paired-comparison analysis with treatment assignment at intervention date, all 6,500 heavy-dose subjects had post-treatment exposure to either (i) name-brand thioridazine only or (ii) generic and/or name-brand thioridazine. Thus, tentatively assume because of this fact that ns and s could be unambiguously identified and so return to the analysis of Section 5 with Y defined as outcome intensities post-intervention-date. Because of the undeniable tendency to accept an analysis of data in the absence of evidence indicating that the analysis is based on incorrect assumptions, we might have been satisfied with this analysis if it were not for the existence of substantial differences in Z (outcomes during the post-intervention pre-switch period), substantial in the sense of what was expected under the unconfounded assignment mechanism (24.1). The observed differences in Z, however, could be due to a confounded assignment mechanism similar to (24.1), but one that involved unrecorded covariates U correlated with outcomes and having different distributions in the switching and nonswitching groups even after matching on X. Appropriate adjustment for such a U would presumably lead to essentially no Z effect and would reveal the real effect of switching on Y.

7.2. A Confounded, Nonignorable Treatment Assignment Mechanism

Continue to accept an unambiguous switcher–nonswitcher designation, and focus on the 6,500 heavy-dose subjects. Suppose that treatment assignment took place on intervention date by independently assigning switch with probability $k(X, U)$ and nonswitch with probability $1 - k(X, U)$:

$$\Pr(\mathbf{W} \mid \mathbf{X}, \mathbf{Y}, \mathbf{U}) = \prod_{i=1}^{6,500} k(X_i, U_i)^{W_i}[1 - k(X_i, U_i)]^{1-W_i}. \qquad (24.6)$$

Matching on X was no mistake, since it balanced X, although matching on (X, U) would have been better, albeit impossible. Our task with U missing is to specify enough about its distribution to allow some adjustment for it and assessment of the precision of the resulting causal-effect estimates.

As in Section 5, let Y be outcome intensities post-intervention-date, and as in Section 6, let Z be outcome intensities post-intervention pre-switch (assigning switch date of the switcher to the nonswitcher). Thus Z for everyone measures post-intervention outcome intensities while on name-brand thioridazine (i.e., before any switching), and Y for everyone measures outcome intensities post-treatment assignment, with the result that the vagaries of switch-date for the nonswitcher are relatively unimportant. Since both Y and Z are post-treatment variables, in general the representation in Table 24.3 needs $\mathbf{Y} = (\mathbf{Y}_s, \mathbf{Y}_{ns})$ and $\mathbf{Z} = (\mathbf{Z}_s, \mathbf{Z}_{ns})$. Now, when all subjects are on name-brand thioridazine, it is reasonable to suppose that there is no systematic effect of treatment on outcome intensities, with the result that $\mathbf{Z}_s = \mathbf{Z}_{ns}$. Thus Z can be used as a covariate, although this does not explain the substantial switcher–nonswitcher differences in Z that have been observed. An unobserved covariate, however, can do this: An unobserved covariate U that is correlated with Z and was used for treatment assignment can lead to the results we have seen.

Parenthetically, an alternative definition of the outcomes presumably unaffected by treatment exposure would be $Z^* =$ "outcome intensities post-intervention while on name-brand thioridazine." This leads to $Z_i^* = Z_i$ when $W_i = s$, and $Z_i^* = Y_i$ when $W = ns$, or more precisely since Z^* is an outcome, $\mathbf{Z}^* = (\mathbf{Z}_s^*, \mathbf{Z}_{ns}^*)$, where $\mathbf{Z}_s^* = \mathbf{Z}_s$ and $\mathbf{Z}_{ns}^* = \mathbf{Y}_{ns}$. With this definition, even if there is no systematic effect of switching on $\mathbf{Z}^*$, we cannot assert $\mathbf{Z}_s^* = \mathbf{Z}_{ns}^*$ for all subjects, because for each subject Z_s and Z_{ns} refer to different periods of time. This formulation is related to "before–after" assessments for causal effects using only a treated group (e.g., see Cochran, 1983, §7.4; Campbell and Stanley, 1963). In fact, the subsequent analyses using $\mathbf{Z}$ were also done using $\mathbf{Z}^*$ with the same practical conclusions.

Adjusting for the covariate U is not direct because it is nowhere observed, but after adjustment for (X, U), the working assumption is that there should be no systematic effect of switching on the outcome Z; that is, Z and W should be conditionally independent given (X, U):

$$\mathbf{Z} \perp \mathbf{W} \mid (\mathbf{X}, \mathbf{U}). \qquad (24.7)$$

Expression (24.7) is easily achieved by setting $U = Z$ but there are many other U's that will do this (e.g., $U = W, U = (Z, V)$ for any V), with different relationships with Y. An honest adjustment for U must reflect uncertainty as to the distributional assumptions.

7.3. Adjustment Assuming $U = Z$

Nevertheless, it is of interest to adjust Y for Z and see what happens. Mechanically, the point estimate that results is that obtained by a covariance adjustment of Y for Z (and X), but the fully-missing-covariate perspective is different from the standard covariance adjustment framework because (i) the covariate is part of the outcome, and (ii) the reported standard errors are lower bounds achieved only under a special implausible assumption. The full perspective offers a whole family of possible corrections consistent with the assumptions (24.6) and (24.7), and will result in larger standard errors than the analysis of covariance. Because we have available many covariates with pre- and post-matching biases and partial regression coefficients with outcomes, there are data available to make sensible expansions of uncertainty, such as the ones proposed in Rubin (1977b) in the context of nonresponse, or at least reasonable sensitivity analyses analogous to those in Rosenbaum and Rubin (1983b), but these are beyond the scope of this article.

Adjusting Y for Z and X leads to smaller, but still nontrivial, effects for dose intensity, which are displayed in Table 24.6 for the heavy-dose 2,000 matched pairs, the 1,000 very-heavy-dose mix, and for the 600 extremely-heavy-dose matched pairs (since neither X nor Z is affected by treatment in this perspective, it is legitimate to stratify on both). In fact, the only effects larger than 1% are for dose intensity, and they suggest that there is a tendency for switchers, especially less frequent users, to obtain higher-dose prescriptions. Since the usual standard errors from the covariance adjustment are misleadingly small, they are suppressed. A proper analysis under this sort of assignment mechanism would provide standard errors that reflected uncertainty due to the adjustment and explore the nature of the nonparallel response surfaces (i.e., try to determine specific X and Z values that predict dose intensity differences, as by a regression of matched-pair Y differences on matched-pair average X values). Such analyses become more and more based on full probability models, and if the data were available, would be worth pursuing.

7.4. A Final Comment on This Assignment Mechanism

A feature of the "treatment assignment at intervention date" formulation that is definitely appealing beyond its statistical simplicity is that it addresses the simple policy question: "If intervention had been postponed from March 18, 1983 until after April 15, 1985, what would be the differences in outcomes for those who switched during the period March 18, 1983–April 15, 1985?" Nevertheless, this formulation of the assignment mechanism remains less than fully satisfactory because of our tentative assumption that the nonswitcher designation is unambiguous. In

Table 24.6. *Summary of analyses for outcomes Y during post-intervention period. Estimated average percentage reduction for nonswitchers obtained from an analysis of covariance of log(Y) on (X, Z) using matched-pair differences*

Number of matched pairs	2,000	1,000	600
Number of pre-intervention prescriptions	$\geqslant 6$	$\geqslant 6$	$\geqslant 10$
Number of post-intervention pre-switch prescriptions		$\geqslant 3$	$\geqslant 5$
Conclusions for nonswitchers on key outcomes			
Fewer thioridazine prescriptions per day	1%	1%	1%
Smaller total thioridazine dose per day	49%	10%	3%
Fewer other prescriptions	<1%	<1%	<1%
Fewer medical encounters	<1%	<1%	<1%

fact, in this formulation, the 4,500 potential controls may still include some early quitting switchers assigned to switch at a date after they quit using thioridazine. This potential problem is less severe than with our previous formulations because there appears to be little evidence that the assumption is incorrect. As Table 24.4 shows, the estimated effect on total number of prescriptions is essentially null so that switchers and their matches have similar numbers of post-treatment prescriptions, as well as similar values for all outcomes except dose intensity. A more realistic view of the assignment mechanism would allow for a factorial structure, where each unit is assigned *s* or *ns* at the time of each prescription with probability a function of previous experience including time to that prescription, previous doses, whether previous prescriptions were generic or name-brand, and possibly unmeasured covariates. This formulation is similar to that for a complicated sequential crossover experiment with many potential crossover times, and has these attendant problems in addition to those arising from unmeasured covariates. In particular, this complex assignment mechanism is confounded, but possibly ignorable since later treatment assignments depend on earlier values of outcomes. Any analysis under such a treatment assignment will be highly model-dependent, at least for some outcomes [see, for example, Robins (1989)].

8. CONCLUSION

Four distinct modes of statistical inference for causal effects can be identified. All accept a common framework for defining causal effects, and all require the specification of a treatment assignment mechanism. Except in the simplest cases, in which randomization tests are unambiguous and elegant, Bayesian predictive inference seems preferable because its basic principles generate a toolkit rich enough to address real-world problems in their full complexity. Repeated-sampling modes of inference should be used to evaluate procedures to ensure their propriety in a broad range of applications, and these generally support the use of carefully formulated Bayesian models. Differences notwithstanding, the most critical issues when applying any mode of causal inference are first to formulate a clear conceptual framework in which causal estimands can be defined, and second to posit plausible assignment mechanisms. Critical discussion of the framework and plausible assignment mechanisms should accompany every analysis of observational data for causal effects.

ACKNOWLEDGMENT

I wish to thank Gregory Maislin for performing data analyses.

25. In Utero Exposure to Phenobarbital and Intelligence Deficits in Adult Men

June Machover Reinisch, Stephanie A. Sanders,
Erik Lykke Mortensen, and Donald B. Rubin

Objective: To test whether exposure to phenobarbital in utero is associated with deficits in intelligence scores in adult men and whether the magnitude of the postnatal effect is mediated by exposure parameters and/or postnatal environmental factors.

Design: Two double-blind studies were conducted on independent samples of adult men prenatally exposed to phenobarbital and matched control samples using different measures of general intelligence. Based on data from control subjects, regression models were built relating intelligence scores to relevant pre-exposure matching variables and age at testing. Models generated predicted scores for each exposed subject. Group mean differences between the individually predicted and observed scores estimated exposure effects.

Setting: Copenhagen, Denmark.

Participants: Exposed subjects were adult men born at the largest hospital in Copenhagen between 1959 and 1961 who were exposed to phenobarbital during gestation via maternal medical treatment and whose mothers had no history of a central nervous system disorder and no treatment during pregnancy with any other psychopharmacological drug. Study 1 included 33 men and study 2, 81 men. Controls were unexposed members of the same birth cohort matched on a wide spectrum of maternal variables recorded prenatally and perinatally. Controls for studies 1 and 2 included 52 and 101 men, respectively.

Main Outcome Measures: In study 1: Wechsler Adult Intelligence Scale (Danish version); in study 2: Danish Military Draft Board Intelligence Test (Børge Priens Prøve).

Result: Men exposed prenatally to phenobarbital had significantly lower verbal intelligence scores (approximately 0.5 SD) than predicted. Lower socioeconomic status and being the offspring of an "unwanted" pregnancy increased the magnitude of the negative effects. Exposure that included the last trimester was the most detrimental.

Conclusion: Phenobarbital exposure during early development can have long-term deleterious effects on cognitive performance. Detrimental environmental conditions can interact with prenatal biological insult to magnify negative outcomes. Physicians are urged to use increased caution in prescribing such medications during pregnancy.

Between 1950 and the late 1970s more than 22 million children were born in the United States to women taking prescribed barbiturates during pregnancy (Reinisch and Sanders, 1982). This estimate does not reflect those who ingested these drugs for nonmedical reasons. Barbiturates were second only to alcohol as drugs of abuse during this period (Wesson and Smith, 1977). Barbiturate-containing medications

Reprinted with permission from *The Journal of the American Medical Association*, 1995, **274**, 1518–1525.

are still administered to pregnant women and neonates in the United States and elsewhere, particularly for the treatment of convulsive disorders of mothers and children (Farwell et al., 1990a,b; Gaily, Kantola-Sorsa, and Granstrom, 1990; Hirtz et al., 1986; Vining et al., 1987), refractory hyperbilirubinemia (Rayburn et al., 1988; Valaes et al., 1980; Wallin and Boreus, 1984), predicted premature delivery (Morales and Koerten, 1986; Rayburn et al., 1989), and neonatal abstinence syndrome (Kaltenbach and Finnegan, 1986). Possible deleterious effects of these exposures to barbiturates have been widely considered (Reinisch and Sanders, 1982; Farwell et al., 1990a,b; Camfield et al., 1979; van der Pol, 1991; Wilson, 1971).

Even in the absence of observable morphological anomalies at birth, permanent alterations in endocrine, neural, and behavioral function have been identified in laboratory animals exposed both prenatally and postnatally to barbiturates, including phenobarbital (Reinisch and Sanders, 1982; Dodson, 1989; Fishman and Yanai, 1983; Martin, 1986; Yanai and Bergman, 1981; Arai and Gorski, 1968; Gorski and Arai, 1968; Yaffe and Dorn, 1990). In rodents, barbiturate exposure late in gestation or perinatally causes deficits in cell proliferation and differentiation concurrent with exposure and appears to negatively affect cells formed prior to exposure (Fishman and Yanai, 1983; Yanai and Bergman, 1981). Behavioral teratological effects of early exposure to barbiturates in laboratory animals include the following: (1) retardation of developmental reflexes, suggesting the possibility of neurological impairment (Zemp and Middaugh, 1975); and (2) long-lasting reduction in adaptive responses to a number of learning and reasoning situations (Reinisch and Sanders, 1982; Dodson, 1989; Fishman and Yanai, 1983; Martin, 1986; Armitage, 1952; Martin et al., 1979; Middaugh, 1986; Pereira de Vasconcelos et al., 1990; Yanai et al., 1989). These deficits appear to increase with greater task complexity (Fishman and Yanai, 1983) and emerge at stages in development much later than the time of exposure (Reinisch and Sanders, 1982; Yaffe and Dorn, 1990; Kilbey and Asghar, 1991).

In humans, negative developmental outcomes, some associated with cognitive development, have been reported in offspring of mothers treated with antiepileptic drugs, including phenobarbital, during pregnancy (Gaily, Kantola-Sorsa, and Granstrom, 1990; Rayburn et al., 1988; van der Pol et al., 1991; Fishman and Yanai, 1983; Sobeyzuk, Dowzenko, and Krasicka, 1977; Jones, Johnson, and Chambers, 1992; Liewendahl, Majuri, and Helenius, 1978; Yaffe, 1980; Friis and Sardemann, 1977; Klipstein, 1964). However, because of a range of confounding factors, it has proven difficult to evaluate definitively the teratogenicity of phenobarbital from such studies. There are also indications of adverse behavioral and cognitive side effects in children treated with phenobarbital for seizure disorders that may persist for several months after termination of treatment (Farwell et al., 1990a,b; Vining et al., 1987; Camfield et al., 1979; Brent et al., 1990). These adverse consequences of phenobarbital treatment reflect the more immediate neurochemical impact of barbiturates on the central nervous system (CNS).

Thus, it is reasonable to suggest that behavior problems, learning disabilities, and deficits in intelligence may be more common among individuals prenatally exposed to barbiturates. We chose to focus on phenobarbital because it is the most commonly prescribed barbiturate and the differences in chemical structure among barbiturates may result in different effects. The present set of studies using adult men tested the hypothesis that prenatal exposure to phenobarbital is associated with deficits in intelligence scores. Study 2 was conducted subsequent to study 1 to independently replicate its findings using a different standard measure of intelligence and an independent sample of subjects. The impact of selected socioenvironmental risk factors and drug administration parameters on the exposure effect were also investigated.

METHODS

Subjects

The subjects participating in the two studies belong to the Danish Perinatal Cohort, comprising the offspring of 9,006 deliveries (Zachau-Christiansen and Ross, 1975). They were born at the largest hospital in Copenhagen, Denmark, between 1959 and 1961. During the establishment of the cohort, demographic, socioeconomic, and medical variables were prospectively recorded prenatally and postnatally. Original medical records were reviewed to confirm exclusion criteria and obtain information on dosage, timing, and duration of exposure. Evaluation of the subjects in study 1 took place from 1983 through 1986. The retrieval of subjects' records for study 2 began in 1988.

SELECTION OF EXPOSED SUBJECTS: GENERAL CRITERIA. Consultation of the data tape constructed when the cohort was established revealed that approximately 10% of the cohort was prenatally exposed to some barbiturate preparation via maternal treatment, often in combination with other medications. During data collection for the original cohort, all drug treatments of the mother of at least 5 days' duration were recorded for each of six gestational periods: first month, second month, third month, second trimester, third trimester, and last month of pregnancy. Subjects were excluded from consideration if they met any of the following criteria: (1) died within the first year of life (there were no significant differences in death rate between barbiturate-exposed and unexposed subjects); (2) were twins or triplets; (3) were offspring of incest; (4) had a gestational length of less than 28 weeks; or (5) had a congenital malformation or Down's syndrome. Offspring were also excluded if their mothers (1) had a history of diabetes; (2) were younger than 16 years at delivery; (3) were diagnosed with polio, encephalitis, meningitis, viral pneumonia, or ornithosis during pregnancy; or (4) were exposed to drugs classified by the physicians who established the cohort as psychopharmacological agents (excluding barbiturates), antiepileptics, or medications for urinary tract infection.

STUDY 1: EXPOSED SUBJECTS. For the first study, the following additional inclusion criteria were applied: at least two periods of exposure to barbiturate and no indication of eclampsia during the pregnancy. As a result of the application of the general and specific selection criteria listed above, 222 men and women exposed to some form of barbiturate during at least two periods were identified from the computer tape. The study was restricted to the 107 men to allow comparison with study 2, where another measure of intelligence was available only for men. Then the original medical records were retrieved from the hospital archives and reviewed. It was only at this stage that the health status of the mother and her fetus and neonate could be confirmed, additional medical factors ascertained, and the specific barbiturate administered identified. Only subjects exposed to phenobarbital and not other barbiturates were included. In addition, dosage, timing, and duration of exposure for each individual could only be determined from these original medical records. Based on these records, 11 cases were eliminated because of notations of maternal syphilis, epilepsy, or psychosis, or lack of sufficient documentation of barbiturate exposure. Referring to the national registries, four cases were eliminated because of death and two because of institutionalization. Nine were unavailable for evaluation as they were living abroad, serving in the military, participating in another institute study, homebound, or unlocatable. Four individuals refused to participate in the research despite multiple recruitment attempts. Eleven were unable to be scheduled during the data collection period, and one did not complete the Wechsler Adult Intelligence Scale (WAIS). Thirty-two of the remaining 65 men were ineligible because of exposure to hormones, barbiturates other than phenobarbital, or phenobarbital in combination with other barbiturates.

Thirty-three men met the selection criteria and were evaluated by us at the Institute for Preventive Medicine. Total dosage of phenobarbital ranged from 900 to 20,250 mg (mean, 7,847 mg; SD, 5,490 mg). Duration of exposure ranged from 10 days to the entire pregnancy (mean, 90 days; SD, 59 days). Thirty subjects were exposed to phenobarbital during at least some part of the third trimester and/or last month of gestation. Of these, 18 had no exposure to phenobarbital during the first or second trimesters. The reason most commonly cited for maternal treatment was hypertension related to pregnancy ($n = 21$). There were no notations on the mothers' records of preexisting hypertension. The following other indications were noted on the medical records: need for hypnotic or sedative ($n = 2$); hyperemesis ($n = 3$); and leg cramps ($n = 1$). For six subjects, records failed to reveal any written indication by the attending physicians of the reason for prescribing phenobarbital. The mean age at testing was 22.79 years (SD, 1.02 years).

STUDY 2: EXPOSED SUBJECTS. For study 2, in addition to the general selection criteria, the following selection criteria were applied: no indication of maternal epilepsy or CNS disorder, no hormone exposure, at least one period of exposure to barbiturates, and not participating in study 1. Based on the general and specific selection criteria, 518 potential index cases were identified from the cohort data tape. Of these, 264 were men. Following a review of the original medical records, 83 cases were eliminated because of exposure to barbiturates other than phenobarbital or phenobarbital in combination with other barbiturate preparations. Records for 14 cases revealed insufficient documentation of phenobarbital exposure, confounding exposures to hormones or psychopharmacological agents, or maternal convulsive disorder. Two cases were eliminated because of a history of hospitalization for mental illness and two because of death. Eleven were abroad or emigrated prior to testing for the Danish military draft. Forty-two cases did not have a Danish Military Draft Board Intelligence Test (Børge Priens Prøve [BPP]) score in the register and 29 were excluded to avoid overlap between the samples in studies 1 and 2.

Study 2 included all 81 men with available outcome data. Total dosage of phenobarbital ranged from 225 to 22,500 mg (mean, 2,894 mg; SD, 3,110 mg). Duration of exposure ranged from 5 to 211 days (mean, 32 days; SD, 36 days). Seventy-seven were exposed to phenobarbital during at least some part of the third trimester and/or last month of gestation. Of these, 72 were not exposed to phenobarbital during the first or second trimester. The most common indication for maternal treatment was hypertension, apparently related to pregnancy (n = 48). For only one of these subjects did medical records note hypertension as a condition that occurred prior to pregnancy. In this instance, however, phenobarbital treatment was limited to 2 weeks in the last month of pregnancy. Other indications included preeclampsia (n = 8), eclampsia (n = 4), need for sedative or hypnotic (n = 3), muscle spasm (n = 1), albuminuria (n = 1), headache (n = 1), vomiting (n = 1), and staining (n = 1). For 13 exposed subjects, medical records lacked any indication by the attending physicians of the reason for prescription of phenobarbital. Mean age at testing by the draft board was 19.37 years (SD, 1.69 years).

Selection of Controls

Except for the absence of prenatal phenobarbital exposure, the same selection criteria applied to both exposed subjects and potential controls. The techniques for matching exposed subjects to the pool of over 3,000 potential cohort controls were developed specifically for this research (Rubin, 1984c). The objective of the matching was to obtain a set of control subjects, approximately the same number as exposed, whose distributions of matching variables were nearly the same as the distributions for exposed subjects. First, for each exposed subject, the 10 statistically best potential matches were identified by computer from original cohort records, using Mahalanobis metric matching within calipers defined by the estimated propensity score (Rosenbaum and Rubin, 1985a,b). The senior author (J.M.R.) then refined the selection of potential matches for inclusion in the study from this list of 10 by examining more subtle trade-offs in deviations across the matching variables. When there was more than one close match, all were designated for potential inclusion. For cases in which none of the 10 computer-selected controls provided a match that was sufficiently close to the exposed subject on all variables, two matches among the 10 were selected to bracket the exposed subject on the less well-matched variables. Final assignment to the selected control group required that there be no overlap of control subjects between studies 1 and 2.

Table 25.1 compares the exposed and selected control subjects for studies 1 and 2 on the matching variables including the pre-exposure variables used in the data analyses and a number of potential confounding variables. Also included for comparison are the data for the pool of all unexposed cohort men from which the controls were selected. These data illustrate that the selected control samples were substantially closer to their respective exposed samples on these matching variables than was the population of 3,308 potential unexposed male control subjects. As a result of our matching procedures, none of the remaining differences in matching variables between the exposed samples and their selected control samples exceeded 0.1 SD, with the exception of the maternal complaint score. Maternal complaint score incorporated the specific symptoms most often cited as reasons for the prescription of phenobarbital in this cohort (see Table 25.1, ninth footnote). As would be expected, the exposed subjects had higher maternal complaint scores than controls. However, because maternal complaint score was included as a matching variable, the selected

Table 25.1. *Distributions of matching variables for exposed subjects, selected controls, and the total pool of unexposed male cohort members*

		Study 1		Study 2	
Matching variable*	Cohort pool of controls (n = 3,308)	Exposed subjects (n = 33)	Selected controls (n = 52)	Exposed subjects (n = 81)	Selected controls (n = 101)
In Prediction Models					
Firstborn, no. (%)*	1,866/3,308 (56.41)	23/33 (69.70)	37/52 (71.15)[†]	41/81 (50.62)	51/101 (50.50)
Unwanted pregnancy, no. (%)[‡]	1,884/3,166 (59.51)	16/32 (50.00)	25/51 (49.02)	36/75 (48.00)[†]	49/101 (48.51)[†]
Abortion attempted, no. (%)[§]	253/3,197 (7.91)	2/32 (6.25)	2/51 (3.92)	5/76 (6.58)	7/101 (6.93)
Single mother, no. (%)	1,352/3,290 (41.09)	6/33 (18.18)[†]	10/52 (19.23)[†]	18/80 (22.50)[†]	23/101 (22.77)[†]
Mean (SD) socioeconomic status‖	4.07 (1.85)	4.87 (1.50)[†]	4.81 (1.53)[†]	4.53 (1.76)[†]	4.47 (1.52)[†]
Mean (SD) breadwinner's education¶	3.39 (0.74)	3.61 (0.76)[#]	3.61 (0.80)[†]	3.44 (0.71)	3.44 (0.71)
Mean (SD) predisposing risk score**	28.14 (17.33)	30.25 (18.16)	29.91 (17.27)	26.52 (18.40)	26.02 (15.57)
Mean (SD) mother's age, y	24.76 (6.37)	29.27 (7.32)[†]	29.10 (6.64)[†]	27.04 (6.65)[†]	26.50 (6.21)[†]
Mean (SD) father's age, y	28.63 (7.77)	33.18 (8.82)[†]	32.80 (8.40)[†]	29.62 (6.93)	29.70 (7.31)
Potential Confounding Variable					
Mean (SD) gestational length, wk	38.59 (2.69)	38.24 (2.05)	38.19 (1.99)	38.73 (1.70)	38.63 (1.69)
Mean (SD) birth weight, g	3,233 (642)	3,336 (596)	3,319 (589)	3219 (586)	3260 (433)
Mean (SD) birth length, cm	51.28 (3.17)	51.80 (2.53)	52.03 (2.80)[#]	51.57 (2.63)	51.64 (2.22)
Mean (SD) no. of cigarettes/day in third trimester	6.40 (7.65)	3.94 (7.01)[#]	4.43 (6.69)[#]	5.03 (7.67)	5.26 (6.61)
Mean (SD) maternal weight gain (kg)/height cubed (m)	26.88 (8.94)	25.64 (8.29)	26.31 (7.23)	27.65 (7.68)	28.18 (8.65)
Mean (SD) maternal complaint score[††]	1.70 (1.88)	3.85 (2.48)[†]	2.57 (2.03)[‡‡]	4.95 (2.28)[†]	3.97 (2.13)[‡‡]

(footnotes on p. 432)

Table 25.1 (*continued*)

* These variables were used for matching in conjunction with the following additional variables: respiratory illness of mother, severe preeclampsia, and the squares of socioeconomic status, education, predisposing risk score, gestation length, number of cigarettes, and maternal complaint score. We chose variables for the linear regression prediction model that could not be affected by phenobarbital exposure. Those listed as potential confounding variables were used for matching, but were not included in the regression model as their values might be affected by phenobarbital exposure.

† Significantly different from the pool of cohort controls, $P < .05$ (two-tailed).

‡ According to Villumsen (1970), as part of the development of the cohort database, during the first talk with the mothers early in pregnancy, an attempt was made to evaluate whether the actual pregnancy had been "desired at the time of conception." The term "wanted" was applied to situations in which the child was wanted and contraception was not used or the child was not unwanted and contraception failed. The term "unwanted" refers to situations in which the child was unwanted and contraception had failed or was not used.

§ In study 1, for each of the two exposed subjects whose mothers attempted abortion, only one well-matched control whose mother also attempted abortion was available. Thus, in study 1 the *number* of abortion attempts in the exposed and selected control samples were equal. In study 2, for two of the five exposed subjects whose mothers attempted abortion, additional good matches whose mothers also attempted abortion were included so that the *proportions* of abortion attempts in the exposed and selected control samples were equivalent.

‖ Family socioeconomic status when the child was 1 year of age. Danish system categorized on a 9-point scale, $1 =$ lowest, $9 =$ highest.

¶ Education was categorized on a 5-point scale, $1 =$ remedial instruction, $5 =$ college.

Trend toward significant difference from the pool of cohort controls, $.5 < P < .10$ (two-tailed).

** The predisposing risk score is a score based on pregravidatis factors concerned with the mother's physical and emotional state prior to the pregnancy. Information includes such items as whether the mother was married when she conceived; whether she had previously had an abortion, a miscarriage, a stillbirth, or neonatal death; her age; her weight; and previous history of central nervous system illness, syphilis, cardiovascular illness, or diabetes. Points indicate that conditions (physical and emotional) were probably "less than optimum" for conception at that time (Zachau-Christiansen and Ross, 1975).

†† The maternal complaint score included the following: severe preeclampsia, hypertension, prescription of diuretics, edema and proteinuria, bleeding/staining, allergies and treatment with antihistamines, and anemia. This score includes the most common indications for treatment with phenobarbital in this cohort, and thus, the maternal complaint score is related to exposure status. Maternal complaint was not always available for use in initial matching from the original cohort data tape or better matching would have been achieved on this variable. It was used when available in subsequent matching after initial matches were chosen and this information could be obtained from original medical records.

‡‡ Significantly greater than the pool of cohort controls ($P < .05$, two-tailed), but still significantly less than respective exposed subjects ($P < .5$, two-tailed).

432

controls were closer to the exposed subjects on this variable than was the cohort pool of unexposed men. It should be noted that the sizes of estimated phenobarbital exposure effects may have been reduced by the inclusion of maternal complaint score as a matching variable: if this score was both affected by exposure and related to IQ, then controlling for the score would remove some of the effect of exposure. The similarity of matching variables between exposed and control groups meant that simple linear regression models could be used in data analysis without excessive concern for sensitivity to nonlinear relationships.

STUDY 1: SELECTED CONTROLS. Fifty-two controls were selected for study 1 using the matching procedures described above. Because of the high cost of testing individual subjects as part of this study, each exposed subject was systematically evaluated prior to recruitment of his controls to avoid the expense of evaluating controls whose exposed subject(s) might not be enlisted as participants during the study period. Therefore, the sample of selected controls was slightly older at the time of testing (mean, 24.10 years, SD, 1.18 years, vs 22.79 years, SD, 1.02 years) than the sample of exposed subjects. Adjustment for this difference in age at testing was incorporated into data analyses. It is important to note that for the distributions in these samples, age is not highly correlated with dependent variables, and so issues of sensitivity to linear extrapolation are not important. Throughout the evaluation phase, both exposed and unexposed subjects were tested concurrently, interspersed with index and control subjects from other ongoing research. Thus, evaluators were blind as to exposure status.

STUDY 2: SELECTED CONTROLS. For study 2, 101 controls were selected. Mean age at testing for the matched controls was 19.52 years (SD, 1.77 years), nearly identical to the exposed subjects (19.37 years, SD, 1.69 years).

Procedures

The procedures used in these studies were reviewed and approved by the institutional review boards of Rutgers, The State University of New Jersey and Indiana University for research involving human subjects. All necessary permissions were obtained from the appropriate Danish registries.

STUDY 1: INSTRUMENT AND DATA COLLECTION. The WAIS (Wechsler, 1955) Danish version (Hess, 1974) is a standard individually administered measure of general intelligence, which yields three IQ scores. The verbal IQ (VIQ) is generated from six subscales primarily measuring verbal comprehension, the performance IQ (PIQ) is derived from five subscales primarily measuring perceptual organization (Hill, Reddon, and Jackson, 1985), and the full scale IQ (FIQ) is based on all 11 subscales and may be considered a measure of general intelligence. These IQ scores form the primary outcome (or dependent) variables for study 1.

The purpose and procedures for the study were explained to participants by a social worker, and an informed consent form was signed. The WAIS was administered by a psychologist as part to double-blind evaluation (Reinisch, Mortensen, and Sanders, 1993). Standard scoring procedures were applied and IQs were derived from norms developed by our laboratory for Danes aged 20 to 26 years old (E. L. Mortensen, G. Høgaard, and J. M. Reinisch, unpublished data, 1995).

STUDY 2: INSTRUMENT AND DATA COLLECTION. The Danish Military Draft Board Intelligence Test, the BPP, is a standardized, 45-minute group intelligence test developed in 1957 for the Danish draft board and administered to nearly all Danish young men as part of the universal military draft evaluation. The test consists of 78 questions including four categories of items: letter matrices, verbal analogies, number series, and geometric figures. The score equals the total number of correct responses, ranging from 0 to 78. This test is highly correlated with the WAIS VIQ (0.78), PIQ (0.71), and FIQ (0.82) (Mortensen, Reinisch, and Teasdale, 1989). This high correlation with the WAIS FIQ indicates that the BPP total is a relatively good measure of general intelligence.

Data Analyses

To adjust for the small differences in age at testing and in the other covariates, least squares linear regression prediction models were developed for each study based on data from the selected control samples. The following variables were used as predictors of the intelligence scores: (1) family socioeconomic status (SES) when the child was 1 year old [Danish system: 1 = lowest, 9 = highest (Zachau-Christiansen and Ross, 1975; General Register Office, 1951; Graffar, 1960)]; (2) breadwinner's education; (3) sibling position, first or latter born; (4) whether the pregnancy was "wanted" (the term "unwanted" refers to situations in which pregnancy was not desired at the time of conception (Zachau-Christiansen and Ross, 1975; Villumsen, 1970) and Table 25.1, second footnote); (5) whether or not the mother attempted an abortion during this pregnancy; (6) maternal marital status; (7) predisposing risk score (see Table 25.1 footnotes) indicating conditions were less than optimum for conception at that time; (8) mother's age; (9) father's age; (10) subject's age at testing; (11) square of the deviation of SES from the mean; and (12) square of the deviation of age at testing from the mean. (These latter two were used to reflect possible nonlinearity in the relation between IQ and SES or age.) As the maximum missing data rate did not exceed 6%, a predicted value for the missing predictor data was substituted and a variable indicating that this substitution had been made was included (Little and Rubin, 1987).

The models, developed separately for each study, were used to predict expected intelligence scores individually for each exposed subject. These analyses did not use indicator variables for pairwise matching of exposed and control subjects because the exact values of the matching variables were available (Cochran and Rubin, 1973). Estimates of SEs correctly reflected the number of subjects in each group. It should be noted that in these models we used only variables that logically could not be affected by phenobarbital exposure. A similar analytic method has been previously used in a comparable study on criminality (Witkin et al., 1976), although in that study there was no matching involved. (The details of the regression models are available from the authors on request.) For such predictive analyses in which the focus is on the adjustment and not the individual regression coefficients, it is neither desirable nor necessary to delete "nonsignificant" variables from the regression. In fact, doing so would falsely increase the reported precision of estimates because such a procedure asserts that the removed variable has a zero regression weight with 0 SE. The low R^2 in these regressions is a good sign in these restricted samples because they indicate that any residual bias due to the matching variables (not taken care of by the matching) is not likely to have a major effect on the validity of estimated exposure effects and can be addressed through the linear adjustments, which are locally valid.

The residuals or differences between observed and predicted scores for each exposed subject served as the measure of the effect of prenatal exposure to phenobarbital. The mean difference between observed and predicted intelligence scores is automatically zero for the group of controls on which the model is based. For each study, the primary hypothesis was that the mean residual (observed minus predicted value) for the group of exposed subjects would be negative and significantly different from zero (indicating an overall negative effect of phenobarbital exposure). A t statistic was applied, which used an adjusted SE term (accounting for both the correlation induced by a common estimated regression coefficient for all predictions as well as the different number of exposed and control subjects) and the appropriate dfs from the error term of the regression model.

Several decades of research indicate that socially, culturally, and psychologically advantaged individuals generally score higher on measures of intelligence (Scarr, 1981). Animal models suggest that subjects exposed during the latter stages of gestation would be particularly vulnerable to the teratogenic effects of phenobarbital exposure (Reinisch and Sanders, 1982) and that although the dose-response relationship may not be simple or linear (Fishman and Yanai, 1983), a larger effect would result from higher dosage exposure. After testing for the overall exposure effect, exposed subjects were subclassified to test the following secondary hypotheses applying adjusted t statistics to the mean residuals. It was specifically hypothesized that phenobarbital effects would be greater among exposed subjects who were (1) from lower, rather than higher, SES; (2) born of unwanted pregnancies; (3) exposed to phenobarbital in the latter months of gestation; or (4) exposed to higher total dosages. One-tailed P values are reported because of the directional nature of all of these a priori hypotheses.

RESULTS

Study 1

Table 25.2 presents the results of data analyses on the WAIS. Phenobarbital-exposed men scored significantly lower on VIQ than was predicted, with the mean difference (observed minus predicted scores) equal to -7.17 (adjusted SE, 3.99; adjusted t, -1.79; df, 37; $P < .04$). The mean observed VIQ for phenobarbital-exposed men was 100.69 (SD, 14.94), whereas the predicted mean was 107.86 (SD, 6.38). This mean difference of -7.17 represents approximately 0.5 of the normalized SD (set at 15) on VIQ. Using the binomial effect size display (BESD) (Rosenthal and Rubin, 1982a,b) to illustrate the phenobarbital effect on VIQ in terms of simple proportions, it could be said that after controlling for the predictor variables used in the regression model, 64% of the exposed subjects scored below the median of all subjects vs only 36% of the controls.

Table 25.2. *Difference in observed and predicted scores on Wechsler Adult Intelligence Scale Verbal Intelligence Quotient (VIQ), Performance IQ (PIQ), and Full Scale IQ (FIQ) for 33 phenobarbital-exposed men and VIQ subclass analyses*

Group	N	Mean observed IQ	Mean predicted IQ	Mean residual difference*	Adjusted standard error of difference	Adjusted t statistic	Adjusted P (one-tailed, df = 37)
Overall Exposure Effects							
All exposed subjects							
VIQ	33	100.69	107.86	−7.17	3.99	−1.79	.04
PIQ	33	99.85	104.77	−4.92	4.18	−1.18	.13
FIQ	33	100.36	106.97	−6.61	4.13	−1.60	.06
VIQ Subclass Analyses†							
Socioeconomic status							
Lower	20	95.35	107.85	−12.50	5.09	−2.46	.01
Higher	11	109.18	109.02	0.16	5.40	0.03	.49
Wanted pregnancy?							
Unwanted	16	93.88	105.85	−11.97	4.78	−2.50	.01
Wanted	16	108.75	109.39	−0.64	5.19	−0.12	.45
Lower socioeconomic status and unwanted pregnancy	10	86.00	106.00	−20.00	5.81	−3.44	.001
Timing of exposure							
Third trimester and/or last month only	18	99.11	106.00	−6.89	4.30	−1.60	.06
Third trimester and earlier	12	100.67	110.89	−10.22	5.18	−1.97	.03
Prior to third trimester only	3	110.33	106.92	3.41	9.81	0.35	.37
Total phenobarbital dosage mg‡							
≤5,000	13	103.31	107.57	−4.26	4.73	−0.90	.19
>5,000	20	99.00	108.05	−9.05	4.74	−1.91	.03

* Observed minus predicted scores.

† Exposed subjects were not included in subclass analyses if there were missing data on that variable.

‡ To permit comparison of study 1 and study 2, 5,000 mg was used as the cutoff.

The difference on PIQ, though in the same direction as the VIQ, was not statistically significant. The difference that was observed in FIQ ($P < .06$) was primarily due to the contribution of the VIQ subscales.

Because the major effect was demonstrated for VIQ, subclass analyses were conducted only with this outcome variable. For simplicity of display, SES was dichotomized as follows: lower (Danish SES categories 1 to 5) and higher (Danish SES categories 6 to 9). Consistent with our hypothesis, the disparity between the observed VIQ of phenobarbital-exposed subjects and predicted levels was clearly negatively related to SES. The mean difference for the lower SES subgroup was -12.50 (adjusted SE, 5.09; adjusted t, -2.46; df, 37; $P < .01$), more than 0.8 SD lower than expected. For the higher SES subgroup, the mean difference was not significant (mean residual, 0.16; adjusted SE, 5.40).

For offspring of unwanted pregnancies, the mean difference between observed and predicted VIQ scores was -11.97 (adjusted SE, 4.78; adjusted t, -2.50; df, 37; $P < .01$), again, close to 0.8 SD lower than expected. Observed IQs for those who were wanted were similar to predicted values, the mean difference being -0.64 (adjusted SE, 5.19).

Adverse cognitive effects were largest among the subgroup with gestational exposure during the third trimester and earlier ($n = 12$). This subgroup had a mean difference between observed and predicted VIQ of -10.22 (adjusted SE, 5.18; adjusted t, -1.97; df, 37; $P < .03$). For those exposed only during the third trimester and/or last month of gestation ($n = 18$), the mean difference was -6.89 (adjusted SE, 4.30; adjusted t, -1.60; df, 37; $P < .06$). An exposure effect was not found for the three subjects exposed only during the first two trimesters.

In this sample, in which the majority of subjects have late gestational exposure, a higher total dosage was associated with larger effect size (Table 25.2).

Study 2

Table 25.3 presents the results from the analyses of the Danish Military Draft Board Intelligence BPP scores. The mean BPP score for prenatally phenobarbital-exposed subjects was 39.58 (SD, 12.80) compared with a predicted group mean of 44.35 (SD, 6.50). Similar to the findings for the WAIS VIQ in study 1, exposed subjects had significantly lower BPP scores than predicted (mean difference, observed minus predicted, -4.77; adjusted SE, 1.63; adjusted t, -2.92; df, 85; $P < .002$). This difference was equivalent to almost 0.5 SD of BPP scores for the control subjects (mean, 43.15; SD, 10.24) or the SD (SD, 11.38) reported for the national sample of subjects born between 1954 and 1958 (Teasdale and Owen, 1987). Using the BESD (Rosenthal and Rubin, 1982a,b) to illustrate the phenobarbital effect in terms of simple proportions, it could be said that after controlling for the predictor variables used in the regression model, 65% of the exposed subjects scored below the median of all subjects, whereas only 35% of the controls did so.

As found in study 1, those phenobarbital-exposed subjects in the lower SES subgroup had significantly lower scores (by more than 0.5 SD) than expected (adjusted t, -3.36; df, 85; $P < .001$). The mean difference between observed and predicted BPP scores for lower and higher SES subgroups were -6.01 (adjusted SE, 1.79) and 2.29 ($P = .23$, adjusted SE, 3.02), respectively. As in study 1, those

Table 25.3. *Difference in observed and predicted scores on Danish Military Board Intelligence Test (Børge Priens Prøve [BPP]) for 81 phenobarbital-exposed men and subclass analyses*

Group	N	Mean observed BPP score	Mean predicted BPP score	Mean residual difference*	Adjusted standard error of difference	Adjusted t statistic	Adjusted P (one-tailed, df = 85)
Overall Exposure Effects							
All exposed subjects	81	39.58	44.35	-4.77	1.63	-2.92	.002
Subclass Analyses†							
Socioeconomic status							
Lower	55	36.24	42.25	-6.01	1.79	-3.36	.001
Higher	21	49.57	47.28	+2.29	3.02	0.76	.23
Wanted pregnancy?							
Unwanted‡	36	36.89	42.01	-5.12	2.45	-2.09	.02
Wanted	39	42.77	45.84	-3.07	2.13	-1.44	.08
Timing of exposure							
Third trimester and/or last month only	72	40.26	44.64	-4.38	1.71	-2.56	.006
Third trimester and earlier	5	23.80	41.22	-17.42	4.52	-3.85	.001
Prior to third trimester only	4	47.00	43.01	+3.99	5.22	0.77	.23
Total phenobarbital dosage, mg§							
≤5,000	71	40.60	44.58	-3.98	1.73	-2.30	.02
>5,000	10	32.30	42.72	-10.42	3.28	-3.17	.001

* Observed minus predicted scores.

† Exposed subjects were not included in subclass analyses if there were missing data on that variable.

‡ Thirty of the 36 unwanted offspring also fell in the lower socioeconomic status category, so there is little difference when looking at the combination of lower socioeconomic status and unwanted pregnancy.

§ To permit comparisons of study 1 and study 2, 5,000 mg was used as the cutoff.

from unwanted pregnancies scored significantly lower than predicted (adjusted t, -2.09; df, 85; $P < .02$). The mean difference between observed and predicted BPP scores for phenobarbital-exposed offspring of unwanted pregnancies was -5.12 (adjusted SE, 2.45), approximately .5 SD of control subjects. For the wanted subgroup, the difference was -3.07 ($P = .08$, adjusted SE, 2.13).

Consistent with the findings in study 1, timing of exposure was related to the magnitude of the cognitive deficit. The five subjects with exposure in the last trimester and earlier had the largest deficit. The mean difference between exposed and unexposed was -17.42 (adjusted SE, 4.52; adjusted t, -3.85; df, 85; $P < .001$) representing a deficit of more than 1.5 SDs. Those exposed only in the third trimester and/or the last month of pregnancy ($n = 72$) had a mean difference of -4.38 (adjusted SE, 1.71; adjusted t, -2.56; df, 85; $P < .006$). The four subjects exposed only prior to the last trimester did not differ from predicted levels. As in study 1, across all subjects, higher total dosage was associated with larger effect size. However, in this study, a significant exposure effect was also demonstrated even for those exposed to lower dosages ($\leq 5,000$ mg: -3.98 mean residual, adjusted SE $= 1.73$, adjusted $t = -2.30$, $P < .02$; $>5,000$ mg: -10.42 mean residual, adjusted SE $= 3.28$, adjusted $t = -3.17$, $P < .001$).

COMMENT

To summarize, study 1 found that observed VIQ, as measured by the WAIS, was significantly lower than predicted for adult men exposed prenatally to phenobarbital. Subclass analyses revealed that the phenobarbital exposure effect was larger among those who were (1) from lower socioeconomic backgrounds, (2) offspring of an unwanted pregnancy, (3) exposed in the third trimester and earlier, and (4) exposed to higher dosages. The results of study 2 corroborate the results of study 1 using a different measure of intelligence and an independent sample.

While the results from the subclass analyses for SES and unwanted pregnancies are strongest, the overall deficits in VIQ and BPP scores are nonetheless significant. The regression equation adjusts for the overall effects of such factors. The subgroup analyses demonstrate the interaction between phenobarbital exposure and being from a lower SES or an unwanted pregnancy above the amount adjusted for by the regression.

In addition, the interactions between the effects of prenatal exposure to phenobarbital, and SES and being wanted, cannot be attributed to differences in total dosage or duration of phenobarbital administered because, within each study, the subgroups used in those analyses did not differ significantly with respect to dosage or duration. Similarly, the effects of timing of exposure were independent of differences in dosage or duration. Despite the comparability of dosage and duration, phenobarbital exposure effects were demonstrated most sharply for those with late gestational exposure. This exposure effect was exacerbated when phenobarbital administration spanned more of gestation than just the third trimester. Thus, timing of exposure and the social variables appear to be more closely related to outcome than other parameters of phenobarbital administration such as duration of exposure.

The individuals exposed to phenobarbital are not mentally retarded nor did they have any obvious physical anomalies. Rather, because of their exposure more than 20 years previously, they ultimately test at approximately 0.5 SD or more lower on measured intelligence than would otherwise have been expected. (As noted elsewhere (Mortensen, Reinisch, and Teasdale, 1989; Teasdale, Owen, and Sorensen, 1988), the observed IQ in this cohort population is significantly higher than that of the general population of Danes in this age group. This cohort derives primarily from Copenhagen and is therefore a more urban and higher SES sample than the country as a whole, thus explaining the generally higher IQ.) Using the BESD (Rosenthal and Rubin, 1982a,b) to illustrate the size of the effect, nearly twice as many phenobarbital-exposed subjects (65%) vs control subjects (35%) scored below the median after controlling for the predictor variables used in the regression models. Significant decrements and increments in IQ scores of similar or smaller magnitudes have been reported for such early exposures as alcohol (Streissguth et al., 1989; Russell et al., 1991), mother's milk (Lucas et al., 1992; Rodgers, 1978), phenylalanine (Smith, Beasley, and Ades, 1991), and such variables as advanced age (Schaie and Hertzog, 1983) and height (Teasdale, Owen, and Sorensen, 1991).

Despite our detailed multidimensional matching of exposed subjects and unexposed controls, it is appropriate to ask whether the pregnant women medically treated with phenobarbital differed from those untreated in ways relevant to offspring intelligence but not captured by our matching variables. It would have been ideal to have some measure of parental intelligence, but such a measure was not available. For differences in parental intelligence to have created the current findings it would be necessary to hypothesize that less intelligent mothers were systematically more likely to be prescribed phenobarbital than those of higher intelligence. There is evidence to suggest that this was not the case. A positive relationship has been demonstrated between education level and/or SES and IQ. The fact that families of exposed offspring tended to have higher SES and education levels than the cohort population of unexposed men (Table 25.1) means the current findings run counter to the possibility that the deficit in intelligence scores is the result of lower parental intelligence. In addition, because (1) hypertension during pregnancy rather than psychiatric symptoms was the most common indication for maternal treatment and (2) maternal pathology does not explain the exposure effect related to timing of phenobarbital treatment, it seems unlikely that maternal rather than drug exposure effects are responsible for the observed deficits.

That the IQ scores were administered 10 to 15 years ago should not affect the relevance of the conclusions for phenobarbital use today, for two reasons. First, IQ scores taken in adulthood tend to remain stable throughout adulthood, ie, the men in the studies should have similar results if retested today. Second, we analyzed differences between predicted and observed scores rather than absolute scores. There is no reason to believe that such differences would change with time.

The current data demonstrate that when exposed to phenobarbital prenatally, coming from lower socioeconomic backgrounds or being the offspring of an unwanted pregnancy increased the deficits in adult intelligence scores. More advantaged circumstances, such as high SES and being a wanted child, were associated with diminished negative phenobarbital effects. These data are consistent with other studies suggesting that enhanced social environmental circumstances can ameliorate to some extent at least some early biological insults, such as low birth weight (Infant Health and Development Program, 1990; Feingold, 1994; Schraeder, 1986; Pfeiffer and Aylward, 1990).

In the United States, lower SES groups and disadvantaged minorities are often subject to negative social, nutritional, and medical conditions far exceeding those found in Denmark. These conditions are likely to exacerbate the effects of developmental insult. In Denmark, racial and socioeconomic environments that are relatively homogeneous and the more supportive lifespan social and health care systems mitigate against negative developmental outcomes. Moreover, the mothers in our sample were medically treated with phenobarbital and were not involved in the use of illegal street-obtained substances, nor were they participants in a subculture of abuse likely to significantly exacerbate any negative effects. Thus, it is likely that our findings from Denmark provide a conservative estimate of the developmental consequences that prenatal phenobarbital exposure would have in more diverse and less socially and economically supportive cultures such as the United States.

It is clearly difficult to evaluate alterations in the CNS that primarily manifest themselves as long-term effects on behavior or performance. Many behavioral patterns that are important to normal adolescent and adult functioning cannot be evaluated during infancy or early childhood, thus potentially obscuring the causative role of prenatal or early postnatal factors. For example, certain cognitive deficits (eg, reading ability) may not be evident until the individual has been challenged with increasing task complexity (Fishman and Yanai, 1983; Kilbey and Asghar, 1991). A child may appear to be normal at birth and during infancy in terms of both morphology and behavior, whereas assessment later in life may nonetheless reveal long-term deficits resulting from prenatal insult to the CNS. The adult intelligence deficits measured in the current studies may reflect damage to the CNS [like that demonstrated in rodents (Reinisch and Sanders, 1982; Fishman and Yanai, 1983; Middaugh, 1986)] as a consequence of phenobarbital exposure during gestation. Systematic behavioral assessment at various points in the lifespan may be necessary to determine teratological effects on the CNS.

The following conclusions can be derived from these findings: (1) behavioral teratological sequelae of prenatal exposure to phenobarbital can continue to be manifest well into adulthood even in the absence of physical anomalies; (2) timing of drug administration to the mother may be a highly significant parameter regarding adverse effects of prenatal exposure; and (3) sociopsychological factors interact with prenatal phenobarbital exposure to affect the magnitude of the cognitive deficit. Physicians should exercise increased caution in prescribing phenobarbital to pregnant women and most likely to premature neonates, particularly to those in less privileged psychological, social, and economic environments. As we have just begun to evaluate the long-term negative consequences of prenatal exposure to phenobarbital, it is prudent to carefully weigh the benefit of treating each pregnant woman against the potential jeopardy to the fetal CNS. Finally, the current findings suggest that significant amelioration of long-term postnatal behavioral and/or cognitive difficulties may be effected through positive intervention in the social-educational environment.

ACKNOWLEDGMENTS

We would like to thank Fini Schulsinger, MD, and Grethe Locke-Winther, MSA, of the Institute of Preventive Medicine, Copenhagen, Denmark, for their sustained participation; Bengt Zachau-Christiansen, MD, and colleagues for the initiation and establishment of the 1959–1961 Danish Cohort; the members of the Danish Perinatal Cohort who served as subjects; Carolyn Kaufman, MA, for project coordination and research assistance in Denmark and the United States; Robert Patrick of Datametrics Research, Inc, Waban, Mass, for computer programming and database design; Thomas Albright, MA, for database management at The Kinsey Institute, Bloomington, Ind; Mary Ziemba-Davis, MA, Kim Sare, and Petra Miskus for their varied and valued contributions; Sarnoff Mednick, PhD, DrMed, for help in initiating and establishing the project in Denmark; the Department of Psychology, Rutgers, The State University of New Jersey, New Brunswick, for its supportive role in the initiation of the project; and John Money, PhD, Leonard Rosenblum, PhD, Thorkild Sorensen, MD, and Brian Sutton-Smith, PhD, for their continued support.

26. Estimating Causal Effects from Large Data Sets Using Propensity Scores

Donald B. Rubin

Abstract: The aim of many analyses of large databases is to draw causal inferences about the effects of actions, treatments, or interventions. Examples include the effects of various options available to a physician for treating a particular patient, the relative efficacies of various health care providers, and the consequences of implementing a new national health care policy. A complication of using large databases to achieve such aims is that their data are almost always observational rather than experimental. That is, the data in most large data sets are not based on the results of carefully conducted randomized clinical trials, but rather represent data collected through the observation of systems as they operate in normal practice without any interventions implemented by randomized assignment rules. Such data are relatively inexpensive to obtain, however, and often do represent the spectrum of medical practice better than the settings of randomized experiments. Consequently, it is sensible to try to estimate the effects of treatments from such large data sets, even if only to help design a new randomized experiment or shed light on the generalizability of results from existing randomized experiments. However, standard methods of analysis using available statistical software (such as linear or logistic regression) can be deceptive for these objectives because they provide no warnings about their propriety. Propensity score methods are more reliable tools for addressing such objectives because the assumptions needed to make their answers appropriate are more assessable and transparent to the investigator.

Many observational studies based on large databases attempt to estimate the causal effects of some new treatment or exposure relative to a control condition, such as the effect of smoking on mortality. In most such studies, it is necessary to control for naturally occurring systematic differences in background characteristics between the treatment group and the control group, such as age or sex distributions, that would not occur in the context of a randomized experiment. Typically, many background characteristics need to be controlled.

Propensity score technology, introduced by Rosenbaum and Rubin (1983a), addresses this situation by reducing the entire collection of background characteristics to a single composite characteristic that appropriately summarizes the collection. This reduction from many characteristics to one composite characteristic allows the straightforward assessment of whether the treatment and control groups overlap enough with respect to background characteristics to allow a sensible estimation

Reprinted from *Annals of Internal Medicine*, 1997, **127**, 757–763, with permission from the American College of Physicians.

of treatment versus control effects from the data set. Moreover, when such overlap is present, the propensity score approach allows a straightforward estimation of treatment versus control effects that reflects adjustment for differences in all observed background characteristics.

SUBCLASSIFICATION ON ONE CONFOUNDING VARIABLE

Before describing the use of propensity scores in the statistical analysis of observational studies with many confounding background characteristics, I begin with an example showing how subclassification adjusts for a single confounding covariate, such as age, in a study of smoking and mortality. I then show how propensity score methods generalize subclassification in the presence of many confounding covariates, such as age, region of the country, and sex.

The potential for a large database to suggest causal effects of treatments is indicated in Table 26.1, adapted from Cochran's work (1968a), which concerns mortality rates per 1000 person-years for nonsmokers, cigarette smokers, and cigar and pipe smokers drawn from three large databases in the United States, the United Kingdom, and Canada. The treatment factor here involves three levels of smoking. The unadjusted mortality rates in Table 26.1 make it seem that cigarette smoking is good for health, especially relative to cigar and pipe smoking; clearly, this result is contrary to current wisdom. A problem with this naive conclusion is exposed in Table 26.1, where the average ages of the subpopulations are given. Age correlates with both mortality rates and smoking behavior. In this example, age is a confounding covariate, and conclusions about the effects of smoking should be adjusted for its effects.

A straightforward way of adjusting for age is to 1) divide the population into age categories of approximately equal size (such as younger and older if two categories are appropriate; younger, middle-aged, and older if three are appropriate; and so on), 2) compare mortality rates within an age category (for example, compare mortality rates for the three treatment groups within the younger population and similarly for the older population), and 3) average the age-group–specific comparisons to obtain overall estimates of the age-adjusted mortality rates per 1000 person-years for each of the three groups. Table 26.1 shows the results for different numbers of age categories where the subclass-age boundaries were defined to have equal numbers of nonsmokers in each subclass. These results align better than the unadjusted mortality rates with our current understanding of the effects of smoking, especially when 9 to 11 subclasses are used. Incidentally, having approximately equal numbers of nonsmokers within each subclass is not necessary, but if the nonsmokers are considered the baseline group, it is a convenient and efficient choice because then the overall estimated effect is the simple unweighted average of the subclass-specific results. That is, the mortality rates in all three groups are being standardized (Finch, 1988) to the age distribution of nonsmokers as defined by their subclass counts.

Cochran (1968a) calls this method *subclassification* and offers theoretical results showing that as long as the treatment and exposure groups overlap in their age distributions (that is, as long as a reasonable number of persons from each treatment group are in each subclass), comparisons using five or six subclasses will typically remove 90% or more of the bias present in the raw comparisons shown in Table 26.1.

Table 26.1. *Comparison of mortality rates for three smoking groups in three databases**

Variable	Canadian study			United Kingdom study			United States study		
	Nonsmokers	Cigarette smokers	Cigar and pipe smokers	Nonsmokers	Cigarette smokers	Cigar and pipe smokers	Nonsmokers	Cigarette smokers	Cigar and pipe smokers
Mortality rates per 1000 person-years, %	20.2	20.5	35.5	11.3	14.1	20.7	13.5	13.5	17.4
Average age, γ	54.9	50.5	65.9	49.1	49.8	55.7	57.0	53.2	59.7
Adjusted mortality rates using subclasses, %									
2 subclasses	20.2	26.4	24.0	11.3	12.7	13.6	13.5	16.4	14.9
3 subclasses	20.2	28.3	21.2	11.3	12.8	12.0	13.5	17.7	14.2
9–11 subclasses	20.2	29.5	19.8	11.3	14.8	11.0	13.5	21.2	13.7

* Adapted from Tables 1–3 in Cochran (1968a).

More than five subclasses were used for the adjusted mortality rates because the large size of the data sets made it possible to do so.

A particular statistical model, such as a linear regression (or a logistic regression model; or in other settings, a hazard model) could have been used to adjust for age, but subclassification has two distinct advantages over such models, at least for offering initial trustworthy comparisons that are easy to communicate. First, if the treatment or exposure groups do not adequately overlap on the confounding covariate age, the investigator will see it immediately and be warned. Thus, if members of one group have ages outside the range of another group's ages, it will be obvious because one or more age-specific subclasses will consist almost solely of members exposed to one treatment. In contrast, nothing in the standard output of any regression modeling software will display this critical fact; the reason is that models predict an outcome (such as death) from regressors (such as age and treatment indicators), and standard regression diagnostics do not include careful analysis of the joint distribution of the regressors (such as a comparison of the distributions of age across treatment groups). When the overlap on age is too limited, the database, no matter how large, cannot support any causal conclusions about the differential effects of the treatments. For example, comparing 5-year survival rates among 70-year-old smokers and 40-year-old nonsmokers gives essentially no information about the effect of smoking or nonsmoking for either 70-year-old or 40-year-old persons.

The second reason for preferring subclassification to models concerns situations such as that found in Table 26.1, in which the groups overlap enough on the confounding covariate to make a comparison possible. Subclassification does not rely on any particular functional form, such as linearity, for the relation between the outcome (death) and the covariate (age) within each treatment group, whereas models do. If the groups have similar distributions of the covariate, such specific assumptions like linearity are usually harmless, but when the groups have different covariate distributions, model-based methods of adjustment are dependent on the specific form of the model (for example, linearity or log linearity) and their results are determined by untrustworthy extrapolations.

If standard models can be so dangerous, why are they commonly used for such adjustments when large databases are examined for estimates of causal effects? One reason is the ease with which automatic data analysis can be done using existing, pervasive software on plentiful, speedy hardware. A second reason is the seeming difficulty of using subclassification when many confounding covariates need adjustment, which is the common case. Standard modeling software can automatically handle many regressor variables and produce results, although they can be remarkably misleading. With many confounding covariates, however, the issues of lack of adequate overlap and reliance on untrustworthy model-based extrapolations are even more serious than with only one confounding covariate. The reason is that small differences in many covariates can accumulate into a substantial overall difference. For example, if members of one treatment or exposure group are slightly older, have slightly higher cholesterol levels, and have slightly more familial history of cancer, that group may be substantially less healthy. Moreover, although standard comparisons of means between the groups like those in Table 26.1, or comparisons of histograms for each confounding covariate among groups are adequate with one covariate, they are inadequate with more than one. The groups may differ in a multivariate direction to an extent that cannot be discerned from separate analyses of each covariate. This multivariate direction is closely related to the statistical concept of the best linear discriminant and intuitively is the single combination of the covariates on which the treatment groups are farthest apart.

Subclassification techniques can be applied with many covariates with almost the same reliability as with only one covariate. The key idea is to use propensity score techniques, as developed by Rosenbaum and Rubin (1983a). These methods can be viewed as important extensions of discriminant matching techniques, which calculate the best linear discriminant between the treatment groups and match on it (Rubin, 1980b).

Since their introduction approximately 15 years ago, propensity score methods have been used in various applied problems in medical and other research disciplines[1] but not nearly as frequently as they should have been relative to model-based methods.

PROPENSITY SCORE METHODS

Propensity score methods must be applied to groups two at a time. Therefore, an example with three treatment or exposure conditions will generally yield three distinct propensity scores, one for each comparison (for the example in Table 26.1, nonsmokers compared with cigarette smokers, nonsmokers compared with cigar and pipe smokers, and cigarette smokers compared with cigar and pipe smokers). To describe the way propensity scores work, I first assume two treatment conditions. Cases with more than two treatment groups are considered later.

The basic idea of propensity score methods is to replace the collection of confounding covariates in an observational study with one function of these covariates, called the propensity score (that is, the propensity to receive treatment 1 rather than treatment 2). This score is then used just as if it were the only confounding covariate. Thus, the collection of predictors is collapsed into a single predictor. The propensity score is found by predicting treatment group membership (that is, the indicator variable for being in treatment group 1 as opposed to treatment group 2) from the confounding covariates, for example, by a logistic regression or discriminant analysis. In this prediction of treatment group measurement, it is critically important that the outcome variable (for example, death) play no role; the prediction of treatment group must involve only the covariates. Each person in the database then has an estimated propensity score, which is the estimated probability (as determined by that person's covariate values) of being exposed to treatment 1 rather than treatment 2. This propensity score is then the single summarized confounding covariate to be used for subclassification.

Subclassification into about five groups on the basis of the propensity score then has the rather remarkable property of adjusting for all of the covariates that went

---

[1] Aiken, Smith, and Lake (1994); Cook and Goldman (1989a,b); Eastwood and Fisher (1988); Fiebach et al. (1990); Harrell et al. (1990); Kane et al. (1991); Lavori, Keller, and Endicott (1988); Lavori and Keller (1988); Myers et al. (1987); Stone et al. (1995); Willoughby et al. (1990); Drake and Fisher (1995); Rosenbaum and Rubin (1984b, 1985a,b); Reinisch et al. (1995); Connors et al. (1996); Gu and Rosenbaum (1993).

into its estimation, no matter how many there are. This is a large-sample claim that relies on certain conditions dealt with in technical statistical publications, but it is nevertheless an extremely useful guide for practice. The intuition behind the validity of this claim is fairly straightforward and proceeds as follows.

If two persons, one exposed to treatment 1 and the other exposed to treatment 2, had the same value of the propensity score, these two persons would then have the same predicted probability of being assigned to treatment 1 or treatment 2. Thus, as far as we can tell from the values of the confounding covariates, a coin was tossed to decide who received treatment 1 and who received treatment 2. Now suppose that we have a collection of persons receiving treatment 1 and a collection of persons receiving treatment 2 and that the distributions of the propensity scores are the same in both groups (as is approximately true within each propensity subclass). In subclass 1, the persons who received treatment 1 were essentially chosen randomly from the pool of all persons in subclass 1, and analogously for each subclass. As a result, within each subclass, the multivariate distribution of the covariates used to estimate the propensity score differs only randomly between the two treatment groups.

The formal proof of this result appears in Rosenbaum and Rubin (1983a). Research on how well this theoretical result is satisfied when using estimated rather than true propensity scores is the topic of technical statistical publications (Rubin, 1984b; Drake, 1993; Rubin and Thomas, 1992a,b, 1996). Generally, the conclusion is that using estimated propensity scores in place of true propensity scores works very well.

PROPENSITY SUBCLASSIFICATION

Several years ago, the U.S. General Accounting Office (1994) summarized results from randomized experiments comparing mastectomy (removal of the breast but not the pectoral muscle with nodal dissection but no radiation) and breast conservation therapy (lumpectomy, nodal dissection, and radiation) for the treatment of breast cancer in node-negative patients. The results, shown in Table 26.2 (U.S. GAO, 1994), provide no evidence of differential treatment effect, at least for the type of women who participated in these informed consent clinical trials and who received the kind of care dispensed at the centers participating in these trials. The question remained, however, how broadly these results could be generalized to other node-negative women and other medical facilities. The U.S. General Accounting Office used the National Cancer Institute's SEER (Surveillance, Epidemiology and End Results) observational database to address this question. Restrictions (including node-negative diagnosis, 70 years of age or younger, and tumor size ≤4 cm [U.S. GAO, 1994]) were applied to match criteria for the randomized experiments. These restrictions reduced the database to 1106 women who received breast conservation therapy and 4220 who received mastectomy, for a total of 5326 women.

The U.S. General Accounting Office used propensity score methods on the SEER database to compare the two treatments for breast cancer. First, approximately 30 potential confounding covariates and interactions were identified: year of diagnosis (1983–1985), age category (4 levels), tumor size, geographical registry

Table 26.2. *Estimated 5-year survival rates for node-negative*
*patients in six randomized experiments**

Study	Treatment	Women n	Estimated survival rate %
US-NCI†	Breast conservation	74	93.9
	Mastectomy	67	94.7
Milanese†	Breast conservation	257	93.5
	Mastectomy	263	93.0
French†	Breast conservation	59	94.9
	Mastectomy	62	95.2
Danish‡	Breast conservation	289	87.4
	Mastectomy	288	85.9
EORTC‡	Breast conservation	238	89.0
	Mastectomy	237	90.0
US-NSABP‡	Breast conservation	330	89.0
	Mastectomy	309	88.0

*Adapted from Table 2 in Finch (1988).
† Single-canter trial.
‡ Multicenter trial.

(9 levels), race (4 levels), marital status (4 levels), and interactions of year and registry. A logistic regression was then used to predict treatment (mastectomy compared with conservation therapy) from these confounding covariates on the basis of data from the 5326 women. Each woman was then assigned an estimated propensity score, which was her probability, on the basis of her covariate values, of receiving breast conservation therapy rather than mastectomy. The group was then divided into five subclasses of approximately equal size on the basis of the womens' individual propensity scores: 1064 in the most mastectomy-oriented subclass, 1070 in the next subclass, 1059 in the middle subclass, 1067 in the next subclass, and 1066 in the most breast conservation–oriented subclass.

Before examining any outcomes (5-year survival results), the subclasses were checked for balance with respect to the covariates. Propensity score theory claims that if the propensity scores are relatively constant within each subclass, then within each subclass, the distribution of all covariates should be approximately the same in both treatment groups. This balance was found to be satisfactory. If important within-subclass differences between treatment groups had been found on some covariates, then either the propensity score prediction model would need to be reformulated or it would have been concluded that the covariate distributions did not overlap sufficiently to allow subclassification to adjust for these covariates. This process of cycling between checking for balance on the covariates and reformulating the propensity score model is described by Rosenbaum and Rubin (1984b) in the context of a study investigating coronary bypass surgery. For example, when the variances of an important covariate were found to differ importantly between treatment and control groups, then the square of that covariate was included in the revised propensity score model. For another example, if the correlations between two important

Table 26.3. *Estimated 5-year survival rates for node-negative patients in the SEER database within each of five propensity score subclasses*[*]

Propensity score subclass	Treatment	Women n	Estimated survival rate for women %	Omitting women whose deaths were unrelated to cancer n	Estimated survival rates omitting women whose deaths were unrelated to cancer %
1	Breast conservation	56	85.6	54	88.8
	Mastectomy	1008	86.7	966	90.5
2	Breast conservation	106	82.8	102	86.0
	Mastectomy	964	83.4	917	87.7
3	Breast conservation	193	85.2	184	89.4
	Mastectomy	866	88.8	841	91.4
4	Breast conservation	289	88.7	279	92.0
	Mastectomy	978	87.3	742	91.5
5	Breast conservation	462	89.0	453	90.7
	Mastectomy	604	88.5	589	90.7

[*] Adapted from Tables 5 and 7 in Finch (1988). SEER = Surveillance, Epidemiology, and End Results.

covariates differed between the groups, then the product of the covariates was added to the propensity score model.

The estimates of 5-year survival rates made on the basis of the resulting propensity score subclassification are given in Table 26.3 (U.S. GAO, 1994). Total rates and rates excluding deaths unrelated to cancer are shown. Several features of Table 26.3 are particularly striking, especially when compared with the results of the randomized experiments shown in Table 26.2. First, the general conclusion of similar performance of both treatments is maintained. Second, although overall survival is similar across treatment groups, the results indicate that survival in general practice may be slightly lower than suggested by data from the population of women and types of clinics participating in the randomized clinical trials, especially in the single-clinic studies.

Third, results slightly indicate that, in general practice, women and their physicians may be making beneficial choices. More precisely, women in propensity subclasses 1 to 3, composed of patients whose characteristics (including age, size of tumor, and region of country) make them relatively more likely to receive mastectomy than breast conservation therapy, seem to show better 5-year survival with mastectomy than with breast conservation therapy. In contrast, for women in propensity subclasses 4 and 5 (whose characteristics make them relatively more likely to receive breast conservation therapy than mastectomy), there seems to be no advantage to mastectomy and possibly a slight advantage to breast conservation therapy. Of course, this last interpretation is subject to two caveats. First, we only adjusted for the covariates that were used to estimate the propensity score; hence, other hidden covariates may alter this interpretation. In a randomized experiment, the effects of these hidden covariates are reflected in the SEs of the estimates, but in an observational study, these effects can create bias not reflected in the SEs. Second, the sampling variability (that is, SEs) of the results do not permit firm conclusions, even if the collection of confounding covariates was sufficient to remove bias in this observational study.

Although there is no randomized assignment in the SEER database, the propensity score analyses seem to provide useful suggestive results, especially when coupled with the results of the randomized experiments, with which they are consistent.

MORE THAN TWO TREATMENT CONDITIONS

With more than two treatment conditions, the propensity score usually differs for each pair of treatment groups being compared (that is, with three treatment groups labelled A, B, and C, there are three propensity scores: A compared with B, A compared with C, and B compared with C). At first, this may seem to be a limitation of propensity score technology relative to a model-based analysis, but in fact it is an important strength and points to further weaknesses in a model-based approach. We show this by exploring a range of hypothetical modifications to Cochran's (1968a) smoking example.

First, consider what we could have learned if the nonsmokers and cigarette smokers had had adequately overlapping age distributions, but the cigar and pipe smokers had been substantially older than persons in either of the other groups, with essentially no overlap with the cigarette smokers or the nonsmokers. Even with only one covariate, with more than two groups, the groups in one two-group comparison (nonsmokers compared with cigarette smokers) may overlap adequately, whereas for all other comparisons (in this example, those involving cigar and pipe smokers), the overlap may be inadequate. A typical model-based analysis would use all the data to provide estimates for all three two-group comparisons, even using the data from the cigar and pipe smokers to influence the comparison between the nonsmokers and the cigarette smokers, with no warning of either the extreme extrapolations involved in two of the three two-group comparisons or the use of data on cigar and pipe smokers to help estimate the comparison of nonsmokers and cigarette smokers.

Let us again modify the Cochran (1968a) smoking example but now include an additional covariate: an index of socioeconomic status. We assume that nonsmokers and cigarette smokers have adequate overlap in their age distributions but not much overlap in their socioeconomic status distributions, with nonsmokers having higher socioeconomic status values. In contrast, we suppose that nonsmokers and cigar and pipe smokers have substantial overlap in their socioeconomic distributions but have essentially no overlap in their age distributions. This scenario illustrates that with more than two groups and more than one covariate, the comparison of one pair of groups can be compromised by one covariate and the comparison of another pair of groups can be compromised by a different covariate. As discussed earlier, typical model-based analyses provide no warning that comparisons may be based on extreme extrapolations, nor do they show that the extrapolations include data from groups that are not in the pair of groups being compared.

Now suppose that the nonsmokers and cigarette smokers have the same age distributions and adequately overlapping socioeconomic status distributions. For this comparison, age needs no adjustment but socioeconomic status does need to be adjusted. The propensity score for the comparison would essentially equal socioeconomic status because it, and not age, would predict being a cigarette smoker as opposed to being a nonsmoker. Thus, for this comparison, adjusting for the propensity score would be the same as adjusting for socioeconomic status. Now also assume that the nonsmokers and cigar and pipe smokers have the same socioeconomic status distributions, so that socioeconomic status needs no adjustment, and have adequately overlapping age distributions that need adjustment. Then the propensity score for this comparison would equal age, and therefore, adjusting for the propensity score would be the same as adjusting for age. Thus, the propensity score for a comparison of one pair of groups generally needs to be different from that for a comparison of a different pair of groups. To complete the current scenario, assume that cigarette smokers and cigar and pipe smokers have adequate overlap in both age and socioeconomic status and that both need adjustment. The propensity score for this comparison would involve both age and socioeconomic status because both help to predict cigarette group membership, as opposed to cigar and pipe smoking group membership, and adjusting for this propensity score would adjust for both age and socioeconomic status. Clearly, different propensity score models are needed to adjust appropriately for different comparisons. Estimating all effects by using one model in our example with three groups and adequate overlap on all covariates can be even more deceptive than estimation in the two-group setting because the model being used to compare one pair of groups (for example, nonsmokers compared with cigarette smokers) is affected by the data from the third group (here cigar and pipe smokers), which probably has covariate values that differ from those in either one of the other two groups being compared.

LIMITATIONS OF PROPENSITY SCORES

Despite the broad utility of propensity score methods, when addressing causal questions from nonrandomized studies, it is important to keep in mind that even propensity score methods can only adjust for observed confounding covariates and not for unobserved ones. This is always a limitation of nonrandomized studies compared with randomized studies, where the randomization tends to balance the distribution of all covariates, observed and unobserved.

In observational studies, confidence in causal conclusions must be built by seeing how consistent the obtained answers are with other evidence (such as results from related experiments) and how sensitive the conclusions are to reasonable deviations from assumptions, as illustrated by Connors and colleagues (1996), who used techniques from Rosenbaum and Rubin's work (1983b). Such sensitivity analyses suppose that a relevant but unobserved covariate has been left out of the propensity score model. By explicating how this hypothetical unmeasured covariate is related to treatment assignment and outcome, we can obtain an estimate of the treatment effect that adjusts for it as well as for measured covariates and hereby investigate how answers might change if such a covariate were available for adjustment. Of course, medical knowledge is needed when assessing whether the posited relations involving the hypothetical unmeasured covariate are realistic or extreme. Clarifications of nomenclature and extended sensitivity analyses reported by Lin and colleagues (1997) moderate the initial conclusions of Connors and colleagues (1996).

Another limitation of propensity score methods is that they work better in larger samples for the following reason. The distributional balance of observed covariates created by subclassifying on the propensity score is an expected balance, just as the balance of all covariates in a randomized experiment is an expected balance. In a small randomized experiment, random imbalances of some covariates can be substantial despite randomization; analogously, in a small observational study, substantial imbalances of some covariates may be unavoidable despite subclassification using a sensibly estimated propensity score. The larger the study, the more minor are such imbalances.

A final possible limitation of propensity score methods is that a covariate related to treatment assignment but not to outcome is handled the same way as a covariate with the same relation to treatment assignment but strongly related to outcome. This feature can be a limitation of propensity scores because inclusion of irrelevant covariates reduces the efficiency of the control on the relevant covariates. However, recent work (Rubin and Thomas, 1996) suggests that, at least in modest or large studies, the biasing effects of leaving out even a weakly predictive covariate dominate the efficiency gains from not using such a covariate. Thus, in practice, this limitation may not be substantial if investigators use some judgment.

CONCLUSION

Large databases have tremendous potential for addressing (although not necessarily settling) important medical questions, including important causal questions involving issues of policy. Addressing these causal questions using standard statistical (or econometric, psychometric, or neural net) models can be fraught with pitfalls because of their possible reliance on unwarranted assumptions and extrapolations without any warning. Propensity score methods are more reliable; they generalize the straightforward technique of subclassification with one confounding covariate to allow simultaneous adjustment for many covariates. One critical advantage of propensity score methods is that they can warn the investigator that, because of inadequately overlapping covariate distributions, a particular database cannot address the causal question at hand without relying on untrustworthy model-dependent extrapolation or restricting attention to the type of person adequately represented in both treatment groups. Because of this advantage, any causal questions put to a large database should be first approached using propensity score methods to see whether the question can be legitimately addressed. If so, subclassification on a well-estimated propensity score can be used to provide reliable results, which are adjusted for the covariates used to estimate the propensity score and which can be clearly displayed. After that, modeling can play a useful role. For example, standard statistical models, such as least-squares regression, can be safely applied within propensity score subclasses to adjust for minor within-subclass differences in covariate distributions between treatment groups. This was done in the example of the study by the U.S. General Accounting Office (1994). Of course, it always must be remembered that propensity scores only adjust for the observed covariates that went into their estimation.

ACKNOWLEDGMENTS

The author thanks Jennifer Hill and Frederick Mosteller for helpful editorial comments on an earlier draft of this article.

27. On Estimating the Causal Effects of DNR Orders

Martin W. McIntosh and Donald B. Rubin

Patients write DNR orders to avoid a vegetative state or an otherwise low-quality life, which they fear will follow resuscitation, but policy makers are concerned that DNR orders hasten death, especially for vulnerable subgroups. Both patients and policy makers would be well informed by knowing how often and for whom obeying a DNR order will cause the early termination of life and how often those terminated lives would have been high quality.

There are two ways to learn about the causal effects of DNR orders: an observational study, such as performed by Shepardson et al. (1999) and a randomized clinical trial (RCT). At the end of our discussion, we consider an alternative to both an observational study and a traditional RCT that involves random assignment, yet may be both ethical and more policy relevant than a traditional RCT.

When treatment evaluation does not benefit from any random assignment, observational study methods must be used to assess treatment effects. Here, the authors use propensity score methods and conclude that a patient's choice to write a DNR order on average hastens death. At first glance, it may seem strange to use formal quantitative methods to evaluate this question. After all, can forbidding "resuscitation to preserve life" do anything but hasten death? However, the authors' give their attention to assessing the magnitude of the effect, which has far more policy relevance than the more mundane question of assessing statistical significance. Also, they mention the important issue of determining which subgroups or types of patients may be most affected by DNR orders, which is an issue to which we return after discussing the fundamental statistical technology of propensity scores.

Propensity score methods are ideally suited to assess observational data for their ability to support causal inferences and to provide effect-size estimates with clear interpretation. Because readers are likely more familiar with regression methods for estimating effects in observational studies, we first briefly discuss propensity score methods before commenting on their use in the specific application of Shepardson et al. (1999). A more complete description of propensity score methods accessible to the practitioner can be found in the article by Rubin (1997).

Randomized clinical trials are undertaken because they eliminate systematic pre-treatment differences between the treatment groups, which makes simple

Reprinted from *Medical Care*, 1999, **37**, 8, 722–726. © 1999 Lippincott Williams & Wilkins, Inc.

comparisons sufficient to estimate causal effects under standard assumptions. Most observational data sets, however, have differences in background characteristics between treatment groups that confound simple treatment group comparisons, and these confounders must be controlled. Despite their popularity, control by linear regression (ordinary least squares or logistic) can be deceptive and inappropriate. Regression methods impose a linear association between the outcomes and confounders in the treatment groups, then estimate the treatment effect by means of interpolation and extrapolation. These methods avoid the dangers of linear extrapolation only when the treatment groups have substantial overlap, or balance, with respect to their potential confounders (i.e., when pre-treatment differences are mild) (Rubin, 1973b; 1977a; 1979b). Typically the adequacy of the data set for the estimation of causal effects, i.e., the overlap in confounder distributions, is not addressed when those methods are applied.

The most natural way to avoid reliance on linearity assumptions is by adequately categorizing, or stratifying, on the potential confounders. However, stratification becomes practically impossible with only a moderate number of confounders. For instance, Shepardson et al. (1999) control for nine potential confounders, which, if each had only two levels (typically inadequate for real control), would imply $2^9 = 512$ strata.

The method of propensity scores, introduced by Rosenbaum and Rubin (1983a), focuses attention initially on assessing the overlap in the distributions of confounders, and when there is adequate overlap, provides a single composite stratifying variable that simultaneously controls for bias due to all confounders. The propensity score is the probability of a subject being exposed to one treatment rather than another as a function of a set of observed covariates. Often, as in the present example, logistic regression is used to estimate the propensity score by predicting the treatment choice (here, DNR order or not) from a collection of potential confounders. The estimated propensity score can be used to help create a matched data set (Rubin, 1979b; Reinisch et al., 1995) or as in Shepardson et al. (1999), to stratify the data, just as if it were the only confounding variable (U.S. GAO, 1994). Strata constructed from propensity scores will balance background characteristics across the treatment groups so that within each stratum, simple comparisons of treated and control subjects will often be sufficient to estimate causal effects. Put another way, within each stratum, the data should look at least as balanced as if they had been collected from an RCT, with respect to the observed confounders used to estimate the score (Rubin and Thomas, 1996).

Unlike regression methods, checking the propensity score model addresses directly the validity of the associated estimated causal effects adjusted for confounders. We simply confirm that within each stratum our data appear as if they were collected from an RCT by checking that background characteristics, whether used to predict the propensity score or not, are balanced across treatment groups. However, even when fully satisfactory balance has not been achieved, residual imbalance may still be reduced to the level that standard linear regression adjustments within each stratum will be robust to their assumptions.

In principle, the propensity scores should be predicted using all potential confounders or characteristics that relate to both the treatment choice (DNR order) and outcome (hospital mortality). In Shepardson et al. (1999), the authors used only those characteristics available at admission and found to be "independently associated with the use of DNR orders" at a significance level of $P = 0.01$. In general, this criterion is too exclusive: it is possible that two or more characteristics together are highly predictive of treatment choice but that none is significant on its own. A better rule is to use forward selection with a liberal inclusion criterion and to use any covariate that improves prediction. Although such stepwise rules will tend to include many predictors in the model, any loss of efficiency will likely be offset by further reduction in bias, and including irrelevant predictors will not compromise the bias reducing property of propensity score methods (Rubin and Thomas, 1996). Moreover, the well known complication of biased coefficients induced by stepwise selection is not a problem when estimating propensity scores because the model is used to create a matching or stratifying variable, which is used to create balance in potential confounders across treatment groups: the number of predictors and the values of their coefficients are not of direct interest.

Despite a generally too exclusive inclusion criterion, Shepardson et al.'s (1999) propensity score model appears to be adequate, at least as determined by their diagnostics: there is substantial balance (according to their chi-square tests) across most of the eight strata, and the imbalances that remain are substantially smaller than with the raw data (see their Table 2). Also, when re-estimating the propensity score using the nineteen individual characteristics that summarize the admission severity score,* they find a minimal change in strata, which provides further evidence that their initial propensity score estimation was successful. More details about fitting and evaluating propensity scores can be found in Rosenbaum and Rubin (1984b, 1985a).

With respect to their substantive results, the authors find, in Tables 4 and 5, that odds ratios of hospital death within their eight strata decrease from 258.9, in the stratum representing the lowest propensity score (lowest probability of DNR), down to 9.8 in the stratum representing the highest propensity score. Another choice of effect estimate, however, suggests a different and more interpretable pattern. An important property of propensity score methods is their ability to estimate any causal estimand, such as risk difference, while controlling for background characteristics. Using Table 4, we compute the risk differences, from lowest to highest stratum, as 23.9%, 34.7%, 24.7%, 29.9%, 31.9%, 27.5%, 32.0%, and 41.0%, revealing a mild increase in effect across increasing propensity score strata, not decreasing as the odds ratios suggest. Risk difference should be considerd the superior measure of effect because, unlike the odds ratio, it has a direct interpretation as a causal effect (i.e., as a comparison of potential outcomes [Rubin, 1990a]). Moreover, we may also interpret the estimated average risk difference, here 34.7%, as an estimated average causal effect, analogous to the intention-to-treat estimate of the average effect in an RCT. Because patients with a DNR order had an average mortality of 40.2%,[†] if the propensity score model has successfully balanced all confounders, then we

---

* The severity of illness summary was estimated using the outcome 'hospital death.' In general practice the outcome variable should be excluded completely from the propensity score fitting.

[†] The mean effect and mortality rate must be computed by weighting each stratum by the number of patients with a DNR order in it. We may compute these weights from Shepardson et al's (1999) Table 4. See references for specific details on how to compute the mean causal effect from propensity score strata.

may interpret this estimated average causal effect as follows: if all patients who wrote a DNR had instead not written it, they would have had an average in-hospital mortality rate near $40.2\% - 34.7\% = 5.5\%$, which has direct policy relevance in contrast with the odds ratio, because it suggests the change in in-hospital mortality this population would experience if DNR orders were eliminated.

This estimated effect size is quite large and Shepardson et al. (1999) suggest that their concern would be especially strong if the effect were "greatest among patients with the best prognosis," but they do not refine their analysis to assess this concern. If some subgroups are of more concern than others, then more refined estimates of effects are desirable and can be obtained by cross-classifying the propensity score strata by subgroups. For example, Shepardson et al. (1999) identify admission with cancer or intracerebral hemorrhage as possible modifiers of a DNR's effect. We can determine if cancer patients have a more or less than typical increased rate of death by cross-classifying the patients by propensity score strata and cancer status and then estimating the causal effect separately by cancer status. We can add intracerebral hemorrhage to the subgroup analysis by cross-classifying on the four possible combinations of these traits to allow for the estimation of separate effects. Continuous covariates, such as Shepardson et al.'s (1999) admission severity score, can be included by cross-classifying with a categorized version of the covariate. There is no need to re-estimate the propensity score model when subgroup analyses are performed, assuming that the variable used to define the subgroups was either included in the propensity score or checked for balance overall. Nevertheless, it is wise to check for chance imbalances within the subgroups; just as in an RCT, with multiple subgroups, chance imbalances can arise, especially in smaller subgroups.

When setting up an observational study, it is important to have in mind a template for an underlying RCT, and observational studies can be misleading when their design deviates from a possible RCT. Shepardson et al. (1999) originally intended to have as a template an RCT that takes the pool of all patients who write a DNR order and randomly deny one-half of them. Reconstructing this experiment requires controlling for all important covariates that antecede writing the DNR order, including post admission pre-DNR experiences of the patients. Covariates describing post-admission and pre-DNR experience can be controlled by using them to match controls without a DNR to those with a DNR (for example, matching on total days in critical care ward, a second stroke, etc., occurring between admission and writing the DNR). Because Shepardson et al. (1999) have covariate information that was recorded on the day of admission only, it makes sense for them to retreat from their original template RCT and to reconstruct a hypothetical RCT that can be controlled using only those covariates available to them.

When estimating the effect of DNRs for patients who wrote them on day 1, we have to choose whom to include in their potential comparison group. Two possible choices are as follows: use all patients who did not write a DNR order on day 1 (i.e., both those who wrote them after day 1 and those who never wrote one) or use only those patients who never wrote a DNR. Both choices may lead to valid causal effect estimates, but for different underlying hypothetical experiments. Both experiments would take a pool of subjects who choose to write the DNR on day 1, and randomly deny one-half, but they differ on whether those initially denied would then be permitted to write a DNR order at a later day (control group = all those who did not write a DNR on day 1) or forbidden from doing so (control group = those who never wrote a DNR). In either case, the hypothetical experiment now differs than Shepardson et al.'s (1999) original experiment, because those who wrote DNRs after day 1 are now either in the control group or in neither group. As a result, the propensity scores should be re-estimated or at least re-checked to confirm that balance has been maintained with the new definition of treatment and control groups.

Shepardson et al. (1999) estimated their day 1 effect by omitting patients who wrote their DNR after day 1 from consideration, but they did not re-estimate the propensity score model, nor did they check the balance. They estimate the day 1 effects, given by Table 5, as being smaller, more believable sizes (from OR = 18.1 down to OR = 2.0) but as still suggestive of increased in-hospital mortality (risk differences cannot be computed from the summary information they provide). However, until the balance has been checked, Shepardson et al. (1999) should be cautious when interpreting these estimates as valid causal effects.

As mentioned in the start of our discussion, an alternative to an observational study and a traditional RCT is a nontraditional RCT; a randomized encouragement design (Holland, 1988). An encouragement RCT of DNRs would encourage a randomly chosen group of doctors to discuss DNRs with their patients. Under typically plausible assumptions, the effect of the random assignment can then be decomposed (Angrist et al., 1996; Frangakis, Rubin, and Zhou, 1998) to reveal the causal effect attributed to writing the DNR order. Moreover, if DNR orders increase the risk of death primarily by withholding resuscitation, then the framework of encouragement designs can be used to assess the length and quality of the lives that could have been resuscitated (McIntosh, 1999). Randomized encouragement trials may also be more policy relevant than traditional RCTs because they estimate the effect of a DNR order on those patients who are indecisive enough about writing one to be influenced by their doctor's discussion. A particularly relevant example of a modern clustered encouragement design and analysis is described by Frangakis et al. (1998).

Clearly, Shepardson et al. (1999) are aware of many of the issues in our comments, and we do not disagree with their main conclusion that DNR orders are potentially causing a large increased risk of in-hospital death, especially in certain subgroups, and, thus, their use deserves further attention. Recent developments in conducting and evaluating both observational studies and randomized experiments can be used to shed light on this question. We think that randomized encouragement designs may be particularly appropriate.

ACKNOWLEDGMENT

We thank the editors for inviting us to comment on this important manuscript by Shepardson et al. (1999).

Conclusion: Advice to the Investigator

Matched samples are usually created to aid in the design of a study to assess the causal effect of some active treatment or intervention relative to some control treatment, based on nonrandomized observational data. Consequently, this summary of advice on matching will also offer some general suggestions for the design of such studies based on my four decades of work on them. The theoretical perspective for this advice, the "Rubin Causal Model" (RCM – Holland, 1986b; Rubin, 2006), has two essential parts: the definition of the scientific situation using "potential outcomes" to define causal effect estimands, and the formulation of a real or hypothetical "assignment mechanism"; and a third optional part, the modeling of the science to produce imputations of missing potential outcomes. Matched sampling is focused on the second step. As stated in the initial introduction, a full-length textbook from this perspective is Imbens and Rubin (2006b); and recent summaries of the RCM appear in Imbens and Rubin (2006a) and Rubin (2006a).

The first part of the RCM implies that we should always start by carefully defining all causal estimands in terms of potential outcomes, which are all values that could be observed in some real or hypothetical experiment comparing the results under an active treatment to the results under a control treatment. That is, causal effects are defined by a comparison of (a) the values that would be observed if the active treatment were applied and (b) the values that would be observed if instead the control treatment were applied. This first effort can be completely abstract and can, and often should, take place before any data are observed or even collected.

The second part of the RCM, the assignment mechanism, implies that given the defined science, we should continue by explicating the design of the study to estimate that science. The assignment mechanism describes why some study units were exposed to the active treatment and why other study units were exposed to the control treatment. If the study is a true experiment, the assignment mechanism can involve the consideration of background (i.e., pretreatment) variables for the purpose of creating strata of similar units to be randomized into treatment and control, thereby improving the balance of treatment and control groups with respect to these background variables (i.e., covariates). A true experiment cannot use any outcome (posttreatment) variables to influence design because they are not yet observed, but this is not necessarily true in an observational study because the assignment mechanism can involve the potential outcomes.

When the study is not a true experiment, but rather an observational study, the guidelines for designing an experiment should, nevertheless, be followed. That is, the guiding principle should be to formulate the study to approximate (or replicate) as closely as possible a true randomized experiment, using the pretreatment covariates to create strata (or matched pairs or groups) within which the treatment and control groups only randomly differ with respect to the observed covariates, but these strata must be created without using, or ideally even seeing, the outcome variables until after the design phase is final. This way, the researcher cannot select a design to produce a desired answer (even unconsciously), because the outcome variables were not available at this step. It is wise to speculate about omitted variables to be used for future sensitivity analyses at this stage too, rather than leaving this entirely until after the investigator sees the results. Classical analyses of the covariates, as if they were outcome variables, are ideal diagnostics for assessing the balance of the covariates in treatment and control groups because, by definition, treatments do not affect pretreatment variables.

The optional third part of the RCM derives inferences for causal effects from the observed data by conceptualizing the problem as one of imputing the missing potential outcomes using Bayesian models, which is beyond the matching step, and so not discussed here.

The general advice on the second step, the design phase for an observational study, is to try to approximate (or replicate) an underlying hypothetical randomized experiment that led to the observed data set. The propensity score is the single most important covariate in the sense that it is the observational study equivalent of randomization in an experiment: in a large simple randomized experiment, the distribution of all covariates will be very similar in the treatment and control groups; in a large observational study with treatment and control groups perfectly matched on the estimated propensity scores, the distribution of all observed covariates that entered that estimated propensity score will be very similar in the treatment and control groups. The purpose of all other stratification and matching in either experiments or observational studies is to increase precision; that is, to reduce the variance of these distributional differences. This is not to say that such increases in precision are not important – they are. And of course, true experiments have the critically important advantage of balancing all covariates, observed and unobserved, whereas the use of propensity scores at best balances only observed covariates.

Thus, propensity score matching is not a panacea for the deficiencies of observational studies, but it is a critical tool contributing to their appropriate design. If balance is achieved in an observational study – that is, if the treatment and control groups have very similar distributions of the observed covariates within blocks (subclasses, matched pairs, etc.) of the propensity score (perhaps crossed by blocks based on critical covariates) – then it really makes no difference, for point estimation of effects, how this balance was achieved. Within blocks balanced on propensity scores, future model-based adjustments for distributional differences between treatment and control groups (e.g., using linear covariance; relative risk models; proportional hazard models – Roseman, 1998) will typically have only minor effects on point estimates, although they can have important effects on estimated precisions, and therefore, on interval estimates.

This advice, however, does not solve the problem of which distributional diagnostics to examine. If the yet to be observed outcome variables (or, e.g., their logits) are thought to be approximately linearly related to the covariates, then it is important that the means of the covariates be assessed for balance. If it is thought instead that the outcomes are approximately linearly related to the logs of these covariates, then the means of the logs should be assessed for balance. If it is thought that such dependencies might also involve second-order terms and interactions among covariates, then the variances and correlations of the covariates should also be checked for balance. If some outcomes are thought to be linearly related to a covariate X and others linearly related to $\log(X)$, then the means of both X and $\log(X)$ should be assessed for balance. And so forth.

Of course, at some point, this sort of assessment must terminate, because no matter how large the samples, the investigator will almost certainly not be able to achieve this balance for all covariates and their interactions simultaneously, and higher order terms in prognostically minor covariates are clearly less important than prognostically important ones, and so scientific judgment must enter the process, just as it does when designing a randomized experiment. If balance cannot be achieved on important covariates in a particular observational data set, then the inescapable conclusion must be that inferences for the treatment-control effect cannot be reliably drawn in the study population, unless either the inferences are restricted to a subpopulation where such balance can be achieved, or heroic assumptions are made (e.g., linearity of outcomes on covariates – which justifies linear extrapolation, or an assertion of irrelevance of previously considered covariates). The conclusion that the data set cannot support any valid conclusion about treatment effects may be the right one.

Regarding specific advice on matched sampling, my current feeling is that some sort of metric matching within propensity score calipers is generally a good choice. The specific metric to use is a more complex issue. With a small number of prognostically important continuous covariates, the Mahalanobis metric seems to work well. With more complex, markedly nonellipsoidally symmetric distributions, other metrics could be superior, but still the metric matching should be within propensity score calipers on a subset of prognostically important covariates. A particular recent application to economic data used a "genetic algorithm" in place of the Mahalanobis metric, and indicated some real promise (Raessler and Rubin, 2005a,b; Diamond and Sekon, 2005).

In observational studies, it must be remembered that the issue of bias reduction nearly always dominates the issue of variance reduction: a precise estimate that is badly biased can be more deceptive than helpful, and matched sampling is a key tool for reducing this bias without compromising the integrity or objectivity of the study's design. In general, the use of model-based adjustments on matched samples improves the quality of the resulting estimates of treatment effects.

References

Abadie, A., and Imbens, G. [2005]. Large sample properties of matching estimators for average treatment effects. *Econometrica,* **74**, 235–267.

Agency for Health Care Policy and Research. [1992]. National medical expenditure survey, calendar year 1987. Center for General Health Services Research, Agency for Health Care Policy and Research, Public Health Service, Rockville, MD.

Aiken, L., Smith, H., and Lake, E. [1994]. Lower Medicare mortality among a set of hospitals known for good nursing care. *Medical Care,* **32**, 771–787.

Althauser, R. P., and Rubin, D. B. [1970]. The computerized construction of a matched sample. *American Journal of Sociology,* **76**, 325–346.

Althauser, R. P., and Rubin, D. B. [1971]. Measurement error and regression to the mean in matched samples. *Social Forces,* **50**, 206–214.

Anderson, S., Auquier, A., Hauck, W. W., Oakes, D., Vandaele, W., and Weisberg, H. I. [1980]. *Statistical Methods for Comparative Studies.* Wiley, New York.

Anderson, T. W. [1958]. *An Introduction to Multivariate Statistics.* Wiley: New York.

Angrist, J. D., Imbens, G. W., and Rubin, D. B. [1996]. Identification of causal effects using instrumental variables. *Journal of the American Statistical Association,* **91**, 444–472.

Arai, Y., and Gorski, R. A. [1968]. Critical exposure time for androgenization of the developing hypothalamus in the female rat. *Endocrinology,* **82**, 1010–1014.

Armitage, S. G. [1952]. The effects of barbiturates on the behavior of rat offspring as measured in learning and reasoning situations. *Journal of Comparative and Physiological Psychology,* **45**, 146–152.

Ashikaga, T., and Chang, P. [1981]. Robustness of Fisher's linear discriminant function under two-component mixed normal models. *Journal of the American Statistical Association,* **76**, 676–680.

Baker, S. G., and Laird, N. M. [1988]. Regression analysis for categorical variables with outcome subject to nonignorable nonresponse. *Journal of the American Statistical Association,* **83**, 62–69.

Barker, F. G. II, Chang, S. M., Gutin, P. H., Malec, M. K., McDermott, M. W., Prados, M. D., and Wilson, C. B. [1998]. Survival and functional status after resection of recurrent glioblastoma multiforme. *Neurosurgery,* **42**, 709–720.

Barnard, J., Frangakis, C. Hill, J., and Rubin, D. B. [2003]. A principal stratification approach to broken randomized experiments: A case study of vouchers in New York City (with discussion and rejoinder). *Journal of the American Statistical Association,* **98**, 299–323.

Barr, M. D., and Bertram, E. G. [1949]. A morphological distinction between neurones of the male and female, and the behavior of the nucleolar satellite during accelerated nucleoprotein synthesis. *Nature*, **163**, 676–677.

Bartlett, D. J., Hurley, W. P., Brand, C. R., and Poole, E. W. [1968]. Chromosomes of male patients in a security prison. *Nature*, **219**, 351–354.

Basu, D. [1980]. Randomization analysis of experimental data: The Fisher Randomization Test. *Journal of the American Statistical Association*, **75**, 575–582.

Baughman, F. A., Jr., and Mann, J. D. [1972]. Ascertainment of seven YY males in a private neurology practice. *Journal of the American Medical Association*, **222**, 446–448.

Beaton, A. E., and Tukey, J. W. [1974]. The fitting of power series, meaning polynomials, illustrated on band-spectroscopic data. *Technometrics*, **16**, 147–185.

Belson, W. A. [1956]. A technique for studying the effects of a television broadcast. *Applied Statistics*, **5**, 195–202.

Benjamin, D. J. [1999]. Does 401(k) eligibility increase net national savings?: Reducing bias in the eligibility effect estimate. A. B. Honors Thesis in Economics, Harvard University, Cambridge, MA.

Benson, H., and McCallie, D. [1979]. Angina pectoris and the placebo effect. *New England Journal of Medicine*, **300**, 1424–1428.

Billewicz, W. Z. [1964]. Matched samples in medical investigations. *British Journal of Preventative Social Medicine*, **18**, 167–173.

Billewicz, W. Z. [1965]. The efficiency of matched samples: An empirical investigation. *Biometrics*, **21**, 623–644.

Bishop, Y. M. M., Fienberg, S. E., and Holland, P. W. [1975]. *Discrete Multivariate Analysis*. MIT Press, Cambridge, MA.

Borgaonkar, D. S., and Shah, S. A. [1974]. The XYY chromosome male – or syndrome? In A. G. Steinberg and A. G. Bearn (eds.), *Progress in Medical Genetics*, **10**. Grune & Stratton: New York. 135–222.

Boue, J., Boue, A., and Lazar, P. [1975]. Respective and prospective epidemiological studies of 1500 karyotyped spontaneous human abortions. *Teratology*, **12**, 11–26.

Brent, D. A., Crumrine, P. K., Varma, R., Brown, R. V., and Allan, M. J. [1990]. Phenobarbital treatment and major depressive disorder in children with epilepsy: A naturalistic follow-up. *Pediatrics*, **85**, 1086–1091.

Breslow, N. E., and Day, N. E. [1980]. *Statistical Methods in Cancer Research. Vol 1: The analysis of case-control studies*. International Agency for Research on Cancer, Lyon, France.

Bross, I. D. J. [1966]. Spurious effects from an extraneous variable. *Journal of Chronic Diseases*, **19**, 637–647.

Bross, I. D. J. [1967]. Pertinency of an extraneous variable. *Journal of Chronic Diseases*, **20**, 487–495.

Bunker, J. P., Forrest, W. H., Mosteller, F., and Vandam, L. D. (eds.). [1969]. *The National Halothane Study*. United States Government Printing Office: Washington, DC.

Camfield, C. S., Chaplin, S., Doyle, A., Shapiro, S. H., Cummings, C., and Camfield, P. R. [1979]. Side effects of phenobarbital in toddlers: Behavioral and cognitive aspects. *Journal of Pediatrics*, **95**, 361–365.

Campbell, D. T., and Erlebacher, A. [1970]. How regression artifacts in quasi-experimental evaluations can mistakenly make compensatory education look harmful. In J. Hellmuth (ed.), *The Disadvantaged Child* (Vol. 3), *Compensatory Education: A National Debate*. Brunner/Mazel: New York.

Campbell, D. T., and Stanley, J. C. [1963]. Experimental and quasi-experimental designs for research on teaching. In N. L. Gage (ed.), *Handbook of Research on Teaching*. Rand McNally: Chicago.

Campbell, D. T., and Stanley, J. C. [1963]. *Experimental and Quasi-Experimental Designs for Research*. Rand McNally: Chicago.

Campbell, D. T., and Stanley, J. C. [1966]. *Experimental and Quasi-Experimental Designs.* Houghton Mifflin: Boston.

Card, D., and Kreuger, A. [1994]. Minimum wages and employment: A case study of the fast food industry in New Jersey and Pennsylvania. *American Economic Review*, **84**, 772–793.

Carpenter, R. G. [1977]. Matching when covariates are normally distributed. *Biometrika*, **64**, 299–307.

Caspersson, T., Zech, L., and Johansson, C. [1970]. Analysis of human metaphase chromosome set by aid of DNA-binding fluorescent agents. *Experimental Cell Research*, **62**, 490–492.

Chambers, J. M., Cleveland, W. S., Kleiner, B., and Tukey, P. A. [1983]. *Graphical Methods for Data Analysis*. Wadsworth: Belmont, CA.

Chapin, F. S. [1947]. *Experimental Designs in Sociological Research*. Harper and Brothers: New York.

Cochran, W. G. [1950]. The comparison of percentages in matched studies. *Biometrika*, **37**, 256–266.

Cochran, W. G. [1952]. An appraisal of the repeated population censuses in the eastern health district, Baltimore. In *Research in Public Health*. Milbank Memorial Fund: New York. 255–265.

Cochran, W. G. [1953a]. Matching in analytical studies. *American Journal of Public Health*, **43**, 684–691.

Cochran, W. G. [1953b]. Analysis of records with a view to their evaluation. *The Family Health Maintenance Demonstration*. Milbank Memorial Fund: New York. 228–236.

Cochran, W. G. [1955]. Research techniques in the study of human beings. *Milbank Memorial Fund Quarterly*, **33**, 121–136.

Cochran, W. G. [1957]. Analysis of covariance: Its nature and uses. *Biometrics*, **13**, 261–281.

Cochran, W. G. [1963]. *Sampling Techniques*. Wiley: New York.

Cochran, W. G. [1965]. The planning of observational studies of human populations (with discussion). *Journal of the Royal Statistical Society*, **A**, **128**, 234–255.

Cochran, W. G. [1967a]. Planning and analysis of non-experimental studies. *Proceedings of the Twelfth Conference on the Design of Experiments in Army Research Development and Testing*, ARO-D Report 67-2, 319–336.

Cochran, W. G. [1967b]. Footnote by William G. Cochran. *Science*, **156**, 1450–1462.

Cochran, W. G. [1968a]. The effectiveness of adjustment by subclassification in removing bias in observational studies. *Biometrics*, **24**, 295–313.

Cochran, W. G. [1968b]. Errors of measurement in statistics. *Technometrics*, **10**, 637–666.

Cochran, W. G. [1969]. The use of covariance in observational studies. *Applied Statistics*, **18**, 270–275.

Cochran, W. G. [1970a]. Performance of a preliminary test of comparability in observational studies. *ONR Technical Report* No. 29. Harvard University: Cambridge, MA.

Cochran, W. G. [1970b]. Some effects of errors of measurement on linear regression. *Proceedings of the 6th Berkeley Symposium*, **1**, 527–539.

Cochran, W. G. [1972]. Observational studies. In T. A. Bancroft (ed.), *Statistical Papers in Honor of George W. Snedecor*. Iowa State University Press: Ames.

Cochran, W. G. [1974]. The vital role of randomization in comparative experimentation. In J. Neyman (ed.), *The Heritage of Copernicus*. MIT Press: Cambridge, MA. 445–463.

Cochran, W. G. [1977]. *Sampling Techniques*. Wiley: New York.

Cochran, W. G. [1978]. Early development of techniques in comparative experimentation. In D. Owen (ed.), *On the History of Statistics and Probability.* Dekker: New York. 2–25.

Cochran, W. G. [1983]. *Planning and Analysis of Observational Studies.* Wiley: New York.

Cochran, W. G., and Cox, G. M. [1957]. *Experimental Designs.* Wiley: New York.

Cochran, W. G., and Rubin, D. B. [1973]. Controlling bias in observational studies: A review. *Sankhya,* **A**, **35**, 417–446.

Cohn, P. F., Harris, P., Barry, W., Rosati, R. A., Rosenbaum, P. R., and Waternaux C. [1981]. Prognostic importance of anginal symptoms in angiographically defined coronary artery disease. *American Journal of Cardiology,* **47**, 233–237.

Coleman, J. S., Campbell, E. Q., Hobson, C. J., McPartland, J., Mood, A. M., Weinfield, F. D., and York, R. L. [1966]. *Equality of Educational Opportunity.* U.S. Office of Education: Washington, DC.

Coleman, J. S., Hoffer, T., and Kilgore, S. [1981]. *Public and Private Schools.* March 1981 Report to the National Center for Educational Statistics: Washington, DC.

Conaway, M. R. [1992]. The analysis of repeated categorical measurements subject to nonignorable nonresponse. *Journal of the American Statistical Association,* **87**, *817–824*.

Connors, A. F., Jr., Speroff, T., Dawson, N. V., Thomas, C., Harrell, F. E., Jr., Wagner, D., Desbiens, N., Goldman, L., Wu, A. W., Califf, R. M., Fulkerson, W. J., Jr., Vidaillet, H., Broste, S., Bellamy, P., Lynn, J., and Knaus, W. A. [1996]. The effectiveness of right heart catheterization in the initial care of critically ill patients. SUPPORT Investigators. *Journal of the American Medical Association,* **276**, 889–897.

Cook, E. F., and Goldman, L. [1989a]. Asymmetric stratification: An outline for an efficient method for controlling confounding in cohort studies. *American Journal of Epidemiology,* **127**, 626–639.

Cook, E. F., and Goldman, L. [1989b]. Performance of tests of significance based on stratification by a multivariate confounder score or by a propensity score. *Journal of Clinical Epidemiology,* **42**, 317–324.

Cook, T. D., and Campbell, D. T. [1979]. *Quasi-Experimentation: Design and Analysis Issues for Field Settings.* Rand McNally: Chicago.

Cornfield, J. [1951]. A method of estimating comparative rates from clinical data, application to cancer of the lung, breast and cervix. *Journal of the National Cancer Institute,* **11**, 1269–1275.

Cornfield, J. [1956]. A statistical problem arising from retrospective studies. Proceedings of the Third Berkeley Symposium, **4**, 135–148.

Cornfield, J., et al. [1959]. Smoking and lung cancer: Recent evidence and a discussion of some questions. *Journal of the National Cancer Institute,* **22**, 173–200.

Cox, D. R. [1951]. Some systematic experimental designs. *Biometrika,* **38**, 312–323.

Cox, D. R. [1957a]. Note on grouping. *Journal of the American Statistical Association,* **52**, 543–547.

Cox, D. R. [1957b]. The use of a concomitant variable in selecting an experimental design. *Biometrika,* **44**, 150–158.

Cox, D. R. [1958]. *The Planning of Experiments.* Wiley: New York.

Cox, D. R. [1970]. *The Analysis of Binary Data.* Methuen: London.

Cox, D. R. [1972]. The analysis of multivariate binary data. *Applied Statistics,* **21**, 113–120.

Cox, D. R. [1986]. Comment on "Statistics and causal inference" by Holland (with discussion and reply). *Journal of the American Statistical Association,* **81**, 945–970.

Cox, D. R., and Hinkley, D. V. [1974]. *Theoretical Statistics.* Chapman and Hill: London.

Crandall, B. F., Carrel, R. E., and Sparkes, R. S. [1972]. Chromosome findings in 700 children referred to a psychiatric clinic. *Journal of Pediatrics,* **80**, 62–68.

Curley, C., McEachern, J. E., and Speroff, T. [1998]. A firm trial of interdisciplinary rounds on the inpatient medical wards: An intervention designed using continuous quality improvement. *Medical Care*, **36**, AS4–12.

Czajka, J. C., Hirabayashi, S. M., Little, R. J. A., and Rubin, D. B. [1992] Projecting from advance data using propensity modeling. *Journal of Business and Economics Statistics*, **10**, 117–131.

D'Agostino, R. B., Jr. [1994]. Estimating propensity scores when covariates have either ignorable or nonignorable missing values. PhD Thesis. Department of Statistics, Harvard University: Cambridge, MA.

D'Agostino, R. B., Jr. [1998]. Propensity score methods for bias reduction in the comparison of a treatment to a nonrandomized control group. *Statistics in Medicine*, **17**, 225–228.

D'Agostino, R. B., and Rubin, D. B. [2000]. Estimation and use of propensity scores with incomplete data. *Journal of the American Statistical Assocation*, **95**, 749–759.

Daudin, J. J. [1986]. Selection of variables in mixed-variable discriminant analysis. *Biometrics*, **42**, 473–481.

Dawid, A. P. [1976]. Properties of diagnostic data distributions. *Biometrics*, **32**, 647–658.

Dawid, A. P. [1979]. Conditional independence in statistical theory (with discussion). *Journal of the Royal Statistical Society*, **B**, **41**, 1–31.

DeBault, L. E., Johnston, E., and Loeffelholz, P. [1972]. Incidence of XYY and XXY individuals in a security hospital population. *Diseases of the Nervous System*, **33**, 590–593.

Dehejia, R. H., and Wahba, S. [1999]. Causal effects in nonexperimental studies: Reevaluating the evaluation of training programs. *Journal of the American Statistical Association*, **94**, 1053–1062.

Dempster, A. P. [1969]. *Elements of Continuous Multivariate Analysis*. Addison-Wesley: Reading, MA.

Dempster, A. P. [1971]. An overview of multivariate analysis. *Journal of Mulivariate Analysis*, **1**, 316–346.

Dempster, A. P. [1973]. Aspects of multinomial logit model. In P. R. Krishnaiah (ed.), *Multivariate Analysis III*. Academic Press: New York. 129–142.

Dempster, A. P., Laird, N., and Rubin, D. B. [1977]. Maximum likelihood from incomplete data via the EM algorithm (with discussion and reply). *Journal of the Royal Statistical Society*, **B**, **39**, 1–38.

Diaconis, P., and Freedman, D. [1984]. Asymptotics of graphical projection pursuit. *Annals of Statistics*, **12**, 793–815.

Diamond, A., and Sekon, J. S. [2005]. Genetic matching for estimating causal effects: A general multivariate matching method for achieving balance in observational studies. Political Methodology, The Society of Political Methodology <polmeth@ARTSCI.WUSTL.EDU>.

Dixon, W., Brown, M. B., Engelman, L., Frane, J. W., Hill, M. A., Jennrich, R. I., and Toporek, J. D. [1981]. *BMD-81: Biomedical Computer Programs*. University of California Press: Berkeley.

Dodson, W. E. [1989]. Deleterious effects of drugs on the developing nervous system. *Neonatal Neurology*, **16**, 339–360.

Dorn, H. F. [1953]. Philosophy of inference from retrospective studies. *American Journal of Public Health*, **43**, 692–699.

Drake, C. [1993]. Effects of misspecification of the propensity score on estimators of treatment effect. *Biometrics*, **49**, 1231–1236.

Drake, C., and Fisher, L. [1995]. Prognostic models and the propensity score. *International Journal of Epidemiology*, **24**, 183–187.

Eastwood, E., and Fisher, G. [1988]. Skills acquisition among matched samples of institutionalized and community-based persons with mental retardation. *American Journal of Mental Retardation*, **93**, 75–83.

Efron, B. [1971]. Forcing a sequential experiment to be balanced. *Biometrika*, **583**, 403–417.

Efron, B. [1975]. The efficiency of logistic regression compared to normal discriminant analysis. *Journal of the American Statistical Association*, **70**, 892–898.

Ekstrom, R. B., French, J. W., and Harman, H. H. [1975]. Technical Report No. 8, Office of Naval Research contract N 00014-71-C-0117, HR 150 329.

Fang, K., Kotz, S., and Ng, K. [1990]. *Symmetric Multivariate and Related Distributions*. Chapman and Hall: London.

Farwell, J. R., Lee, Y. J., Hirtz, D. G., Sulzbacher, S. I., Ellenberg, J. H., and Nelson, K. B. [1990a]. Phenobarbital for febrile seizures: Effects on intelligence and on seizure recurrence. *New England Journal of Medicine*, **322**, 364–369.

Farwell, J. R., Lee, Y. J., Hirtz, D. G., Sulzbacher, S. I., Ellenberg, J. H., and Nelson, K. B. [1990b]. Phenobarbital for febrile seizures. *New England Journal of Medicine*, **323**, 485–486.

Feingold, C. [1994]. Correlates of cognitive development in low-birth-weight infants from low-income families. *Journal of Pediatric Nursing*, **9**, 91–97.

Feller, W. [1966]. *An Introduction to Probability Theory and Its Applications,* Vol. 2. Wiley: New York.

Fiebach, N. H., Cook, E. F., Lee, T. H., Brand, D. A., Rouan, G. W., Weisberg, M., et al. [1990]. Outcomes in patients with myocardial infarction who are initially admitted to stepdown units: Data from the Multicenter Chest Pain Study. *American Journal of Medicine*, **89**, 15–20.

Finch, P. E. [1988]. Standardization. In S. Kotz and N. L. Johnson (eds.), *Encyclopedia of Statistical Sciences* (Vol. 8). Wiley: New York. 629–632.

Finley, W. H., McDanal, C. E., Jr., Finley, S. C., and Rosecrans, C. J. [1973]. Prison survey for the XYY karyotype in tall inmates. *Behavior Genetics*, **3**, 97–100.

Finney, D. J. [1957]. Stratification, balance, and covariance. *Biometrics*, **13**, 373–386.

Fisher, R. A. [1925]. *Statistical Methods for Research Workers*. Oliver and Boyd: Edinburgh.

Fisher, R. A. [1935]. *The Design of Experiments*. Oliver and Boyd: Edinburgh.

Fishman, R. H. B., and Yanai, J. [1983]. Long-lasting effects of early barbiturates on central nervous system and behavior. *Neuroscience & Biobehavioral Reviews*, **7**, 19–28.

Forssman, H. [1967]. Epilepsy in XYY man. *Lancet*, 1389.

Frangakis, C. E., and Rubin, D. B. [2002]. Principal stratification in causal inference. *Biometrics*, **58**, 21–29.

Frangakis, C. E., Rubin, D. B., and Zhou, X-H. [1998]. The clustered encouragement design. *Proceedings of the Biometrics Section of the American Statistical Association*, 71–79.

Friis, B., and Sardemann, H. [1977]. Neonatal hypocalcemia after intrauterine exposure to anticonvulsant drugs. *Archives of Disease in Childhood*, **52**, 239–241.

Gail, M. H., Wieand, S., and Piantadosi, S. [1984]. Biased estimates of treatment effect in randomized experiments and nonlinear regressions and omitted covariates. *Biometrika*, **71**, 431–444.

Gaily, E., Kantola-Sorsa, E., and Granstrom, M. L. [1990]. Specific cognitive dysfunction in children with epileptic mothers. *Developmental Medicine and Child Neurology*, **32**, 403–414.

Gelfand, A. E., and Smith, A. F. M. [1990]. Sampling-based approaches to calculating marginal densities. *Journal of the American Statistical Association*, **85**, 972–985.

General Register Office. [1951]. *Classification of Occupations 1950*. His Majesty's Statistical Office: London. 3–13.

Gilbert, J. P., Light, R. J., and Mosteller, F. [1975]. Assessing social innovation: An empirical base for policy. In A. R. Lumsdaine and C. A. Bennett (eds.), *Some Critical Issues in Assessing Social Programs*. Academic Press: New York.

Glynn, R. J., Laird, N. M., and Rubin, F. B. [1993]. Multiple imputation in mixture models for nonignorable nonresponse with follow-ups. *Journal of the American Statistical Association*, **88**, 984–993.

Goldberger, A. S. [1972a]. *Some selection bias in evaluating treatment effects: Some formal illustrations*. Discussion paper. University of Wisconsin Institute for Research on Poverty: Madison, WI.

Goldberger, A. S. [1972b]. *Some selection bias in evaluating treatment effects: The case of interaction*. Discussion paper. University of Wisconsin Institute for Research on Poverty: Madison, WI.

Goodman, L. A. [1968]. The analysis of cross-classified data: Independence, quasi-independence, and interactions in contingency tables with or without missing entries. *Journal of the American Statistical Association*, **63**, 1091–1131.

Goodman, R. M., Smith, W. S., and Migeon, C. J. [1967]. Sex chromosome abnormalities. *Nature*, **216**, 942–943.

Gorski, R. A., and Arai, Y. [1968]. Protection against the organizing effect of exogenous androgen in the neonatal female rat. *Endocrinology*, **82**, 1005–1009.

Graffar, M. [1960]. Social study of samples. *Modern Problems in Pediatrics*, **5**, 30–42.

Granger, C. W. J. [1969]. Investigating causal relations by econometric models and cross-spectral methods. *Econometrica*, **37**, 424–438.

Greenberg, B. G. [1953]. The use of covariance and balancing in analytical surveys. *American Journal of Public Health*, **43**, 692–699.

Greenlees, J. S., Reece, W. S., and Zieschang, K. D. [1982]. Imputation of missing values when the probability of nonresponse depends upon the variable being imputed. *Journal of the American Statistical Association*, **77**, 251–261.

Greenwood, E. [1945]. *Experimental Sociology: A Study in Method*. Kings Crown Press: New York.

Gu, X. S., and Rosenbaum, P. R. [1993]. Comparison of multivariate matching methods: Structures, distances, and algorithms. *Journal of Computational and Graphical Statistics*, **2**, 405–420.

Haggstrom, G. [1976]. The pitfalls of manpower experimentation. In H. W. Sinaido and L. A. Broedling (eds.), *Perspectives on Attitude Assessment: Surveys and Their Alternatives*. Pendleton: Champaign, IL. 228–231.

Hahn, J. [1998]. On the role of the propensity score in efficient semiparametric estimation of average treatment effects. *Econometrica*, **66**, 315–331.

Hamilton, M. A. [1979]. Choosing a parameter for 2 x 2 table or 2 x 2 x 2 table analysis. *American Journal of Epidemiology*, **109**, 362–375.

Hansen, B. B. [2004]. Full matching in an observational study of coaching for the SAT. *Journal of the American Statistical Association*, **99**, 609–618.

Harrell, F. E., Jr., Marcus, S. E., Layde, P. M., Broste, S. K., Cook, E. F., Wagner, D. P., et al. [1990]. Statistical methods in SUPPORT. *Journal of Clinical Epidemiology*, **43** (Supplement), 89S–98S.

Harrison, G. W. [1998] Expert report, April 27, 1998: "Health care expenditures attributable to smoking in Oklahoma." *The State of Oklahoma, ex rel., et al., Plaintiffs, vs. Reynolds Tobacco Co., et al., Defendants.* Case no. CJ-96-1499-L. District Court of Cleveland County, Oklahoma.

Harter, H. L. [1960]. Expected values of normal order statistics. Aeronautical Research Laboratories Technical Report, 60–292.

Heckman, J. J. [1976]. The common structure of statistical models of truncation, sample selection, and limited dependent variables and a simple estimator for such models. *Annals of Economic and Social Measurement*, **5**, 475–492.

Heckman, J. J., and Hotz, V. J. [1989]. Choosing among alternative nonexperimental methods for estimating the impact of social programs: The case of manpower training. *Journal of the American Statistical Association*, **84**, 862–880.

Heckman, J. J., Ichimura, H., Smith, J., and Todd, P. [1996]. Sources of selection bias in evaluating social programs: An interpretation of conventional measures and evidence on the effectiveness of matching as a program evaluation method. *Proceedings of the National Academy of Sciences of the United States of America*, **93**, 13416–13420.

Hess, G. [1974]. *WAIS Anvendt på 698 50-årige*. Akademisk Forlag: Copenhagen, Denmark.

Hill, A. B., and Knowelden, J. [1950]. Inoculation and poliomyelitis: A statistical investigation in England and Wales in 1949. *British Medical Journal*, ii: 1–16.

Hill, J., Rubin, D. B., and Thomas, N. [1999]. The design of the New York school choice scholarship program evaluation. In L. Bickman (ed.), *Research Designs: Donald Campbell's Legacy*. Sage: London. 155–180.

Hill, J. L., Reiter, J. P., and Zanutto, E. [2004]. A comparison of experimental and observational date analyses. In A. Gelman and X.L. Meng (eds.), *Applied Bayesian Modeling and Causal Inference from Incomplete-Data Perspectives*. Wiley: New York. 49–60.

Hill, T. D., Reddon, J. R., and Jackson, D. N. [1985]. The factor structure of the Wechsler scales: A brief review. *Clinical Psychology Review*, **5**, 287–306.

Hirano, K., Imbens, G., and Ridder, G. [2003]. Efficient estimation of average treatment effects using the estimated propensity score. *Econometrica*, **71**, 1161–1189.

Hirschhorn, K. [1965]. In D. B. Amos (ed.), *Histocompatibility Testing*. National Academy of Sciences – National Research Council: Washington, DC. 177–178.

Hirtz, D. G., Lee, Y. J., Ellenberg, J. H., and Nelson, K. B. [1986]. Survey on the management on febrile seizures. *American Journal of Diseases of Children*, **140**, 909–914.

Holland, P. W. [1986a]. Which comes first, cause or effect? *New York Statistician*, **38**, 1–6.

Holland, P. W. [1986b]. Statistics and causal inference (with discussion and reply). *Journal of the American Statistical Association*, **81**, 945–970.

Holland, P. W. [1988]. Causal inference, path analysis, and recursive structural equations models. *Sociological Methodology*, 449–493.

Holland, P. W., and Rubin, D. B. [1983]. On Lord's Paradox. In H. Wainer and S. Messick (eds.), *Principles of Modern Psychological Measurement*. Lawrence Erlbaum, Hillsdale, NJ.

Holland, P. W., and Rubin, D. B. [1988]. Causal inference in retrospective studies. *Evaluation Review*, **12**, 203–231.

Hook, E. B. [1973]. Behavioral implications of the human XYY genotype. *Science*, **179**, 139–150.

Hook, E. B., and Kim, D. S. [1971]. Height and antisocial behavior in XY and XYY boys. *Science*, **172**, 284–286.

Horvitz, D. G., and Thompson, D. J. [1952]. A generalization of sampling without replacement from a finite universe. *Journal of the American Statistical Association*, **47**, 663–685.

Imbens, G. W. [2000]. The role of the propensity score in estimating dose-response functions. *Biometrika*, **87**, 706–710.

Imbens, G. W. [2004]. Nonparametric estimation of average treatment effects under exogeneity: A review. *Review of Economics and Statistics*, **86**, 4–29.

Imbens, G. W., and Rubin, D. [2006a]. Rubin Causal Model. In S. Durlauff and L. Blume (eds.), *Palgrave Dictionary of Economics*. Palgrave Macmillan: London.

Imbens, G. W., and Rubin, D. B. [2006b]. *Causal Inference in Statistics and the Medical and Social Sciences*. Cambridge University Press: Cambridge, UK.

Infant Health and Development Program. [1990]. Enhancing the outcomes of low-birth-weight infants from low-income families. *Journal of the American Medical Association*, **263**, 3035.

Jacobs, P. A., Brunton, M., Melville, M. M., Brittain, R. P., and McClemont, W. F. [1965]. Aggressive behavior, mental, sub-normality and the XYY male. *Nature*, **208**, 1351–1352.

Jacobs, P. A., Price, W. H., Richmond, S., and Ratcliff, R. A. W. [1971]. Chromosome surveys in penal institutions and approved schools. *Journal of Medical Genetics*, **8**, 49–58.

Jarvik, L. F., Klodin, V., and Matsuyama, S. S. [1973]. Human aggression and the extra Y chromosome: Fact or fantasy? *American Psychologist*, **28**, 674–682.

Jick, H., et al. [1973]. Coffee and myocardial infarction. *New England Journal of Medicine*, **289**, 63–67.

Johnson, N. L., and Kotz, S. [1971a]. *Continuous Univariate Distributions – 1*. Houghton Mifflin: Boston.

Johnson, N. L., and Kotz, S. [1971b]. *Continuous Univariate Distributions – 2*. Houghton Mifflin: Boston.

Jones, K. L., Johnson, K. A., and Chambers, C. C. [1992]. Pregnancy outcome in women treated with phenobarbital monotherapy. *Teratology*, **45**, 452–510.

Kaltenbach, K., and Finnegan, L. P. [1986]. Neonatal abstinence syndrome, pharmacotherapy, and developmental outcomes. *Neurobehavioral Toxicology and Teratology*, **8**, 353–355.

Kalton, G. [1968]. Standardization: A technique to control for extraneous variables. *Applied Statistics*, **16**, 118–136.

Kane, R., Garrad, J., Buchanon, J., Rosenfeld, A., Skay, C., and McDermott, S. [1991]. Improving primary care in nursing homes. *Journal of the American Geriatric Society*, **39**, 359–367.

Kaplan, E. L., and Meier, P. [1958]. Nonparametric estimation from incomplete observations. *Journal of the American Statistical Association*, **53**, 457–481.

Kelly, S., Almy, R., and Bernard, M. [1967]. Another XYY phenotype. *Nature*, **215**, 405.

Kempthorne, O. [1952]. *The Design and Analysis of Experiments*. Wiley: New York.

Kempthorne, O. [1976]. Discussion of "On rereading R. A. Fisher" by Leonard J. Savage. *Annals of Statistics*, **4**, 495–497.

Kendall, M. G., and Buckland, W. R. [1971]. *A Dictionary of Statistical Terms*. Oliver and Boyd: London.

Kenney, D. A. [1975]. A quasi-experimental approach to assessing treatment effects in the nonequivalent control group design. *Psychological Bulletin*, **82**, 345–362.

Kihlberg, J. K., and Narragon, E. A. [1964]. A failure of the accident severity classification. *Cornell Aeronautical Laboratory Report* No. VJ-1823-R8, 62–70.

Kihlberg, J. K., and Robinson, S. J. [1968]. Seat belt use and injury patterns in automobile accidents. *Cornell Aeronautical Laboratory Report* No. VJ-1823-R30.

Kilbey, M. M., and Asghar, K. (eds.). [1991]. *Methodological Issues in Controlled Studies on Effects of Prenatal Exposure to Drug Abuse*. Research report DHHS-NIDA 114 ADM 9-1837. U.S. Department of Health and Human Services, National Institute on Drug Abuse: Rockville, MD.

Klipsten, F. A. [1964]. Subnormal serum folate and macrocytosis associated with anticonvulsant drug therapy. *Blood*, **23**, 68–86.

Knuth, D. E. [1969]. *Seminumerical Algorithms* (Vol. 2). Addison-Wesley: Reading, MA.

Kruskal, W. [1980]. The significance of Fisher. *Journal of the American Statistical Association*, **75**, 1019–1030.

Krzanowski, W. J. [1975]. Discrimination and classification using both binary and continuous variables. *Journal of the American Statistical Association*, **70**, 782–790.

Krzanowski, W. J. [1980]. Mixtures of continuous and categorical variables in discriminant analysis. *Biometrics*, **36**, 486–499.

Krzanowski, W. J. [1982]. Mixtures of continuous and categorical variables in discriminant analysis: A hypothesis testing approach. *Biometrics*, **38**, 991–1002.

Lalonde, R. [1986]. Evaluating the econometric evaluations of training programs with experimental data. *American Economic Review*, **76**, 604–620.

Lavori, P. W., and Keller. M. B. [1988]. Improving the aggregate performance of psychiatric diagnostic methods when not all subjects receive the standard test. *Statistics in Medicine*, **7**, 727–737.

Lavori, P. W., Keller, M. B., and Endicott, J. [1988]. Improving the validity of FH-RDC diagnosis of major affective disorder in uninterviewed relatives in family studies: A model based approach. *Journal of Psychiatric Research*, **22**, 249–259.

Lechner, M. [2002]. Some practical issues in the evaluation of heterogeneous labour market programmes by matching methods. *Journal of the Royal Statistical Society, A*, **165**, 59–82.

Lederberg, J. [1973]. The genetics of human nature. *Social Research*, **43**, 375–406.

Li, K. C. [1991]. Sliced inverse regression for dimension reduction. *Journal of the American Statistical Association*, **86**, 316–342.

Lieberman, E., Cohen, A., Lang, J. M., D'Agostino, R. B., Jr., Datta, S., and Frigoletto, F. D., Jr. [1996]. The association of epidural anesthesia with caesarian delivery in nulliparas. *Obstetrics and Gynecology*, **88**, 993–1000.

Liewendahl, K., Majuri, H., and Helenius, T. [1978]. Thyroid function tests in patients on long-term treatment with various anticonvulsant drugs. *Clinical Endocrinology*, **8**, 185–191.

Light. J. R., Mosteller, F., and Winokur, H. S. [1971]. Using controlled field study to improve public policy. *Federal Statistics* (report of the President's Commission), **11**, 367–402.

Lin, D. Y., Psaty, B. M., and Kronal, R. A. [1997]. Assessing the sensitivity of regression results to unmeasured confounders in observational studies. Technical Report No. 144. University of Washington School of Public Health: Seattle.

Lindley, D. V. [1947]. Regression lines and the linear functional relationship. *Journal of the Royal Statistical Society, B*, **9**, 218–224.

Lindley, D. V., and Novick, M. R. [1981]. The role of exchangeability in inference. *Annals of Statistics*, **9**, 45–58.

Lindsley, D. B. [1939]. A longitudinal study of the occipital alpha rhythm in normal children: Frequency and amplitude standards. *Journal of Genetic Psychology*, **55**, 197–213.

Little, R. J. A. [1993]. Pattern-mixture models for multivariate incomplete data. *Journal of the American Statistical Association*, **88**, 125–134.

Little, R. J. A., and Rubin, D. B. [1987]. *Statistical Analysis with Missing Data*. Wiley: New York.

Little, R. J. A., and Schlucter, M. D. [1985]. Maximum likelihood estimation for mixed continuous and categorical data with missing values. *Biometrika*, **72**, 497–512.

Liu, C., and Rubin, D. B. [1995]. ML estimation of the *t* distribution using EM and its extensions, ECM and ECME. *Statistic Sinica*, **5**, 19–39.

Liu, C., and Rubin, D. B. [1998]. Ellipsoidally symmetric extensions of the general location model for mixed categorical and continuous data. *Biometrika*, **85**, 673–688.

Lord, F. M. [1960]. Large-sample covariance analysis when the control variable is fallible. *Journal of the American Statistical Association*, **55**, 307–321.

Lord, F. M. [1967]. A paradox in the interpretation of group comparisons. *Psychological Bulletin*, **68**, 304–305.

Lucas, A., Morley, R., Cole, T. J., Lister, G., and Leeson-Paynee, C. [1992]. Breast milk and subsequent intelligence quotient in children born preterm. *Lancet*, **339**, 261–264.

Lytle, B. W., Blackstone, E. H., Loop, F. D., Hotalling, P. L., Arnold, J. H., McCarthy, P. M., and Cosgrove, D. M. [1999]. Two internal thoracic artery grafts are better than one. *Journal of Thoracic and Cardiovascular Surgery*, **117**, 855–872.

Mahalanobis, P. C. [1927]. Analysis of race mixture in Bengal. *Journal of the Asiatic Society of Bengal*, **23**, 301–333.

Martin, J. C. [1986]. Irreversible changes in mature and aging animals following intrauterine drug exposure. *Neurobehavioral Toxicology and Teratology*, **8**, 335–343.

Martin, J. C., Martin, D. C., Lamire, R., and Mackler, B. [1979]. Effects of maternal absorption of phenobarbital upon rat offspring development and function. *Neurobehavioral Toxicology*, **1**, 49–55.

Matousek, M., and Petersen, I. [1973]. Frequency analysis of the EEG in normal children and adolescents. In P. Kellaway and I. Petersen (eds.), *Automation of Clinical Electroencephalography*. Raven Press: New York. 75–102.

Maxwell, S. E., and Jones, L. V. [1976]. Female and male admission to graduate school: An illustrative inquiry. *Journal of Educational Statistics*, **1**, 1–37.

McIntosh, M. W. [1999]. Instrumental variables and cancer screening trials: Estimating the effect of detecting cancer by screening. *Statistics in Medicine*, **18**, 2775–2794.

McIntosh, M., and Rubin, D. B. [1999]. On estimating the causal effects of do not resuscitate orders. *Medical Care*, **37**, 722–726.

McKerracher, D. W. [1971]. Psychological aspects of a sex chromatin abnormality. *Canadian Psychology*, **12**, 270.

McKinlay, S. M. [1973]. An assessment of the relative effectiveness of several measures of association in removing bias from a comparison of qualitative variables. (unpublished).

McKinlay, S. M. [1974]. The expected number of matches and its variance for matched-pair designs. *Applied Statistics*, **23**, 372–383.

McKinlay, S. M. [1975a]. The design and analysis of observational studies – A review. *Journal of the American Statistical Association*, **70**, 503–520.

McKinlay, S. M. [1975b]. The effect of bias on estimators of relative risk for pair-matched and stratified samples. *Journal of the American Statistical Association*, **70**, 859–864.

McKinlay, S. M. [1977]. Pair matching: A reappraisal of a popular technique. *Biometrics*, **33**, 725–735.

Mednick, S. A., Mura, E., Schulsinger, F., and Mendick, B. [1971]. Prenatal conditions and infant development in children with schizophrenic parents. *Social Biology*, **18**, 5103–5113.

Meier, P. [1978]. The biggest public health experiment ever: The 1954 trial of the Salk poliomyelitis vaccine. In J. M. Tanur, et al. (eds.), *Statistics: A Guide to the Unknown* Holden Day: San Francisco, 3–14.

Meng, X.-L., and Rubin, D. B. [1993]. Maximum likelihood estimation via the ECM algorithm: A general framework. *Biometrika*, **80**, 267–278.

Messick, S. [1980]. *The Effectiveness of Coaching for SAT: Review and Reanalysis of Research from the Fifties to the FTC*. Educational Testing Service: Princeton, NJ.

Middaugh, L. D. [1986]. Phenobarbital during pregnancy in mouse and man. *Neurotoxicology*, **7**, 287–302.

Miettinen, O. [1976]. Stratification by a multivariate confounder score. *American Journal of Epidemiology*, **104**, 609–620.

Ming, K., and Rosenbaum, P. [2000]. Substantial gains in bias reduction from matching with a variable number of controls. *Biometrics*, **42**, 109–142.

Moorhead, P. S., Nowell, P. C., Mellman, W. J., Battips, D. M., and Hungerford, D. A. [1960]. Chromosome preparations of leukocytes cultured from human peripheral blood. *Experimental Cell Research*, **20**, 163–166.

Morales, W. J., and Koerten, J. [1986]. Prevention of intraventricular hemorrhage in very low birth weight infants by maternally administered phenobarbital. *Obstetrics and Gynecology*, **68**, 295–299.

Morris, C. [1979]. A finite selection model for experimental design of the health insurance study. *Journal of Econometrics*, **11**, 43–61.

Mortensen, E. L., Reinisch, J. M., and Teasdale, T. W. [1989]. Intelligence as measured by the WAIS and a military draft board group test. *Scandinavian Journal of Psychology*, **30**, 315–318.

Mosteller, C. F., and Tukey, J. W. [1977]. *Data Analysis and Regression*. Addison-Wesley: Reading, MA.

Murphy, M. L., Hultgren, H. N., Detre, K., Rhomsen, P. H. J., Takaro, T., and participants of the Veterans Administration Cooperative Study. [1977]. Treatment of chronic stable angina. *New England Journal of Medicine*, **297**, 621–627.

Myers, W. O., Gersh, B. J., Fisher, L. D., Mock, M. B., Holmes, D. R., Schaff, H. V., et al. [1987]. Medical versus early surgical therapy in patients with triple-vessel disease and mild angina pectoris: A CASS registry study of survival. *Annals of Thoracic Surgery*, **44**, 471–486.

Nakamura, Y., Moss, A. J., Brown, M. W., Kinoshita, M., and Kawai, C. [1999]. Long-term nitrate use may be deleterious in ischemic heart disease: A study using the databases from two large-scale postinfarction studies. Multicenter Myocardial Ischemia Research Group. *American Heart Journal*, **138**, 577–585.

Neyman, J. [1990]. On the application of probability theory to agricultural experiments: Essay on principles. Translated by D. M. Dabrowska and edited by T. P. Speed. *Statistical Science*, **5**, 465–472.

Nielsen, J. [1971]. Prevalence and a 2 years incidence of chromosome abnormalities among all males in a forensic psychiatric clinic. *British Journal of Psychiatry* **119**, 503–512.

Nielsen, J., and Christensen, A. L. [1974]. Thirty-five males with double Y chromosomes. *Psychological Medicine*, **4**, 28–37.

Noël, B., Dupont, J. P., Revil, D., Dussuyer, I., and Quack, B. [1974]. The XYY syndrome: Reality or myth? *Clinical Genetics*, **5**, 387–394.

Ogawa, J. [1951]. Contributions to the theory of systematic statistics. *Osaka Mathematical Journal*, **4**, 175–213.

Olkin, I., and Tate, R. F. [1961]. Multivariate correlation models with mixed discrete and continuous variables. *Annals of Mathematical Statistics*, **32**, 448–465.

Owen, D. R. [1972]. The XYY male: A review. *Psychological Bulletin*, **78**, 209–233.

Park, T., and Brown, M. B. [1994]. Models for categorical data with nonignorable nonresponse. *Journal of the American Statistical Association*, **89**, 44–52.

Pearson, P. L., and Bobrow, M. J. [1970]. Fluorescent staining of the Y chromosome in meiotic stages of the human male. *Journal of Reproduction and Fertility*, **22**, 177–179.

Pereira De Vasconcelos, A., Colin, C., Desor, D., Divry, M., and Nehlig, A. [1990]. Influence of early neonatal phenobarbital exposure on cerebral energy metabolism and behavior. *Experimental Neurology*, **108**, 176–187.

Persson, T. [1967]. An XYY man and his relatives. *Journal of Mental Deficiency Research*, **11**, 239–245.

Peters, C. C. [1941]. A method of matching groups for experiment with no loss of population. *Journal of Educational Research*, **34**, 606–612.

Peters, C. C., and Van Voorhis, W. R. [1940]. *Statistical Procedures and Their Mathematical Bases*. McGraw-Hill: New York.

Pfeiffer, S. I., and Aylward, G. P. [1990]. Outcome for preschoolers of very low birthweight: Sociocultural and environmental influences. *Perceptual and Motor Skills*, **70**, 1367–1378.

Philip, J., Lundsteen, C., Owen, D., and Hirschhorn, K. [1976]. The frequency of chromosome aberrations in tall men with special reference to 47, XYY and 47, XXY. *American Journal of Human Genetics*, **28**, 404–411.

Pitman, E. J. G. [1937]. Significance tests which may be applied to samples from any populations. *Biometrika*, **29**, 322–335.

Pocock, S. J., and Simon, R. [1975]. Sequential treatment assignment with balancing for prognostic factors in the controlled clinical trial. *Biometrics*, **31**, 103–115.

Pratt, J. W., and Schlaifer, R. [1988]. On the interpretation and observation of laws. *Journal of Econometrics*, **39**, 23–52.

Price, W. H., and Whatmore, P. B. [1967]. Criminal behavior and the XYY male. *Nature*, **213**, 815.

Raessler, S., and Rubin, D. B. [2005a]. The use of multiple imputation to create a nulldata set from nonrandomized job training data. *Proceedings of the International Statistical Institute.*

Raessler, S., and Rubin, D. B. [2005b]. Complications when using nonrandomized job training data to draw causal inferences. *Proceedings of the International Statistical Institute.*

Rao, C. R. [1973]. *Linear Statistical Inference and Its Applications.* 2nd edition. Wiley: New York.

Rao, P. S. R. S., and Sedransk, J. (eds.) [1984]. *W. G. Cochran's Impact on Statistics.* Wiley: New York.

Rayburn, W., Donn, S., Compton, A., and Piehl, E. [1989]. Oral phenobarbital given antenatally to reduce neonatal intraventricular hemorrhage: A comparison between maternal and umbilical cord serum levels at delivery. *Journal of Perinatology*, **9**, 268–270.

Rayburn, W., Donn, S., Piehl, E., and Compton, A. [1988]. Antenatal phenobarbital and bilirubin metabolism in the very low birth weight infant. *American Journal of Obstetrics and Gynecology*, **159**, 1491–1493.

Raynor, W. J. [1983]. Caliper pair-matching on a continuous variable in case-control studies. *Communications in Statistics: Theory and Methods*, **12**, 1499–1509.

Raynor, W. J., and Kupper, L. L. [1981]. Category-matching of continuous variables in case-control studies. *Biometrics*, **37**, 811–818.

Reinisch, J. M., and Karow, W. G. [1977]. Prenatal exposure to synthetic progestins and estrogens: Effects on human development. *Archives of Sexual Behavior*, **6**, 257–288.

Reinisch, J. M., Mortensen, E. L., and Sanders, S. A. [1993]. The Prenatal Development Project. *Acta Psychiatrica Scandinavica*, **370** (Supplement), 54–61.

Reinisch, J. M., and Sanders, S. A. [1982]. Early barbiturate exposure: The brain, sexually dimorphic behavior, and learning. *Neuroscience & Biobehavioral Reviews*, **6**, 311–319.

Reinisch, J. M., Sanders, S. A., Lykke-Mortensen, E., and Rubin, D. B. [1995]. In utero exposure to phenobarbital and intelligence deficits in adult men. *Journal of the American Medical Association*, **274**, 1518–1525.

Rich, S. S. [1998]. Analytic options for asthma genetics. *Clinical and Experimental Allergy*, **28**, 108–110.

Richards, B. W., and Stewart, A. [1966]. The YY syndrome. *Lancet*, 984–985.

Robins, J. M. [1989]. The control of confounding by intermediate variables. *Statistics in Medicine*, **8**, 679–701.

Rodgers, B. [1978]. Feeding in infancy and later ability and attainment. *Developmental Medicine and Child Neurology*, **20**, 421–426.

Roseman, L. [1998]. Reducing bias in the estimate of the difference in survival in observational studies using subclassification on the propensity score. PhD thesis, Department of Statistics, Harvard University: Cambridge, MA.

Rosenbaum, P. R. [1984a]. Conditional permutation tests and the propensity score in observational studies. *Journal of the American Statistical Association*, **79**, 565–574.

Rosenbaum, P. R. [1984b]. From association to causation in observational studies: The role of tests of strongly ignorable treatment assignment. *Journal of the American Statistical Association*, **79**, 41–48.

Rosenbaum, P. R. [1984c]. The consequences of adjustment for a concomitant variable that has been affected by the treatment. *Journal of the Royal Statistical Society*, A, **147**, 656–666.

Rosenbaum, P. R. [1986]. Dropping out of high-school in the United States: An observational study. *Journal of Educational Statistics*, **11**, 207–224.

Rosenbaum, P. R. [1987]. Model-based direct adjustment. *Journal of the American Statistical Association*, **82**, 387–394.

Rosenbaum, P. R. [1988]. Permutation tests for matched pairs with adjustments for covariates. *Applied Statistics*, **37**, 401–411.

Rosenbaum, P. R. [1989]. Optimal matching for observational studies. *Journal of the American Statistical Association*, **84**, 1024–1032.

Rosenbaum, P. R. [1991]. A characterization of optimal designs for observational studies. *Journal of the Royal Statistical Society,* B, **53**, 597–610.

Rosenbaum, P. R. [1995]. *Observational Studies*. Springer-Verlag: New York.

Rosenbaum, P. R. [2002]. *Observational Studies,* 2nd ed. Springer-Verlag: New York.

Rosenbaum, P. R., and Rubin, D. B. [1983a]. The central role of the propensity score in observational studies for causal effects. *Biometrika*, **70**, 41–55.

Rosenbaum, P. R., and Rubin, D. B. [1983b]. Assessing sensitivity to an unobserved binary covariate in an observational study with binary outcome. *Journal of the Royal Statistical Society*, B, **45**, 212–218.

Rosenbaum, P. R., and Rubin, D. B. [1984a]. Estimating the effects caused by treatments: Discussion of a paper by Pratt and Schlaiffer. *Journal of the American Statistical Association*, **79**, 26–28.

Rosenbaum, P. R., and Rubin, D. B. [1984b]. Reducing bias in observational studies using subclassification on the propensity score. *Journal of the American Statistical Association*, **79**, 516–524.

Rosenbaum, P. R., and Rubin, D. B. [1985a]. Constructing a control group by multivariate matched sampling methods that incorporate the propensity score. *The American Statistician*, **39**, 33–38.

Rosenbaum, P. R., and Rubin, D. B. [1985b]. The bias due to incomplete matching. *Biometrics*, **41**, 103–116.

Rosenthal, R., and Rubin, D. B. [1982a]. A simple, general purpose display of the magnitude of experimental effect. *Journal of Educational Psychology*, **74**, 166–169.

Rosenthal, R., and Rubin, D. B. [1982b]. Further meta-analytic procedures for assessing cognitive gender differences. *Journal of Educational Psychology*, **74**, 708–712.

Rubin, D. B. [1970]. *The Use of Matched Sampling and Regression Adjustment in Observational Studies*. PhD thesis. Department of Statistics, Harvard University: Cambridge, MA.

Rubin, D. B. [1972]. *Estimating Causal Effects of Treatments in Experimental and Observational Studies*. Educational Testing Service: Princeton, NJ.

Rubin, D. B. [1973a]. Matching to remove bias in observational studies. *Biometrics,* **29**, 159–183. Correction note [1974]: *Biometrics*, **30**, 728.

Rubin, D. B. [1973b]. The use of matched sampling and regression adjustment to remove bias in observational studies. *Biometrics,* **29**, 185–203.

Rubin, D. B. [1974]. Estimating causal effects of treatments in randomized and nonrandomized studies. *Journal of Educational Psychology*, **66**, 688–701.

Rubin, D. B. [1975]. Bayesian inference for causality: The importance of randomization. *Proceedings of the Social Statistics Section of the American Statistical Association*, 233–239.

Rubin, D. B. [1976a]. Inference and missing data (with discussion). *Biometrika*, **63**, 581–592.

Rubin, D. B. [1976b]. Multivariate matching methods that are equal percent bias reducing, I: Some examples. *Biometrics*, **32**, 109–120. Printer's correction note, p. 955.

Rubin, D. B. [1976c]. Multivariate matching methods that are equal percent bias reducing, II: Maximums on bias reduction for fixed sample sizes. *Biometrics*, **32**, 121–132. Printer's correction note, p. 955.

Rubin, D. B. [1977a]. Assignment to treatment group on the basis of a covariate. *Journal of Educational Statistics*, **2**, 1–26. Printer's correction note, **3**, p. 384.

Rubin, D. B. [1977b]. Formalizing subjective notions about the effect of nonrespondents in sample surveys. *Journal of the American Statistical Association*, **72**, 538–543.

Rubin, D. B. [1978a]. Bayesian inference for causal effects: The role of randomization. *Annals of Statistics*, **6**, 34–58.

Rubin. D. B. [1978b]. *Bias Reduction Using Mahalanobis Metric Matching*, Research Bulletin 78–17. Educational Testing Service: Princeton, NJ.

Rubin, D. B. [1978c]. Multiple imputations in sample surveys: A phenomenological Bayesian approach to nonresponse. *Proceedings of Survey Research Methods Section of the American Statistical Association*. 20–28.

Rubin, D. B. [1979a]. Discussion of 'Conditional independence in statistical theory,' by A. P. Dawid. *Journal of the Royal Statistical Society, Series*, **B**, **41**, 27–28.

Rubin, D. B. [1979b]. Using multivariate matched sampling and regression adjustment to control bias in observational studies. *Journal of the American Statistical Association*, **74**, 318–328.

Rubin, D. B. [1980a]. Discussion of "Randomization analysis of experimental data in the Fisher randomization test" by D. Basu. *Journal of the American Statistical Association*, **75**, 591–593.

Rubin, D. B. [1980b]. Bias reduction using Mahalanobis metric matching. *Biometrics*, **36**, 293–298. Printer's correction p. 296 ((5,10) = 75%).

Rubin, D. B. [1981]. Estimation in parallel randomized experiments. *Journal of Educational Statistics*, **6**, 377–400.

Rubin, D. B. [1983]. Comment: Probabilities of selection and their role for Bayesian modeling in sample surveys (discussion of Hansen, Madow, and Tepping). *Journal of the American Statistical Association*, **78**, 803–805.

Rubin, D. B. [1984a]. Bayesianly justifiable and relevant frequency calculations for the applied statistician. *Annals of Statistics*, **12**, 1151–1172.

Rubin, D. B. [1984b]. Comment: Assessing the fit of logistic regressions using the implied discriminant analysis (discussion of "Graphical Methods for Assessing Logistic Regression Models" by Landwehr, Pregibon, and Shoemaker). *Journal of the American Statistical Association*, **79**, 79–80.

Rubin, D. B. [1984c]. William G. Cochran's contributions to the design, analysis, and evaluation of observational studies. In P. S. R. S. Rao and J. Sedransk (eds.), *W. G. Cochran's Impact on Statistics*. Wiley: New York. 37–69.

Rubin, D. B. [1986]. Which ifs have causal answers? Discussion of Holland's "Statistics and causal inference." *Journal of the American Statistical Association*, **81**, 961–962.

Rubin, D. B. [1987]. *Multiple Imputation for Nonresponse in Surveys*. Wiley: New York.

Rubin, D. B. [1990a]. Formal modes of statistical inference for causal effects. *Journal of Statistical Planning and Inference*, **25**, 279–292.

Rubin, D. B. [1990b]. Neyman [1923] and causal inference in experiments and observational studies. *Statistical Science*, **5**, 472–480.

Rubin, D. B. [1991a]. EM and beyond. *Psychometrika*, **56**, 241–254.

Rubin, D. B. [1991b]. Practical implications of modes of statistical inference for causal effects. *Biometrics*, **4**, 1213–1234.

Rubin, D. B. [1997]. Estimating causal effects from large data sets using propensity scores. *Annals of Internal Medicine*, **127**, 757–763.

Rubin, D. B. [2000a]. Statistical issues in the estimation of the causal effects of smoking due to the conduct of the tobacco industry. In J. Gastwirth (ed.), *Statistical Science in the Classroom*. Springer-Verlag: New York. 321–351.

Rubin, D. B. [2000b]. Statistical assumptions in the estimation of the causal effects of smoking due to the conduct of the tobacco industry. In J. Blasius, J. Hox, E. de Leeuw, and P. Schmidt (eds.), *Social Science Methodology in the New Millennium* Proceedings of the Fifth International Conference on Logic and Methodology. October 6, 2002. Cologne, Germany. 1–22.

Rubin, D. B. [2001a]. Estimating the causal effects of smoking. *Statistics in Medicine*, **20**, 1395–1414.

Rubin, D. B. [2001b]. Using propensity scores to help design observational studies: Application to the tobacco litigation. *Health Services & Outcomes Research Methodology*, **2**, 169–188.

Rubin, D. B. [2002]. The ethics of consulting for the tobacco industry. Special Issue on 'Ethics, Statistics and Statisticians'. *Statistical Methods in Medical Research*, **11**, 373–380.

Rubin, D. B. [2005]. Causal inference using potential outcomes: Design, modeling, decisions. 2004 Fisher Lecture. *Journal of the American Statistical Association*, **100**, 322–331.

Rubin, D. B. [2006]. Statistical inference for causal effects, with emphasis on applications in psychometrics and education. In C. R. Rao and S. Sinharay (eds.), *Handbook of Statistics, Psychometrics*. Elsevier: North Holland, Amsterdam.

Rubin, D. B., Schafer, J. L., and Schenker, N. [1988]. Imputation strategies for missing values in post-enumeration surveys. *Survey Methodology*, **14**, 209–221.

Rubin, D. B., and Stuart, E. A. [2006]. Affinely invariant matching methods with discriminant mixtures of ellipsoidally symmetric distributions. To appear in the *Annals of Statistics*, **34**.

Rubin, D. B., and Thomas, N. [1992a]. Affinely invariant matching methods with ellipsoidal distributions. *Annals of Statistics*, **20**, 1079–1093.

Rubin, D. B., and Thomas, N. [1992b]. Characterizing the effect of matching using linear propensity score methods with normal distributions. *Biometrika*, **79**, 797–809.

Rubin, D. B., and Thomas, N. [1996]. Matching using estimated propensity scores: Relating theory to practice. *Biometrics*, **52**, 249–264.

Rubin, D. B., and Thomas, N. [2000]. Combining propensity score matching with additional adjustments for prognostic covariates. *Journal of the American Statistical Association*, **95**, 573–585.

Russell, M., Czarnecki, D. M., Cowan, R., McPherson, E., and Mudar, P. J. [1991]. Measures of maternal alcohol use as predictors of development in early childhood. *Alcoholism: Clinical and Experimental Research*, **15**, 991–1000.

Sarhan, A. E., and Greenberg, B. G. [1962]. *Contributions to Order Statistics*. Wiley: New York.

Scarr, S. [1981]. *Race, Social Class, and Individual Differences in I.Q.* Lawrence Erlbaum Associates: Hillsdale, NJ.

Schafer, J. L. [1997]. *Analysis of Incomplete Multivariate Data*. CRC Press: New York.

Schaie, K. W., and Hertzog, C. [1983]. Fourteen-year cohort-sequential analyses of adult intellectual development. *Developmental Psychology*, **19**, 531–543.

Schlesselman, J. J. [1978]. Assessing the effects of confounding variables. *Americn Journal of Epidemiology*, **108**, 3–8.

Schraeder, B. D. [1986]. Developmental progress in very low birth weight infants during the first year of life. *Nursing Research*, **35**, 237–242.

Seabright, M. [1971]. A rapid banding technique for human chromosomes. *Lancet*, **2**, 971–972.

Seltser, R., and Sartwell, P. E. [1965]. The influence of occupational exposure to radiation on the mortality of American radiologists and other medical specialists. *American Journal of Epidemiology*, **81**, 2–22.

Shah, S. A. [1970]. *Report on the XYY Chromosomal Abnormality*. Public Health Service Publication No. 2103.

Shah, S. A., and Borgaonkar, D. S. [1974]. *American Psychologist*, **29**, 357.

Shepardson, L. B., Youngner, S. J., Speroff, T., and Rosenthall, G. E. [1999]. Increased risk of death in patients with do not resuscitate orders. *Medical Care*, **37**, 727–737.

Siegel, D. G., and Greenhouse, S. W. [1973]. Validity in estimating relative risk in case-control studies. *Journal of Chronic Diseases*, **26**, 219–225.

Smith, I., Beasley, M. G., and Ades, A. E. [1991]. Effect on intelligence of relaxing the low phenylalanine diet in phenylketonuria. *Archives of Disease in Childhood*, **66**, 311–316.

Smith, N. L., Reiber, G. E., Psaty, B. M., Heckbert, S. R., Siscovick, D. S., Ritchie, J. L., Every, N. R., and Koepsell, T. D. [1998]. Health outcomes associated with beta-blocker and diltiazem treatment of unstable angina. *Journal of the American College of Cardiology*, **32**, 1305–1311.

Smith, J., and Todd, P. [2001]. Reconciling conflicting evidence on the performance of propensity score matching methods. *American Economic Review Papers and Proceedings*, **91**, 112–118.

Snedecor, G. W., and Cochran, W. G. [1967]. *Statistical Methods*, 6th ed. Iowa State University Press: Ames.

Snedecor, G. W., and Cochran, W. G. [1974]. *Statistical Methods*, 6th ed. Iowa State University Press: Ames.

Snedecor, G. W., and Cochran, W. G. [1980]. *Statistical Methods*, 7th ed. Iowa State University Press: Ames.

Sobcyzuk, W., Dowzenko, A., and Krasicka, J. [1977]. Study of children of mothers treated with anticonvulsants during pregnancy. *Polish Journal of Neurology and Neurosurgery*, **11**, 59–63.

Stone, R. A., Obrosky, D. S., Singer, D. E., Kapoor, W. N., and Fine, M. J. [1995]. Propensity score adjustment for pretreatment differences between hospitalized and ambulatory patients with community-acquired pneumonia. Pneumonia Patient Outcomes Research Team (PORT) Investigators. *Medical Care*, **33** (Supplement), AS56–AS66.

Street, D. R. K., and Watson, R. A. [1969]. In D. J. West (ed.), *Criminological Implications of Chromosome Abnormalities*. Cropwood Round-Table Conference, Institute of Criminology, University of Cambridge: Cambridge, UK. 61–67.

Streissguth, A. P., Barr, H. M., Sampson, P. D., et al. [1989]. IQ at age 4 in relation to maternal alcohol use and smoking during pregnancy. *Developmental Psychology*, **25**, 3–11.

Student [1937]. Comparison between balanced and random arrangements of field plots. *Biometrika*, **29**, 363–379.

Svalastoga, K. [1959]. *Prestige, Class and Mobility*. Gyldenhal: Copenhagen, Denmark.

Takizawa, T., Haga, M., Yagi, N., Terashima, M., Uehara, H., Yokoyama, A., and Kurita, Y. [1999]. Pulmonary function after segmentectomy for small peripheral carcinoma of the lung. *Journal of Thoracic and Cardiovascular Surgery*, **118**, 536–541.

Tanner, M., and Wong, M. [1987]. The calculation of posterior distributions by data augmentation. *Journal of the American Statistical Association*, **82**, 528–558.

Teasdale, T. W., and Owen, D. R. [1987]. National secular trends in intelligence and education: A twenty-year cross-sectional study. *Nature*, **325**, 119–121.

Teasdale, T. W., Owen, D. R., and Sørensen, T. I. [1988]. Regional differences in intelligence and educational level in Denmark. *British Journal of Educational Psychology*, **58**, 307–314.

Teasdale, T. W., Owen, D. R., and Sørensen, T. I. [1991]. Intelligence and educational levels in adult males at the extremes of stature. *Human Biology*, **63**, 19–30.

Tukey, J. W. [1977]. *Exploratory Data Analysis*. Addison-Wesley: Reading, MA.

U.S. General Accounting Office (GAO). [1994]. Breast conservation versus mastectomy: Patient survival in day-to-day medical practice and in randomized studies. Report to the Chairman, Subcommittee on Human Resources and Intergovernmental Relations, Committee on Government Operations, House of Representatives. Report No. GAO-PEMD-95-9. U.S. General Accounting Office: Washington, DC.

U.S. Supreme Court Decision. [2002]. *Zelman et al. v. Simmons-Harris, et al.* Nos. 00-1751, 00-1777 and 00-1779.

U.S. Surgeon General's Committee. [1964]. *Smoking and Health*. United States Government Printing Office: Washington, DC.

Valaes, T., Kipouros, K., Petmezaki, S., Solman, M., and Doxiadis, S. A. [1980]. Effectiveness and safety of prenatal phenobarbital for the prevention of neonatal jaundice. *Pediatric Research*, **14**, 947–952.

Van der Pol, M. C., Hadders-Algra, M., Huises, H. J., and Touwen, B. C. L. [1991]. Antiepileptic medication in pregnancy: Late effects on the children's central nervous system development. *American Journal of Obstetrics and Gynecology*, **164**, 121–128.

Vianna, A. M., Froto-Pessoa, O., Lion, M. F., and Decourt, L. [1972]. Searching for XYY males through electrocardiograms. *Journal of Medical Genetics*, **9**, 165–167.

Villumsen, A. L. [1970]. *Environmental Factors in Congenital Malformations: A Prospective Study of 9006 Human Pregnancies*. F.A.D.L.S. Forlag: Copenhagen, Denmark. 124–125.

Vining, E. P. G., Mellitis, E. D., Dorsen, M. M., et al. [1987]. Psychologic and behavioral effects of antiepileptic drugs in children: A double-blind comparison between phenobarbital and valproic acid. *Pediatrics*, **80**, 165–174.

Volavka, J., Mednick, S. A., Sergeant, J., and Rasmussen, L. [1976]. EEGs of XYY and XXY men found in a large birth cohort. In S. A. Mednick and K. O. Christiansen (eds.), *Biosocial Bases of Criminal Behavior*. Gardner Press: New York.

Wallin, A., and Boreus, L. O. [1984]. Phenobarbital prophylaxis for hyperbilirubinemia in preterm infants: A controlled study of bilirubin disappearance and infant behavior. *Acta Paediatrica Scandinavica*, **73**, 488–497.

Wechsler, D. [1955]. *Manual for the Wechsler Adult Intelligence Scale*. The Psychological Corp.: New York.

Welch, B. L. [1937]. On the z-test in randomized block and Latin squares. *Biometrika*, **29**, 21–52.

Wesson, D. R., and Smith, D. E. [1977]. *Barbiturates: Their Use, Misuse, and Abuse*. Human Science Press: New York.

Wilks, S. S. [1932]. On the distribution of statistics in samples from a normal population of two variables with matched sampling of one variable. *Metron*, **9**, 87–126.

Willoughby, A., Graubard, B. I., Hocker, A., Storr, C., Vietze, P., Thackaberry, J. M., et al. [1990]. Population-based study of the developmental outcome of children exposed to chloride-deficient infant formula. *Pediatrics*, **85**, 485–490.

Wilson, J. T. [1971]. Caution with phenobarbital. *Clinical Pediatrics*, **10**, 684–687.

Witkin, H. A., Mednick, S. A., Schulsinger, F., Bakkestrom, E., Christiansen, K. O., Goodenough, D. R., Hirschhorn, K., Lundsteen, C., Owen, D. R., Philip, J., Rubin, D. B., and Stocking, M. [1976]. Criminality in XYY and XXY men. *Science*, **193**, 547–555.

Yaffe, S. J. [1980]. Drug and chemical risks to the fetus and newborn, summary: Pediatrician's view. *Progress in Clinical and Biological Research*, **36**, 157–161.

Yaffe, S. J., and Dorn, L. D. [1990]. Effects of prenatal treatment with phenobarbital. *Developmental Pharmacology and Therapeutics*, **15**, 213–223.

Yanai, J., and Bergman, A. [1981]. Neuronal deficits after neonatal exposure to phenobarbital. *Experimental Neurology*, **73**, 199–208.

Yanai, J., Fares, F., Gavish, M., et al. [1989]. Neural and behavioral alterations after early exposure to phenobarbital. *Neurotoxicology*, **10**, 543–554.

Yinger, J., Milton, I. K., and Laycock, F. [1967]. Treating matching as a variable in sociological experiment. *American Sociological Review*, **32**, 801–812.

Zachau-Christiansen, B., and Ross, E. M. [1975]. *Babies: Human Development During the First Year.* Wiley, New York.

Zeger, S. L., Wyant, T., Miller, L., and Samet, J. [2000]. Statistical testimony on damages in *Minnesota v. Tobacco Industry*. In J. Gastwirth (ed.), *Statistical Science in the Classroom.* Springer-Verlag: New York. 303–320.

Zemp, J. W., and Middaugh, L. D. [1975]. Some effects of prenatal exposure to D-amphetamine sulfate and phenobarbital on developmental neurochemistry and on behavior. *Journal of Addictive Diseases*, **2**, 307–331.

Zeuthen, S., and Nielsen, J. [1973]. Prevalence of chromosome abnormalities among males examined for military service. *Clinical Genetics*, **4**, 422–428.

Zhao, Z. [2004]. Using matching to estimate treatment effects: Data requirements, matching metrics, and monte carlo evidence. *The Review of Economics and Statistics*, **86**, 91–107.

Author Index

Subject Index